Neurobiology of Autism Spectrum Disorders

Abdeslem El Idrissi • Dan McCloskey
Editors

Neurobiology of Autism Spectrum Disorders

Springer

Editors
Abdeslem El Idrissi
Center for Developmental
Neuroscience and Department of Biology
College of Staten Island
City University of New York
New York, NY, USA

Graduate Center
Program Biology-Neuroscience
City University of New York
New York, NY, USA

Dan McCloskey
College of Staten Island
City University of New York
New York, NY, USA

ISBN 978-3-031-42382-6 ISBN 978-3-031-42383-3 (eBook)
https://doi.org/10.1007/978-3-031-42383-3

This Springer imprint is published by the registered company Springer Nature Switzerland AG
The registered company address is: Gewerbestrasse 11, 6330 Cham, Switzerland

Paper in this product is recyclable.

Preface

The human brain is a complex self-organizing system which depends on precise timing of genes, circuits, experiences, and behaviors to function. Through neurodiversity, we learn the different ways that the brain can self-organize with varying degrees of impact on overall brain function. Of the neurodevelopmental disorders which demonstrate neurodiversity, none are more informative than autism spectrum disorder (ASD). Understanding ASD and its diverse manifestations not only helps us to unravel the basic principles underlying the organization and function of the human brain, but also helps us to find ways to improve the quality of life of those that are negatively impacted by it.

ASD is a neurodevelopmental condition that affects millions of individuals worldwide, encompassing a broad range of symptoms and challenges. It can profoundly impact an individual's social interaction, communication abilities, and sensory processing, often accompanied by repetitive behaviors and restricted interests. Despite its prevalence and profound impact on individuals and their families, the underlying mechanisms and causes of ASD remain a subject of intense scientific investigations and debates.

This book, *Neurobiology of Autism Spectrum Disorder*, delves deep into the intricate interplay between genetics, brain structure and function, and environmental factors that shape the development and expression of ASD. It was designed to provide a comprehensive overview of the current state of research, synthesizing findings from various disciplines such as epidemiology, neurobiology, psychology, genetics, and clinical studies.

Within these pages, we gather the expertise of leading researchers and clinicians in the field of autism research to present a comprehensive overview of our current understanding of ASD. We offer a diverse range of perspectives, theories, and empirical evidence that shed light on the intricate mechanisms underlying this complex disorder. We aim to bridge the gap between cutting-edge research and practical knowledge, providing a resource that is accessible to both professionals and individuals seeking to deepen their understanding of autism spectrum disorder. Throughout this book, we recognize the immense impact that ASD can have on individuals, families, and communities worldwide. Our ultimate goal is to contribute to the advancement of evidence-based practices that can improve the lives of individuals on the autism spectrum.

While this book does not claim to provide all the answers, it serves as a valuable resource for researchers, clinicians, educators, and anyone seeking a deeper understanding of the neurobiological foundations of ASD. By synthesizing cutting-edge research and presenting it in a clear and accessible manner, we hope to inspire further investigation and foster collaborative efforts in unraveling the complexities of this enigmatic condition.

We extend our gratitude to all the authors who have contributed their expertise and insights to this volume. We would also like to express our appreciation to the countless individuals and families affected by ASD, whose lived experiences continually remind us of the importance of our work.

May this book serve as a stepping stone towards greater understanding, compassion, and progress in the field of autism spectrum disorder research.

New York, NY, USA

Abdeslem El Idrissi
Dan McCloskey

Contents

Chapter 1
Dysfunctional Circuit Mechanisms of Sensory Processing in FXS and ASD: Insights from Mouse Models

Anubhuti Goel

The art of being wise is knowing what to overlook

– William James

1.1 Sensory Processing and Decision Making

Deciding if one should smile or be sober and if the constant noise in the background is not relevant to the current conversation is imperative for any social interaction or cognition. There is no doubt that decisions and choices are heavily influenced by sensory input, and thus arises the question of how do the building blocks in the brain decide which sensory input is relevant to the current context and which one to overlook (not be distracted by). Imagine a scene at a restaurant where a group of friends are enjoying a meal or a late Saturday morning at the park, with kids running around and playing. In both scenarios, during a social interaction at dinner or at play, the decision to show joy or sympathy is dependent on whether the person we are interacting with is smiling or sad–thus processing the visual input and other sensory information is key to deciding the appropriate social response and any subsequent actions. Therefore, discriminating between sensory stimuli across modalities is an important prerequisite to most behaviors. Understanding how sensory stimuli shape decisions and judgments is particularly critical in the context of Autism Spectrum Disorders (ASDs) and Fragile X Syndrome (FXS), where sensory issues such as hypersensitivity to sounds, touch and sights are the most common and pervasive symptom. Sensory atypicalities in ASD and FXS range from sensory seeking (taking pleasure from activities such as fidgeting or repeating noises, sensory distortions in the perception of physical objects, sensory tune-outs or blanking of sound or vision, overload of senses, difficulties in processing stimulation of more than one of

---

A. Goel (✉)
University of California, Riverside, Riverside, CA, USA
e-mail: anubhuti.goel@ucr.edu

A. El Idrissi, D. McCloskey (eds.), *Neurobiology of Autism Spectrum Disorders*,
https://doi.org/10.1007/978-3-031-42383-3_1

the senses at any given time, difficulty in recognizing the channel through which the stimulation is received and suppressing targets in the visually attended area. Further, atypical sensory experience in individuals with ASD spans across sensory modalities including taste, touch, audition, smell and vision (Robertson & Baron-Cohen, 2017). While disorders associated with ASD are typically described as difficulties in establishing eye contact and social interactions, repetitive behaviors and an atypical reduction in broader interests, an emerging view that is gaining support is that these cognitive impairments are likely due to atypical processing and discriminating of sensory stimuli. For example, if an individual with ASD or FXS perceives normal stimuli as overwhelming, or is unable to tune out irrelevant stimuli then, he/she could limit social interactions (avoiding eye contact or hugging), try to control sensory inputs through rituals and experience delays in learning and adapting to changes in the environment. Therefore, the consequence of altered sensory processing is debilitating, resulting in impairment in sensory discrimination and an inability to ignore irrelevant sensory stimuli such as innocuous sounds, smells, sights, or touches. This hyperarousal to sensory stimuli leads to a commonly reported symptoms–hypersensitivity, tactile defensiveness and ADD, all of which eventually impair learning. Indeed, several recent studies suggest that sensory issues and atypical sensory processing can be predictive of and contribute to abnormal anxiety and other cognitive and social deficits (Wheeler et al., 2016; Kojovic et al., 2019; Tavassoli et al., 2014; Robertson & Simmons, 2013; Green & Ben-Sasson, 2010). At this time our understanding of how disruption in neuronal communication results in the diverse symptoms of ASD and FXS is severely lacking. In this chapter I will discuss what we have learned from using animal models of FXS and ASD to understand the neural impairments contributing to atypical sensory processing and highlight the translational potential of this approach to hopefully solving FXS and ASD.

1.2 Fragile X Syndrome

Extensive research in the field of ASD has revolved around Fragile X Syndrome (FXS). Symptoms such as abnormal sensory sensitivity and processing is common in humans with FXS and ASD, manifesting usually as sensory hyperarousal, hypersensitivity and Attention Deficit Disorder (ADD) (Marco et al., 2011; Ethridge et al., 2016, 2017; Miller et al., 1999; Sullivan et al., 2006). FXS, the leading known genetic cause of atypical behaviors associated with autism spectrum disorders (ASD) (Chudley & Hagerman, 1987; Niu et al., 2017; Yu & Berry-Kravis, 2014), arises due to the reduced expression or loss of the Fragile X Messenger Ribonucleoprotein 1 Protein (FMRP). FMRP is an mRNA binding protein that, in response to neuronal activity, negatively regulates the translation of many mRNAs that encode proteins important for neuronal development and synaptic function (Darnell et al., 2011; Bassell & Warren, 2008). Hence, FMRP is important for synapse maturation and experience-dependent rewiring of brain circuits through its regulation of protein synthesis (Kooy et al., 2000; Ronesi & Huber, 2008).

Interestingly, some of the signaling pathways regulated by FMRP are also implicated in other ASDs (Parikshak et al., 2013) and alterations in neuronal and circuit excitability, a hallmark of FXS (Contractor et al., 2015), may be a key to understand autism more generally (Rubenstein & Merzenich, 2003). This argument suggests that elucidating circuit-level alterations in FXS may not only accelerate the development of therapeutics in FXS but may also be of broad importance to other types of ASD and intellectual disability.

1.3 Mouse Models of Fragile X Syndrome and ASD

There is a significant overlap in the attentional impairments between FXS and ASD symptoms (Hagerman & Hagerman, 2002). Further there is a growing consensus that these attentional impairments could arise from sensory processing deficits. However, the underlying neural causes of the sensory and attentional deficits are unknown. To dissect the circuit mechanisms of sensory processing and how this sensory deficit contributes to learning, attention and decision making, research has utilized the potential of animal models of FXS and ASD. One well established model–the *Fragile X* Messenger Ribonucleoprotein 1 *gene* (*Fmr1)* knockout (KO) mouse, is popular for many reasons as discussed here. The mouse *Fmr1* gene product shares 97% homology with human FMRP including a conservation of the CGG repeats (Ashley et al., 1993). *Fmr1* KO mice show functional alterations that are similar to humans and hypersensitivity phenotypes in mice resonate with human symptoms (Consortium, D.-B.F.X, 1994; Kazdoba et al., 2014). There is evidence that abnormal sensory processing in animal models of ASD, can also contribute to abnormal anxiety and social impairments (Orefice et al., 2016). Further several symptoms such as auditory startle, audiogenic seizure, visual deficits such as discriminating between orientations and contrasts, have been reliably observed in mouse models of FXS (Razak et al., 2021).

Several studies in humans have examined the oscillatory dynamics associated with FXS, and found a reduction in alpha power and enhancement in theta power in resting state EEG data (Van der Molen et al., 2012). The reduction in alpha power is suggestive of a disruption in inhibition that impairs task irrelevant information from being actively processed–leading to attentional deficits, hypersensitivity and distractibility. Further a disruption in oscillatory dynamics is interpreted as an impairment in the Excitation-Inhibition (E-I) balance. Single-cell mRNA expression studies have shown that many genes that are differentially expressed in human ASD are interneuron-specific genes, including PVALB, SST and VIP (which are all downregulated in ASD) (Polioudakis et al., 2019; Velmeshev et al., 2019)–again suggestive of a disruption in inhibition. The huge explosion of recording and imaging techniques combined with mouse genetics over the last decade as well as optogenetic manipulation of specific subsets of cells in mice allow examination of dysfunctional interneurons and E-I balance in sensory processing. Importantly, similar EEG phenotypes of sensory abnormalities such as increased magnitude and reduced habituation of sound-evoked responses (Lovelace et al., 2016), are seen across species in FXS (Lovelace et al., 2018). Therefore, for several reasons, *Fmr1* KO mouse in particular, has emerged as a popular model system to study the molecular and cellular pathogenesis of FXS. And given the overlap in sensory issues across FXS and ASD, dissecting the dysfunctional synaptic and circuit impairments in FXS will shed light on defects in circuit mechanisms in ASD in general.

Currently, there exists neither a cure for FXS nor any therapy that can reverse its core pathogenic mechanisms. The development of viable and effective therapies in FXS and ASD is hampered by a lack of understanding in: (1) circuit mechanisms of atypical sensory processing in FXS and ASD, (2) the relationship of impairments in synaptic and cellular physiology to atypical circuit dynamics, and (3) the relationship of impaired sensory processing (at the level of synapses and circuits) to learning, social processing and other cognition impaired in FXS and ASD (Fig. 1.1). Various research groups, highlight parallel alterations in neural responses to sensory stimuli in FXS mice and humans (Ethridge et al., 2016; Lovelace et al., 2018; Goel et al., 2018), thus insights from mouse models of FXS will provide the basis for future translational studies and pave the way for more viable therapies in FXS and potentially other ASDs (Fig. 1.1)

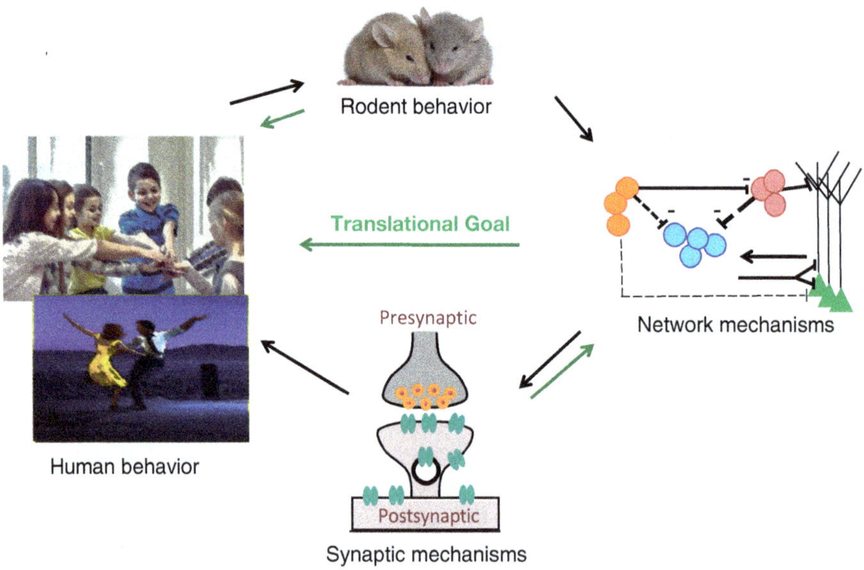

Fig. 1.1 Linking sensory discrimination to complex behavior. Several components are critical to generating complex behavior: genetic, synaptic and network. Studies in mouse models of FXS and ASD will allow probing/understanding of different components so that we can delineate how behavior emerges from neuronal communication. The ultimate goal of FXS research is to use insights from mouse models to understand mechanisms linked to human behavior and find solutions for neurodevelopmental disorders

1.4 Sensory Processing Deficits in the Context of Impaired Inhibition

Sensory cortex undergoes a maturation and refinement during development which allows emergence of a functional architecture between the different cell types and emergence of sensory maps. These cortical dynamics are then integral to sensory discrimination, perceptual learning and behavior throughout adulthood (Froemke, 2015). The emerging cortical dynamics have to be orchestrated across multiple synapses in order to maintain the general neural coding mechanisms and strategies, including a balance between excitation and inhibition. Maturation of inhibition is synergistic with excitation and has several roles to play – (1) regulating the generation of spikes at cells, (2) shaping responses of excitatory cells to specific stimulus features, (3) regulating the brain state and, importantly, (4) defining the closure of the critical period (Isaacson & Scanziani, 2011; Fishell & Rudy, 2011). Critical periods during early postnatal development allow plasticity and refinement of cortical circuits. Several studies across sensory modalities have shown that closure of critical periods during development adults reduces the available plasticity mechanisms, thus limiting the modification of circuits and synapses in adult cortex. The closure poses a problem for neurodevelopmental disorders, where enhancing plasticity in adulthood can be beneficial and improve the efficacy of therapeutic strategies by overcoming early abnormal development and closure of critical period. Hence given the importance of inhibition in shaping cortical dynamics during development and through adulthood, any disruptions in inhibition in neurodevelopmental disorders such as ASD and FXS is a problem. This realization led to the idea that a defect in inhibition is a central problem in ASD and FXS. Further there is mounting evidence from clinicians, psychiatrists and scientists that individuals with ASD and FXS experience atypical sensory sensitivity across different sensory modalities. There is a growing consensus that sensory issues and atypical sensory processing can be predictive of and contribute to abnormal anxiety, learning and other cognitive and social deficits (Wheeler et al., 2016; Kojovic et al., 2019; Tavassoli et al., 2014; Robertson & Simmons, 2013; Green & Ben-Sasson, 2010). And as I explained at the start of this chapter the influence of sensory processing is crucial prerequisite to behavior and cognition. These different ideas have resulted in the redirection of ASD and FXS research on understanding the role of inhibition and excitation-inhibition balance in sensory processing and resulting behavior. While excitation-inhibition balance is a broad overarching concept that defines brain function and behavior to a large extent, how does the complex interplay between excitatory and inhibitory neurons generate a fine orchestration? How is this orchestration disrupted in ASD and FXS to result in atypical learning, social interactions and cognition?

Further adding to the complexity of these questions is the fact that the inhibitory system is highly complex. The complexity arises from the large variety of interneuron subtypes (Petilla Interneuron Nomenclature, G, 2008), cortical inhibitory cells can synapse onto different compartments of excitatory cells, such as soma, axonal hillock and dendrites, and interneurons innervate different layers of the cortex. Further, interneurons can control the output of excitatory cells as well as inhibitory cells. In addition to the dynamic role of inhibition, studies have shown that anatomical constraints, spatially restrict the inhibitory influence on cortical networks (Froemke, 2015). This was first observed in the hippocampus where a conserved ratio of excitatory and inhibitory synapses was found on the dendrite of the hippocampal neurons (Liu, 2004). More recently, a study using visual cortical slices showed that ratio of evoked inhibition scaled proportionally to the magnitude of excitation (Xue et al., 2014). This underscores the idea that influence of inhibition on circuit function and behavior is profound, therefore any disruptions in inhibition in neurodevelopmental disorders such as ASD need to be thoroughly examined.

1.5 Inhibition in the Neurotypical Sensory Cortex

Inhibition in the sensory cortex is mediated by three populations of interneurons–Parvalbumin (PV), Somatostatin (SST) and Vasoactive Intestinal Peptide (VIP) neurons (Gonchar & Burkhalter, 1997) (Fig. 1.2). PV interneurons, the most prevalent inhibitory neuron in the sensory cortex, synapses on the soma and axonal hillock of pyramidal cells (Gonchar & Burkhalter, 1997; Wood et al., 2017). PV cells are not selective and exhibit very broad orientation tuning by simply responding to all orientations, since they receive local input from a wide range of orientation tuned pyramidal cells (Zariwala et al., 2012; Hofer et al., 2011; Runyan & Sur, 2013). Furthermore, selective stimulation of PV cells in V1 with channelrhodopsin-2 leads to improved feature selectivity and visual discrimination (Lee et al., 2012). SST neurons contact the distal dendrite of excitatory cells and exhibit broader spikes and lower firing rates, compared to PV cells. Further, SST cells exhibit a broad range of receptive field properties, orientation and direction tuning (Ma et al., 2010). VIP neurons inhibit PV and SST interneurons, thereby reducing tonic inhibition of target pyramidal neurons. Through this disinhibitory circuit, VIP cells can increase the gain of pyramidal neurons during reinforcement learning (Letzkus et al., 2011; Pi et al., 2013; Jiang et al., 2013) by reducing tonic inhibition of target pyramidal neurons. Recent work shows that a small subset of VIP cells exhibit more canonical inhibitory properties by directly inhibiting the pyramidal neurons (Guet-McCreight et al., 2020; Harris et al., 2018; Bezaire & Soltesz, 2013). Each interneuron type has a specific role in shaping selectivity and fine-tuning excitatory output. VIP interneurons in the sensory cortex are also under the influence of cholinergic input from subcortical areas, such as the nucleus basalis. Subcortical neuromodulation is a robust mechanism that desynchronizes the pyramidal cell activity in the visual cortex, enhances the magnitude and reliability of visually evoked responses, thus aiding detection of sensory stimuli and making learned associations (Bennett et al., 2013; Fu et al., 2014; Pinto et al., 2013; Carcea & Froemke, 2013; Lee et al., 2014; Goard & Dan, 2009; Pafundo

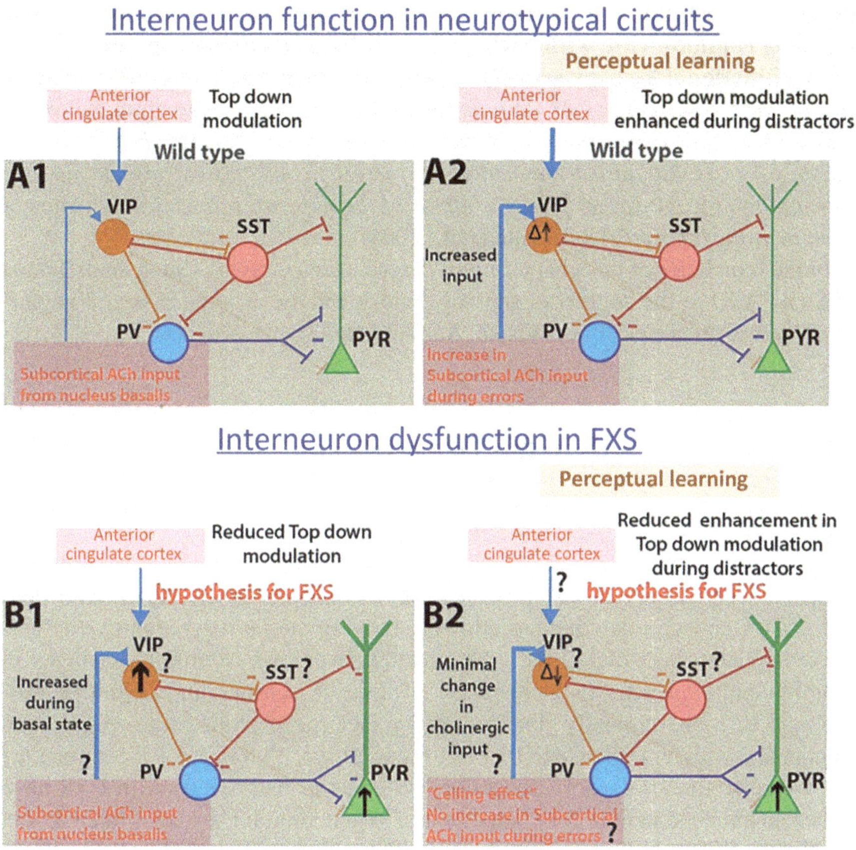

Fig. 1.2 Potential dysfunctional sensory cortical mechanisms associated with FXS and ASD. Perceptual learning and selectivity to sensory stimuli in layer 2/3 sensory cortex can be modulated by local interneuron dynamics as well as by cholinergic input from nucleus basalis (NB) and top-down input from anterior cingulate cortex (ACC). (**A1**) Neurotypical circuits under basal/unaroused conditions have a low cholinergic tone. (**A2**) During perceptual learning long range input to sensory cortex increases thus modulating VIP cells and the dynamics involving other cell types. This allows selective modulation of PYR cells. (**B1**) *Hypothesis:* Circuits in FXS have an elevated basal cholinergic tone that disrupts basal VIP activity and local interneuron dynamics, thus reducing the dynamic range and selectivity of PYR cells. (**B2**) *Hypothesis:* In FXS, disruption in long range inputs to sensory cortex, for example elevated cholinergic tone, renders the cells unable to respond to any further enhancement in cholinergic tone. Further diminished input from ACC in FXS, reduces selectivity and increases the susceptibility to distractors. These hypotheses remain to be tested. VIP: vasoactive intestinal polypeptide, PV: parvalbumin, SST: somatostatin, PYR: pyramidal

et al., 2016). Therefore, in summary VIP neurons can dynamically regulate sensory responses and plasticity as a function of the behavioral brain state and arousal of the animal (Fig. 1.2). **Hence, while the field realizes that the dysfunctional inhibition is the underlying cause of many symptoms associated with FXS**

and ASD, investigating how inhibitory neurons (and by association excitatory neurons) communicate with each other will be instrumental in understanding dysfunctional sensory sensitivity, arousal, learning and decision making in FXS and ASD (Fig. 1.2). This is underscored by the quote from Belmonte et al., 2004, "*the heterogeneity of neuropatho-logical and genetic observations in autism suggests that autism's essential characteristic may not be any specific cellular pathology, but rather a perturbation of the network properties that emerge when neurons interact*" (Belmonte et al., 2004)

In the remainder of this chapter I will describe the current research landscape in FXS and ASD in the context of sensory deficits and the associated neural impairments across three modalities–Visual, Auditory and Somatosensory.

1.6 Sensory Deficits in the Visual Domain

1.6.1 Humans

Over the decades several visual sensory deficits have been associated with FXS and ASD. One technique that has allowed studying the sensory-cognitive defects is event-related potential (ERP). This technique allows examining changes in neural population activity in response to specific sensory and cognitive processes. ERPs are typically detected using electroencephalograms (EEG) and magnetoencephalograms (MEG). EEG data consist of N1 and P2 components and are reflective of sensory processing. Studies in individuals with FXS have shown an enhancement in N1 component of visual stimuli (Van der Molen & Van der Molen, 2013), indicative of hyperexcitability and a disruption in E-I balance. Visual-motor dysfunction in ASD include disruption in drawing, constructing abstract designs (Crowe & Hay, 1990; Freund & Reiss, 1991). Other studies of children with autism showed a deficit in the ability to detect coherent motion (Milne et al., 2002) and contrast detection (Farzin et al., 2008). Children with ASD also showed reduced "spatial suppression", which is an enhancement in perceiving motion of high contrast stimuli, again indicative of disruption in excitatory-inhibitory balance (Foss-Feig et al., 2013). Another common deficit is the stronger image center bias irrespective of object distribution. Here individuals with ASD were shown images of different scenes, for example a desk or a bedroom. Using gaze tracking researchers found that, compared to neurotypical controls, individuals with ASD fixated on the center of the image, irrespective of the components in the visual scene (Wang et al., 2015). Therefore, a combination of ERP and gaze tracking methods have revealed several sensory processing issues in FXS and ASD that are manifested in a range of different symptoms. The neural mechanisms, however, remain largely unknown and this has necessitated research using animal models.

1.6.2 Mice

Some of the early studies in *Fmr1* KO mice examining the neural underpinnings of lack of *Fmr*1 were focused on examining a disruption in mGluR5 signaling. mGluR5 signaling is downregulated during development in the neurotypical cortex. However *Fmr1* KO mice showed enhanced mGluR5 signaling (Dudek & Bear, 1989). Some of the early studies linking lack of *Fmr1* to behavior, found that FMRP and mGluR5 was required for ocular dominance plasticity, a classic experience dependent plasticity in the visual system, which allows development of binocular vision. In normal mice, the responsiveness of neurons in the binocular zone of primary visual cortex (V1b) is dominated by input from the contralateral eye. However, depriving the contralateral eye of input for 3–4 d by monocular deprivation (MD) during the critical period for the visual domain (~postnatal day (P)19–P32 in the mouse), causes a shift in V1b responsiveness toward the non-deprived ipsilateral eye (Gordon & Stryker, 1996). Studies showed that FMRP constrains OD plasticity and an absence of FMRP in *Fmr1* KO mice showed aberrant OD plasticity (Dolen & Bear, 2008; Dolen et al., 2007). As the visual system matures, refines and develops through visual experience, an important milestone is keeping track of familiar features that allow recognition of familiar objects. An inability to discriminate between familiar and novel objects can contribute to disruption in higher order social cognitions, as is often observed in FXS humans and in *Fmr1* KO mice. One study found that visual familiarity is encoded in visually evoked low frequency oscillations (Kissinger et al., 2018). In *Fmr1* KO mice, these oscillations were reduced in power and shorter in duration, and mediated by a reduced enhancement in synaptic strength from layer 5 pyramidal cells onto layer 4 fast spiking cells (inhibitory cells), indicating a disruption in neural mechanisms that encode for familiarity (Kissinger et al., 2020).

Studies in humans with FXS and *Fmr1* KO mice showed that maturations and refinement of dendritic spines required FMRP and lack of FMRP in visual cortex of *Fmr1* KO resulted in immature spines. Dolen et al. 2007 found that *Fmr1* KO exhibited a reduction in spine density and this could be rescued by a 50% reduction in mGluR5 expression (Dolen et al., 2007). Therefore, these studies set the stage for the role of FMRP and atypical enhancement of mGluR5 as the neural contributors to visual processing and impairments. However, how these molecules cause dysfunctional circuits and ultimately atypical behavior is unknown.

Since inhibition has a powerful influence on maturation of cortical circuits, an investigation of the role of inhibition in FXS and ASD showed a ~50% reduction in the mRNA of α1, α3, α4, β1, β2, γ1, and γ2 GABA receptor subunits and a decrease in α1, β2, and δ GABA receptor subunits (Adusei et al., 2010; El Idrissi et al., 2005; D'Hulst et al., 2006). Further several lines of converging evidence from human and animal studies support a role for PV interneuron hypofunction in the pathogenesis of autism: (1) reduced density of PV interneurons and lower expression levels of the protein (2) the density of PNNs around interneurons is decreased. In the neurotypical cortex PV cells exhibit very broad orientation tuning (Zariwala et al., 2012;

Hofer et al., 2011; Runyan & Sur, 2013), contribute to the selectivity of pyramidal cells (Lee et al., 2012) (although SST cells have also been show to influence pyramidal cell tuning) and cortical gain of V1 (Atallah et al., 2012). Therefore given the mounting evidence that sensory processing is heavily influenced by PV cells, an emerging idea is that a disruption in PV function, leads to atypical communication, "handshake" between PV cells and pyramidal cells contributing to many symptoms in FXS and potentially ASD (Contractor et al., 2021). More generally several studies have shown that the influence of lack of *Fmr1* on inhibitory and excitatory function is varied and complex. While studies have shown a disruption in PV cells, an investigation of the contribution of this disruption to FXS symptoms is more recent (Contractor et al., 2015; Cea-Del Rio & Huntsman, 2014). In the visual cortex hypoactivity of parvalbumin (PV) expressing cortical interneurons leads to impaired perceptual learning in *Fmr1* KO mice and restoration of PV function using a DREADD approach rescues learning (Goel et al., 2018). Hypoactivity of PV cells was accompanied by a reduction in percentage of orientation tuned cells, broader tuning of existing pyramidal cells and correlated with impaired learning. Chemogenetic rescue of PV function also had ameliorative effects on orientation selectivity and tuning (Goel et al., 2018).

PV cells interact with different inhibitory and excitatory cells in a dynamic fashion and thus the influence of PV cells on the circuit and behavior is complex (Fig. 1.2). PV cells are shown to be modulated by other interneurons such as VIP cells and SST cells, as well pyramidal cells, which are excitatory. While this has not been tested, the prediction is that atypical sensory processing could result from a disruption at any, or a combination of the loci in the complex circuit (Fig. 1.2). Further, recent studies have shown that PV and pyramidal cells do not develop in isolation, rather the development and maturation of both cell types is a synergistic process requiring both their functional contribution during early postnatal development (Wong et al., 2018)– contributing to the "handshake". During neurotypical neocortical development in mice, pruning of interneuron population occurs where excess neurons are eliminated by pyramidal neuron activity (Wong & Marin, 2019; Southwell et al., 2012). In juvenile mice, during the first postnatal week, both inhibitory and pyramidal neurons collaborate in synchronous bursts of activity (Golshani et al., 2009). Collectively studies in mice suggest that development of cortical network activity and establishment of an E-I balance is dependent on the regulation of interneurons in an activity dependent manner (Wong et al., 2018; Wong & Marin, 2019; Dana et al., 2019). Therefore, reduced PV cell density in ASD could be caused by disruptions in early cortical network activity, or in their ability to establish functional connections with pyramidal neurons. Further any disruption in the "handshake" can influence development and maturations of the circuits in the sensory cortex, thus affecting sensory processing.

As shown in Fig. 1.2, PV cell function is heavily influenced by other interneurons in the circuit. In particular, the influence of VIP interneurons on PV cells, SST cells and sensory processing has received a lot of recent attention. There are many reasons for this.

VIP cells are modulated by local intracortical circuits as well as two well studied long-range afferents–top-down inputs from anterior cingulate cortex (ACC) and subcortical cholinergic inputs from nucleus basalis (NB). These long range afferents have been shown to modulate sensory processing in primary visual cortex (V1) of wild type (WT) mice. The following discussion highlights the importance of long range and local inputs to VIP cells in the context of FXS and ASD.

1. *Influence of long range afferents on VIP cells.* Top-down modulation from frontal cortical areas is important in perceptual learning by allowing visual cortical neurons to maximize their responses to behaviorally relevant stimuli and discard input from competing distractors (Desimone & Duncan, 1995; Gilbert & Li, 2013; Li et al., 2004). One specific frontal cortical area–anterior cingulate cortex has been implicated in tasks involving attention, detecting change in stimuli, error detection and contribute to emergence of stimulus selective responses (Posner & Petersen, 1990; Garavan et al., 2002; Menon et al., 2001; Fiser et al., 2016). Recent studies in mice have shown that stimulation of ACC in mice, selectively modulates visual processing (Zhang et al., 2014) and viral tracing studies identified projection neurons in ACC that densely innervate V1 (Zhang et al., 2016). Studies in humans with FXS and ASD have shown disruptions in ACC activity (Minshew & Keller, 2010), reduced AAC activity during attentive tasks (Chan et al., 2011) and reduced glutamate metabolism in ACC (Tebartz van Elst et al., 2014). Therefore, ACC improves selectivity to behaviorally relevant stimuli and FXS and ASD is associated with hypersensitivity to sensory stimuli and impairments in ACC function. However, the contribution of ACC afferent on sensory processing in V1 or its influence on VIP cells in FXS remains to be investigated. Subcortical neuromodulation to V1 enhances the magnitude and reliability of visually evoked responses, thus aiding detection of sensory stimuli and making learned associations (Bennett et al., 2013; Fu et al., 2014; Pinto et al., 2013; Jimenez-Martin et al., 2021; Hasselmo & Giocomo, 2006). Specifically, basal forebrain stimulation enhanced magnitude and reliability of orientation specific responses in V1 (Goard & Dan, 2009; Pafundo et al., 2016) and optogenetic activation of cholinergic input to V1 improved visual discrimination learning by reducing the variation over trials (Pinto et al., 2013). Several studies in *Fmr1* KO mice show overactive cholinergic signaling (Volk et al., 2007; Veeraragavan et al., 2012; D'Antuono et al., 2003) including deficits in learning (Volk et al., 2007) which can contribute to anxiety, repetitive behavior and hypersensitivity.
2. *Local effects of VIP cell dynamics:* VIP cells mediate a disinhibitory circuit, which has emerged as an intracortical mechanism for providing cholinergic modulation to neural circuits in different parts of the brain. Sensory responses by cortical neurons are modulated by different behavioral states, and cholinergic modulation has been shown to improve discrimination by reducing trial to trial variability (Goard & Dan, 2009; Pafundo et al., 2016; Metherate et al., 1992).

Visual and auditory responses are accompanied by robust, phasic activity in VIP neurons (Pi et al., 2013; Reimer et al., 2014; Fu et al., 2014), which are strongly driven by basal forebrain stimulation (Pinto et al., 2013). Thus, VIP cells can relay cholinergic inputs to excitatory neurons in order to dynamically regulate their sensory responses as a function of the behavioral brain state of the animal (Fig. 1.2). This line of reasoning is supported by the fact that VIP cells mediate the enhancement of visually-evoked responses by locomotion, i.e., a heightened state of arousal (Fu et al., 2014). Ultimately, VIP neurons inhibit PV and SST interneurons, thereby reducing tonic inhibition of target pyramidal neurons. Through this disinhibitory circuit, VIP cells can increase the gain of pyramidal neurons during reinforcement learning (Letzkus et al., 2011; Pi et al., 2013). Also, in mice performing an auditory detection task that required sustained attention, cholinergic neurons responded most strongly to the unexpectedness of the reinforcement (Hangya et al., 2015). Collectively these studies underscore the influence of VIP cells in sensory processing and suggest that disruption of VIP function can have a lasting impact on cortical dynamics and perceptual learning.

Recognizing the importance of VIP cells, recent work has examined the role of developmental disruption of VIP function. One study used a genetic approach to produce a disruption of VIP function during development (Batista-Brito et al., 2017). The genetic manipulation involved a disruption in ErbB4-Nrg1signalling, which is an important pathway in interneuron development. They found that disrupting ErbB4-Nrg1signalling did not change overall density of VIP, PV or SST cells, however dramatic impairments were reported in the cortical dynamics– further highlighting the importance of examining circuit deficits associated with interneuron dysfunction. Extracellular recordings from awake mice showed an elevation in firing, reduced bursting and decreased firing rate variability in regular spiking cells (putative excitatory cells). Further, there was a drastic reduction in phase locking of regular spiking (excitatory) cells to low frequency and gamma oscillations. Optimal sensory processing and learning requires temporal synchrony and organization is cortical circuits and this feature of cortical dynamics was disrupted by developmental disruption of VIP function. VIP cells can relay cholinergic inputs to excitatory neurons in order to dynamically regulate their sensory responses as a function of the behavioral brain state of the animal. Therefore, it was not surprising that atypical VIP development resulted in an absence of cortical state transitions, reduced orientation selectivity of RS neurons and deficits in performing a visual discrimination task (Batista-Brito et al., 2017). Investigation of VIP function deficits in a mouse model of Rett Syndrome also showed impaired cortical dynamics, state modulation and deficits in performing social tasks (Mossner et al., 2020)

Future research needs to be targeted towards improving our understanding of dysfunctional long-range influence on VIP cells and whether the link between VIP, PV, SST neurons and arousal/reinforcement is disrupted in FXS.

1.7 Sensory Deficits in the Auditory Domain

Atypical exaggerated responses to sounds and an inability to habituate to irrelevant sounds is a hallmark of the auditory hypersensitivity phenotype, often reported in FXS. It is measured using a pre-pulse inhibition (PPI) assay (Frankland et al., 2004). In this assay subjects are presented with a less intense auditory stimuli (pre-pulse stimulus) followed by a louder startle stimulus. The pre-pulse stimulus reduces the response to the following startle stimulus and is measured as PPI. Reduced PPI has been observed in individuals with FXS but not in ASD. PPI is a standard and reliable outcome measure that is often used to assay the efficacy of treatments, both in mice and humans. Further the magnitude of PPI response is correlated with deficits in attention, autism and adaptive behaviors (Frankland et al., 2004). *Fmr1* KO mice show a similar robust enhancement in PPI[109]. Auditory hypersensitivity can be associated with an atypical enhancement auditory cortex function. Indeed some of the early reports showed an enhancement in responses to sounds and broader frequency tuning of adult cortical neurons (Rotschafer & Razak, 2013). This disruption of cortical function is conserved across modalities– visual cortical neurons (Goel et al., 2018) and neurons in the somatosensory cortex (Arnett et al., 2014) also exhibit broader tuning. Collectively, studies across modalities show that neurons in the sensory cortex of *Fmr1* KO mice have reduced stimulus selectivity and hence, sensory stimuli recruit a larger proportion of cortical neurons and a potential disruption in inhibition results in a sustained enhancement of cortical activity.

Several studies in humans using EEG have shown an enhancement in the magnitude of N1 component (Van der Molen et al., 2012; Van der Molen & Van der Molen, 2013; Rojas et al., 2001; Castren et al., 2003) and a reduction in N1 habituation (Castren et al., 2003). Further, a study using MEG found an enlargement and reduced latency of the N100m[112]. Pioneering work from the Razak lab, using EEG recordings in mice, has shown parallel EEG phenotypes in *Fmr1* KO mice and individuals with FXS (Lovelace et al., 2016, 2018). Atypical EEG phenotypes include enhanced EEG gamma band power, reduced cross frequency coupling, reduced sound-evoked synchrony of neural response at gamma band frequencies and audiogenic seizures (Lovelace et al., 2018, 2020) and a lack of modulation in the auditory startle response, irrespective to stimulus intensity. This lack of modulation in the behavioral response is another feature that is observed consistently in *Fmr1* KO mice across modalities (He et al., 2017; Nielsen et al., 2002). Reduced stimulus selectivity, enhanced non phase locked EEG responses are indicative of a reduced signal to noise ratio, which can lead to a disruption in sensory discrimination. Cortical activity in the gamma range is associated with a range of sensory and cognitive processes, therefore disruption in gamma activity can contribute to many of the symptoms associated with FXS and ASD. Indeed, enhanced PPI and gamma activity is also reported in *Shank3*B null mutant mouse, a mouse model of ASD (Dhamne et al., 2017), that also causes the rare neurodevelopmental disorder Phelan McDermid Syndrome (PMS)

Aberrant gamma function can also be explained by deficits in PV function. For example, a single PV neuron synapses onto multiple neighboring pyramidal cells

(Runyan & Sur, 2013) providing synchronous inhibition that can coordinate generation of gamma activity (Cardin et al., 2009; Sohal et al., 2009). Although somatostatin cells in the visual cortex have also shown to be required for context dependent gamma rhythm generation (Veit et al., 2017). However, the contribution of any disruptions in SOM cells to sensory processing or hypersensitivity impairments in FXS remains to be investigated.

Specialized extracellular matrix structures called the perineuronal nets (PNNs) also influence PV function and observed reduction in PNN ensheathed PV cells in *Fmr1* KO have been associated with disruption in fear conditioning (Reinhard et al., 2019).

1.8 Sensory Deficits in Somatosensory Domain

The sensory deficits observed in Fmr1 KO mice prompted researchers to identify a reduction in the number of PV cells in somatosensory cortical tissue early on (Selby et al., 2007). Further somatosensory cortical neurons of *Fmr1* KO, also showed a decrease in fast spiking inhibitory neuron activity and an increase in excitatory function (Gibson et al., 2008; Hays et al., 2011; Paluszkiewicz et al., 2011), resulting in a disruption in network activity, measured as long UP states (Gibson et al., 2008; Hays et al., 2011) and decrease in synchrony of gamma waves (Gibson et al., 2008). Long UP states were specific to a deletion of *Fmr*1 in glutamatergic cells and not GABAergic (inhibitory) cells, highlighting the complexity of the interaction between excitatory and inhibitory cells. Prolonged UP states were attributed to the hyperfunction of mGluR5 receptors on excitatory neurons (Hays et al., 2011). Up states are a well-studied signature of cortical network dynamics. It is a network wide phenomenon where a large population of neurons mutually contribute to a sustained depolarization. Up states influence memory consolidation, homeostatic plasticity as well as some interpretations suggest that Up states contribute to the decorrelation of activity during arousal. Therefore, a disruption in Up states can have significant impact on sensory processing and perceptual learning. Indeed, a lack of decorrelation in somatosensory cortical network activity has been reported in *Fmr1* KO mice (Golshani et al., 2009; Goncalves et al., 2013) and this decorrelation is attributed to deficits in selective processing of stimuli and learning. A study using an ex vivo model system (organotypic slice cultures) showed a delay in emergence of Up states. This resulted in enhanced variability in the spatiotemporal structure of Up states. The variability prevented circuits from learning a temporal pattern (Motanis & Buonomano, 2020).

Reduced feedforward inhibition, mediated by PV cells, has also been observed in several other mouse models of ASD such as *Cntnap2*$^{-/-}$, *16p11.2*$^{del/+}$, *Tsc2*$^{+/-}$) (Antoine et al., 2019). Disruption in network synchrony has also been observed in Cntnap2$^{-/-}$ mice (Penagarikano et al., 2011). Ca^{2+} imaging studies during development have shown that in juvenile mice, somatosensory cortical neurons of *Fmr1* KO mice, show increased synchrony of cortical activity and an elevation in firing probability during Up states (Goncalves et al., 2013). Further young and adult *Fmr1* KO mice exhibited tactile defensiveness, a form of sensory hypersensitivity as measured by enhanced motor responses to innocuous whisker stimulation (He et al., 2017). Ca^{2+} imaging revealed that the hypersensitivity to whisker stimulation was accompanied by fewer time locked or stimulus selective layer 2/3 neurons in the somatosensory cortex as well as a reduced neuronal adaptation to the innocuous stimuli (He et al., 2017). Impaired network adaption correlated with a reduction in time locked cells. These studies reveal a reduced range of selectivity available in circuits of *Fmr1* KO mice, reducing the flexibility and potentially the number of computations that can be performed.

1.9 Conclusion

Choices are the hinges of destiny – Pythagoras.

Dissecting how the brain interrogates the environment and chooses/discriminates between stimuli of different modalities will provide an important piece to the puzzle of FXS and ASD. Research using mouse models of FXS and ASD have identified several cellular, synaptic and circuit level defects in the associated dysfunctional circuits (Fig. 1.3). In addition,

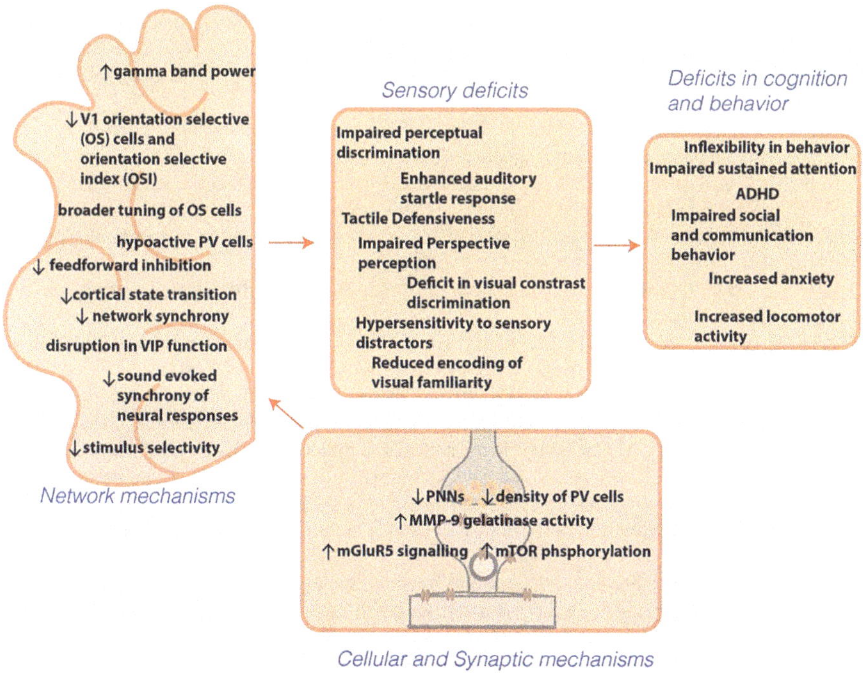

Fig. 1.3 Summary schematic shows the synaptic and network anomalies identified in mouse models of FXS and ASD, across modalities. These deficits contribute to atypical sensory issues and cognition

mouse models have provided a valuable characterization of sensory, learning and behavioral defects that resonate with human symptoms. An immediate future need in the field of FXS research is designing parallel behavior paradigms in mice and humans, that capture sensory issues, and elucidating associated circuit dysfunction in mice to provide a translation- and clinically-relevant read out with objectively measurable biomarkers.

References

Adusei, D. C., Pacey, L. K., Chen, D., & Hampson, D. R. (2010). Early developmental alterations in gabaergic protein expression in fragile x knockout mice. *Neuropharmacology, 59*, 167–171.

Antoine, M. W., Langberg, T., Schnepel, P., & Feldman, D. E. (2019). Increased excitation-inhibition ratio stabilizes synapse and circuit excitability in four autism mouse models. *Neuron, 101*, 648–661.e644.

Arnett, M. T., Herman, D. H., & McGee, A. W. (2014). Deficits in tactile learning in a mouse model of fragile x syndrome. *PLoS One, 9*, e109116.

Ashley, C. T., et al. (1993). Human and murine fmr-1: Alternative splicing and translational initiation downstream of the cgg-repeat. *Nature Genetics, 4*, 244–251.

Atallah, B. V., Bruns, W., Carandini, M., & Scanziani, M. (2012). Parvalbumin-expressing interneurons linearly transform cortical responses to visual stimuli. *Neuron, 73*, 159–170.

Bassell, G. J., & Warren, S. T. (2008). Fragile x syndrome: Loss of local mrna regulation alters synaptic development and function. *Neuron, 60*, 201–214.

Batista-Brito, R., et al. (2017). Developmental dysfunction of vip interneurons impairs cortical circuits. *Neuron, 95*, 884–895 e889.

Belmonte, M. K., et al. (2004). Autism and abnormal development of brain connectivity. *The Journal of Neuroscience, 24*, 9228–9231.

Bennett, C., Arroyo, S., & Hestrin, S. (2013). Subthreshold mechanisms underlying state-dependent modulation of visual responses. *Neuron, 80*, 350–357.

Bezaire, M. J., & Soltesz, I. (2013). Quantitative assessment of ca1 local circuits: Knowledge base for interneuron-pyramidal cell connectivity. *Hippocampus, 23*, 751–785.

Carcea, I., & Froemke, R. C. (2013). Cortical plasticity, excitatory-inhibitory balance, and sensory perception. *Progress in Brain Research, 207*, 65–90.

Cardin, J. A., et al. (2009). Driving fast-spiking cells induces gamma rhythm and controls sensory responses. *Nature, 459*, 663–667.

Castren, M., Paakkonen, A., Tarkka, I. M., Ryynanen, M., & Partanen, J. (2003). Augmentation of auditory n1 in children with fragile x syndrome. *Brain Topography, 15*, 165–171.

Cea-Del Rio, C. A., & Huntsman, M. M. (2014). The contribution of inhibitory interneurons to circuit dysfunction in fragile x syndrome. *Frontiers in Cellular Neuroscience, 8*, 245.

Chan, A. S., et al. (2011). Abnormalities in the anterior cingulate cortex associated with attentional and inhibitory control deficits: A neurophysiological study on children with autism spectrum disorders. *Research in Autism Spectrum Disorders, 5*, 254–266.

Chudley, A. E., & Hagerman, R. J. (1987). Fragile x syndrome. *The Journal of Pediatrics, 110*, 821–831.

Consortium, D.-B.F.X. (1994). Fmr1 knockout mice: A model to study fragile x mental retardation. The dutch-belgian fragile x consortium. *Cell, 78*, 23–33.

Contractor, A., Klyachko, V. A., & Portera-Cailliau, C. (2015). Altered neuronal and circuit excitability in fragile x syndrome. *Neuron, 87*, 699–715.

Contractor, A., Ethell, I. M., & Portera-Cailliau, C. (2021). Cortical interneurons in autism. *Nature Neuroscience, 24*, 1648–1659.

Crowe, S. F., & Hay, D. A. (1990). Neuropsychological dimensions of the fragile x syndrome: Support for a non-dominant hemisphere dysfunction hypothesis. *Neuropsychologia, 28*, 9–16.
D'Antuono, M., Merlo, D., & Avoli, M. (2003). Involvement of cholinergic and gabaergic systems in the fragile x knockout mice. *Neuroscience, 119*, 9–13.
D'Hulst, C., et al. (2006). Decreased expression of the GABAA receptor in fragile x syndrome. *Brain Research, 1121*, 238–245.
Dana, H., et al. (2019). High-performance calcium sensors for imaging activity in neuronal populations and microcompartments. *Nature Methods, 16*, 649–657.
Darnell, J. C., et al. (2011). Fmrp stalls ribosomal translocation on mrnas linked to synaptic function and autism. *Cell, 146*, 247–261.
Desimone, R., & Duncan, J. (1995). Neural mechanisms of selective visual attention. *Annual Review of Neuroscience, 18*, 193–222.
Dhamne, S. C., et al. (2017). Replicable in vivo physiological and behavioral phenotypes of the shank3b null mutant mouse model of autism. *Molecular Autism, 8*, 26.
Dolen, G., & Bear, M. F. (2008). Role for metabotropic glutamate receptor 5 (mglur5) in the pathogenesis of fragile x syndrome. *The Journal of Physiology, 586*, 1503–1508.
Dolen, G., et al. (2007). Correction of fragile x syndrome in mice. *Neuron, 56*, 955–962.
Dudek, S. M., & Bear, M. F. (1989). A biochemical correlate of the critical period for synaptic modification in the visual cortex. *Science, 246*, 673–675.
El Idrissi, A., et al. (2005). Decreased gaba(a) receptor expression in the seizure-prone fragile x mouse. *Neuroscience Letters, 377*, 141–146.
Ethridge, L. E., et al. (2016). Reduced habituation of auditory evoked potentials indicate cortical hyper-excitability in fragile x syndrome. *Translational Psychiatry, 6*, e787.
Ethridge, L. E., et al. (2017). Neural synchronization deficits linked to cortical hyper-excitability and auditory hypersensitivity in fragile x syndrome. *Molecular Autism, 8*, 22.
Farzin, F., Whitney, D., Hagerman, R. J., & Rivera, S. M. (2008). Contrast detection in infants with fragile x syndrome. *Vision Research, 48*, 1471–1478.
Fiser, A., et al. (2016). Experience-dependent spatial expectations in mouse visual cortex. *Nature Neuroscience, 19*, 1658–1664.
Fishell, G., & Rudy, B. (2011). Mechanisms of inhibition within the telencephalon: "Where the wild things are". *Annual Review of Neuroscience, 34*, 535–567.
Foss-Feig, J. H., Tadin, D., Schauder, K. B., & Cascio, C. J. (2013). A substantial and unexpected enhancement of motion perception in autism. *The Journal of Neuroscience, 33*, 8243–8249.
Frankland, P. W., et al. (2004). Sensorimotor gating abnormalities in young males with fragile x syndrome and fmr1-knockout mice. *Molecular Psychiatry, 9*, 417–425.
Freund, L. S., & Reiss, A. L. (1991). Cognitive profiles associated with the fra(x) syndrome in males and females. *American Journal of Medical Genetics, 38*, 542–547.
Froemke, R. C. (2015). Plasticity of cortical excitatory-inhibitory balance. *Annual Review of Neuroscience, 38*, 195–219.
Fu, Y., et al. (2014). A cortical circuit for gain control by behavioral state. *Cell, 156*, 1139–1152.
Garavan, H., Ross, T. J., Murphy, K., Roche, R. A., & Stein, E. A. (2002). Dissociable executive functions in the dynamic control of behavior: Inhibition, error detection, and correction. *NeuroImage, 17*, 1820–1829.
Gibson, J. R., Bartley, A. F., Hays, S. A., & Huber, K. M. (2008). Imbalance of neocortical excitation and inhibition and altered up states reflect network hyperexcitability in the mouse model of fragile x syndrome. *Journal of Neurophysiology, 100*, 2615–2626.
Gilbert, C. D., & Li, W. (2013). Top-down influences on visual processing. *Nature Reviews. Neuroscience, 14*, 350–363.
Goard, M., & Dan, Y. (2009). Basal forebrain activation enhances cortical coding of natural scenes. *Nature Neuroscience, 12*, 1444–1449.
Goel, A., et al. (2018). Impaired perceptual learning in a mouse model of fragile x syndrome is mediated by parvalbumin neuron dysfunction and is reversible. *Nature Neuroscience, 21*, 1404–1411.

Golshani, P., et al. (2009). Internally mediated developmental desynchronization of neocortical network activity. *The Journal of Neuroscience, 29*, 10890–10899.
Goncalves, J. T., Anstey, J. E., Golshani, P., & Portera-Cailliau, C. (2013). Circuit level defects in the developing neocortex of fragile x mice. *Nature Neuroscience, 16*, 903–909.
Gonchar, Y., & Burkhalter, A. (1997). Three distinct families of gabaergic neurons in rat visual cortex. *Cerebral Cortex, 7*, 347–358.
Gordon, J. A., & Stryker, M. P. (1996). Experience-dependent plasticity of binocular responses in the primary visual cortex of the mouse. *The Journal of Neuroscience, 16*, 3274–3286.
Green, S. A., & Ben-Sasson, A. (2010). Anxiety disorders and sensory over-responsivity in children with autism spectrum disorders: Is there a causal relationship? *Journal of Autism and Developmental Disorders, 40*, 1495–1504.
Guet-McCreight, A., Skinner, F. K., & Topolnik, L. (2020). Common principles in functional organization of vip/calretinin cell-driven disinhibitory circuits across cortical areas. *Frontiers in Neural Circuits, 14*, 32.
Hagerman, R. J., & Hagerman, P. J. (2002). *Fragile x syndrome: Diagnosis, treatment, and research* (3rd ed.). Johns Hopkins University Press.
Hangya, B., Ranade, S. P., Lorenc, M., & Kepecs, A. (2015). Central cholinergic neurons are rapidly recruited by reinforcement feedback. *Cell, 162*, 1155–1168.
Harris, K. D., et al. (2018). Classes and continua of hippocampal ca1 inhibitory neurons revealed by single-cell transcriptomics. *PLoS Biology, 16*, e2006387.
Hasselmo, M. E., & Giocomo, L. M. (2006). Cholinergic modulation of cortical function. *Journal of Molecular Neuroscience, 30*, 133–135.
Hays, S. A., Huber, K. M., & Gibson, J. R. (2011). Altered neocortical rhythmic activity states in fmr1 ko mice are due to enhanced mglur5 signaling and involve changes in excitatory circuitry. *The Journal of Neuroscience, 31*, 14223–14234.
He, C. X., et al. (2017). Tactile defensiveness and impaired adaptation of neuronal activity in the fmr1 knock-out mouse model of autism. *The Journal of Neuroscience, 37*, 6475–6487.
Hofer, S. B., et al. (2011). Differential connectivity and response dynamics of excitatory and inhibitory neurons in visual cortex. *Nature Neuroscience, 14*, 1045–1052.
Isaacson, J. S., & Scanziani, M. (2011). How inhibition shapes cortical activity. *Neuron, 72*, 231–243.
Jiang, X., Wang, G., Lee, A. J., Stornetta, R. L., & Zhu, J. J. (2013). The organization of two new cortical interneuronal circuits. *Nature Neuroscience, 16*, 210–218.
Jimenez-Martin, J., Potapov, D., Potapov, K., Knopfel, T., & Empson, R. M. (2021). Cholinergic modulation of sensory processing in awake mouse cortex. *Scientific Reports, 11*, 17525.
Kazdoba, T. M., Leach, P. T., Silverman, J. L., & Crawley, J. N. (2014). Modeling fragile x syndrome in the fmr1 knockout mouse. *Intractable & Rare Diseases Research, 3*, 118–133.
Kissinger, S. T., Pak, A., Tang, Y., Masmanidis, S. C., & Chubykin, A. A. (2018). Oscillatory encoding of visual stimulus familiarity. *The Journal of Neuroscience, 38*, 6223–6240.
Kissinger, S. T., et al. (2020). Visual experience-dependent oscillations and underlying circuit connectivity changes are impaired in fmr1 ko mice. *Cell Reports, 31*, 107486.
Kojovic, N., Ben Hadid, L., Franchini, M., & Schaer, M. (2019). Sensory processing issues and their association with social difficulties in children with autism spectrum disorders. *Journal of Clinical Medicine, 8*, 1508.
Kooy, R. F., Willemsen, R., & Oostra, B. A. (2000). Fragile x syndrome at the turn of the century. *Molecular Medicine Today, 6*, 193–198.
Lee, S. H., et al. (2012). Activation of specific interneurons improves v1 feature selectivity and visual perception. *Nature, 488*, 379–383.
Lee, A. M., et al. (2014). Identification of a brainstem circuit regulating visual cortical state in parallel with locomotion. *Neuron, 83*, 455–466.
Letzkus, J. J., et al. (2011). A disinhibitory microcircuit for associative fear learning in the auditory cortex. *Nature, 480*, 331–335.

Li, W., Piech, V., & Gilbert, C. D. (2004). Perceptual learning and top-down influences in primary visual cortex. *Nature Neuroscience, 7*, 651–657.

Liu, G. (2004). Local structural balance and functional interaction of excitatory and inhibitory synapses in hippocampal dendrites. *Nature Neuroscience, 7*, 373–379.

Lovelace, J. W., et al. (2016). Matrix metalloproteinase-9 deletion rescues auditory evoked potential habituation deficit in a mouse model of fragile x syndrome. *Neurobiology of Disease, 89*, 126–135.

Lovelace, J. W., Ethell, I. M., Binder, D. K., & Razak, K. A. (2018). Translation-relevant eeg phenotypes in a mouse model of fragile x syndrome. *Neurobiology of Disease, 115*, 39–48.

Lovelace, J. W., et al. (2020). Deletion of fmr1 from forebrain excitatory neurons triggers abnormal cellular, eeg, and behavioral phenotypes in the auditory cortex of a mouse model of fragile x syndrome. *Cerebral Cortex, 30*, 969–988.

Ma, W. P., et al. (2010). Visual representations by cortical somatostatin inhibitory neurons--Selective but with weak and delayed responses. *The Journal of Neuroscience, 30*, 14371–14379.

Marco, E. J., Hinkley, L. B., Hill, S. S., & Nagarajan, S. S. (2011). Sensory processing in autism: A review of neurophysiologic findings. *Pediatric Research, 69*, 48R–54R.

Menon, V., Adleman, N. E., White, C. D., Glover, G. H., & Reiss, A. L. (2001). Error-related brain activation during a go/nogo response inhibition task. *Human Brain Mapping, 12*, 131–143.

Metherate, R., Cox, C. L., & Ashe, J. H. (1992). Cellular bases of neocortical activation: Modulation of neural oscillations by the nucleus basalis and endogenous acetylcholine. *The Journal of Neuroscience, 12*, 4701–4711.

Miller, L. J., et al. (1999). Electrodermal responses to sensory stimuli in individuals with fragile x syndrome: A preliminary report. *American Journal of Medical Genetics, 83*, 268–279.

Milne, E., et al. (2002). High motion coherence thresholds in children with autism. *Journal of Child Psychology and Psychiatry, 43*, 255–263.

Minshew, N. J., & Keller, T. A. (2010). The nature of brain dysfunction in autism: Functional brain imaging studies. *Current Opinion in Neurology, 23*, 124–130.

Mossner, J. M., Batista-Brito, R., Pant, R., & Cardin, J. A. (2020). Developmental loss of mecp2 from vip interneurons impairs cortical function and behavior. *eLife, 9*, e55639.

Motanis, H., & Buonomano, D. (2020). Decreased reproducibility and abnormal experience-dependent plasticity of network dynamics in fragile x circuits. *Scientific Reports, 10*, 14535.

Nielsen, D. M., Derber, W. J., McClellan, D. A., & Crnic, L. S. (2002). Alterations in the auditory startle response in fmr1 targeted mutant mouse models of fragile x syndrome. *Brain Research, 927*, 8–17.

Niu, M., et al. (2017). Autism symptoms in fragile x syndrome. *Journal of Child Neurology, 32*, 903–909.

Orefice, L. L., et al. (2016). Peripheral mechanosensory neuron dysfunction underlies tactile and behavioral deficits in mouse models of asds. *Cell, 166*, 299–313.

Pafundo, D. E., Nicholas, M. A., Zhang, R., & Kuhlman, S. J. (2016). Top-down-mediated facilitation in the visual cortex is gated by subcortical neuromodulation. *The Journal of Neuroscience, 36*, 2904–2914.

Paluszkiewicz, S. M., Olmos-Serrano, J. L., Corbin, J. G., & Huntsman, M. M. (2011). Impaired inhibitory control of cortical synchronization in fragile x syndrome. *Journal of Neurophysiology, 106*, 2264–2272.

Parikshak, N. N., et al. (2013). Integrative functional genomic analyses implicate specific molecular pathways and circuits in autism. *Cell, 155*, 1008–1021.

Penagarikano, O., et al. (2011). Absence of cntnap2 leads to epilepsy, neuronal migration abnormalities, and core autism-related deficits. *Cell, 147*, 235–246.

Petilla Interneuron Nomenclature, G, et al. (2008). Petilla terminology: Nomenclature of features of gabaergic interneurons of the cerebral cortex. *Nature Reviews. Neuroscience, 9*, 557–568.

Pi, H. J., et al. (2013). Cortical interneurons that specialize in disinhibitory control. *Nature, 503*, 521–524.

Pinto, L., et al. (2013). Fast modulation of visual perception by basal forebrain cholinergic neurons. *Nature Neuroscience, 16*, 1857–1863.

Polioudakis, D., et al. (2019). A single-cell transcriptomic atlas of human neocortical development during mid-gestation. *Neuron, 103*, 785–801.e788.

Posner, M. I., & Petersen, S. E. (1990). The attention system of the human brain. *Annual Review of Neuroscience, 13*, 25–42.

Razak, K. A., Binder, D. K., & Ethell, I. M. (2021). Neural correlates of auditory hypersensitivity in fragile x syndrome. *Frontiers in Psychiatry, 12*, 720752.

Reimer, J., et al. (2014). Pupil fluctuations track fast switching of cortical states during quiet wakefulness. *Neuron, 84*, 355–362.

Reinhard, S. M., et al. (2019). Reduced perineuronal net expression in fmr1 ko mice auditory cortex and amygdala is linked to impaired fear-associated memory. *Neurobiology of Learning and Memory, 164*, 107042.

Robertson, C. E., & Baron-Cohen, S. (2017). Sensory perception in autism. *Nature Reviews. Neuroscience, 18*, 671–684.

Robertson, A. E., & Simmons, D. R. (2013). The relationship between sensory sensitivity and autistic traits in the general population. *Journal of Autism and Developmental Disorders, 43*, 775–784.

Rojas, D. C., et al. (2001). Auditory evoked magnetic fields in adults with fragile x syndrome. *Neuroreport, 12*, 2573–2576.

Ronesi, J. A., & Huber, K. M. (2008). Metabotropic glutamate receptors and fragile x mental retardation protein: Partners in translational regulation at the synapse. *Science Signaling, 1*, pe6.

Rotschafer, S., & Razak, K. (2013). Altered auditory processing in a mouse model of fragile x syndrome. *Brain Research, 1506*, 12–24.

Rubenstein, J. L., & Merzenich, M. M. (2003). Model of autism: Increased ratio of excitation/inhibition in key neural systems. *Genes, Brain, and Behavior, 2*, 255–267.

Runyan, C. A., & Sur, M. (2013). Response selectivity is correlated to dendritic structure in parvalbumin-expressing inhibitory neurons in visual cortex. *The Journal of Neuroscience, 33*, 11724–11733.

Selby, L., Zhang, C., & Sun, Q. Q. (2007). Major defects in neocortical gabaergic inhibitory circuits in mice lacking the fragile x mental retardation protein. *Neuroscience Letters, 412*, 227–232.

Sohal, V. S., Zhang, F., Yizhar, O., & Deisseroth, K. (2009). Parvalbumin neurons and gamma rhythms enhance cortical circuit performance. *Nature, 459*, 698–702.

Southwell, D. G., et al. (2012). Intrinsically determined cell death of developing cortical interneurons. *Nature, 491*, 109–113.

Sullivan, K., et al. (2006). Adhd symptoms in children with fxs. *American Journal of Medical Genetics. Part A, 140*, 2275–2288.

Tavassoli, T., Miller, L. J., Schoen, S. A., Nielsen, D. M., & Baron-Cohen, S. (2014). Sensory overresponsivity in adults with autism spectrum conditions. *Autism, 18*, 428–432.

Tebartz van Elst, L., et al. (2014). Disturbed cingulate glutamate metabolism in adults with high-functioning autism spectrum disorder: Evidence in support of the excitatory/inhibitory imbalance hypothesis. *Molecular Psychiatry, 19*, 1314–1325.

Van der Molen, M. J., & Van der Molen, M. W. (2013). Reduced alpha and exaggerated theta power during the resting-state eeg in fragile x syndrome. *Biological Psychology, 92*, 216–219.

Van der Molen, M. J., et al. (2012). Auditory and visual cortical activity during selective attention in fragile x syndrome: A cascade of processing deficiencies. *Clinical Neurophysiology, 123*, 720–729.

Veeraragavan, S., et al. (2012). Genetic reduction of muscarinic m4 receptor modulates analgesic response and acoustic startle response in a mouse model of fragile x syndrome (fxs). *Behavioural Brain Research, 228*, 1–8.

Veit, J., Hakim, R., Jadi, M. P., Sejnowski, T. J., & Adesnik, H. (2017). Cortical gamma band synchronization through somatostatin interneurons. *Nature Neuroscience, 20*, 951–959.

Velmeshev, D., et al. (2019). Single-cell genomics identifies cell type-specific molecular changes in autism. *Science, 364*, 685–689.

Volk, L. J., Pfeiffer, B. E., Gibson, J. R., & Huber, K. M. (2007). Multiple gq-coupled receptors converge on a common protein synthesis-dependent long-term depression that is affected in fragile x syndrome mental retardation. *The Journal of Neuroscience, 27*, 11624–11634.

Wang, S., et al. (2015). Atypical visual saliency in autism spectrum disorder quantified through model-based eye tracking. *Neuron, 88*, 604–616.

Wheeler, A. C., Raspa, M., Bishop, E., & Bailey, D. B., Jr. (2016). Aggression in fragile x syndrome. *Journal of Intellectual Disability Research, 60*, 113–125.

Wong, F. K., & Marin, O. (2019). Developmental cell death in the cerebral cortex. *Annual Review of Cell and Developmental Biology, 35*, 523–542.

Wong, F. K., et al. (2018). Pyramidal cell regulation of interneuron survival sculpts cortical networks. *Nature, 557*, 668–673.

Wood, K. C., Blackwell, J. M., & Geffen, M. N. (2017). Cortical inhibitory interneurons control sensory processing. *Current Opinion in Neurobiology, 46*, 200–207.

Xue, M., Atallah, B. V., & Scanziani, M. (2014). Equalizing excitation-inhibition ratios across visual cortical neurons. *Nature, 511*, 596–600.

Yu, T. W., & Berry-Kravis, E. (2014). Autism and fragile x syndrome. *Seminars in Neurology, 34*, 258–265.

Zariwala, H. A., et al. (2012). A cre-dependent gcamp3 reporter mouse for neuronal imaging in vivo. *The Journal of Neuroscience: The Official Journal of the Society for Neuroscience, 32*, 3131–3141.

Zhang, S., et al. (2014). Selective attention. Long-range and local circuits for top-down modulation of visual cortex processing. *Science, 345*, 660–665.

Zhang, S., et al. (2016). Organization of long-range inputs and outputs of frontal cortex for top-down control. *Nature Neuroscience, 19*, 1733–1742.

Chapter 2
Theory of Mind in Autism

Bertram O. Ploog

2.1 Introduction

Premack and Woodruff (1978) coined the term "Theory of Mind" (ToM) in their pioneering paper entitled "Does the chimpanzee have a theory of mind?" This paper inspired a lot of animal and human research in cognitive science, neuroscience, developmental psychology, comparative psychology, anthropology, behavior analysis, and other fields. Premack and Woodruff (1978) defined ToM as follows: "An individual has a theory of mind if he imputes mental states to himself and others." In other words, an individual has ToM, if this individual has the ability to take another individual's perspective into account and thereby can make a number of predictions about the other's intentions, beliefs, emotions, thinking, perceptions, or other mental processes. Perhaps because the term ToM has frequently been misunderstood as implying that this is about a general, scientific theory of the workings of the mind, ToM has been referred to as "mind reading" (e.g., Golan et al., 2006, 2007; Heyes, 2015; Krupenye & Call, 2019; Rutherford et al., 2002; Trope & Gaunt, 2010; Udell et al., 2011) – a concept that might be intuitive even for laypersons.

In Premack's and Woodruff's definition, it is not specified which species, other than chimpanzee, might possess ToM, but clearly it is assumed that humans do. Subsequent research, some discussed below when presenting landmark research in ToM, has included a number of different species in addition to human and chimpanzee (e.g., other primate, bird, dog, pig, goat, dolphin, and elephant). However, there is still controversy of whether ToM is uniquely human (Krupenye & Call, 2019; Wimmer & Perner, 1983). Perhaps ToM is only possible with bona fide, naturally acquired, human language but this seems curious given the fact that Premack and

B. O. Ploog (✉)
Department of Psychology, College of Staten Island and Graduate Center of The City University of New York, Staten Island, NY, USA
e-mail: bertram.ploog@csi.cuny.edu

A. El Idrissi, D. McCloskey (eds.), *Neurobiology of Autism Spectrum Disorders*,
https://doi.org/10.1007/978-3-031-42383-3_2

Woodruff (1978) conducted their pioneering work with a chimpanzee and coined the term based on that research. Perhaps ToM can only exist in a species that one assumes possesses some rudimentary form of language (e.g., see Premack, 1971, for language/chimpanzee; Kimhi, 2014, for relationship between language and ToM) as a prerequisite to perform the high-order cognitive operations such as second-order representations (Baron-Cohen et al., 1985) that underly ToM. Even with the claim of ToM being uniquely human, chances are that nonhuman animals do have something that is comparable to human ToM, at least in rudimentary form. It is therefore worthwhile to continue human and nonhuman ToM research to explore perhaps precursors to ToM or ToM-like behaviors (e.g., joint attention, Sodian & Kristen-Antonow, 2015; executive function and central coherence, Kimhi, 2014; perspective-taking in ASD, Pearson et al., 2015, and Udell et al., 2011, in dog; deception, Semple & McComb, 1996, Woodruff & Premack, 1979) that may turn out to be critical components of bona fide ToM.

As will be discussed more in context, ToM research is extremely challenging for several reasons. The two most irking issues are: Firstly, there is little consensus on a scientifically sound operational definition. Being a first, Premack's and Woodruff's original definition was still understandably vague because "imputing mental states" may encompass a large number of different phenomena, skills, abilities, or processes – each of them possibly requiring different paradigms to be applied when investigating them (for a review of different ToM paradigms, see Baron-Cohen, 2001). Secondly, there are epistemological, even ideological, differences among the researchers that study ToM. A term that may be acceptable to a cognitive scientist (e.g., "mental representation") may not pass muster for a behavioral scientist who might point out that the term "mental representation" is merely a description but not an explanatory concept in itself. There may even be ideological differences, not specific to ToM, within a discipline. Haith (1998), for example, wondered "how much of cognition is in the head of the infant and how much in the mind of the theoretician" (p. 167). Thus, Haith made a distinction between "rich" vs. "simple" interpretations of empirical data. The unfortunate consequence of all this is that there is no agreement on any particular paradigm that would scientifically, unequivocally, and universally be accepted as the definitive test of ToM, let alone as an agreement on how to interpret the empirical data obtained with any given paradigm. For more discussion on this topic, see Krupenye and Call (2019) who thoroughly reviewed these challenges and suggest a more standardized way of testing ToM, at least in nonhuman animals, in the hope of making comparisons of research findings across different fields possible.

In the following sections, just a few of the ToM landmark studies will be presented. What follows is not an exhaustive literature review of the relationship between ToM and ASD. For a review of that literature from 1985 through 2000, see Baron-Cohen (2001). For a review through 2014, see Kimhi (2014). Instead, where appropriate, a discussion of the interpretations of data will be included to further illustrate the challenges a ToM researcher might face when studying the relationship between ToM and ASD.

2.2 Critical and Landmark Studies of ToM

2.2.1 Premack and Woodruff (1978)

As stated before, Premack and Woodruff (1978) published the first paper on ToM involving a chimpanzee, Sarah. With this, for decades to come, they initiated an entire new line of human and nonhuman research in the behavioral, clinical/medical, and social sciences. In one example, in a remarkable display of complex reasoning, Sarah chose a picture of a plugged-in over an unplugged phonograph, which Premack and Woodruff (1978) took as evidence that Sarah put herself in the human's shoes who wanted to listen to music. In their conclusion, Premack and Woodruff (1978) stated that ToM is truly a "theory" because the mental "states are not directly observable [...] and can be used to make predictions." They further ask the rhetorical question: "Having decided that behaviorism is unnatural because it requires suppressing primitive inferences, whereas theories of mind are natural, can we conclude that mentalism is, therefore, preferable, and more likely to lead to valid theories?" They answer their question: "The ape could only be a mentalist. Unless we are badly mistaken, he is not intelligent enough to be a behaviorist." They specifically *rejected* explanations of ToM in terms of learning or ontogenetic experience (that a behaviorist might attempt to address) and instead proposed a purely mental, possibly nativist, explanation of ToM.

Two points are offered here for consideration: If it was true, as Premack and Woodruff (1978) claim, that Sarah *had no prior experience relevant to the testing situation* that could explain her choice but instead used ToM, that is, "read the mind" of the experimenter who wanted to listen to music, the only explanations for Sarah's behavior would be either telepathy (and not only intra- but even inter-species telepathy with high-level of abstraction from photographs or video tapes, not from real-life situation), or, alternatively, that Sarah had some knowledge of an electrician or of phonographs without ever having learned about phonographs during her lifetime. Both explanations can safely be rejected. The only other possibility is that Sarah did have some relevant experience in her life (i.e., she *learned* something in the past that she could then generalize to a novel setting). To analyze the learning (i.e., benefitting from past experience) that is involved in an individual's performance is exactly the focus of behavior-analytical research, some of which is done by behaviorists. To study material/physical phenomena in an individual's life is also consistent with the approach of other scientists who reject the term "mind" as an *explanation* for non-material, purely mental phenomena even though they might accept the term "mind" as a *description* to refer to phenomena whose existence is not denied but that are not directly observable (e.g., ToM). Regardless, it is fair to say that any explanation of unobservable phenomena remains a practical and philosophical challenge in science. An attempt to address this challenge will be offered below.

The second point for consideration is Premack's and Woodruff's (1978) apparent use of reification and circular reasoning. Most agree that Sarah was capable of truly amazing reasoning that can be *described* as ToM. But when using ToM as an explanatory concept, one violates a cardinal rule of science by engaging in reification and circular reasoning. Reification refers to "making a thing" out of something. Premack and Woodruff (1978) state that "[a]n individual has a theory of mind" where "has" implies that Sarah possesses a thing called ToM. Circular reasoning refers to first merely describing a phenomenon, then engaging in reification, and finally using the reified object as an explanation for the phenomenon under study. Applied to Sarah's case: Premack and Woodruff (1978) demonstrate a series of Sarah's remarkable behaviors (which are not denied) and *describe* them as Sarah's ability to make inferences about the human's intentions (i.e., the man wants to listen to music). These descriptions are then taken as evidence that Sarah "has" ToM (reification). Then ToM is offered as an *explanation* for the phenomenon that was to be studied originally (circular reasoning).

Another example of reification and circular reasoning specifically in relation to ToM and ASD research is provided by Lartseva et al. (2015) who write (annotation in brackets added): "One of the influential cognitive models of ASD is the Theory of Mind account. It assumes that people normally develop an ability to attribute mental states to themselves and to other people [...; here, ToM is still purely descriptive]. This ability is also called Theory of Mind reasoning or mentalizing. Mentalizing is required to reason about thoughts and desires of people, infer their feelings and beliefs, and predict their actions [now ToM is a putative explanation for people's ability to exhibit ToM] (p. 17)".

Even though not all ToM research is necessarily marred by such fallacies, a good explanation in terms of the determinants that result in ToM is still lacking.

2.2.2 Perner and Wimmer (1983)

The study by Wimmer and Perner (1983) represents the next landmark that has greatly shaped ToM research, even though they never referred to the term ToM in their General Discussion, and they avoided the use of ToM as an explanatory term throughout. Rather, Wimmer and Perner (1983) described ToM as "beliefs about beliefs" and other phenomena that appear to be related to ToM (e.g., the ability to deceive, also shown in chimpanzee, Woodruff & Premack, 1979). Through several experiments, they were able to show that the ability to have beliefs about beliefs and the ability to deceive were correlated with chronological age in typically developing 3- to 9-year-old children.

The basic paradigm was as follows: The participating child watches Maxi (doll) hiding some chocolate in the blue cupboard. Then, while Maxi is absent, Mother (doll) appears at the scene and misplaces the chocolate in the green cupboard. Then Maxi returns and wants to retrieve the chocolate. The participating child is asked by the experimenter (person) "Where do you think Maxi will look for the chocolate?" Children who answered "green cupboard" failed this test because they failed to take Maxi's perspective into account who did not observe Mother changing the location of the chocolate. The deception task is an extension of this paradigm: Maxi's Big

Brother (doll) – a competitor who also wants the chocolate – appears at the scene before Maxi returns. Upon Maxi's return, the child is asked what will Maxi tell Big Brother where the chocolate is hidden (given that Maxi does not want to share with Big Brother). If the child understands Maxi's motivation, the child should predict that Maxi will deceive Big Brother in order to save the chocolate for himself. Thus, children who predicted "blue cupboard" failed this deception test.

This paradigm relies on at least two assumptions: (1) The participating child can generalize from doll-enactments to real-life situations and (2) the child's linguistic skills are sufficient to understand receptively a complex narrative that accompanied the doll-enactment. In an additional step, when the child is asked to verbalize the rationale for their choice, also a high level of expressive language is required. These two assumptions are not problematic with typically developing children even though one can perhaps assume that linguistic skills are more limited in younger children, and the results may have less to do with testing ToM directly. This test, however, may have greater limitations when applied to children with developmental disabilities whose linguistic skills may be lacking. This type of test is clearly not useful for participants who have linguistic deficits or are even nonverbal as many children with ASD are. Therefore, this test has been modified over the years in order to make it applicable for a wider population of children with varying levels of linguistic and intellectual abilities.

2.2.3 *Baron-Cohen, Leslie, and Frith (1985)*

Baron-Cohen et al. (1985) employed the fundamentals of the Perner and Wimmer paradigm to studying the social and communication deficits that are common in people with ASD (cf. American Psychiatric Association, 2013). Baron-Cohen et al.'s (1985) approach has become known as the Sally-and-Anne test which goes like this: Sally (doll) places a marble in a basket, then leaves the scene. Anne (doll) appears and transfers the marble to a box. Upon Sally's return, the child is asked "Where will Sally look for her marble?" (Belief Question). Children who answer "In the basket" pass the test. In a second trial, the general procedure is repeated but now the marble is transferred to the experimenter's (person) pocket.

Baron-Cohen et al. (1985) employed three groups of children (autistic, typical, and with Downs syndrome), roughly matched by mental age and verbal skill. If anything, the autistic group had a higher mental age on the verbal scale than the Downs syndrome group, and the typical group, because of their younger chronological and corresponding mental age, had a lower mental age than the other two groups. In other words, if the autistic children failed the test more often than the other two groups (which is what happened), then the argument that ToM is a reflection of a higher IQ or of better verbal skills is undermined. Instead, the failure on this test can be viewed as specific to autism. Baron-Cohen et al. (1985) also employed three additional control questions: "Which doll is Sally/Anne?" (Naming Question); "Where is the marble really?" (Reality Question); and "Where was the marble in the beginning?" (Memory Question). Every child in all groups answered these three questions correctly, which indicates that the results (i.e., children with ASD underperformed on the Belief Question) were not due to confusion, lack of memory, or deficit in linguistic skills. The inclusion of a third location on the second trial is also evidence that the autistic children, when they failed the test, did not simply choose a "wrong" location but rather pointed to the actual location of the marble. This is indication of specifically having failed to take Sally's belief into account. Another strength of this study was that it did not rely on any long, complex narrative provided by the experimenter even though some level of linguistic competence was still required. Thus, inclusion of low-functioning autistic children or nonverbal children would require further modification of this paradigm. (For an example of a paradigm that allows for inclusion of nonverbal or low-functioning autistic participants – albeit, for now, not investigating ToM but perception of prosody and emotion in speech – see Brooks & Ploog, 2013; Brooks et al., 2018; Ploog et al., 2009, 2014.)

While these results were very clear and strong as far as describing the autistic children's inability to take another's belief into account is concerned (certainly a valuable and acceptable *description* of a ToM behavior), Baron-Cohen et al.'s (1985) *explanation* of the results in terms of ToM seem less convincing because of the embedded circularity of their argument. In their Discussion, they argue that their results support the hypothesis that autistic children fail to employ ToM and thereby they answered the question posed in their paper's title in the negative: Autistic children do not *have* a theory of mind. This failure is then explained by these children's inability to represent mental states – a skill that defines ToM. Earlier in their paper, in the Introduction, they state that a lack of "second-order representations" of mental states may lead to a lack of ToM with ToM being defined as "ability to impute mental states" to self and/or others (cf. Premack & Woodruff, 1978). Beyond the very clear and valuable description of the phenomenon that appears specific to autism, one learns little about the underlying determinants of either the failure to form second-order representations, the inability to impute mental states, or the failure to acquire ToM. This absence of explanatory power has to do with the operationally undefined term "second-order representation" especially when one has to assume that representations have to be somehow acquired (learned) during an individual's lifetime but no account is offered of how they are learned.

Unfortunately, with this line of reasoning, the determinants that lead to the successful or unsuccessful development of second-order representations, or of ToM itself, remain unexplained. It also remains unexplained why a lack of ToM might be specific to autism.

There are many other paradigms that have been employed for investigating ToM – each possibly addressing a different facet of ToM depending on how it is conceptualized. In addition to the false-belief and deception paradigms, Baron-Cohen (2001) discusses paradigms that address "Seeing Leads to Knowing" tests, tests of recognizing mental-state words, the ability to engage in pretend play, understanding complex emotions, perspective taking, understanding humor, sarcasm, and irony, and tests of imagination. As stated previously, which paradigm a researcher might select depends much on a specific working definition of ToM and on the researcher's epistemological view.

2.2.4 *Mirror-Neuron Research*

In an attempt to investigate actual underlying determinants of ToM and its links to autism beyond a pure description of the phenomenon itself, research in the neurosciences has focused, among other topics, on investigating so-called mirror neurons that have been identified in humans in their relation to autism (e.g., Dapretto et al., 2006; Oberman et al., 2005; Williams et al., 2001, 2006). For a comprehensive and current review of the relationship between ToM, autism, and mirror neurons, see Andreou and Skrimpa (2020). While it is beyond the scope of this chapter to cover in depth this line of research, suffice it here to describe the rationale for these investigations. Presumably, an individual's mirror neurons are activated by the individual's own action as well as by observation of another individual's same action. It has been suggested that mirror neurons are acquired through visual-motor learning (e.g., Heyes & Catmur, 2022). If indeed "acquisition" (learning) is involved, behavior analysis should be useful to identify the conditions under which mirror neurons are acquired or activated and, if mirror neurons are deficient in some way, as hypothesized in autism, behavior analysis may also offer therapeutic approaches such as applied behavior analysis (ABA) to normalize the number or the workings of mirror neurons in autistic people.

It is intuitively plausible that mirror neurons play a critical role in imitation, for example, which in turn is thought to be a possible component of ToM. Note, however, that so far mirror neurons have only been implicated in visual-motor learning. It is on shaky grounds to hypothesize that mirror neurons may be involved in allowing an individual to impute mental states to other individuals, that is, observe mental states in others and simultaneously recognize these states in self, thus showing evidence of ToM. (Outside neuroscience, this "do as others do" has been proposed also in "simulation theory", e.g., Gordon, 1986, which unfortunately does not go beyond just a description of the phenomenon that individuals possess the ability to "do as others do)." But if true, according to the mirror-neuron theory, observing the manifestations of the other individual's action (like expressing an intention or emotion) would activate neurons in the individual mirroring the same mental states.

While mirror neurons may play an important role in ToM intuitively, the actual findings linking mirror neurons and autism have been somewhat disappointing (Heyes & Catmur, 2022). These authors write that after a peak in mirror neuron research around 2013 (measured by number of research publications), a sharp decline occurred, and mirror neuron research in reference to autism research has "failed to find evidence supporting the 'broken-mirror theory' of autism."

2.2.5 Studies to Investigate Precursors or Prerequisites of ToM

Many studies have also looked into investigating precursors of Tom or the partial skills that are considered prerequisites for ToM. Even though these studies so far also have not provided a convincing analysis of what the actual determinants of ToM are, they have shed light on the next deeper level of description of ToM, which ultimately may lead to a discovery of the underlying causes that lead to the development of ToM, or a weakness or the lack thereof, in autism. One conceptual framework is provided by the Weak Central Coherence Hypothesis (WCC). WCC posits that a person has a deficit in understanding of context or, figuratively, an inability to "see the bigger picture." For example, Burnette et al. (2005) linked WCC to a lack of ToM in autism. In this study, 31 children (mean age 11.25 years) with high-functioning autism participated in a series of ToM (first- and second-order false-belief) and WCC (block design, differential abilities, embedded figures, and modified homograph) tasks. (For details and references, see Burnette et al., 2005.) One control group consisted of 17 children with learning disabilities and another control group consisted of 16 typically developing children. The results were mixed. However, there was evidence of language-specific WCC (i.e., in the homograph tasks but not in the visual-spatial measures) in the ASD group. These measures were also correlated with the ToM measures. This outcome makes sense considering that in order to be able to impute mental states to another, one has to simultaneously attend to a multitude of contextual and person-specific information (stimuli) where attention is a function of motivation and many contextual contingencies and conditions. For example, in the Sally-and-Anne task, only looking at the marble, doll, or box, or only listening to parts of the narrative or question will cause failure on this test.

Ploog (2020) compared and contrasted WCC with the stimulus-overselectivity hypothesis (Lovaas et al., 1971). In Lovaas et al.'s (1971) landmark study, coining the concept of "stimulus overselectivity," three groups of children participated: Children with autism, children with intellectual impairment, and typically developing children. All children were trained to respond to a compound stimulus comprising an auditory (tone), a visual (light), and tactile (pressure cuff) component. (Also, a temporal component was included but was not attended to by any of the children, therefore it was not considered further in the analysis.) The relevant results were that when tested with isolated components, the typical children usually attended to all three components. The children with intellectual impairments usually attended to two of the three components, whereas the autistic children attended to only one component (with no clear preference for a given sensory modality). This study inspired decades of research in stimulus overselectivity, which appears to be common in but not unique to autism. For a review of this literature see Ploog (2010). It is obvious how both, WCC and stimulus overselectivity, may touch on similar aspects of the processing of multi-dimensional and contextual information that is critical for the development of ToM. Also note that atypical selective attention has been shown in autistic children in their perception of prosody, affect, and other linguistic aspects such as statement vs. question sentences (Brooks & Ploog, 2013; Brooks et al., 2018; Ploog et al., 2009, 2014) under consideration that linguistic skills and consequently social skills may possibly be crucial for the development of ToM.

2.3 A Challenge to Study Theory of Mind: The Problem of Investigating the Unobservable in Science

When ToM research is considered in its entirety, it becomes clear that the seemingly unsurmountable hurdle is that ToM appears to be a function of inherently unobservable events, and to go beyond a pure description of ToM, and an identifaction of the causes, roots, or determinants of it, have to remain elusive. Thus, to revert to the term "mind reading" as a description is excusable. However, the observability of a phenomenon and the assumption that one deals with the material/physical world, even in the investigation of human behavior (cognition included), are critical in a true science. The neuroscience approach to identifying the role of mirror neurons in the development of ToM was one attempt to do so but as stated above seems to have ended in a dead-end.

The debate between cognitivists and behaviorists has focused for decades on how to study the "mind" – the prototypical example of a seemingly unobservable phenomenon. It has been a misrepresentation, or at least misunderstanding, that behaviorists presumably refuse to deal with the unobservable. This is not correct, certainly not for radical behaviorists. Almost 80 years ago, Skinner (1945) addressed the definitions of "psychological terms." Skinner (1957), in his *Verbal Behavior*, included a chapter on thinking (Chap. 19). Watson (1970/1930) wrote about emotions (Chap. 8) and about talking and thinking (Chap. 10). Many other behaviorists or behavior analysts have written about so-called "private events" which are not directly observable (e.g., Catania, 1995; Lowenkron, 2004; Skinner, 1974, p. 23; Tourinho, 2006a, b).

What follows is an offer to explain the investigation of "private events" (not directly observable) from a behavior analyst's perspective. Figure 2.1 (legend for terms provided in figure) shows an example of how a child and a parent may learn about private events that are not directly accessible by an outside observer. The child experiences nausea (US; private). The US elicits crying and vomiting (UR; public). Conjointly, with the US, nausea also functions as an S^D (private) which sets the occasion for the child to gesture and vocalize (R; public). The parent does not have access to the child's US or S^D, but does have access to the child's UR and R, which function as an S^D (public) for the parent. This S^D sets the occasion for the parent to respond to the child's obvious distress by comforting the child and saying "Your tummy hurts!" (R; public). The parent's comforting serves as an S^R (positive reinforcement) for the child's making her distress publicly known. That is, this S^R makes it more likely for the child in the future to indicate publicly her distress by

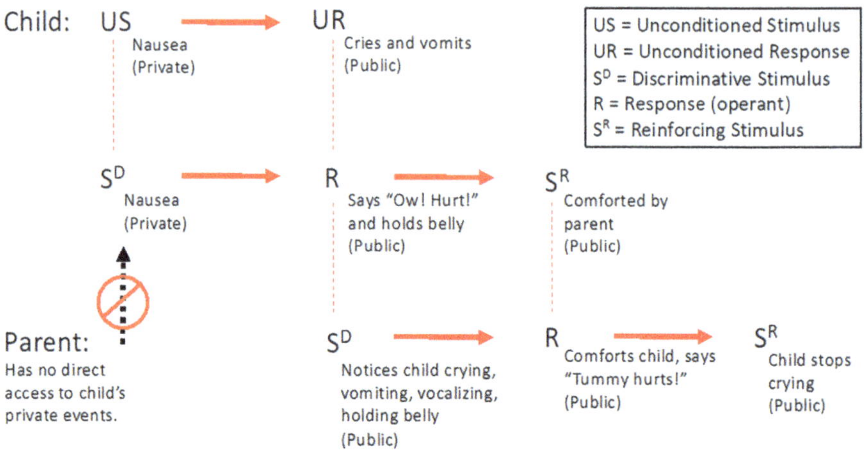

Fig. 2.1 Example of "Private Event" analysis: The child learns to say "Tummy hurts!" (public) in response to her internal state (private); the parent learns to identify the child's internal state (private) even though it is not directly observable. Note: These terms describe behavioral contingencies (psychological phenomena), not "things" such as mental schemata or representations that are said to be located inside a person. This is not to say that the child's physiology and anatomy do not provide the "hardware" for these phenomena

labeling in gesture and words her private event (i.e., experiencing nausea). When the child responds to the parent's comforting by calming down and not crying anymore, the parent's verbalizing and comforting (R) is reinforced by the child's noticeably (public) reduced distress (negative reinforcement). Note that the child's verbalizations (first perhaps only nonlinguistic vocalizations "ow, ow") are shaped to became more and more effective and efficient verbalizations (e.g., from crying and simple "Ow, ow!" to "Ow! Hurt!" to "Mom, my tummy really hurts!" in contrast to learning to say "Mom, I have a headache!") because the more the child's verbalizations approximate adult language and match the child's private event, the faster and more effective the parent can comfort the child (e.g., give warm tea for nausea as opposed to acetaminophen for a headache). An analysis of verbal behavior was offered by Skinner (1957) with critique by Chomsky (1959), and rebuttals and support by MacCorquodale (1969, 1970) and Stemmer (1990).

2.4 Conclusion

ToM research is valuable because regardless of what ToM is, how it is defined, or what scientific framework one employs, it will help improve our knowledge of possible deficits associated with autism and how to offer appropriate support, accommodations, and interventions if desired. Even at a purely descriptive level, not knowing the actual underlying causes of ToM, scientific research may inform us how to help people who have deficits to acquire ToM skills. More generally, we might glean an increased knowledge about human behavior which can be employed to improve society at large. What if a typical adult person were to show a lack of ToM (e.g., a politician's lack of empathy or a lack of understanding of a political opponent's motivations)? It would be helpful if one could develop policies (e.g., in school curricula) to incorporate ToM in the general socialization that typically occurs in our society throughout a person's lifespan. Lastly, as argued by comparative psychologists, it is also clear that nonhuman ToM research remains valuable. In nonhuman research, it is possible to conduct true experiments with their controlled environment (often impossible with humans because of limited experimental control and ethical considerations) to study partial aspects of what ToM might be, and to test to what extent these findings might be relevant for human ToM. In summary, ToM research continues to strive and continues to have value for many different fields far beyond research in autism.

References

American Psychiatricl Association. (2013). *Diagnostic and statistical manual of mental disorders* (5th ed.). Author.

Andreou, M., & Skrimpa, V. (2020). Theory of mind deficits and neurophysiological operations in autism spectrum disorders: A review. *Brain Sciences, 10*, 393.

Baron-Cohen, S. (2001). Theory of mind and autism: A review. In L. M. Glidden (Ed.), *International review of research in mental retardation: Autism* (Vol. 23, pp. 169–184). Academic.

Baron-Cohen, S., Leslie, A. M., & Frith, U. (1985). Does the autistic child have a "theory of mind"? *Cognition, 21*, 37–46.

Brooks, P. J., & Ploog, B. O. (2013). Attention to emotional tone of voice in speech perception in children with autism. *Research in Autism Spectrum Disorders, 7*, 845–857.

Brooks, P. J., Gaggi, N. L., & Ploog, B. O. (2018). Generalization of content and emotional prosody across speakers varying in gender in youth with Autism Spectrum Disorder. *Research in Developmental Disabilities, 83*, 57–68.

Burnette, C. P., Mundy, P. C., Meyer, J. A., Sutton, S. K., Vaughan, A. E., & Charak, D. (2005). Weak central coherence and its relations to theory of mind and anxiety in autism. *Journal of Autism and Developmental Disorders, 35*, 63–73.

Catania, A. C. (1995). Higher-order behavior classes: Contingencies, beliefs, and verbal behavior. *Journal of Behavior Therapy and Experimental Psychiatry, 26*, 191–200.

Chomsky, N. (1959). A review of B. F. Skinner's verbal behavior. *Language, 35*, 26–58.

Dapretto, M., Davies, M. S., Pfeifer, J. H., Scott, A. A., Sigman, M., Bookheimer, S. Y., & Iacoboni, M. (2006). Understanding emotions in others: Mirror neuron dysfunction in children with autism spectrum disorder. *Nature Neuroscience, 9*, 28–30.

Golan, O., Baron-Cohen, S., Hill, J. J., & Golan, Y. (2006). The "reading the mind in films" task: Complex emotion recognition in adults with and without autism spectrum conditions. *Social Neuroscience, 1*, 111–123.

Golan, O., Baron-Cohen, S., Hill, J. J., & Rutherford, M. D. (2007). The "Reading the Mind in the Voice" test-revised: A study of complex emotion recognition in adults with and without autism spectrum conditions. *Journal of Autism and Developmental Disorders, 37*, 1096–1106.

Gordon, R. M. (1986). Folk psychology as simulation. *Mind & Language, 2*, 158–171.

Haith, M. M. (1998). Who put the cog in infant cognition? Is rich interpretation too costly? *Infant Behavior and Development, 21*, 167–179.

Heyes, C. (2015). Animal mindreading: What's the problem? *Psychonomic Bulletin & Review, 22*, 313–327.
Heyes, C., & Catmur, C. (2022). What happened to mirror neurons? *Perspectives on Psychological Science, 17*, 153–168.
Kimhi, Y. (2014). Theory of mind abilities and deficits in autism spectrum disorders. *Topics in Language Disorders, 34*, 329–343.
Krupenye, C., & Call, J. (2019). Theory of mind in animals: Current and future directions. *WIREs Cognitive Science, 10*, e1503.
Lartseva, A., Dijkstra, T., & Buitelaar, J. K. (2015). Emotional language processing in autism spectrum disorders: A systematic review. *Frontiers in Human Neuroscience, 8*, 991.
Lovaas, O. I., Schreibman, L., Koegel, R. L., & Rehm, R. (1971). Selective responding by autistic children to multiple sensory input. *Journal of Abnormal Psychology, 77*, 211–222.
Lowenkron, B. (2004). Meaning: A verbal behavior account. *The Analysis of Verbal Behavior, 20*, 77–97.
MacCorquodale, K. (1969). B. F. Skinner's verbal behavior: A retrospective appreciation. *Journal of the Experimental Analysis of Behavior, 12*, 831–841.
MacCorquodale, K. (1970). On Chomsky's review of Skinner's verbal behavior. *Journal of the Experimental Analysis of Behavior, 13*, 83–99.
Oberman, L. M., Hubbard, E. M., McCleery, J. P., Altschuler, E. L., Ramachandran, V. S., & Pineda, J. A. (2005). EEG evidence for mirror neuron dysfunction in autism spectrum disorders. *Cognitive Brain Research, 24*, 190–198.
Pearson, A., Ropar, D., & Hamilton, A. F. d. C. (2015). A review of visual perspective taking in autism spectrum disorder. *Frontiers in Human Neuroscience, 7*, 652.
Ploog, B. O. (2010). Stimulus overselectivity four decades later: A review of the literature and its implications for current research in autism spectrum disorder. *Journal of Autism and Developmental Disorders, 40*, 1332–1349.
Ploog, B. O. (2020). Selective attention. In F. M. Volkmar (Ed.), *Encyclopedia of autism spectrum disorders* (2nd ed.). Springer.
Ploog, B. O., Banerjee, S., & Brooks, P. J. (2009). Attention to prosody (intonation) and content in children with autism and in typical children using spoken sentences in a computer game. *Research in Autism Spectrum Disorders, 3*, 743–758.
Ploog, B. O., Brooks, P. J., Scharf, A., & Aum, S. (2014). Perception of the prosody and content of sentences in an unfamiliar language in children with autism spectrum disorders. *Research in Autism Spectrum Disorders, 8*, 775–787.
Premack, D. (1971). Language in chimpanzee? *Science, 172*, 808–822.
Premack, D., & Woodruff, G. (1978). Does the chimpanzee have a theory of mind? *The Behavioral & Brain Sciences, 4*, 515–526.
Rutherford, M. D., Baron-Cohen, S., & Wheelwright, S. (2002). Reading the mind in the voice: A study with normal adults and adults with Asperger Syndrome and high functioning autism. *Journal of Autism and Developmental Disorders, 32*, 189–194.
Semple, S., & McComb, K. (1996). Behavioural deception. *Trends in Ecology & Evolution, 11*, 434–437.
Skinner, B. F. (1945). The operational analysis of psychological terms. *Psychological Review, 52*, 270–277.
Skinner, B. F. (1957). *Verbal behavior*. Appleton-Century-Crofts, Inc.
Skinner, B. F. (1974). *About behaviorism*. Vintage Books: A Division of Random House.
Sodian, B., & Kristen-Antonow, S. (2015). Declarative joint attention as a foundation of theory of mind. *Developmental Psychology, 51*, 1190–1200.
Stemmer, N. (1990). Skinner's verbal behavior, Chomsky's review, and mentalism. *Journal of the Experimental Analysis of Behavior, 54*, 307–315.
Tourinho, E. Z. (2006a). On the distinction between private events and the physiology of the organism. *The Behavior Analyst Today, 7*, 548–559.

Tourinho, E. Z. (2006b). Private stimuli, covert responses, and private events: Conceptual remarks. *The Behavior Analyst, 29*, 13–31.
Trope, Y., & Gaunt, R. (2010). Attribution and person perception. In M. A. Hogg & J. Cooper (Eds.), *The SAGE handbook of social psychology* (pp. 176–194). Sage.
Udell, M. A. R., Dorey, N. R., & Wynne, C. D. L. (2011). Can your dog read your mind? Understanding the causes of canine perspective taking. *Learning & Behavior, 39*, 289–302.
Watson, J. B. (1970). *Behaviorism [Reprint of Watson, 1930]*. W. W. Norton & Company, Inc.
Williams, J. H. G., Whiten, A., Suddendorf, T., & Perrett, D. I. (2001). Imitation, mirror neurons and autism. *Neuroscience & Biobehavioral Reviews, 25*, 287–295.
Williams, J. H. G., Waiter, G. D., Gilchrist, A., Perrett, D. I., Murray, A. D., & Whiten, A. (2006). Neural mechnanisms of imitation and "mirror neuron" functioning in autistic spectrum disorder. *Neuropsychologia, 44*, 610–621.
Wimmer, H., & Perner, J. (1983). Beliefs about beliefs: Representation and constraining function of wrong beliefs in young children's understanding of deception. *Cognition, 13*, 103–128.
Woodruff, G., & Premack, D. (1979). Intentional communication in the chimpanzee: The development of deception. *Cognition, 7*, 333–362.

Chapter 3
Prenatal and Early Life Environmental Stressors: Chemical Moieties Responsible for the Development of Autism Spectrum Disorder

Kanishk Luhach, Poonam Sharma, Niti Sharma, Neerupma Dhiman, Harsha Kharkwal, and Bhupesh Sharma

Abbreviations

5-HT	5-Hydroxytryptamine
5-MT	5-Methoxytryptamine
ASD	Autism Spectrum Disorders
BDNF	Brain Derived Neurotrophic Factor
BPA	Bisphenol-A
CNS	Central Nervous System
DNA	Deoxyribonucleic Acid
EDC	Endocrine Disruptor Chemicals
EED	Estrogenic Endocrine Disruptor

K. Luhach
Department of Pharmacology, Amity Institute of Pharmacy, Amity University Noida, Noida, Uttar Pradesh, India

Vytals Wellbeing India Private Limited, Gurugram, Haryana, India

P. Sharma · B. Sharma (✉)
Department of Pharmacology, Amity Institute of Pharmacy, Amity University Noida, Noida, Uttar Pradesh, India
e-mail: bsharma5@amity.edu

N. Sharma
Department of Pharmacology, Llyod School of Pharmacy, Knowledge Park - II, Greater Noida, Uttar Pradesh, India

N. Dhiman
Department of Chemistry, Amity Institute of Pharmacy, Amity University Noida, Noida, Uttar Pradesh, India

H. Kharkwal
Amity Institute of Phytochemistry and Phytomedicine, Amity University Noida, Noida, Uttar Pradesh, India

A. El Idrissi, D. McCloskey (eds.), *Neurobiology of Autism Spectrum Disorders*,
https://doi.org/10.1007/978-3-031-42383-3_3

IL	Interleukin
PCB	Polychlorinated Biphenyl
SSRIs	Selective Serotonin Receptor Inhibitors
TNF	Tumor Necrosis Factor
VDRs	Vitamin D Receptors
VPA	Valproic Acid

3.1 Introduction

Autism spectrum disorder (ASD) is considered a group of neurodevelopmental disorders characterized by persistent impairment in social communication/interaction and restricted/repetitive pattern of behavior (Lai et al., 2014). Several co-morbid traits are associated with ASD, such as anxiety, sleep problems, gastrointestinal issues and epilepsy, to name a few commonly occurring traits (Volkmar & Pauls, 2003). The current prevalence rate of ASD has been estimated at worldwide 1%, i.e., one in every four individual out of a sample size of 10,000 people (Pardo & Eberhart, 2007). A thorough review conducted by the Center for Disease Control and Prevention in 2014 suggests an incidence rate of 17 in 1000 individuals per year along with an increased rate of incidence from 2010 to 2014 (Christensen et al., 2019). It is suggested that the increased diagnosis might be a result of environment and the increased exposure to various environmental stressors (Christensen & Zubler, 2020; Keyes et al., 2012).

ASD presents with high heterogeneity in both phenotypical and genotypical makeup (Eapen et al., 2017). Early twin studies suggested high heritability of ASD, (Bailey et al., 1995) and genetic experiments identified gene mutations, probably resulting in altered neuronal developmental pathways and structural changes commonly associated with ASD (Luo et al., 2018). Although, a key etiological factor but genetic factors alone have not been able to completely explain the heterogeneity presented in ASD. Besides the genetics, early life environmental factors play a significant role in driving ASD pathology (M. R. Herbert et al., 2006). Studies with dizygotic twins (50% similarity in DNA sequence) and monozygotic twins (100% DNA similarity) have revealed that environment play a significant role in deciding disease outcome/development in ASD (Hegarty et al., 2020; Hertz-Picciotto et al., 2006).

Also, it is noted that close to 40–50% genetic/epigenetic variance observed in ASD might be a result of environmental influences (Edelson & Saudino, 2009; Gaugler et al., 2014; Hallmayer et al., 2011).

Environmental stressors acting prenatally/perinatally play a significant role in deciding developmental trajectories in clinical and preclinical ASD. Chemical stressors such as prenatal exposure to valproic acid (VPA) (Christianson et al., 1994), serotonin agonist/hyperserotonemia (Montgomery et al., 2018), cocaine, alcohol (A. Ornoy et al., 2015), nicotine (L. Wang et al., 2015), heavy metal (Yassa, 2014), pesticides (Shelton et al., 2012), bisphenol-A (BPA) (Kalkbrenner et al., 2014) and immune reactions/maternal infections (Shuid et al., 2021) all have been implicated in development of ASD pathology. In probability the interactions between multiple genes, and changes in gene expression as a consequence of epigenetic factors and susceptibility towards varied environmental factors are liable to cause ASD (Muhle et al., 2004).

This chapter will highlight several notable environmental stressors and present a summary of findings till date.

3.2 Prenatal/Perinatal Exposure to Environmental Stressors

The action and reaction caused by environmental factors may begin at a prenatal, and early postnatal stages of cerebral development (Watts, 2008). The environmental stimuli strongly affect the structural and functional development of brain and spinal cord along with psychopathological development (Julio-Pieper et al., 2014). The long list of environmental risk factors consists of neurotoxins such as mercury (A. V. Skalny et al., 2018), enteric bacteria (clostridia, desulfovibrio, bacteroides) metabolic products (MacFabe et al., 2007), food containing a short chain fatty acid, such as propionic acid (Foley et al., 2014), organophosphate insecticide/pesticide (De Felice et al., 2015), and maternal immune system activators such as lipopolysaccharide (LPS) or polyinosinic: polycytidylic acid (Poly IC) (Kirsten et al., 2015). Further, gestational factors like maternal hyperandrogenism (Xu et al., 2015), viral or bacterial infection (respiratory infection & rubella) (Ohkawara et al., 2015), teratogenic effects producing drugs like valproic acid (Roux & Bossu, 2018), thalidomide (Miyazaki et al., 2005), ethanol, misoprostol (Eliasen et al., 2010; A. Ornoy et al., 2015), fetal hypoxia (McGinnis et al., 2013), Chorioamnionitis (inflammation of the fetal membranes due to bacterial infection) (Cordeiro et al., 2015), and prenatal stress (Walder et al., 2014), all have been implicated the development of ASD neurobiology. These factors potentially trigger brain injury, threaten the structural and functional aspects of CNS development.

Moreover, recently few studies suggested that air pollutants specifically oxides of nitrogen (NOx), ozone (O3), particulate matter <2.5 μm (PM2.5) and PM10 (Volk et al., 2013), diesel PM, lead, manganese, nickel, and sulfur dioxide (SO2) prenatal/early postnatal exposure might increase the probability of development of ASD, later in life (Suades-González et al., 2015). Several drug molecules/small molecules of interest have been identified which have shown to introduce pathology related to ASD.

3.2.1 Valproic Acid

Valproic acid (VPA), or 2-propylpentanoic acid, is a short-chained fatty acid widely used as an antiepileptic drug. VPA readily crosses the blood brain barrier and has prominent CNS effects (Ghodke-Puranik et al., 2013). VPA at therapeutic levels, acts as a histone deacetylase inhibitor, causing hyperacetylation of histones in cultured cells (Phiel et al., 2001). Accumulated evidence indicates that in utero exposure to VPA is associated with increased risk of ASD occurrence (Dietert et al., 2011; Hertz-Picciotto et al., 2006). VPA is a potent teratogen and exposure during the first trimester is known to result in neural tube defects (Hughes et al., 2018). An association between in utero exposure to VPA and autism has been confirmed by larger cohort studies (Bromley et al., 2008; Dean et al., 2002; Moore et al., 2000; Rasalam et al., 2005), where 8.9% of the studied children exposed to VPA monotherapy were later diagnosed with either autism or Asperger's syndrome. More recently, studies have provided further evidence of an association between prenatal valproate exposure and increased risk of autism (Honybun et al., 2021). Research in neuroanatomy and neuroimaging have repeatedly shown that autism sufferers have "increased cerebral-volume" and enlarged ventricles. Others have identified irregularities in serotonin production, neuron electrophysiology, and brain chemistry through the results of neuroimaging (Ecker & Murphy, 2014; Herbert, 2005; Petropoulos et al., 2006).

Literature also points at increased chances of developing cognitive problems with first trimester exposure to VPA. A study reported that children of women exposed to VPA monotherapy in the United States and the United Kingdom had lower IQ scores at 3 years of age compared to children exposed to other common anti-epileptic drugs (Eppolito et al., 2010; Wood et al., 2015).

The prenatal exposure of Valproic acid exerts various effects such as reduced social behavioral interaction, increased restricted repetitive behaviors (Kumar et al., 2015; Luhach et al., 2021c; Schneider & Przewłocki, 2005), increased anxiety, spatial learning disability (Bambini-Junior et al., 2011; Hara et al., 2015; Luhach et al., 2021c; Mirza & Sharma, 2019), impairment in the regulation of cholinergic neuronal development, acetylcholinesterase (AChE) system (prefrontal cortex region) (C. S. Choi et al., 2014; Eissa et al., 2020), hyposensitive to pain (Schneider et al., 2008), and reduction in number of purkinje cells in the cerebellum (Morakotsriwan et al., 2016). Further, key symptomatology such as reduced preference for social novelty, hyperactivity, high chance for seizure occurrence (Romoli et al., 2018; Zamberletti et al., 2019), decreased in dendritic spine density (prefrontal cortex region & hippocampal CA1 region) (Hara et al., 2015; Takuma et al., 2014), altered BDNF-AKT-Bcl-2 signaling pathway, activated the caspase-3 activity (Guo et al., 2018), neuroinflammation (both in brain and ileum myeloperoxidase activity) (Lucchina & Depino, 2014), altered serotonin homeostasis (amygdala) (Veenstra-VanderWeele et al., 2012), reduction in cortical thickness (Mychasiuk et al., 2012), increased in utero synaptic plasticity (mPFC) (Sui & Chen, 2012), have been implicated in ASD. Also, altered Wnt1 signaling, the GSK-3β/β-catenin pathway (Go et al., 2012), Nissl-positive cell loss (prefrontal and somatosensory cortices) (Hara et al., 2015), increased brain oxidative stress (thiobarbituric acid reactive species, glutathione, catalase), nitrosative stress (nitrite/nitrate), altered calcium level and impaired blood-brain barrier permeability (Kumar et al., 2015; Kumar & Sharma, 2016; Luhach et al., 2021a) are among the most common phenotype associated with prenatal exposure to VPA.

Prenatal stress associated with VPA exposure is known to generate the pathological state associated with ASD, both clinically (Christianson et al., 1994) and in preclinical animal models (Luhach et al., 2021c; Schneider et al., 2001). Further, VPA is also known to generate epigenetic modifications associated with ASD (Nicolini & Fahnestock, 2018; Ornoy et al., 2019).

Evidence supports the hypothesis that HDAC inhibition during a critical window of embryonic brain development contributes to autism pathogenesis (Krämer et al., 2003). VPA exposure could result in elevated GABA levels, as VPA inhibits an enzyme GABA transaminase, while enhancing the activity of the GABA synthesizing enzyme glutamic acid decarboxylase (Kumamaru et al., 2014). Further, several key biochemical changes are also observed in the VPA model of ASD, such as reduced levels of phosphorylated – cAMP response element binding protein (pCREB), brain derived neurotropic factor (BDNF) both clinically and experimentally (Fatemi et al., 2010; Lee et al., 2016; Wu et al., 2017). This might explain the loss of cAMP driven neuroplasticity, neuronal survival, and neurogenesis in VPA model of ASD (Contestabile & Sintoni, 2013).

The precise mechanism through which valproic acid exposure during pregnancy causes autistic-like symptoms in mice, rats, zebrafish, and humans is unknown. In the embryonic brain, prenatal exposure to VPA results in temporary hyperacetylation of H3 and H4 histones, which induces postnatal behavioral impairments resembling autism (Phiel et al., 2001). On the other hand, prenatal exposure to valpromide, a VPA analogue lacking histone deacetylase inhibitory effect, does not cause behavioral alterations and does not cause a brief increase in the amounts of acetylated histones in embryonic mouse brains (Kataoka et al., 2013). This information supports the hypothesis that HDAC suppression during a critical stage of embryonic brain development contributes to the genesis of autism.

3.2.2 *Hyperserotonemia*

5-HT is essential for proper brain development, acting not only as a neurotransmitter but also as a neurotrophic factor (Gaspar et al., 2003; Jaiswal et al., 2015). Serotonergic neurons in the human brain first appear at 5 weeks of gestation (Sundström et al., 1993), and they rapidly expand by the 10th week (Kontur et al., 1993; Levallois et al., 1997; Shen et al., 1989). By week 15 of pregnancy, serotonin cell bodies are often organized into raphe nuclei (Whitaker-Azmitia, 2005). A consistent finding in clinical as well as experimental ASD is the increased levels of peripheral circulating serotonin and decreased function of serotonin in the brain (Madden & Zup, 2014; Tanaka et al., 2018). Behavioral and cellular alterations seen in autism may be caused by the high levels of circulating in certain autistic children's blood. High blood serotonin levels can reach a growing fetus's brain at an early stage of development and lead to the loss of serotonin terminals through a known negative feedback function of serotonin throughout development, finally manifesting in ASD (Whitaker-Azmitia, 2005).

Recently, Mostafa and colleagues have reported that autistic children with peripheral hyperserotonemia, presented with higher levels of NGF (nerve growth factor) than their control counter parts (Mostafa et al., 2021). Furthermore, a 2006 study found that moms of autistic children had lower plasma serotonin levels than mothers of generally developing children, raising the possibility that low maternal plasma serotonin may harm fetal brain development (Connors et al., 2006). Fluoxetine, a SSRI in prenatal/perinatal (first in utero, followed by early postnatal exposure via breast milk) exposure has been linked to an increased prevalence of ASD and has been shown to cause behavioral and psychological abnormalities, including depression in adulthood, in both human studies and animal models (Pedersen, 2015; Rai et al., 2013).

In conditions such as ASD the lack of serotonergic function in brain is evident from many studies, whereby substances increasing the levels of serotonin directly or by indirect means help ameliorate symptoms associated with lack of it. This finding stands true in 30% of all autistics regardless the etiology (Cai et al., 2019; Chakraborti et al., 2019; Montgomery et al., 2018; Veenstra-VanderWeele et al., 2012). Gestational use/exposure of non-selective serotonin receptor agonist such as 5 – MT (Luhach et al., 2021b), selective serotonin re-uptake inhibitor (Kobayashi et al., 2016), and tryptophan rich diet (Hranilovic et al., 2011), during the peak serotonergic development stage in mammalian organisms is known to cause pervasive harm to the serotonergic system (Kahne et al., 2002; McNamara et al., 2008; Musumeci et al., 2014). This may lead to development of symptoms associated with ASD such as, deficits in social interactions, presence of repetitive behavior, spontaneous hyperlocomotion (McNamara et al., 2008; Tanaka et al., 2018), anxiety, and lack of exploratory activity (Yang et al., 2014; Zangrossi & Graeff, 2014) in the offspring. Phenotypical features such as sexual dimorphic character of ASD, alterations to oxytocin system (McNamara et al., 2008), serotonin transporters (SERT), 5-Hydroxytryptamine (5-HT1 and 5-HT2) receptor system (Lawson et al., 2016) and increased peripheral serotonin (Muller et al., 2016) along with reduced central levels of 5-HT expression, suggests similarity of DHS model with other well established models mimicking ASD symptomatology (Kahne et al., 2002; Kumar & Sharma, 2016; McNamara et al., 2008).

The levels of BDNF and 5-HT are also inter-dependent and alterations in this system result in altered cerebral development (Popova & Naumenko, 2019). The expression of hyperlocomotive trait in animals is associated with the decrease in the levels of BDNF (Simchon Tenenbaum et al., 2015). Yoo et al. (2014) and colleagues suggested that neuronal health markers such as pCREB and BDNF, are pivotal to synapse function, neuroplasticity, and neuronal survival. Reduction in both of these markers may precipitate repetitive behavior/stereotypy in rodents (Yoo et al., 2014). The activation of microglia can result in an increased level of pro-inflammatory mediators such as TNF-α, IL-1β, IL-6 and oxidative stress markers such as TBARS, GSH, and SOD (Luhach et al., 2021b). Also, 5-HT levels in the brain regulate the immune function, thus regulating release of cytokines and inflammation (Goeden et al., 2016; Wu et al., 2019). The developmental changes induced by 5-MT are known to alter several key areas of the brain such as frontal cortex, hippocampus, cerebellum, striatum, and amygdala. It is important to note that these are also the critical areas which are innervated by serotonergic pathways directly or indirectly (Courchesne et al., 2011; Kahne et al., 2002; Pierce & Courchesne, 2001; Whitaker-Azmitia, 2005).

3.2.3 *Maternal Infections*

Borna disease virus (BDV), a non-typical, non-segmented, noncytolytic, extremely neurotropic and negative, single strand enveloped RNA virus. BDV is the prototypic member of the family Bornaviridae due to its unique biologic and genetic characteristics. The BDV virions are spherical in shape with 70–130 nm in diameter, the inner electron-dense core of 50–60 nm, outer membrane enveloped surface covered with 7 nm long spikes and nucleocapsid of 4 nm in width (De la Torre, 2002).

The intrauterine or perinatal exposures immune, infections (rubella), or environmental elements are suggested for brain damage associated with autism spectrum disorder (Mahic et al., 2017). Viruses are referred as CNS teratogens. As they can destroy the cells in the brain by apoptosis (directly) and immune-mediated (indirectly) mechanisms. The immature nervous system is highly prone to virus-induced injury or destruction by the above mechanisms (Carbone et al., 2002).

Neonatal BDV infection is known to induce autism in the rats. Neonatal BDV virus-induced neuroanatomical and behavioral changes like ASD (Hornig & Ian Lipkin, 2001).

It is the substance present in the outer membrane of gram-negative bacteria and acts as an endotoxin. It is a highly immunogenic antigen that activates the immune system to induce in vivo antibody responses. Prenatal LPS exposed mice exhibited elevated anxiety-like behavior, reduced social interaction and impaired learning and memory. And prenatal LPS exposed rats exhibited reduced pre-pulse inhibition in male offspring. The LPS exposed rodents exhibited increased excitability of pyramidal neurons, postsynaptic glutaminergic responses to NMDA induced synaptic plasticity and increased cell density. These findings suggested that LPS induced maternal immune activation (MIA) can cause ASD symptoms in offspring (Carlezon et al., 2019).

3.2.4 *Endocrine Disruptors*

The endocrine disruptor chemicals (EDCs) are exogenous chemicals, or a combination of chemicals, which can negatively impact neurodevelopment by altering endocrine function (Schug et al., 2015). EDCs can inhibit or simulate the natural hormones responsible for maintaining various functions organs of the human or animal body. EDCs are a class of compounds that have detrimental effects on human and animal cerebral health (Hamra et al., 2019; Vitor-Vieira et al., 2021; Zlatnik,

2016). The consumption and dispersion of these chemicals in many fruit and vegetable crops have now been discovered, indicating a significant potential human exposure (Lu et al., 2015). These chemicals have also been found in processed food, packaging material etc., as bisphenol A (BPA) migrates from packaging to food items (Rubin, 2011), Polychlorinated biphenyls (PCBs) have been shown to be used in several industrial applications (Ampleman et al., 2015).

3.2.4.1 Polychlorinated Biphenyls

PCBs are a group of 209 congeners—chemicals with similar structures—that are composed of a biphenyl with varied quantities of chlorine substitutions at various locations on the benzene rings. Lower-chlorinated PCBs are congeners with fewer than four chlorine substituents, whereas higher-chlorinated PCBs are those with more than four chlorine substituents. Higher-chlorinated PCBs tend to bioaccumulate, whereas lower-chlorinated PCBs are often more volatile and less harmful. PCBs were first manufactured in considerable quantities in the late 1920s, primarily for use as stabilizers, lubricants, and coolants in various industrial and commercial applications (Grimm et al., 2015). However, additional PCB manufacturing was prohibited in 1979 in response to mounting evidence of their environmental persistence and potential carcinogenicity. Humans are still exposed to legacy PCBs despite the end of large-scale PCB due to older material still being used and continued usage by unregulated markets (Koh et al., 2016). Furthermore, data that have been accumulating over the past 10 years show that people have been exposed to both "modern" PCB congeners and historical PCB congeners, both of which were inadvertent consequences of mainly pigment production (Grimm et al., 2015; S. Sethi et al., 2017a). The most plausible source of these non-legacy PCBs, which have been found in indoor and outdoor surroundings as well as in human tissues, is off gassing from typical household paints, varnishes, and caulking (Hu & Hornbuckle, 2010; S. Sethi et al., 2017a).

PCBs are easily transmitted through breast milk and the placenta (Pessah et al., 2019). Therefore, for the rapidly growing brain PCB exposure is of great concern due to the high likelihood of being accumulated in the cerebral region, especially during the first trimester and the early neonatal state. Data from the National Health and Nutrition Examination Survey (NHANES) show that women of reproductive age have been exposed to PCBs often (Thompson & Boekelheide, 2013), including the less chlorinated modern PCBs (Koh et al., 2016; Marek et al., 2013; Sunjay Sethi et al., 2017b). Pregnant women show blood concentrations of 34 total PCB congeners ranging from 16.1 to 148 ng/g of lipid (Eskenazi et al., 2017), whereas the concentration of 17 PCB congeners was 33.6–62.34 ng/g lipid (Granillo et al., 2019). Around the world, other nations have reported outcomes that are comparable. The sum of the PCBs (13 tested) in the blood of pregnant women in two distinct Japanese cohorts was 17.8–362 ng/g lipid and 20–210 ng/g lipid, respectively (Mitoma et al., 2015). In Greenland, pregnant mothers' blood levels of PCBs, averaged 107 ng/g lipid whereas in the Ukraine, it was just 27 ng/g lipid; the latter was linked to cognitive impairments in the offspring (Rosenquist et al., 2017). PCBs have also been found in human umbilical cord blood, human fetal and neonatal brain tissue (Kania-Korwel et al., 2012). It's interesting to note that post-mortem brain tissue from children with ASD contained considerably greater amounts of PCB95 than did brain tissue from control (Mitchell et al., 2012).

Most of the research discovered that prenatal PCB exposures were linked to decreased cognitive function and behavioral issues (Pessah et al., 2019). A recent pilot study found positive association between PCB 153 and ASD in people with a deletion mutation in the glutathione transferase gene but not in people without this mutation, indicating that the genetic substrate likely also affects the impact of PCBs on ASD-relevant outcomes (Bach et al., 2020). Similarly, according to a 2019 study there is positive association between plasma PCB concentrations during gestation and increased likelihood of ASD diagnosis in children aged 3–4 years old (Bernardo et al., 2019).

Preclinical research in non-human primates and rodents shows that developmental PCB exposures, predominantly through nursing or direct oral exposure in young animals, have a deleterious influence on cognitive function even when there are no negative impacts on outcomes of conception or reproduction (Gore et al., 2019; Ulbrich & Stahlmann, 2004; Winneke, 2011). Recently, the evidence on animals illustrating how developmental PCB exposure affects social behavior is of interest (Gore et al., 2019; Klocke et al., 2020). Several studies demonstrate PCB impacts on several sociability measures have been published: one in mice (Karkaba et al., 2017) and four in rats (Bell et al., 2016; Jolous-Jamshidi et al., 2010; Reilly et al., 2015; Topper et al., 2019).

3.2.4.2 Bisphenol – A

Bisphenol A (BPA) a disruptor of endocrine system is classified as an estrogenic endocrine disruptor (EEDs), is usually found to be present in polycarbonate plastics, food can lining, and micro/nano plastic based materials (Brown, 2009; Kaur et al., 2014). Leeching of BPA has been known to occur at high temperature, and acid/basic conditions (Brotons et al., 1995; Krishnan et al., 1993). Exposure of humans to BPA have increased since the dawn of plastic and recent studies have clearly indicated the same (Vandenberg et al., 2007; Welch & Mulligan, 2022). Alarmingly, when BPA levels were measured in the urine of 2500 participants (U.S.A population) who were older than 6-years-old and found that BPA was detectable in as much as 92.6% of the population (Calafat et al., 2008). BPA concentrations have been in the detectable range for maternal serum, urine, neonatal urine, and breast milk (Lee et al., 2018), suggesting perinatal exposure pattern for BPA. Studies in rodents and in vitro cell cultures have revealed that BPA can circulate throughout the body and readily crosses the placenta (Adu-Gyamfi et al., 2022; Balakrishnan et al., 2010) and blood-brain barrier (Mesnil et al., 2021; Sun et al., 2002).

Stein et al., 2015 collected samples from 46 ASD children and 52 controls. The assessment of BPA & BPA metabolites in urine samples were measure and concluded that individuals with ASD have higher urine metabolites for BPA as compared to control group, suggesting increased exposure. Also, correlation analysis proposed a higher likelihood of ASD with BPA (Stein et al., 2015). Further, another study investigated 49 children with ASD and 40 comparable age and sex matched controls indicating that both BPA and 8-oxodG levels were significantly higher in children with autism than control. The study implicates increased oxidative stress and resulting mitochondrial dysfunction due to BPA exposure to have resulted in behavioral changes seen ASD (Metwally et al., 2018). Even at low exposure levels, 85.3% maternal urine samples were found to be positive of BPA (1.2 ng/mL median value). Perinatal exposure of BPA was still found to be significant in causing ASD pathology (Hansen et al., 2021). Interestingly, prenatal BPA exposure has been shown to cause social behavior deficits in girls (4 year old) than age matched boys (Lim et al., 2017). Prenatal exposure to BPA (0.8 ng/mL, range < LOD-79.1 ng/mL, maternal urine), range, and cognitive development. No relationship between exposure and full-scale IQ, was observed at the age of 3 years in a multicenter Canadian study of 812 participants (Braun et al., 2011).

Early developmental exposures to BPA may also affect endocrine system resulting in severe and chronic neurological, reproductive, cardiovascular, metabolic, and immune effects in humans. A study has revealed that serum levels of FSH, INHB, and E2 were lower in ASD (n = 49, children) group than that of control (n = 40, children), results also revealed that BPA causes endocrine dysfunction in ASD children (Metwally et al., 2020). BPA has been shown to inhibit estrogen and testosterone-mediated synaptogenesis, possibly explaining the sex-specific findings (Johnson et al., 2016; Krishnan et al., 1993). Social and sexual behavioral developmental study in juvenile cynomolgus monkeys (Macaca fascicularis), have revealed social deficits along with demasculinization of key sexually dimorphic behaviors with prenatal exposure to BPA (Negishi et al., 2014). Lymphoblastoid cells from autistic subjects and age-matched unaffected sibling controls were exposed to BPA in vitro, an increase in the generation of reactive oxygen species (ROS) and a decrease in mitochondrial membrane potential was seen, suggesting BPA exposure results in increased oxidative stress and mitochondrial dysfunction in cells derived of autistic subjects (Kaur et al., 2014).

Further, BPA is a known disruptor of ASD-candidate genes that are involved in neuronal viability, neurite outgrowth, and learning/memory (Carr et al., 2003; Manshack et al., 2016; Zhang et al., 2020) and altered gene expression is correlated with neuronal characteristics and behaviors disrupted in response to BPA in a sex-dependent manner. Transcriptome profiling of hippocampal tissues isolated from neonatal rat offspring showed that four ASD candidate genes (Cux1, Kdm5c, Arhgap32, and Itga4) were significantly upregulated in males prenatally exposed to BPA compared to the control (Thongkorn et al., 2019, 2021). Similarly, it has been reported that mice exposed to prenatal BPA show changes in gene expression in the anterior hypothalamus, the basal nucleus of the stria terminalis, and hippocampus.

These mice were exposed in utero to human-relevant doses of BPA, resulting in a total of 259 genes that were deferentially expressed between BPA exposed and control mice. Genes that were significantly affected and are of interest were, huntingtin, betaestradiol, alpha-synuclein, Creb1, and estrogen receptor (ER) (Goldsby et al., 2017; Rosenfeld, 2017; Xu et al., 2010). Similarly, expression of corticosterone receptors, their regulator Fkbp5, corticotropin releasing hormone and its receptor, oxytocin and its receptor, estrogen receptor beta, serotonin receptors (Htr1a, Htr2a) and glutamate receptor subunit Grin2b, were modified in the limbic system of C57BL animals exposed developmentally to BPA/mix of EED, in a regionspecific, sexually-dimorphic manner suggesting developmental deficits involved in ASD later in life (Miodovnik et al., 2011; Repouskou et al., 2020).

Female CD1 mice when treated perinatally with BPS (2 or 200 μg BPS/kg/day on a wafer) from GD 9 through PND20 and showed compromised maternal care, such as time spent in the nest, time spent nest building, latency to retrieve pups, or latency to retrieve the entire litter, was evident in F0 (directly exposed) and F1 adult females exposed during the perinatal period (Catanese & Vandenberg, 2017). Wistar albino dams exposed perinatally to BPA (5 mg/kg) till after third day of parturition. Anxiety like behavior and thigmotaxis along with reduction in serum lactate concentrations were observed in pups exposed to BPA. Also, decline in neuronal number and decreased astrocytic population with reduced dendritic spines were revealed by H&E and immunohistochemical analysis, respectively, in right frontal cortices. Over expression of p21 was also detected in BPA exposed offspring. Even postnatal BPA exposure have shown to introduce ASD related behavioral changes (social deficits, anxiety, and depressive behavior), in rodents (Singha et al., 2022).

These observations are not strictly limited to mammalian species, BPA exposure can result in complex gene-environmental interactions resulting in mutation of fragile X mental retardation-1 (dFmr1)-the ortholog of human FMR1 gene, resulting in ASD (Morin et al., 2021). Embryonic exposure of Drosophila melanogaster to BPA (1 mM), have been indicated to precipitate social deficits, hyperactivity, increased grooming (repetitive behavior) (Nguyen et al., 2021), aggression along with decreases in dopamine levels, tyrosine hydroxylase activity (Kaur et al., 2015), and oxidative stress (Musachio et al., 2021). The fact that higher serum BPA were found levels in the autism spectrum disorder group compared to healthy controls suggests that EEDs may have a role in the pathogenesis of ASD (Preciados et al., 2016).

3.2.5 Pesticides

In agriculture, pesticides are frequently employed to eradicate pests that harm crops or other vegetal products. They are also utilized in public health to eliminate infectious disease vectors (i.e., mosquitoes carrying plasmodium malaria, etc.). The number and volume of pesticides used to protect plants and crops have increased along with agricultural output, endangering human health and contaminating the environment globally (Kardas et al., 2016; Rahbar et al., 2017). On an annual basis, according to estimates about 2.27 billion kilogram of pesticides is used (Shokrzadeh & Saravi, 2011).

Several experimental and epidemiological studies have highlighted the association between neurodevelopmental disorders and pre/postnatal or chronic exposure to pesticides. A recent systematic review of human and preclinical models has suggested that the gestational exposure to certain organophosphate agents could be linked to the clinical signs of ASD (Grube et al., 2011). Several case–control studies showed a positive association between the exposure to pyrethroids during pregnancy or early life and ASD (Arab & Mostafalou, 2021; Biosca-Brull et al., 2021). Moreover, a cohort study, highlighted an association between pyrethroid exposure during pregnancy or early life and ASD (He et al., 2022; Miani et al., 2021; Schmidt et al., 2017; Shelton & Hertz-Picciotto, 2015; Von Ehrenstein et al., 2019). A recent comprehensive analysis of human and preclinical models raised the possibility that clinical indications of ASD may be related to fetal exposure to certain organophosphate chemicals (Barkoski et al., 2021). Due to the few studies that have been done and the minimal current use of these pesticides, there is no evidence linking organochlorine pesticides and ASD. There were only two research that looked at this relationship. One case-control study revealed a link between organochlorine pesticides and ASD (Biosca-Brull et al., 2021), but the second case-control study found no link between organochlorine persticide exposure and an elevated risk of ASD (Roberts et al., 2007). Maternal exposure to organophosphorus pesticides, such as chlorpyrifos, was positively linked with ASD (Lyall et al., 2017).

Numerous epidemiological studies have revealed a connection between various pesticides, including diazinon, malathion, myclobutanil, imidacloprid, and dichlorodiphenyltrichloroethane, and the emergence of ASD (Biosca-Brull et al., 2021; Shelton et al., 2014).

3.2.6 Heavy Metals

Numerous neurotoxic environmental variables that may lead to neurodevelopmental impairment and ASD have been found based on epidemiological investigations. Lead, methylmercury, polychlorinated biphenyls, organophosphorus pesticides, organochlorine pesticides, asphaltites, car exhaust fumes, polycyclic aromatic hydrocarbons, polybrominated diphenyl ethers, and perfluorinated compounds are listed as xenobiotics that can cause ASD (Grandjean & Landrigan, 2006). Most epidemiological research show that rising environmental contamination, primarily from heavy metals, trichloroethylene, or vinyl chloride, is linked to an increase in ASD incidence (Alampi et al., 2021; Roberts & English, 2013; Shrader-Frechette & ChoGlueck, 2017).

The efficient functioning of the body's key organs and homeostasis may be seriously hampered by the buildup of hazardous metals (Kinney et al., 2010; Rossignol et al., 2014). High amounts of heavy metals including lead, mercury, and cadmium have been identified in the hair of autistic children. According to Tabatadze et al. (2015), children with ASD had higher levels of lead (78% and 16%), mercury (43% and 10%), and cadmium (38% and 8%), respectively, than do children without ASD. The statistical findings of the study also show a lack of trace elements including zinc, manganese, molybdenum, and selenium in the hair, which is unmistakably linked to the development of ASD (Tabatadze et al., 2015).

It has been suggested that young children are more at risk from hazardous metals even in very low quantities (Grandjean & Landrigan, 2006). Heavy metals can substantially contribute to the etiology of autism because they interfere with the operation of various enzymes, alter cell signaling mechanisms, and cause oxidative stress that results in apoptosis (J. Choi et al., 2017; Hsueh et al., 2017). High toxic loading with metals such as aluminium, cadmium, and lead were found in the hair of patients with ASD in percentages of 17.2%, 8.5%, and 4.8%, respectively. Mercury and arsenic contamination were found in percentages of 2.8% and 2.6% of patients with ASD, respectively (Fujiwara et al., 2016; Modabbernia et al., 2017). The majority of research confirms that hazardous metal levels in children with autism are increased on average and closely connected with the severity of autistic symptoms (Yasuda et al., 2013). There is evidence from several research linking mercury exposure to ASD (Adams et al., 2013). According to Vojdani et al. (2003), mercury can bind to tissue enzymes and lymphocyte receptors and trigger an autoimmune response. This theory was supported by the research of Mostafa and colleagues (Adams et al., 2007; Kern et al., 2017), who demonstrated that the level of mercury in children with ASD correlates with the level of serum anti-myelin basic protein auto-antibodies, which have the potential to alter brain functioning (Mostafalou & Abdollahi, 2017).

In a research of metal content in hair conducted in Japan in 2013 by Yasuda, 2013. on a sample of 1967 children with ASD, it was discovered that the highest load levels of arsenic in these children were 33.5 times higher than the reference level (Yasuda et al., 2013). However, some research indicates no correlation between metal exposure and ASD occurrence. More than 90% of the blood samples tested in Romania by Hessabi et al. (2019) on a cohort of 60 children with ASD between 2015 and 2017 showed As and Cd levels below detection thresholds (Elamin & Al-Ayadhi, 2014; Piras et al., 2021). Only 1.7% of children with ASD had higher cadmium levels, and none of the kids had elevated arsenic levels. Further, brain tissue of people with ASD, aluminium was found in high concentrations (Mold et al., 2018). Aluminum was located using sensitive fluorescence microscopy. The findings of additional research examining the relationship between aluminium exposure and the development of ASD appear to be equivocal (Hessabi et al., 2019).

3.2.7 Nutritional Deficiency

The development of the embryonic brain depends on maternal nutrition. Prior to conception, nutrition plays a key role in aiding developmental processes (Mohamed et al., 2015; Skalny et al., 2017). Nutrition plays a crucial role in gamete function and placental growth throughout gestation (King, 2016). The embryo goes through coordinated processes of neuronal proliferation and migration, synapse formation, myelination, and death to create the fetal brain beginning 2–3 weeks after conception (Stephenson et al., 2018). The fetus at this stage of rapid development may be predisposed to neurodevelopmental insults, as a result of increased sensitivity of the developing system, leading to disorders such as ASD (Tau & Peterson, 2010). Overall, the nutritional profile during gestation can alter fetal developmental trajectories and can also result in introduction of epigenetic changes in the progeny leading to increased probability of ASD diagnosis (Linnet et al., 2003).

The relationship between mother nutrition and offspring cognitive and behavioral capacities has been supported by mounting evidence from human epidemiological research over the past 10 years (Bale et al., 2010; Prado & Dewey, 2014). The risk of clinically diagnosed neurodevelopmental disorders, is not yet elucidated, and it is uncertain how/if maternal diet plays a role in pathogenesis of ASD. It has been established that several nutrients are essential for prenatal neuronal development. For instance, folate is a crucial co-factor in single-carbon metabolism, which is involved in the processes of DNA and RNA synthesis as well as DNA methylation. These activities are crucial during times of fast growth and development. Particularly in animal studies, inadequate folate consumption has been associated with alterations in DNA methylation (Devilbiss et al., 2015; M. Wang et al., 2017) and impaired embryonic brain development (H. S. Lee, 2015; Steegers-Theunissen et al., 2009). The synthetic form of folate, folic acid, has been demonstrated to be helpful in reducing neural tube abnormalities before conception and during the first trimester of pregnancy (Craciunescu et al., 2004; Freedman et al., 2021; Jadavji et al., 2015).

Polyunsaturated fatty acids (PUFAs), notably docosahexaenoic acid (DHA) of the omega-3 family and arachidonic acid (AA) of the omega-6 family, are structural elements of neuronal membrane phospholipids and the myelin sheath insulating the neuronal axons in addition to folate (Pitkin, 2007). Beginning in the third trimester and continuing into the early postnatal period, PUFAs accumulate quickly in the brain (Liu et al., 2010; McNamara et al., 2018), and the membrane PUFA composition is mostly reliant on the amount of DHA consumed by the mother (Martinez, 1992). In animal research, limiting DHA and its precursors during the prenatal period has a detrimental effect on cognitive and behavioral results (Lauritzen et al., 2016). Also, studies have revealed that, in comparison to children without ASD, children with ASD have greater frequencies of aberrant amino acids and lower blood levels of choline (Carlson, 2001; McCann & Ames, 2005), B6 (Freedman et al., 2021), B12 (Martineau et al., 1985; Robea et al., 2020), and folate (Robea et al., 2020; Yektaş et al., 2019). Children with ASD have been shown to have significantly lowered plasma levels of the aforementioned nutrients, which may be related to their reported significantly decreased dietary consumption of these nutrients (Jennings & Basiri, 2022).

3.2.7.1 Vitamin D

Ergocalciferol and cholecalciferol are two examples of the steroid-like compounds that fall under the umbrella of vitamin D, a fat-soluble vitamin (Petruzzelli et al., 2020; Shan et al., 2016). The skin produces vitamin D3 because of 7-dehydrocholesterol reacting with UV-B light (Grant, 2016). It then goes through a two-step hydroxylation process in the liver and kidneys to create 1,25(OH)D and 1,25(OH)2D3, which attach to vitamin D receptors (VDRs) and produce biological effect. The presence of 25-(OH)D in the blood has been proposed as the best single indication of vitamin D status since 1,25(OH)2D3 production is strictly controlled. According to Garcia-Serna and Morales (2020), vitamin D's primary function is to control the metabolism of calcium and phosphorus, which has an impact on bone formation and growth. According to recent research, 1-hydroxylase, a major enzyme for vitamin D production, as well as VDRs, are abundant in brain tissue, and vitamin D is essential for brain development (Hoel & de Gruijl, 2018). Vitamin D modulates synaptic plasticity, neuronal differentiation, oxidative stress, neuroinflammation and the dopaminergic system, having significant impacts on brain development and function (Gáll & Székely, 2021).

Numerous studies looking at the vitamin D status of children and teenagers with ASD from various nations and racial groups found that these individuals had lower vitamin D levels (Lisi et al., 2020). Several studies have indicated that children and adolescents with ASD had significantly lowered vitamin D concentrations than participants in the control group (Altun et al., 2018; Alzghoul et al., 2020; Arastoo et al., 2018; Basheer et al., 2017; Şengenç et al., 2020). Numerous research has looked at the effects of vitamin D intake during pregnancy on brain development.

According to a systematic review and meta-analysis low prenatal 25-(OH)D levels (20 ng/mL) had more ASD-related symptoms, more behavioral issues, and lower social skills at age 5 (Z. Wang et al., 2021). Further, researchers measured maternal vitamin D levels in a study involving 4229 children in the second trimester and at birth, it was concluded that vitamin D deficiency during pregnancy increased the risk of ASD-like symptoms in childhood (López-Vicente et al., 2019). Also, meta-analysis offered evidence that prenatal exposure to increased 25-(OH)D levels is associated with improved cognitive development and a reduced risk of attention-deficit/hyperactivity disorder and ASD related traits later in life (Vinkhuyzen et al., 2018).

In rats, a severe vitamin D shortage during pregnancy can result in pathological alterations such as ventricular and brain volume expansion and have an impact on neuron differentiation and axonal connection (Du et al., 2017). Young children with ASD and their parents' offspring exhibit behaviors that are comparable to those of animals lacking in vitamin D. It has been demonstrated that an animal model of vitamin D insufficiency may replicate the phenotype associated with ASD in the field of neuroanatomy (Ali et al., 2019; Principi & Esposito, 2020). In the meanwhile, some researchers have found that vitamin D can treat or prevent ASD in rats that have been exposed to propionic acid-induced ASD (Ali et al., 2019).

Recent genetic research on ASD has revealed that afflicted infants have several neonatal mutations (Berk et al., 2013; Du et al., 2017). Through several ways, vitamin D may carry out DNA maintenance and repair tasks. Currently, it has been established that more than five vitamin D-dependent genes produce proteins that repair DNA damage on a constant basis (De Rubeis et al., 2014). Additionally, research has shown that low vitamin D levels cause the DNA repair enzyme poly (adenosine diphosphate ribose) polymerase to overreact and damage nearby DNA. Daily supplementation with low doses of vitamin D3 has been shown to increase Bax levels, promote apoptosis, and guard against gene mutations (Fleet et al., 2012). Excessive neurons in the anterior prefrontal and dorsolateral prefrontal cortices play a significant role in the enlargement of the ASD brain at the cellular level (Fedirko et al., 2009). According to research, vitamin D can stop cells from multiplying by promoting the growth of the cyclin-dependent kinase inhibitors p21 and p27.

Additionally, it can stop the production of other proteins necessary for the cell cycle, such cyclin D1 and the proliferating cell nuclear antigen (Hazlett et al., 2011). According to research, chronic vitamin D administration to rodents can enhance the production of GABA, dopamine metabolism, restore glial cell activity in several important regions associated with ASD (Marini et al., 2010; Staal et al., 2012). Also, vitamin D has been shown to reduce cerebral inflammation by reducing inflammatory cytokines (TNF-α, IL-6 and IL-1β) and oxidative stress (Pertile et al., 2018).

Therefore, increased neuronal cell proliferation, reduced inflammation, and oxidative stress that results in early-stage brain enlargement and pathological changes when vitamin D levels are low may be related to the development of autism spectrum disorders (ASD).

3.2.7.2 Amino Acids

Non-essential amino acids can be produced by the body via conversion processes, but essential amino acids must be supplied through diet. The CNS is known to be significantly influenced by amino acids levels, as amino acids such as tyrosine and tryptophan are precursors to neurotransmitters such as serotonin and dopamine (El-Sharkawy & Malki, 2020; Harutyunyan et al., 2021). Serotonin has a role in brain development by influencing synaptogenesis, cortical plasticity, migration, cell division, and cell proliferation while influencing mood, memory, motivation, and reward processes (Celada et al., 2013; Gaspar et al., 2003; Hellmer & Nyström, 2017). Dopamine controls movement, motivation, focus, and the way rewards are processed (Jenkins et al., 2021; Rose'Meyer, 2013). Dopamine and serotonin levels are known to be diminished in autistic individuals (Doya, 2000; Schultz, 1997), therefore appropriate amounts of tryptophan and tyrosine may maintain normal levels of dopamine and serotonin. Tyrosine deficiency has been linked to a loss in attention and an increase in hyperactivity in children without ASD (Farook et al., 2012). Differences in the plasma levels of amino acids among individuals with ASD and individuals without ASD have been documented.

Plasma levels have been used to support high amounts of the amino acid lysine (Petersen & Posner, 2012; Strömland et al., 1994). High quantities of the crucial amino acid lysine have been linked to behavioral or intellectual problems (Aldred et al., 2003; Tirouvanziam et al., 2012). An earlier study (Iovcheva & Dessev, 1980) has also shown that children with ASD are deficient in the amino acid lysine. Agitation can result from low amounts of the important amino acid lysine (Arnold et al., 2003). This might be because glutamate, which is necessary to produce the neurotransmitter GABA, is not synthesized enough (Arum et al., 2017). The primary inhibitory neurotransmitter is GABA, and mood disorders are linked to low levels of GABA (Papes et al., 2001). Researchers found elevated tryptophan and phenylalanine levels to be associated with increased hyperactivity (Petty, 1995). Also, Low levels of tyrosine can contribute to ASD symptoms, such as the inability to concentrate, focus, and increased levels of hyperactivity (Arum et al., 2017). Not only amino acids are involved in the pathophysiology of ASD, B-vitamins are implicated.

3.2.7.3 B-Vitamins

Body receives B vitamins via the diet and supplements, such as fortification and enrichment. Numerous bodily activities, including those involving the central nervous system (CNS), oxygen transport, blood cell creation, and the synthesis and conversion of amino acids, depend on B vitamins (Arnold et al., 2003). It is well known that folate helps homocysteine turn into methionine (Adams & Holloway, 2004; Hunt et al., 2014; Robea et al., 2020). Through the transport of nitrogen, vitamin B6 aids in the conversion and breakdown of amino acids. Hemoglobin, glutathione, and neurotransmitters (dopamine and serotonin) are also produced as a result (Agam et al., 2020). ASD symptoms may improve with vitamin B6 and magnesium supplementation, according to earlier study (Sato, 2018). Each of these vitamins has a unique impact on how the CNS works. A folate shortage may result in altered behavior and cognitive impairment, whereas a vitamin B6 deficiency may produce irritability (Martineau et al., 1985; Rimland, 1988). Symptoms of ASD, including as irritability, motor difficulties, cognitive impairment, and brain cell death, are all often associated with vitamin B12 insufficiency (Hunt et al., 2014).

According to research, getting enough B vitamins is crucial for preventing behavioral and cognitive impairments, which are serious issues for people with ASD. Schmidt et al. investigated the connections between maternal vitamin supplement use before conception and during fetal development and autism, in several studies (Hunt et al., 2014; Kennedy, 2016). Another study determined if multivitamin supplementation during pregnancy and maternal levels of plasma folate and B12 had an association with the incidence of ASD (Schmidt et al., 2011, 2012,

2017). The risk of having a kid with autistic features was not significantly decreased by folic acid intake, according to the study (Raghavan et al., 2018). Contrarily, another research with a bigger sample size (n = 85,176) discovered that folic acid lowers the risk of ASD. Surén et al. investigated the relationship between folic acid supplementation before pregnancy and the likelihood that a kid will experience ASD (Steenweg-de Graaff et al., 2015).

3.3 Conclusion

Several molecules and factors can affect CNS development at various stages of gestational, neonatal, and postnatal development. Besides fetal, maternal exposure to these untoward environmental factors can lead to altered developmental trajectories in the fetus resulting in precipitation or expression of traits associated with neurodevelopmental disorders, especially ASD. Several clinical studies pointed at the involvement of various prenatal factors and their association with onset of ASD. Also, a lot of evidence points at the interplay of genes and their expression with environmental stressors in animal models of ASD utilizing these stressors. However, exact mechanism resulting in the development of ASD are still baffling to say the least. Considering current evidence, it is pertinent to say that environmental stressors play a major role, when describing the increase in the incidence rate of ASD diagnosis later in life. As a result of clinical evidence, susceptible populations are at increased risk of developing ASD and related neurodevelopmental disorders upon birth. Also, preclinical models utilizing these stressors have resulted in accelerated understanding of the basic pathology at play, while bringing us one step closer to possible treatment and prevention of ASD occurrence in a susceptible population.

Better understanding of how these environmental stressors can cause ASD would result in development of strategies and advancing treatment and prevention methodologies. Likely reducing the ASD related burden in society.

References

Adams, J. B., & Holloway, C. (2004). Pilot study of a moderate dose multivitamin/mineral supplement for children with autistic spectrum disorder. *Journal of Alternative and Complementary Medicine, 10*(6), 1033–1039. https://doi.org/10.1089/ACM.2004.10.1033

Adams, J. B., Romdalvik, J., Ramanujam, V. M. S., & Legator, M. S. (2007). Mercury, lead, and zinc in baby teeth of children with autism versus controls. *Journal of Toxicology and Environmental Health. Part A, 70*(12), 1046–1051. https://doi.org/10.1080/15287390601172080

Adams, J. B., Audhya, T., McDonough-Means, S., Rubin, R. A., Quig, D., Geis, E., Gehn, E., Loresto, M., Mitchell, J., Atwood, S., Barnhouse, S., & Lee, W. (2013). Toxicological status of children with autism vs. neurotypical children and the association with autism severity. *Biological Trace Element Research, 151*(2), 171–180. https://doi.org/10.1007/S12011-012-9551-1

Adu-Gyamfi, E. A., Rosenfeld, C. S., & Tuteja, G. (2022). The impact of bisphenol A on the placenta. *Biology of Reproduction, 106*(5), 826–834. https://doi.org/10.1093/BIOLRE/IOAC001

Agam, G., Taylor, Z., Vainer, E., & Golan, H. M. (2020). The influence of choline treatment on behavioral and neurochemical autistic-like phenotype in Mthfr-deficient mice. *Translational Psychiatry, 10*(1), 316. https://doi.org/10.1038/S41398-020-01002-1

Alampi, J. D., Lanphear, B. P., Braun, J. M., Chen, A., Takaro, T. K., Muckle, G., Arbuckle, T. E., & Mccandless, L. C. (2021). Association between gestational exposure to toxicants and autistic behaviors using Bayesian quantile regression. *American Journal of Epidemiology, 190*(9), 1803–1813. https://doi.org/10.1093/AJE/KWAB065

Aldred, S., Moore, K. M., Fitzgerald, M., & Waring, R. H. (2003). Plasma amino acid levels in children with autism and their families. *Journal of Autism and Developmental Disorders, 33*(1), 93–97. https://doi.org/10.1023/A:1022238706604

Ali, A., Vasileva, S., Langguth, M., Alexander, S., Cui, X., Whitehouse, A., McGrath, J. J., & Eyles, D. (2019). Developmental vitamin D deficiency produces behavioral phenotypes of relevance to autism in an animal model. *Nutrients, 11*(5), 1187. https://doi.org/10.3390/NU11051187

Altun, H., Kurutaş, E. B., Şahin, N., Güngör, O., & Fındıklı, E. (2018). The levels of vitamin D, vitamin D receptor, homocysteine and complex B vitamin in children with autism spectrum disorders. *Clinical Psychopharmacology and Neuroscience, 16*(4), 383–390. https://doi.org/10.9758/CPN.2018.16.4.383

Alzghoul, L., Al-Eitan, L. N., Aladawi, M., Odeh, M., & Abu Hantash, O. (2020). The association between serum vitamin D3 levels and autism among Jordanian boys. *Journal of Autism and Developmental Disorders, 50*(9), 3149–3154. https://doi.org/10.1007/S10803-019-04017-W

Ampleman, M. D., Martinez, A., DeWall, J., Rawn, D. F. K., Hornbuckle, K. C., & Thorne, P. S. (2015). Inhalation and dietary exposure to PCBs in urban and rural cohorts via congener-specific measurements. *Environmental Science & Technology, 49*(2), 1156–1164. https://doi.org/10.1021/ES5048039

Arab, A., & Mostafalou, S. (2021). Neurotoxicity of pesticides in the context of CNS chronic diseases. *International Journal of Environmental Health Research, 32*, 2718–2755. https://doi.org/10.1080/09603123.2021.1987396

Arastoo, A. A. et al. (2018). Evaluation of serum 25-Hydroxy vitamin D levels in children with autism Spectrum disorder. *Italian Journal of Pediatrics, 44*(1), 150. https://doi.org/10.1186/s13052-018-0587-5

Arnold, G. L., Hyman, S. L., Mooney, R. A., & Kirby, R. S. (2003). Plasma amino acids profiles in children with autism: Potential risk of nutritional deficiencies. *Journal of Autism and Developmental Disorders, 33*(4), 449–454. https://doi.org/10.1023/A:1025071014191

Arum, P., Amareta, D. I., & Zannah, F. (2017). Phenylalanine and tryptophan intake of hyperactive children with autism. *Journal of Biomedical and Translational Research, 3*(2), 34. https://doi.org/10.14710/JBTR.V3I2.1744

Bach, M. A., Samms-Vaughan, M., Hessabi, M., Bressler, J., Lee, M. J., Zhang, J., Shakespeare Pellington, S., Grove, M. L., Loveland, K. A., & Rahbar, M. H. (2020). Association of polychlorinated biphenyls and organochlorine pesticides with autism spectrum disorder in Jamaican children. *Research in Autism Spectrum Disorder, 76*, 101587. https://doi.org/10.1016/J.RASD.2020.101587

Bailey, A. et al. (1995). Autism as a strongly genetic disorder: evidence from a British twin study. *Psychological Medicine, 25*(1), 63–77. https://doi.org/10.1017/s0033291700028099

Balakrishnan, B., Henare, K., Thorstensen, E. B., Ponnampalam, A. P., & Mitchell, M. D. (2010). Transfer of bisphenol A across the human placenta. *American Journal of Obstetrics and Gynecology, 202*(4), 393.e1–393.e7. https://doi.org/10.1016/J.AJOG.2010.01.025

Bale, T. L., Baram, T. Z., Brown, A. S., Goldstein, J. M., Insel, T. R., McCarthy, M. M., Nemeroff, C. B., Reyes, T. M., Simerly, R. B., Susser, E. S., & Nestler, E. J. (2010). Early life programming and neurodevelopmental disorders. *Biological Psychiatry, 68*(4), 314–319. https://doi.org/10.1016/J.BIOPSYCH.2010.05.028

Bambini-Junior, V., Rodrigues, L., Behr, G. A., Moreira, J. C. F., Riesgo, R., & Gottfried, C. (2011). Animal model of autism induced by prenatal exposure to valproate: Behavioral changes and liver parameters. *Brain Research, 1408*, 8–16. https://doi.org/10.1016/j.brainres.2011.06.015

Barkoski, J. M., Philippat, C., Tancredi, D., Schmidt, R. J., Ozonoff, S., Barr, D. B., Elms, W., Bennett, D. H., & Hertz-Picciotto, I. (2021). In utero pyrethroid pesticide exposure in relation to autism spectrum disorder (ASD) and other neurodevelopmental outcomes at 3 years in the MARBLES longitudinal cohort. *Environmental Research, 194*, 110495. https://doi.org/10.1016/J.ENVRES.2020.110495

Basheer, S., Natarajan, A., van Amelsvoort, T., Venkataswamy, M. M., Ravi, V., Srinath, S., Girimaji, S. C., & Christopher, R. (2017). Vitamin D status of children with autism spectrum disorder: Case-control study from India. *Asian Journal of Psychiatry, 30*, 200–201. https://doi.org/10.1016/J.AJP.2017.10.031

Bell, M. R., Thompson, L. M., Rodriguez, K., & Gore, A. C. (2016). Two-hit exposure to polychlorinated biphenyls at gestational and juvenile life stages: 1. Sexually dimorphic effects on social and anxiety-like behaviors. *Hormones and Behavior, 78*, 168–177. https://doi.org/10.1016/J.YHBEH.2015.11.007

Berk, M., Williams, L. J., Jacka, F. N., O'Neil, A., Pasco, J. A., Moylan, S., Allen, N. B., Stuart, A. L., Hayley, A. C., Byrne, M. L., & Maes, M. (2013). So depression is an inflammatory disease, but where does the inflammation come from? *BMC Medicine, 11*(1), 200. https://doi.org/10.1186/1741-7015-11-200

Bernardo, B. A., Lanphear, B. P., Venners, S. A., Arbuckle, T. E., Braun, J. M., Muckle, G., Fraser, W. D., & McCandless, L. C. (2019). Assessing the relation between plasma PCB concentrations and elevated autistic behaviours using bayesian predictive odds ratios. *International Journal of Environmental Research and Public Health, 16*(3), 457.

Biosca-Brull, J., Pérez-Fernández, C., Mora, S., Carrillo, B., Pinos, H., Conejo, N. M., Collado, P., Arias, J. L., Martín-Sánchez, F., Sánchez-Santed, F., & Colomina, M. T. (2021). Relationship between autism spectrum disorder and pesticides: A systematic review of human and preclinical models. *International Journal of Environmental Research and Public Health, 18*(10), 5190. https://doi.org/10.3390/IJERPH18105190

Braun, J. M., Kalkbrenner, A. E., Calafat, A. M., Yolton, K., Ye, X., Dietrich, K. N., & Lanphear, B. P. (2011). Impact of early-life bisphenol A exposure on behavior and executive function in children. *Pediatrics, 128*(5), 873–882. https://doi.org/10.1542/PEDS.2011-1335

Bromley, R. L. et al. (2008). Autism spectrum disorders following in utero exposure to antiepileptic drugs, *Neurology, 71*(23), 1923–1924. https://doi.org/10.1212/01.wnl.0000339399.64213.1a

Brotons, J. A., Olea-Serrano, M. F., Villalobos, M., Pedraza, V., & Olea, N. (1995). Xenoestrogens released from lacquer coatings in food cans. *Environmental Health Perspectives, 103*(6), 608–612. https://doi.org/10.1289/EHP.95103608

Brown, J. S. (2009). Effects of bisphenol-A and other endocrine disruptors compared with abnormalities of schizophrenia: An endocrine-disruption theory of schizophrenia. *Schizophrenia Bulletin, 35*(1), 256–278. https://doi.org/10.1093/SCHBUL/SBM147

Cai, Y. et al. (2019). Citalopram attenuates social behavior deficits in the BTBR T+Itpr3tf/J mouse model of autism, *Brain Research Bulletin, 150*, pp. 75–85. https://doi.org/10.1016/j.brainresbull.2019.04.026

Calafat, A. M., Ye, X., Wong, L. Y., Reidy, J. A., & Needham, L. L. (2008). Exposure of the U.S. population to bisphenol A and 4-tertiary-octylphenol: 2003-2004. *Environmental Health Perspectives, 116*(1), 39–44. https://doi.org/10.1289/EHP.10753

Carbone, K. M., Rubin, S. A., & Pletnikov, M. (2002). Borna disease virus (BDV)-induced model of autism: application to vaccine safety test design. *Molecular Psychiatry, 7*(2), S36–37. https://doi.org/10.1038/sj.mp.4001174

Carlson, S. E. (2001). Docosahexaenoic acid and arachidonic acid in infant development. *Seminars in Neonatology, 6*(5), 437–449. https://doi.org/10.1053/SINY.2001.0093

Carlezon, W. A. et al. (2019). Maternal and early postnatal immune activation produce sex-specific effects on autism-like behaviors and neuroimmune function in mice. *Scientific Reports, 9*(1), 16928. https://doi.org/10.1038/s41598-019-53294-z

Carr, R. L., Bertasi, F. R., Betancourt, A. M., Bowers, S. D., Gandy, B. S., Ryan, P. L., & Willard, S. T. (2003). Effect of neonatal rat bisphenol a exposure on performance in the Morris water maze. *Journal of Toxicology and Environmental Health. Part A, 66*(21), 2077–2088. https://doi.org/10.1080/713853983

Catanese, M. C., & Vandenberg, L. N. (2017). Bisphenol S (BPS) alters maternal behavior and brain in mice exposed during pregnancy/lactation and their daughters. *Endocrinology, 158*(3), 516–530. https://doi.org/10.1210/EN.2016-1723

Celada, P., Victoria Puig, M., & Artigas, F. (2013). Serotonin modulation of cortical neurons and networks. *Frontiers in Integrative Neuroscience, 7*, 25. https://doi.org/10.3389/FNINT.2013.00025

Chakraborti, B. et al. (2019). Gender-Specific Effect of 5-HT and 5-HIAA on Threshold Level of Behavioral Symptoms and Sex-Bias in Prevalence of Autism Spectrum Disorder. *Frontiers in Neuroscience, 13*, 1375. https://doi.org/10.3389/fnins.2019.01375

Choi, C. S., Shin, C. Y., Kwon, K. J., Seung, H., Kim, J.-W. W., Lee, J., Bahn, G. H., Oh, H. A., Han, S. H., Kim, K. C., You, J. S., Yang, S. M., Ko, M. J., Cheong, J. H., Gonzales, E. L., Choi, D. H., Lee, E. J., Seung, H., Kwon, K. J., et al. (2014). Subchronic treatment of donepezil rescues impaired social, hyperactive, and stereotypic behavior in valproic acid-induced animal model of autism. *PLoS One, 9*(8), e104927. https://doi.org/10.1371/journal.pone.0104927

Choi, J., Chang, J. Y., Hong, J., Shin, S., Park, J. S., & Oh, S. (2017). Low-level toxic metal exposure in healthy weaning-age infants: Association with growth, dietary intake, and iron deficiency. *International Journal of Environmental Research and Public Health, 14*(4), 388. https://doi.org/10.3390/IJERPH14040388

Christianson, A. L., Chester, N., & Kromberg, J. G. R. (1994). Fetal valproate syndrome: Clinical and neuro-developmental features in two sibling pairs. *Developmental Medicine and Child Neurology, 36*(4), 361–369. https://doi.org/10.1111/J.1469-8749.1994.TB11858.X

Christensen, D. L. et al. (2019). Prevalence and Characteristics of Autism Spectrum Disorder Among Children Aged 4 Years - Early Autism and Developmental Disabilities Monitoring Network, Seven Sites, United States, 2010, 2012, and 2014. *Morbidity and Mortality Weekly Report. Surveillance Summaries (Washington, D.C.: 2002), 68*(2), 1–19. https://doi.org/10.15585/mmwr.ss6802a1

Christensen, D., & Zubler, J. (2020). CE: From the CDC: Understanding Autism Spectrum Disorder. *The American Journal of Nursing, 120*(10), 30–37. https://doi.org/10.1097/01.NAJ.0000718628.09065.1b

Connors, S. L., Matteson, K. J., Sega, G. A., Lozzio, C. B., Carroll, R. C., & Zimmerman, A. W. (2006). Plasma serotonin in autism. *Pediatric Neurology, 35*(3), 182–186. https://doi.org/10.1016/j.pediatrneurol.2006.02.010

Contestabile, A., & Sintoni, S. (2013). Histone acetylation in neurodevelopment. *Current Pharmaceutical Design, 19*(28), 5043–5050. https://doi.org/10.2174/1381612811319280003

Cordeiro, C. N., Tsimis, M., & Burd, I. (2015). Infections and brain development. *Obstetrical & Gynecological Survey, 70*(10), 644–655. https://doi.org/10.1097/OGX.0000000000000236

Courchesne, E. et al. (2011). Neuron number and size in prefrontal cortex of children with autism. *JAMA, 306*(18), 2001–2010. https://doi.org/10.1001/jama.2011.1638

Craciunescu, C. N., Brown, E. C., Mar, M. H., Albright, C. D., Nadeau, M. R., & Zeisel, S. H. (2004). Folic acid deficiency during late gestation decreases progenitor cell proliferation and increases apoptosis in fetal mouse brain. *The Journal of Nutrition, 134*(1), 162–166. https://doi.org/10.1093/JN/134.1.162

De Felice, A. et al. (2015). Prenatal exposure to a common organophosphate insecticide delays motor development in a mouse model of idiopathic autism', PloS One, 10(3), p. e0121663. Available at: https://doi.org/10.1371/journal.pone.0121663

de la Torre, J. C. (2002). Bornavirus and the brain. *The Journal of Infectious Diseases, 186*(2), S241–247. https://doi.org/10.1086/344936

De Rubeis, S., He, X., Goldberg, A. P., Poultney, C. S., Samocha, K., Cicek, A. E., Kou, Y., Liu, L., Fromer, M., Walker, S., Singh, T., Klei, L., Kosmicki, J., Fu, S. C., Aleksic, B., Biscaldi, M., Bolton, P. F., Brownfeld, J. M., Cai, J., et al. (2014). Synaptic, transcriptional and chromatin genes disrupted in autism. *Nature, 515*(7526), 209–215. https://doi.org/10.1038/nature13772

Dean, J. C. S. et al. (2002). Long term health and neurodevelopment in children exposed to antiepileptic drugs before birth. *Journal of Medical Genetics, 39*(4), 251–259. https://doi.org/10.1136/jmg.39.4.251

Devilbiss, E. A., Gardner, R. M., Newschaffer, C. J., & Lee, B. K. (2015). Maternal folate status as a risk factor for autism spectrum disorders: A review of existing evidence. *The British Journal of Nutrition, 114*(5), 663–672. https://doi.org/10.1017/S0007114515002470

Dietert, R. R., Dietert, J. M., & DeWitt, J. C. (2011). Environmental risk factors for autism. *Emerging Health Threats Journal, 4*(1), 7111. https://doi.org/10.3402/EHTJ.V4I0.7111

Doya, K. (2000). Complementary roles of basal ganglia and cerebellum in learning and motor control. *Current Opinion in Neurobiology, 10*(6), 732–739. https://doi.org/10.1016/S0959-4388(00)00153-7

Du, L., Zhao, G., Duan, Z., & Li, F. (2017). Behavioral improvements in a valproic acid rat model of autism following vitamin D supplementation. *Psychiatry Research, 253*, 28–32. https://doi.org/10.1016/j.psychres.2017.03.003

Eapen, V. et al. (2017). Current status of biological treatment options in Autism Spectrum Disorder. *Asian Journal of Psychiatry, 30*, 1–10. https://doi.org/10.1016/j.ajp.2017.07.025

Ecker, C., & Murphy, D. (2014). Neuroimaging in autism-from basic science to translational research. *Nature Reviews. Neurology, 10*(2), 82–91. https://doi.org/10.1038/NRNEUROL.2013.276

Edelson, L. R., & Saudino, K. J. (2009). Genetic and environmental influences on autistic-like behaviors in 2-year-old twins. *Behavior Genetics, 39*(3), 255–264. https://doi.org/10.1007/S10519-009-9270-3

Eissa, N., Azimullah, S., Jayaprakash, P., Jayaraj, R. L., Reiner, D., Ojha, S. K., Beiram, R., Stark, H., Łażewska, D., Kieć-Kononowicz, K., & Sadek, B. (2020). The dual-active histamine H3 receptor antagonist and acetylcholine esterase inhibitor E100 alleviates autistic-like behaviors and oxidative stress in valproic acid induced autism in mice. *International Journal of Molecular Sciences, 21*(11), 1–22. https://doi.org/10.3390/ijms21113996

Elamin, N. E., & Al-Ayadhi, L. Y. (2014). Brain autoantibodies in autism spectrum disorder. *Biomarkers in Medicine, 8*(3), 345–352. https://doi.org/10.2217/BMM.14.1

Eliasen, M., Tolstrup, J. S., Andersen, A. M. N., Grønbæk, M., Olsen, J., & Strandberg-Larsen, K. (2010). Prenatal alcohol exposure and autistic spectrum disorders-a population-based prospective study of 80 552 children and their mothers. *International Journal of Epidemiology, 39*(4), 1074–1081. https://doi.org/10.1093/IJE/DYQ056

El-Sharkawy, A., & Malki, A. (2020). Vitamin D signaling in inflammation and cancer: Molecular mechanisms and therapeutic implications. *Molecules, 25*(14), 3219. https://doi.org/10.3390/MOLECULES25143219

Eppolito, A. K., Bachus, S. E., McDonald, C. G., Meador-Woodruff, J. H., & Smith, R. F. (2010). Late emerging effects of prenatal and early postnatal nicotine exposure on the cholinergic system and anxiety-like behavior. *Neurotoxicology and Teratology, 32*(3), 336–345. https://doi.org/10.1016/j.ntt.2009.12.009

Eskenazi, B., Rauch, S. A., Tenerelli, R., Huen, K., Holland, N. T., Lustig, R. H., Kogut, K., Bradman, A., Sjödin, A., & Harley, K. G. (2017). In utero and childhood DDT, DDE, PBDE and PCBs exposure and sex hormones in adolescent boys: The CHAMACOS study. *International Journal of Hygiene and Environmental Health, 220*(2), 364–372. https://doi.org/10.1016/J.IJHEH.2016.11.001

Farook, M. F., DeCuypere, M., Hyland, K., Takumi, T., LeDoux, M. S., & Reiter, L. T. (2012). Altered serotonin, dopamine and norepinepherine levels in 15q duplication and Angelman syndrome mouse models. *PLoS One, 7*(8), e43030. https://doi.org/10.1371/journal.pone.0043030

Fatemi, S. H. et al. (2010). Levels of phosphodiesterase 4A and 4B are altered by chronic treatment with psychotropic medications in rat frontal cortex. *Synapse (New York, N.Y.), 64*(7), 550–555. https://doi.org/10.1002/syn.20762

Fedirko, V., Bostick, R. M., Flanders, W. D., Long, Q., Shaukat, A., Rutherford, R. E., Daniel, C. R., Cohen, V., & Dash, C. (2009). Effects of vitamin D and calcium supplementation on markers of apoptosis in normal colon mucosa: A randomized, double-blind, placebo-controlled clinical trial. *Cancer Prevention Research, 2*(3), 213–223. https://doi.org/10.1158/1940-6207.CAPR-08-0157

Fleet, J. C., Desmet, M., Johnson, R., & Li, Y. (2012). Vitamin D and cancer: A review of molecular mechanisms. *Biochemical Journal, 441*(1), 61–76. https://doi.org/10.1042/BJ20110744

Foley, K. A. et al. (2014). Sexually dimorphic effects of prenatal exposure to propionic acid and lipopolysaccharide on social behavior in neonatal, adolescent, and adult rats: implications for autism spectrum disorders. *International Journal of Developmental Neuroscience: The Official Journal of the International Society for Developmental Neuroscience, 39*, 68–78. https://doi.org/10.1016/j.ijdevneu.2014.04.001

Freedman, R., Hunter, S. K., Law, A. J., Clark, A. M., Roberts, A., & Hoffman, M. C. (2021). Choline, folic acid, vitamin D, and fetal brain development in the psychosis spectrum. *Schizophrenia Research, 247*, 16–25. https://doi.org/10.1016/J.SCHRES.2021.03.008

Fujiwara, T., Morisaki, N., Honda, Y., Sampei, M., & Tani, Y. (2016). Chemicals, nutrition, and autism spectrum disorder: A mini-review. *Frontiers in Neuroscience, 10*, 174. https://doi.org/10.3389/FNINS.2016.00174

Gáll, Z., & Székely, O. (2021). Role of vitamin d in cognitive dysfunction: New molecular concepts and discrepancies between animal and human findings. *Nutrients, 13*(11), 3672. https://doi.org/10.3390/NU13113672

García-Serna, A. M., & Morales, E. (2020). Neurodevelopmental effects of prenatal vitamin D in humans: systematic review and meta-analysis. *Molecular Psychiatry, 25*(10), 2468–2481. https://doi.org/10.1038/s41380-019-0357-9

Gaspar, P., Cases, O., & Maroteaux, L. (2003). The developmental role of serotonin: News from mouse molecular genetics. *Nature Reviews. Neuroscience, 4*(12), 1002–1012. https://doi.org/10.1038/NRN1256

Gaugler, T., Klei, L., Sanders, S. J., Bodea, C. A., Goldberg, A. P., Lee, A. B., Mahajan, M., Manaa, D., Pawitan, Y., Reichert, J., Ripke, S., Sandin, S., Sklar, P., Svantesson, O., Reichenberg, A., Hultman, C. M., Devlin, B., Roeder, K., & Buxbaum, J. D. (2014). Most genetic risk for autism resides with common variation. *Nature Genetics, 46*(8), 881–885. https://doi.org/10.1038/NG.3039

Ghodke-Puranik, Y., Thorn, C. F., Lamba, J. K., Leeder, J. S., Song, W., Birnbaum, A. K., Altman, R. B., & Klein, T. E. (2013). Valproic acid pathway: Pharmacokinetics and pharmacodynamics. *Pharmacogenetics and Genomics, 23*(4), 236–241. https://doi.org/10.1097/FPC.0B013E32835EA0B2

Go, H. S., Kim, K. C., Choi, C. S., Jeon, S. J., Kwon, K. J., Han, S. H., Lee, J., Cheong, J. H., Ryu, J. H., Kim, C. H., Ko, K. H., & Shin, C. Y. (2012). Prenatal exposure to valproic acid increases the neural progenitor cell pool and induces macrocephaly in rat brain via a mechanism involving the GSK-3β/β-catenin pathway. *Neuropharmacology, 63*(6), 1028–1041. https://doi.org/10.1016/j.neuropharm.2012.07.028

Goeden, N. et al. (2016). Maternal Inflammation Disrupts Fetal Neurodevelopment via Increased Placental Output of Serotonin to the Fetal Brain. *The Journal of Neuroscience: The Official Journal of the Society for Neuroscience, 36*(22), 6041–6049. https://doi.org/10.1523/JNEUROSCI.2534-15.2016

Goldsby, J. A., Wolstenholme, J. T., & Rissman, E. F. (2017). Multi- and transgenerational consequences of bisphenol a on sexually dimorphic cell populations in mouse brain. *Endocrinology, 158*(1), 21–30. https://doi.org/10.1210/EN.2016-1188

Gore, A. C., Krishnan, K., & Reilly, M. P. (2019). Endocrine-disrupting chemicals: Effects on neuroendocrine systems and the neurobiology of social behavior. *Hormones and Behavior, 111*, 7–22. https://doi.org/10.1016/J.YHBEH.2018.11.006

Grandjean, P., & Landrigan, P. (2006). Developmental neurotoxicity of industrial chemicals. *Lancet, 368*(9553), 2167–2178. https://doi.org/10.1016/S0140-6736(06)69665-7

Granillo, L., Sethi, S., Keil, K. P., Lin, Y., Ozonoff, S., Iosif, A. M., Puschner, B., & Schmidt, R. J. (2019). Polychlorinated biphenyls influence on autism spectrum disorder risk in the MARBLES cohort. *Environmental Research, 171*, 177–184.

Grant, W. B. (2016). The role of geographical ecological studies in identifying diseases linked to UVB exposure and/or vitamin D. *Dermato-Endocrinology, 8*(1), e1137400. https://doi.org/10.1080/19381980.2015.1137400

Grimm, F. A., Hu, D., Kania-Korwel, I., Lehmler, H. J., Ludewig, G., Hornbuckle, K. C., Duffel, M. W., Bergman, Å., & Robertson, L. W. (2015). Metabolism and metabolites of polychlorinated biphenyls. *Critical Reviews in Toxicology, 45*(3), 245–272. https://doi.org/10.3109/10408444.2014.999365

Grube, A., Donaldson, D., Kiely, T., & Wu, L. (2011). *Pesticides industry sales and usage*. Swap. Stanford.Edu. https://swap.stanford.edu/20140417081610/http://www.epa.gov/opp00001/pestsales/07pestsales/market_estimates2007.pdf

Guo, Q., Yin, X., Qiao, M., Jia, Y., Chen, D., Shao, J., Lebaron, T. W., Gao, Y., Shi, H., & Jia, B. (2018). Hydrogen-rich water ameliorates autistic-like behavioral abnormalities in valproic acid-treated adolescent mice offspring. *Frontiers in Behavioral Neuroscience, 12*, 1–9. https://doi.org/10.3389/fnbeh.2018.00170

Hallmayer, J., Cleveland, S., Torres, A., Phillips, J., Cohen, B., Torigoe, T., Miller, J., Fedele, A., Collins, J., Smith, K., Lotspeich, L., Croen, L. A., Ozonoff, S., Lajonchere, C., Grether, J. K., & Risch, N. (2011). Genetic heritability and shared environmental factors among twin pairs with autism. *Archives of General Psychiatry, 68*(11), 1095–1102. https://doi.org/10.1001/ARCHGENPSYCHIATRY.2011.76

Hamra, G. B., Lyall, K., Windham, G. C., Calafat, A. M., Sjödin, A., Volk, H., & Croen, L. A. (2019). Prenatal exposure to endocrine-disrupting chemicals in relation to autism spectrum disorder and intellectual disability. *Epidemiology, 30*(3), 418–426. https://doi.org/10.1097/EDE.0000000000000983

Hansen, J. B., Bilenberg, N., Timmermann, C. A. G., Jensen, R. C., Frederiksen, H., Andersson, A. M., Kyhl, H. B., & Jensen, T. K. (2021). Prenatal exposure to bisphenol A and autistic- and ADHD-related symptoms in children aged 2 and 5 years from the Odense Child Cohort. *Environmental Health, 20*(1), 24. https://doi.org/10.1186/S12940-021-00709-Y

Hara, Y., Takuma, K., Takano, E., Katashiba, K., Taruta, A., Higashino, K., Hashimoto, H., Ago, Y., & Matsuda, T. (2015). Reduced prefrontal dopaminergic activity in valproic acid-treated mouse autism model. *Behavioural Brain Research, 289*, 39–47. https://doi.org/10.1016/j.bbr.2015.04.022

Harutyunyan, A. A., Harutyunyan, H. A., & Yenkoyan, K. B. (2021). Novel probable glance at inflammatory scenario development in autistic pathology. *Frontiers in Psychiatry, 12*, 788779. https://doi.org/10.3389/FPSYT.2021.788779

Hazlett, H. C., Poe, M. D., Gerig, G., Styner, M., Chappell, C., Smith, R. G., Vachet, C., & Piven, J. (2011). Early brain overgrowth in autism associated with an increase in cortical surface area before age 2 years. *Archives of General Psychiatry, 68*(5), 467–476. https://doi.org/10.1001/ARCHGENPSYCHIATRY.2011.39

He, X., Tu, Y., Song, Y., Yang, G., & You, M. (2022). The relationship between pesticide exposure during critical neurodevelopment and autism spectrum disorder: A narrative review. *Environmental Research, 203*, 111902. https://doi.org/10.1016/J.ENVRES.2021.111902

Hegarty, J. P., Lazzeroni, L. C., Raman, M. M., Pegoraro, L. F. L., Monterrey, J. C., Cleveland, S. C., Hallmayer, J. F., Wolke, O. N., Phillips, J. M., Reiss, A. L., & Hardan, A. Y. (2020). Genetic and environmental influences on lobar brain structures in twins with autism. *Cerebral Cortex, 30*(3), 1946–1956. https://doi.org/10.1093/cercor/bhz215

Hellmer, K., & Nyström, P. (2017). Infant acetylcholine, dopamine, and melatonin dysregulation: Neonatal biomarkers and causal factors for ASD and ADHD phenotypes. *Medical Hypotheses, 100*, 64–66. https://doi.org/10.1016/J.MEHY.2017.01.015

Herbert, M. R. (2005). Large brains in autism: The challenge of pervasive abnormality. *The Neuroscientist, 11*(5), 417–440. https://doi.org/10.1177/0091270005278866

Herbert, M. R., Russo, J. P., Yang, S., Roohi, J., Blaxill, M., Kahler, S. G., Cremer, L., & Hatchwell, E. (2006). Autism and environmental genomics. *Neurotoxicology, 27*(5), 671–684. https://doi.org/10.1016/J.NEURO.2006.03.017

Hertz-Picciotto, I., Croen, L. A., Hansen, R., Jones, C. R., van de Water, J., & Pessah, I. N. (2006). The CHARGE study: An epidemiologic investigation of genetic and environmental factors contributing to autism. *Environmental Health Perspectives, 114*(7), 1119–1125. https://doi.org/10.1289/ehp.8483

Hessabi, M., Rahbar, M. H., Dobrescu, I., Bach, M. A., Kobylinska, L., Bressler, J., Grove, M. L., Loveland, K. A., Mihailescu, I., Nedelcu, M. C., Moisescu, M. G., Matei, B. M., Matei, C. O., & Rad, F. (2019). Concentrations of lead, mercury, arsenic, cadmium, manganese, and aluminum in blood of Romanian children suspected of having autism spectrum disorder. *International Journal of Environmental Research and Public Health, 16*(13), 2303. https://doi.org/10.3390/IJERPH16132303

Hoel, D. G., & de Gruijl, F. R. (2018). Sun exposure public health directives. *International Journal of Environmental Research and Public Health, 15*(12), 2794. https://doi.org/10.3390/IJERPH15122794

Honybun, E., Thwaites, R., Malpas, C. B., Rayner, G., Anderson, A., Graham, J., Hitchcock, A., O'Brien, T. J., Vajda, F. J. E., & Perucca, P. (2021). Prenatal valproate exposure and adverse neurodevelopmental outcomes: Does sex matter? *Epilepsia, 62*(3), 709–719. https://doi.org/10.1111/epi.16827

Hornig, M., & Ian Lipkin, W. (2001). Infectious and immune factors in the pathogenesis of neurodevelopmental disorders: Epidemiology, hypotheses, and animal models. *Mental Retardation and Developmental Disabilities Research Reviews, 7*(3), 200–210. https://doi.org/10.1002/mrdd.1028

Hranilovic, D., Blazevic, S., Ivica, N., Cicin-Sain, L., & Oreskovic, D. (2011). The effects of the perinatal treatment with 5-hydroxytryptophan or tranylcypromine on the peripheral and central serotonin homeostasis in adult rats. *Neurochemistry International, 59*(2), 202–207. https://doi.org/10.1016/j.neuint.2011.05.003

Hsueh, Y. M., Lee, C. Y., Chien, S. N., Chen, W. J., Shiue, H. S., Huang, S. R., Lin, M. I., Mu, S. C., & Hsieh, R. L. (2017). Association of blood heavy metals with developmental delays and health status in children. *Scientific Reports, 7*, 43608. https://doi.org/10.1038/SREP43608

Hu, D., & Hornbuckle, K. C. (2010). Inadvertent polychlorinated biphenyls in commercial paint pigments. *Environmental Science & Technology, 44*(8), 2822–2827. https://doi.org/10.1021/ES902413K

Hughes, A., Greene, N. D. E., Copp, A. J., & Galea, G. L. (2018). Valproic acid disrupts the biomechanics of late spinal neural tube closure in mouse embryos. *Mechanisms of Development, 149*, 20–26. https://doi.org/10.1016/J.MOD.2017.12.001

Hunt, A., Harrington, D., & Robinson, S. (2014). Vitamin B12 deficiency. *BMJ, 349*, g5226. https://doi.org/10.1136/BMJ.G5226

Iovcheva, C., & Dessev, G. N. (1980). Interaction between lysine-rich histones and DNA. *Molecular Biology Reports, 6*(1), 21–25.

Jadavji, N. M., Deng, L., Malysheva, O., Caudill, M. A., & Rozen, R. (2015). MTHFR deficiency or reduced intake of folate or choline in pregnant mice results in impaired short-term memory and increased apoptosis in the hippocampus of wild-type offspring. *Neuroscience, 300*, 1–9. https://doi.org/10.1016/J.NEUROSCIENCE.2015.04.067

Jaiswal, P., Mohanakumar, K. P., & Rajamma, U. (2015). Serotonin mediated immunoregulation and neural functions: Complicity in the aetiology of autism spectrum disorders. In *Neuroscience and biobehavioral reviews* (Vol. 55, pp. 413–431). Elsevier Ltd. https://doi.org/10.1016/j.neubiorev.2015.05.013

Jenkins, D. D., Moss, H. G., Brown, T. R., Yazdani, M., Thayyil, S., Montaldo, P., Vento, M., Kuligowski, J., Wagner, C., Hollis, B. W., & Wiest, D. B. (2021). NAC and vitamin D improve cns and plasma oxidative stress in neonatal hie and are associated with favorable long-term outcomes. *Antioxidants, 10*(9), 1344. https://doi.org/10.3390/ANTIOX10091344

Jennings, L., & Basiri, R. (2022). Amino acids, B vitamins, and choline may independently and collaboratively influence the incidence and core symptoms of autism spectrum disorder. *Nutrients, 14*(14), 2896. https://doi.org/10.3390/NU14142896

Johnson, S. A., Javurek, A. B., Painter, M. S., Ellersieck, M. R., Welsh, T. H., Camacho, L., Lewis, S. M., Vanlandingham, M. M., Ferguson, S. A., & Rosenfeld, C. S. (2016). Effects of developmental exposure to bisphenol A on spatial navigational learning and memory in rats: A CLARITY-BPA study. *Hormones and Behavior, 80*, 139–148. https://doi.org/10.1016/J.YHBEH.2015.09.005

Jolous-Jamshidi, B., Cromwell, H. C., McFarland, A. M., & Meserve, L. A. (2010). Perinatal exposure to polychlorinated biphenyls alters social behaviors in rats. *Toxicology Letters, 199*(2), 136–143.

Julio-Pieper, M. et al. (2014). Review article: intestinal barrier dysfunction and central nervous system disorders--a controversial association. *Alimentary Pharmacology & Therapeutics, 40*(10), 1187–1201. https://doi.org/10.1111/apt.12950

Kahne, D. et al. (2002). Behavioral and magnetic resonance spectroscopic studies in the rat hyperserotonemic model of autism. *Physiology & Behavior, 75*(3), 403–410. https://doi.org/10.1016/s0031-9384(01)00673-4

Kalkbrenner, A. E., Schmidt, R. J., & Penlesky, A. C. (2014). Environmental chemical exposures and autism spectrum disorders: A review of the epidemiological evidence. *Current Problems in Pediatric and Adolescent Health Care, 44*(10), 277–318. https://doi.org/10.1016/j.cppeds.2014.06.001.Environmental

Kania-Korwel, I., Barnhart, C. D., Stamou, M., Truong, K. M., El-Komy, M. H. M. E., Lein, P. J., Veng-Pedersen, P., & Lehmler, H. J. (2012). 2,2′,3,5′,6-Pentachlorobiphenyl (PCB 95) and its hydroxylated metabolites are enantiomerically enriched in female mice. *Environmental Science & Technology, 46*(20), 11393–11401. https://doi.org/10.1021/ES302810T

Kardas, F., Bayram, A. K., Demirci, E., Akin, L., Ozmen, S., Kendirci, M., Canpolat, M., Oztop, D. B., Narin, F., Gumus, H., Kumandas, S., & Per, H. (2016). Increased serum phthalates (MEHP, DEHP) and bisphenol A concentrations in children with autism spectrum disorder. *Journal of Child Neurology, 31*(5), 629–635. https://doi.org/10.1177/0883073815609150

Karkaba, A., Soualeh, N., Soulimani, R., & Bouayed, J. (2017). Perinatal effects of exposure to PCBs on social preferences in young adult and middle-aged offspring mice. *Hormones and Behavior, 96*, 137–146. https://doi.org/10.1016/J.YHBEH.2017.09.002

Kataoka, S., Takuma, K., Hara, Y., Maeda, Y., Ago, Y., & Matsuda, T. (2013). Autism-like behaviours with transient histone hyperacetylation in mice treated prenatally with valproic acid. *The International Journal of Neuropsychopharmacology, 16*(1), 91–103. https://doi.org/10.1017/S1461145711001714

Kaur, K., Chauhan, V., Gu, F., & Chauhan, A. (2014). Bisphenol A induces oxidative stress and mitochondrial dysfunction in lymphoblasts from children with autism and unaffected siblings. *Free Radical Biology & Medicine, 76*, 25–33. https://doi.org/10.1016/J.FREERADBIOMED.2014.07.030

Kaur, K., Simon, A. F., Chauhan, V., & Chauhan, A. (2015). Effect of bisphenol A on Drosophila melanogaster behavior – A new model for the studies on neurodevelopmental disorders. *Behavioural Brain Research, 284*, 77–84. https://doi.org/10.1016/J.BBR.2015.02.001

Kennedy, D. O. (2016). B vitamins and the brain: Mechanisms, dose and efficacy—A review. *Nutrients, 8*(2), 68. https://doi.org/10.3390/NU8020068

Kern, J. K., Geier, D. A., Deth, R. C., Sykes, L. K., Hooker, B. S., Love, J. M., Bjørklund, G., Chaigneau, C. G., Haley, B. E., & Geier, M. R. (2017). Systematic assessment of research on autism spectrum disorder (ASD) and mercury reveals conflicts of interest and the need for transparency in autism research. *Science and Engineering Ethics, 23*(6), 1691–1718.

Keyes, K. M. et al. (2012). Cohort effects explain the increase in autism diagnosis among children born from 1992 to 2003 in California. *International Journal of Epidemiology, 41*(2), 495–503. https://doi.org/10.1093/ije/dyr193

King, J. C. (2016). A summary of pathways or mechanisms linking preconception maternal nutrition with birth outcomes. *The Journal of Nutrition, 146*(7), 1437S–1444S. https://doi.org/10.3945/JN.115.223479

Kinney, D. K., Barch, D. H., Chayka, B., Napoleon, S., & Munir, K. M. (2010). Environmental risk factors for autism: Do they help cause de novo genetic mutations that contribute to the disorder? *Medical Hypotheses, 74*(1), 102–106. https://doi.org/10.1016/J.MEHY.2009.07.052

Kirsten, T. B., Chaves-Kirsten, G. P., Bernardes, S., Scavone, C., Sarkis, J. E., Bernardi, M. M., & Felicio, L. F. (2015). Lipopolysaccharide exposure induces maternal hypozincemia, and prenatal zinc treatment prevents autistic-like behaviors and disturbances in the striatal dopaminergic and mtor systems of offspring. *PLoS One, 10*(7), e0134565. https://doi.org/10.1371/JOURNAL.PONE.0134565

Klocke, C., Sethi, S., & Lein, P. J. (2020). The developmental neurotoxicity of legacy vs. contemporary polychlorinated biphenyls (PCBs): Similarities and differences. *Environmental Science and Pollution Research, 27*(9), 8885–8896. https://doi.org/10.1007/S11356-019-06723-5

Kobayashi, T., Matsuyama, T., Takeuchi, M., & Ito, S. (2016). Autism spectrum disorder and prenatal exposure to selective serotonin reuptake inhibitors: A systematic review and metaanalysis. *Reproductive Toxicology, 65*, 170–178. https://doi.org/10.1016/J.REPROTOX.2016.07.016

Koh, W. X., Hornbuckle, K. C., Marek, R. F., Wang, K., & Thorne, P. S. (2016). Hydroxylated polychlorinated biphenyls in human sera from adolescents and their mothers living in two U.S. Midwestern communities. *Chemosphere, 147*, 389–395. https://doi.org/10.1016/J.CHEMOSPHERE.2015.12.113

Kontur, P. J., Leranth, C., Redmond, D. E., Roth, R. H., & Robbins, R. J. (1993). Tyrosine hydroxylase immunoreactivity and monoamine and metabolite levels in cryopreserved human fetal ventral mesencephalon. *Experimental Neurology, 121*(2), 172–180. https://doi.org/10.1006/EXNR.1993.1084

Krämer, O.H. et al. (2003). The histone deacetylase inhibitor valproic acid selectively induces proteasomal degradation of HDAC2. *The EMBO journal, 22*(13), 3411–3420. https://doi.org/10.1093/emboj/cdg315

Krishnan, A. V., Stathis, P., Permuth, S. F., Tokes, L., & Feldman, D. (1993). Bisphenol-A: An estrogenic substance is released from polycarbonate flasks during autoclaving. *Endocrinology, 132*(6), 2279–2286. https://doi.org/10.1210/ENDO.132.6.8504731

Kumamaru, E. et al. (2014). Valproic acid selectively suppresses the formation of inhibitory synapses in cultured cortical neurons. *Neuroscience Letters, 569*, 142–147. https://doi.org/10.1016/j.neulet.2014.03.066

Kumar, H., & Sharma, B. (2016). Minocycline ameliorates prenatal valproic acid induced autistic behaviour, biochemistry and blood brain barrier impairments in rats. *Brain Research, 1630*, 83–97. https://doi.org/10.1016/j.brainres.2015.10.052

Kumar, H., Sharma, B. M., & Sharma, B. M. (2015). Benefits of agomelatine in behavioral, neurochemical and blood brain barrier alterations in prenatal valproic acid induced autism spectrum disorder. *Neurochemistry International, 91*, 34–45. https://doi.org/10.1016/j.neuint.2015.10.007

Lai, M. C., Lombardo, M. V., & Baron-Cohen, S. (2014). Autism. *Lancet, 383*(9920), 896–910. https://doi.org/10.1016/S0140-6736(13)61539-1

Lauritzen, L., Brambilla, P., Mazzocchi, A., Harsløf, L. B. S., Ciappolino, V., & Agostoni, C. (2016). DHA effects in brain development and function. *Nutrients, 8*(1), 6. https://doi.org/10.3390/NU8010006

Lawson, S. K., Gray, A. C., & Woehrle, N. S. (2016). Effects of oxytocin on serotonin 1B agonist-induced autism-like behavior in mice. *Behavioural Brain Research, 314*, 52–64. https://doi.org/10.1016/j.bbr.2016.07.027

Lee, H. S. (2015). Impact of maternal diet on the epigenome during in utero life and the developmental programming of diseases in childhood and adulthood. *Nutrients, 7*(11), 9492–9507. https://doi.org/10.3390/NU7115467

Lee, S.-M. et al. (2016). Music application alleviates short-term memory impairments through increasing cell proliferation in the hippocampus of valproic acid-induced autistic rat pups. *Journal of Exercise Rehabilitation, 12*(3), 148–155. https://doi.org/10.12965/jer.1632638.319

Lee, J., Choi, K., Park, J., Moon, H. B., Choi, G., Lee, J. J., Suh, E., Kim, H. J., Eun, S. H., Kim, G. H., Cho, G. J., Kim, S. K., Kim, S., Kim, S. Y., Kim, S., Eom, S., Choi, S., Kim, Y. D., & Kim, S. (2018). Bisphenol A distribution in serum, urine, placenta, breast milk, and umbilical cord serum in a birth panel of mother–neonate pairs. *Science of the Total Environment, 626*, 1494–1501. https://doi.org/10.1016/J.SCITOTENV.2017.10.042

Levallois, C., Valence, C., Baldet, P., & Privat, A. (1997). Morphological and morphometric analysis of serotonin-containing neurons in primary dissociated cultures of human rhombencephalon: A study of development. *Brain Research. Developmental Brain Research, 99*(2), 243–252. https://doi.org/10.1016/S0165-3806(97)00026-6

Lim, Y. H., Bae, S., Kim, B. N., Shin, C. H., Lee, Y. A., Kim, J. I., & Hong, Y. C. (2017). Prenatal and postnatal bisphenol A exposure and social impairment in 4-year-old children. *Environmental Health, 16*(1), 79. https://doi.org/10.1186/S12940-017-0289-2

Linnet, K. M., Dalsgaard, S., Obel, G., Wisborg, K., Henriksen, T. B., Rodriguez, A., Kotimaa, A., Moilanen, I., Thomsen, P. H., Olsen, J., & Jarvelin, M. R. (2003). Maternal lifestyle factors in pregnancy risk of attention deficit hyperactivity disorder and associated behaviors: Review of the current evidence. *American Journal of Psychiatry, 160*(6), 1028–1040. https://doi.org/10.1176/APPI.AJP.160.6.1028

Lisi, G., Ribolsi, M., Siracusano, A., & Niolu, C. (2020). Maternal vitamin D and its role in determining fetal origins of mental health. *Current Pharmaceutical Design, 26*(21), 2497–2509. https://doi.org/10.2174/1381612826666200506093858

Liu, R. Z., Mita, R., Beaulieu, M., Gao, Z., & Godbout, R. (2010). Fatty acid binding proteins in brain development and disease. *The International Journal of Developmental Biology, 54*(8–9), 1229–1239. https://doi.org/10.1387/IJDB.092976RL

López-Vicente, M., Sunyer, J., Lertxundi, N., González, L., Rodríguez-Dehli, C., Espada SáenzTorre, M., Vrijheid, M., Tardón, A., Llop, S., Torrent, M., Ibarluzea, J., & Guxens, M. (2019). Maternal circulating vitamin D3 levels during pregnancy and behaviour across childhood. *Scientific Reports, 9*(1), 14792. https://doi.org/10.1038/S41598-019-51325-3

Lu, M. et al. (2015). Endocrine disrupting potential of fipronil and its metabolite in reporter gene assays. *Chemosphere, 120*, 246–251. https://doi.org/10.1016/j.chemosphere.2014.07.015

Lucchina, L., & Depino, A. M. (2014). Altered peripheral and central inflammatory responses in a mouse model of autism. *Autism Research, 7*(2), 273–289. https://doi.org/10.1002/AUR.1338

Luhach, K., Kulkarni, G. T., Singh, V. P., & Sharma, B. (2021a). Cilostazol attenuated prenatal valproic acid-induced behavioural and biochemical deficits in a rat model of autism spectrum disorder. *The Journal of Pharmacy and Pharmacology, 73*, 1460–1469. https://doi.org/10.1093/JPP/RGAB115

Luhach, K., Kulkarni, G. T., Singh, V. P., & Sharma, B. (2021b). Effect of papaverine on developmental hyperserotonemia induced autism spectrum disorder related behavioural phenotypes by altering markers of neuronal function, inflammation, and oxidative stress in rats. *Clinical and Experimental Pharmacology & Physiology, 48*(4), 614–625. https://doi.org/10.1111/1440-1681.13459

Luhach, K., Kulkarni, G. T., Singh, V. P., & Sharma, B. (2021c). Vinpocetine amended prenatal valproic acid induced features of ASD possibly by altering markers of neuronal function, inflammation, and oxidative stress. *Autism Research, 14*(11), 2270–2286. https://doi.org/10.1002/aur.2597

Luo, W. et al. (2018). Systematic reconstruction of autism biology from massive genetic mutation profiles. *Science Advances, 4*(4), e1701799. https://doi.org/10.1126/sciadv.1701799

Lyall, K., Croen, L. A., Sjödin, A., Yoshida, C. K., Zerbo, O., Kharrazi, M., & Windham, G. C. (2017). Polychlorinated biphenyl and organochlorine pesticide concentrations in maternal midpregnancy serum samples: Association with autism spectrum disorder and intellectual disability. *Environmental Health Perspectives, 125*(3), 474–480. https://doi.org/10.1289/EHP277

MacFabe, D. F. et al. (2007). Neurobiological effects of intraventricular propionic acid in rats: possible role of short chain fatty acids on the pathogenesis and characteristics of autism spectrum disorders. *Behavioural Brain Research, 176*(1), 149–169. https://doi.org/10.1016/j.bbr.2006.07.025

Madden, A. M. K., & Zup, S. L. (2014). Effects of developmental hyperserotonemia on juvenile play behavior, oxytocin and serotonin receptor expression in the hypothalamus are age and sex dependent. *Physiology & Behavior, 128*, 260–269. https://doi.org/10.1016/j.physbeh.2014.01.036

Mahic, M., Mjaaland, S., Bøvelstad, H. M., Gunnes, N., Susser, E., Bresnahan, M., Øyen, A.-S., Levin, B., Che, X., Hirtz, D., Reichborn-Kjennerud, T., Schjølberg, S., Roth, C., Magnus, P., Stoltenberg, C., Surén, P., Hornig, M., & Lipkin, W. I. (2017). Maternal immunoreactivity to herpes simplex virus 2 and risk of autism spectrum disorder in male offspring. *mSphere, 2*(1), e00016-17. https://doi.org/10.1128/MSPHERE.00016-17

Manshack, L. K., Conard, C. M., Johnson, S. A., Alex, J. M., Bryan, S. J., Deem, S. L., Holliday, D. K., Ellersieck, M. R., & Rosenfeld, C. S. (2016). Effects of developmental exposure to bisphenol A and ethinyl estradiol on spatial navigational learning and memory in painted turtles (Chrysemys picta). *Hormones and Behavior, 85*, 48–55. https://doi.org/10.1016/J.YHBEH.2016.07.009

Marek, R. F., Thorne, P. S., Wang, K., Dewall, J., & Hornbuckle, K. C. (2013). PCBs and OH-PCBs in serum from children and mothers in urban and rural U.S. communities. *Environmental Science & Technology, 47*(7), 3353–3361. https://doi.org/10.1021/ES304455K

Marini, F., Bartoccini, E., Cascianelli, G., Voccoli, V., Baviglia, M. G., Magni, M. V., Garcia-Gil, M., & Albi, E. (2010). Effect of 1α,25-dihydroxyvitamin D3 in embryonic hippocampal cells. *Hippocampus, 20*(6), 696–705. https://doi.org/10.1002/HIPO.20670

Martineau, J., Barthelemy, C., Garreau, B., & Lelord, G. (1985). Vitamin B6, magnesium, and combined B6-Mg: Therapeutic effects in childhood autism. *Biological Psychiatry, 20*(5), 467–478. https://doi.org/10.1016/0006-3223(85)90019-8

Martinez, M. (1992). Tissue levels of polyunsaturated fatty acids during early human development. *The Journal of Pediatrics, 120*(4 Part 2), S129–S138. https://doi.org/10.1016/S0022-3476(05)81247-8

McCann, J. C., & Ames, B. N. (2005). Is docosahexaenoic acid, an n−3 long-chain polyunsaturated fatty acid, required for development of normal brain function? An overview of evidence from cognitive and behavioral tests in humans and animals. *The American Journal of Clinical Nutrition, 82*(2), 281–295. https://doi.org/10.1093/AJCN/82.2.281

McGinnis, W. R., Audhya, T., & Edelson, S. M. (2013). Proposed toxic and hypoxic impairment of a brainstem locus in autism. *International Journal of Environmental Research and Public Health, 10*(12), 6955–7000.

McNamara, I. M. et al. (2008). Further studies in the developmental hyperserotonemia model (DHS) of autism: social, behavioral and peptide changes. *Brain Research, 1189*, 203–214. https://doi.org/10.1016/j.brainres.2007.10.063

McNamara, R. K., Asch, R. H., Lindquist, D. M., & Krikorian, R. (2018). Role of polyunsaturated fatty acids in human brain structure and function across the lifespan: An update on neuroimaging findings. *Prostaglandins, Leukotrienes, and Essential Fatty Acids, 136*, 23–34. https://doi.org/10.1016/J.PLEFA.2017.05.001

Mesnil, M., Defamie, N., Naus, C., & Sarrouilhe, D. (2021). Brain disorders and chemical pollutants: A gap junction link? *Biomolecules & Therapeutics, 11*(1), 1–64. https://doi.org/10.3390/BIOM11010051

Metwally, F. M., Rashad, H., Zeidan, H. M., Kilany, A., & Abdol Raouf, E. R. (2018). Study of the effect of bisphenol a on oxidative stress in children with autism spectrum disorders. *Indian Journal of Clinical Biochemistry, 33*(2), 196–201. https://doi.org/10.1007/S12291-017-0667-0

Metwally, F. M., Rashad, H., Zeidan, H. M., & Hashish, A. F. (2020). Impact of bisphenol A on gonadotropic hormone levels in children with autism spectrum disorders. *Indian Journal of Clinical Biochemistry, 35*(2), 205–210. https://doi.org/10.1007/s12291-018-0801-7

Miani, A., Imbriani, G., De Filippis, G., De Giorgi, D., Peccarisi, L., Colangelo, M., Pulimeno, M., Castellone, M. D., Nicolardi, G., Logroscino, G., & Piscitelli, P. (2021). Autism spectrum disorder and prenatal or early life exposure to pesticides: A short review. *International Journal of Environmental Research and Public Health, 18*(20), 10991. https://doi.org/10.3390/IJERPH182010991

Miodovnik, A., Engel, S. M., Zhu, C., Ye, X., Soorya, L. V., Silva, M. J., Calafat, A. M., & Wolff, M. S. (2011). Endocrine disruptors and childhood social impairment. *Neurotoxicology, 32*(2), 261–267. https://doi.org/10.1016/J.NEURO.2010.12.009

Mirza, R., & Sharma, B. (2019). Beneficial effects of pioglitazone, a selective peroxisome proliferator-activated receptor-γ agonist in prenatal valproic acid-induced behavioral and biochemical autistic like features in Wistar rats. *International Journal of Developmental Neuroscience, 76*, 6–16. https://doi.org/10.1016/j.ijdevneu.2019.05.006

Mitchell, M. M., Woods, R., Chi, L. H., Schmidt, R. J., Pessah, I. N., Kostyniak, P. J., & Lasalle, J. M. (2012). Levels of select PCB and PBDE congeners in human postmortem brain reveal possible environmental involvement in 15q11-q13 duplication autism spectrum disorder. *Environmental and Molecular Mutagenesis, 53*(8), 589–598. https://doi.org/10.1002/EM.21722

Mitoma, C., Uchi, H., Tsukimori, K., Yamada, H., Akahane, M., Imamura, T., Utani, A., & Furue, M. (2015). Yusho and its latest findings-a review in studies conducted by the Yusho Group. *Environment International, 82*, 41–48. https://doi.org/10.1016/J.ENVINT.2015.05.004

Miyazaki, K., Narita, N., & Narita, M. (2005). Maternal administration of thalidomide or valproic acid causes abnormal serotonergic neurons in the offspring: Implication for pathogenesis of autism. *International Journal of Developmental Neuroscience, 23*(2–3), 287–297. https://doi.org/10.1016/j.ijdevneu.2004.05.004

Modabbernia, A., Velthorst, E., & Reichenberg, A. (2017). Environmental risk factors for autism: An evidence-based review of systematic reviews and meta-analyses. *Molecular Autism, 8*(1), 13. https://doi.org/10.1186/S13229-017-0121-4

Mohamed, F. E. B., Zaky, E. A., El-Sayed, A. B., Elhossieny, R. M., Zahra, S. S., Salah Eldin, W., Youssef, W. Y., Khaled, R. A., & Youssef, A. M. (2015). Assessment of hair aluminum, lead, and mercury in a sample of autistic Egyptian children: Environmental risk factors of heavy metals in autism. *Behavioural Neurology, 2015*, 1–9. https://doi.org/10.1155/2015/545674

Mold, M. et al. (2018). Aluminium in brain tissue in autism. *Journal of trace elements in medicine and biology: organ of the Society for Minerals and Trace Elements (GMS), 46*, 76–82. https://doi.org/10.1016/j.jtemb.2017.11.012

Montgomery, A. K., Shuffrey, L. C., Guter, S. J., Anderson, G. M., Jacob, S., Mosconi, M. W., Sweeney, J. A., Turner, J. B., Sutcliffe, J. S., Cook, E. H., & Veenstra-VanderWeele, J. (2018). Maternal serotonin levels are associated with cognitive ability and core symptoms in autism spectrum disorder. *Journal of the American Academy of Child and Adolescent Psychiatry, 57*(11), 867–875. https://doi.org/10.1016/j.jaac.2018.06.025

Moore, S. J. et al. (2000). A clinical study of 57 children with fetal anticonvulsant syndromes. *Journal of Medical Genetics, 37*(7), 489–497. https://doi.org/10.1136/jmg.37.7.489

Morakotsriwan, N., Wattanathorn, J., Kirisattayakul, W., & Chaisiwamongkol, K. (2016). Autistic-like behaviors, oxidative stress status, and histopathological changes in cerebellum of valproic acid rat model of autism are improved by the combined extract of purple rice and silkworm pupae. *Oxidative Medicine and Cellular Longevity, 2016*, 1–10. https://doi.org/10.1155/2016/3206561

Morin, A., Van de Beeck, L., Person, E., & Plamondon, H. (2021). Adult male rats show resilience to adolescent bisphenol A effects on hormonal and behavioral responses while co-exposure with hop extracts supports synergistic actions. *Frontiers in Toxicology, 3*, 639820. https://doi.org/10.3389/FTOX.2021.639820

Mostafa, G. A., Meguid, N. A., Shehab, A. A. S., Elsaeid, A., & Maher, M. (2021). Plasma levels of nerve growth factor in Egyptian autistic children: Relation to hyperserotonemia and autoimmunity. *Journal of Neuroimmunology, 358*, 577638. https://doi.org/10.1016/J.JNEUROIM.2021.577638

Mostafalou, S., & Abdollahi, M. (2017). Pesticides: An update of human exposure and toxicity. *Archives of Toxicology, 91*(2), 549–599. https://doi.org/10.1007/S00204-016-1849-X

Muhle, R., Trentacoste, S. V., & Rapin, I. (2004). The genetics of autism. *Pediatrics, 113*(5), e472–486. https://doi.org/10.1542/peds.113.5.e472

Muller, C. L., Anacker, A. M. J., & Veenstra-VanderWeele, J. (2016). The serotonin system in autism spectrum disorder: From biomarker to animal models. *Neuroscience, 321*, 24–41. https://doi.org/10.1016/j.neuroscience.2015.11.010

Musachio, E. A. S., de Freitas Couto, S., Poetini, M. R., Bortolotto, V. C., Dahleh, M. M. M., Janner, D. E., Araujo, S. M., Ramborger, B. P., Rohers, R., Guerra, G. P., & Prigol, M. (2021). Bisphenol A exposure during the embryonic period: Insights into dopamine relationship and behavioral disorders in Drosophila melanogaster. *Food and Chemical Toxicology, 157*, 112526. https://doi.org/10.1016/J.FCT.2021.112526

Musumeci, G. et al. (2014). Serotonin (5HT) expression in rat pups treated with high-tryptophan diet during fetal and early postnatal development. *Acta Histochemica, 116*(2), 335–343. https://doi.org/10.1016/j.acthis.2013.08.011

Mychasiuk, R., Richards, S., Nakahashi, A., Kolb, B., & Gibb, R. (2012). Effects of rat prenatal exposure to valproic acid on behaviour and neuro-anatomy. *Developmental Neuroscience, 34*(2–3), 268–276. https://doi.org/10.1159/000341786

Negishi, T., Nakagami, A., Kawasaki, K., Nishida, Y., Ihara, T., Kuroda, Y., Tashiro, T., Koyama, T., & Yoshikawa, Y. (2014). Altered social interactions in male juvenile cynomolgus monkeys prenatally exposed to bisphenol A. *Neurotoxicology and Teratology, 44*, 46–52. https://doi.org/10.1016/J.NTT.2014.05.004

Nguyen, U., Tinsley, B., Sen, Y., Stein, J., Palacios, Y., Ceballos, A., Welch, C., Nzenkue, K., Penn, A., Murphy, L., Leodones, K., Casiquin, J., Ivory, I., Ghenta, K., Danziger, K., Widman, E., Newman, J., Triplehorn, M., Hindi, Z., & Mulligan, K. (2021). Exposure to bisphenol A differentially impacts neurodevelopment and behavior in Drosophila melanogaster from distinct genetic backgrounds. *Neurotoxicology, 82*, 146–157. https://doi.org/10.1016/J.NEURO.2020.12.007

Nicolini, C., & Fahnestock, M. (2018). The valproic acid-induced rodent model of autism. *Experimental Neurology, 299*(Pt A), 217–227. https://doi.org/10.1016/j.expneurol.2017.04.017

Ohkawara, T., Katsuyama, T., Ida-Eto, M., Narita, N., & Narita, M. (2015). Maternal viral infection during pregnancy impairs development of fetal serotonergic neurons. *Brain Dev, 37*(1), 88–93. https://doi.org/10.1016/J.BRAINDEV.2014.03.007

Ornoy, A., Weinstein-Fudim, L., & Ergaz, Z. (2015). Prenatal factors associated with autism spectrum disorder (ASD). *Reproductive Toxicology, 56*, 155–169. https://doi.org/10.1016/j.reprotox.2015.05.007

Ornoy, A., Weinstein-Fudim, L., Tfilin, M., Ergaz, Z., Yanai, J., Szyf, M., & Turgeman, G. (2019). S-adenosyl methionine prevents ASD like behaviors triggered by early postnatal valproic acid exposure in very young mice. *Neurotoxicology and Teratology, 71*, 64–74. https://doi.org/10.1016/j.ntt.2018.01.005

Pardo, C. A., & Eberhart, C. G. (2007). The neurobiology of autism. *Brain Pathology (Zurich, Switzerland), 17*(4), 434–447. https://doi.org/10.1111/j.1750-3639.2007.00102.x

Papes, F., Surpili, M. J., Langone, F., Trigo, J. R., & Arruda, P. (2001). The essential amino acid lysine acts as precursor of glutamate in the mammalian central nervous system. *FEBS Letters, 488*(1–2), 34–38. https://doi.org/10.1016/S0014-5793(00)02401-7

Pedersen, L. H. (2015). Prenatal antidepressant exposure and childhood autism spectrum disorders: Cause for concern? *Pediatric Drugs, 17*(6), 443–448. https://doi.org/10.1007/s40272-015-0141-5

Pertile, R. A. N., Cui, X., Hammond, L., & Eyles, D. W. (2018). Vitamin D regulation of GDNF/ret signaling in dopaminergic neurons. *The FASEB Journal, 32*(2), 819–828. https://doi.org/10.1096/FJ.201700713R

Pessah, I. N., Lein, P. J., Seegal, R. F., & Sagiv, S. K. (2019). Neurotoxicity of polychlorinated biphenyls and related organohalogens. *Acta Neuropathologica, 138*(3), 363–387.

Petersen, S. E., & Posner, M. I. (2012). The attention system of the human brain: 20 years after. *Annual Review of Neuroscience, 35*, 73–89. https://doi.org/10.1146/ANNUREV-NEURO062111-150525

Petropoulos, H., Friedman, S. D., Shaw, D. W. W., Artru, A. A., Dawson, G., & Dager, S. R. (2006). Gray matter abnormalities in autism spectrum disorder revealed by T2 relaxation. *Neurology, 67*(4), 632–636. https://doi.org/10.1212/01.WNL.0000229923.08213.1E

Petruzzelli, M. G., Marzulli, L., Margari, F., De Giacomo, A., Gabellone, A., Giannico, O. V., Margari, L., De Giacomo, A., Gabellone, A., Giannico, O. V., & Margari, L. (2020). Vitamin D deficiency in autism spectrum disorder: A cross-sectional study. *Disease Markers, 2020*, 1–5. https://doi.org/10.1155/2020/9292560

Petty, F. (1995). GABA and mood disorders: A brief review and hypothesis. *Journal of Affective Disorders, 34*(4), 275–281. https://doi.org/10.1016/0165-0327(95)00025-I

Phiel, C. J., Zhang, F., Huang, E. Y., Guenther, M. G., Lazar, M. A., & Klein, P. S. (2001). Histone deacetylase is a direct target of valproic acid, a potent anticonvulsant, mood stabilizer, and teratogen. *The Journal of Biological Chemistry, 276*(39), 36734–36741. https://doi.org/10.1074/JBC.M101287200

Pierce, K., & Courchesne, E. (2001). Evidence for a cerebellar role in reduced exploration and stereotyped behavior in autism. *Biological Psychiatry, 49*(8), 655–664. https://doi.org/10.1016/s0006-3223(00)01008-8

Piras, I. S., Gabriele, S., Altieri, L., Lombardi, F., Sacco, R., Lintas, C., Manzi, B., Curatolo, P., Nobile, M., Rigoletto, C., Molteni, M., & Persico, A. M. (2021). Reevaluation of serum arylesterase activity in neurodevelopmental disorders. *Antioxidants, 10*(2), 1–14. http://www.ncbi.nlm.nih.gov/pubmed/33499329

Pitkin, R. M. (2007). Folate and neural tube defects. *The American Journal of Clinical Nutrition, 85*(1), 285S–288S. https://doi.org/10.1093/AJCN/85.1.285S

Popova, N. K., & Naumenko, V. S. (2019). Neuronal and behavioral plasticity: the role of serotonin and BDNF systems tandem. *Expert Opinion on Therapeutic Targets, 23*(3), 227–239. https://doi.org/10.1080/14728222.2019.1572747

Prado, E. L., & Dewey, K. G. (2014). Nutrition and brain development in early life. *Nutrition Reviews, 72*(4), 267–284. https://doi.org/10.1111/NURE.12102

Preciados, M., Yoo, C., & Roy, D. (2016). Estrogenic endocrine disrupting chemicals influencing NRF1 regulated gene networks in the development of complex human brain diseases. *International Journal of Molecular Sciences, 17*(12), 2086. https://doi.org/10.3390/ijms17122086

Principi, N., & Esposito, S. (2020). Vitamin D deficiency during pregnancy and autism spectrum disorders development. *Frontiers in Psychiatry, 10*, 987. https://doi.org/10.3389/FPSYT.2019.00987

Raghavan, R., Riley, A. W., Volk, H., Caruso, D., Hironaka, L., Sices, L., Hong, X., Wang, G., Ji, Y., Brucato, M., Wahl, A., Stivers, T., Pearson, C., Zuckerman, B., Stuart, E. A., Landa, R., Fallin, M. D., & Wang, X. (2018). Maternal multivitamin intake, plasma folate and vitamin B12 levels and autism spectrum disorder risk in offspring. *Paediatric and Perinatal Epidemiology, 32*(1), 100–111. https://doi.org/10.1111/PPE.12414

Rahbar, M. H., Swingle, H. M., Christian, M. A., Hessabi, M., Lee, M., Pitcher, M. R., Campbell, S., Mitchell, A., Krone, R., Loveland, K. A., & Patterson, D. G. (2017). Environmental exposure to dioxins, dibenzofurans, bisphenol a, and phthalates in children with and without autism spectrum disorder living near the gulf of Mexico. *International Journal of Environmental Research and Public Health, 14*(11), 1425. https://doi.org/10.3390/IJERPH14111425

Rai, D., Lee, B. K., Dalman, C., Golding, J., Lewis, G., & Magnusson, C. (2013). Parental depression, maternal antidepressant use during pregnancy, and risk of autism spectrum disorders: Population based case-control study. *BMJ, 346*(7907), f2059. https://doi.org/10.1136/bmj.f2059

Rasalam, A. D. et al. (2005). Characteristics of fetal anticonvulsant syndrome associated autistic disorder. *Developmental Medicine and Child Neurology, 47*(8), 551–555. https://doi.org/10.1017/s0012162205001076

Reilly, M. P., Weeks, C. D., Topper, V. Y., Thompson, L. M., Crews, D., & Gore, A. C. (2015). The effects of prenatal PCBs on adult social behavior in rats. *Hormones and Behavior, 73*, 47–55. https://doi.org/10.1016/J.YHBEH.2015.06.002

Repouskou, A., Papadopoulou, A. K., Panagiotidou, E., Trichas, P., Lindh, C., Bergman, Å., Gennings, C., Bornehag, C. G., Rüegg, J., Kitraki, E., & Stamatakis, A. (2020). Long term transcriptional and behavioral effects in mice developmentally exposed to a mixture of endocrine disruptors associated with delayed human neurodevelopment. *Scientific Reports, 10*(1), 9367. https://doi.org/10.1038/S41598-020-66379-X

Rimland, B. (1988). Controversies in the treatment of autistic children: Vitamin and drug therapy. *Journal of Child Neurology, 3*(Suppl), S68–S72. https://doi.org/10.1177/0883073888003001S13

Robea, M. A., Luca, A. C., & Ciobica, A. (2020). Relationship between vitamin deficiencies and cooccurring symptoms in autism spectrum disorder. *Medicina (Lithuania), 56*(5), 245. https://doi.org/10.3390/MEDICINA56050245

Roberts, E. M., & English, P. B. (2013). Bayesian modeling of time-dependent vulnerability to environmental hazards: An example using autism and pesticidedata. *Statistics in Medicine, 32*(13), 2308–2319. https://doi.org/10.1002/SIM.5600

Roberts, E. M., English, P. B., Grether, J. K., Windham, G. C., Somberg, L., & Wolff, C. (2007). Maternal residence near agricultural pesticide applications and autism spectrum disorders among children in the California Central Valley. *Environmental Health Perspectives, 115*(10), 1482–1489. https://doi.org/10.1289/ehp.10168

Romoli, M., Mazzocchetti, P., D'Alonzo, R., Siliquini, S., Rinaldi, V. E., Verrotti, A., Calabresi, P., & Costa, C. (2018). Valproic acid and epilepsy: From molecular mechanisms to clinical evidences. *Current Neuropharmacology, 17*(10), 926–946. https://doi.org/10.2174/1570159X17666181227165722

Rose'Meyer, R. (2013). A review of the serotonin transporter and prenatal cortisol in the development of autism spectrum disorders. *Molecular Autism, 4*(1), 37. https://doi.org/10.1186/2040-2392-4-37

Rosenfeld, C. S. (2017). Neuroendocrine disruption in animal models due to exposure to bisphenol A analogues. *Frontiers in Neuroendocrinology, 47*, 123–133. https://doi.org/10.1016/J.YFRNE.2017.08.001

Rosenquist, A. H., Høyer, B. B., Julvez, J., Sunyer, J., Pedersen, H. S., Lenters, V., Jönsson, B. A. G., Bonde, J. P., & Toft, G. (2017). Prenatal and postnatal PCB-153 and p,p′-DDE exposures and behavior scores at 5–9 years of age among children in Greenland and Ukraine. *Environmental Health Perspectives, 125*(10), 107002. https://doi.org/10.1289/EHP553

Rossignol, D. A., Genuis, S. J., & Frye, R. E. (2014). Environmental toxicants and autism spectrum disorders: A systematic review. *Translational Psychiatry, 4*(2), e360. https://doi.org/10.1038/TP.2014.4

Roux, S., & Bossu, J. L. (2018). Valproic acid and autism spectrum disorder: From clinical observations to animal studies. *Current Trends in Neurology, 11*, 53–61.

Rubin, B. S. (2011). Bisphenol A: An endocrine disruptor with widespread exposure and multiple effects. *The Journal of Steroid Biochemistry and Molecular Biology, 127*(1–2), 27–34. https://doi.org/10.1016/J.JSBMB.2011.05.002

Sato, K. (2018). Why is vitamin B6 effective in alleviating the symptoms of autism? *Medical Hypotheses, 115*, 103–106. https://doi.org/10.1016/j.mehy.2018.04.007

Schmidt, R. J., Hansen, R. L., Hartiala, J., Allayee, H., Schmidt, L. C., Tancredi, D. J., Tassone, F., & Hertz-Picciotto, I. (2011). Prenatal vitamins, one-carbon metabolism gene variants, and risk for autism. *Epidemiology, 22*(4), 476–485. https://doi.org/10.1097/EDE.0B013E31821D0E30

Schmidt, R. J., Tancredi, D. J., Ozonoff, S., Hansen, R. L., Hartiala, J., Allayee, H., Schmidt, L. C., Tassone, F., & Hertz-Picciotto, I. (2012). Maternal periconceptional folic acid intake and risk of autism spectrum disorders and developmental delay in the CHARGE (CHildhood Autism

Risks from Genetics and Environment) case-control study. *The American Journal of Clinical Nutrition, 96*(1), 80–89. https://doi.org/10.3945/AJCN.110.004416

Schmidt, R. J., Kogan, V., Shelton, J. F., Delwiche, L., Hansen, R. L., Ozonoff, S., Ma, C. C., McCanlies, E. C., Bennett, D. H., Hertz-Picciotto, I., Tancredi, D. J., & Volk, H. E. (2017). Combined prenatal pesticide exposure and folic acid intake in relation to autism spectrum disorder. *Environmental Health Perspectives, 125*(9), 097007. https://doi.org/10.1289/EHP604

Schneider, T., & Przewłocki, R. (2005). Behavioral alterations in rats prenatally to valproic acid: Animal model of autism. *Neuropsychopharmacology, 30*(1), 80–89. https://doi.org/10.1038/sj.npp.1300518

Schneider, T., Labuz, D., & Przewłocki, R. (2001). Nociceptive changes in rats after prenatal exposure to valproic acid. *Polish Journal of Pharmacology, 53*(5), 531–534. http://www.ncbi.nlm.nih.gov/pubmed/11990073

Schneider, T., Roman, A., Basta-Kaim, A., Kubera, M., Budziszewska, B., Schneider, K., & Przewłocki, R. (2008). Gender-specific behavioral and immunological alterations in an animal model of autism induced by prenatal exposure to valproic acid. *Psychoneuroendocrinology, 33*(6), 728–740. https://doi.org/10.1016/j.psyneuen.2008.02.011

Schug, T. T., Blawas, A. M., Gray, K., Heindel, J. J., & Lawler, C. P. (2015). Elucidating the links between endocrine disruptors and neurodevelopment. *Endocrinology, 156*(6), 1941–1951. https://doi.org/10.1210/EN.2014-1734

Schultz, W. (1997). Dopamine neurons and their role in reward mechanisms. *Current Opinion in Neurobiology, 7*(2), 191–197. https://doi.org/10.1016/S0959-4388(97)80007-4

Şengenç, E., Kıykım, E., & Saltik, S. (2020). Vitamin D levels in children and adolescents with autism. *The Journal of International Medical Research, 48*(7), 030006052093463. https://doi.org/10.1177/0300060520934638

Sethi, S., Chen, X., Kass, P. H., & Puschner, B. (2017a). Polychlorinated biphenyl and polybrominated diphenyl ether profiles in serum from cattle, sheep, and goats across California. *Chemosphere, 181*, 63–73. https://doi.org/10.1016/J.CHEMOSPHERE.2017.04.059

Sethi, S., Keil, K. P., Chen, H., Hayakawa, K., Li, X., Lin, Y., Lehmler, H. J., Puschner, B., & Lein, P. J. (2017b). Detection of 3,3'-dichlorobiphenyl in human maternal plasma and its effects on axonal and dendritic growth in primary rat neurons. *Toxicological Sciences, 158*(2), 401–411. https://doi.org/10.1093/TOXSCI/KFX100

Shan, L., Hu, X. L., Wang, B., & Jia, F. Y. (2016). Research advances in the role of vitamin D in autism spectrum disorders. *Chinese Journal Of Contemporary Pediatrics, 18*(2), 183–188. https://doi.org/10.7499/J.ISSN.1008-8830.2016.02.016

Shelton, J. F., & Hertz-Picciotto, I. (2015). Neurodevelopmental disorders and agricultural pesticide exposures: Shelton and Hertz-Picciotto respond. *Environmental Health Perspectives, 123*(4), A79–A80. https://doi.org/10.1289/EHP.1409124R

Shelton, J. F., Hertz-Picciotto, I., & Pessah, I. N. (2012). Tipping the balance of autism risk: Potential mechanisms linking pesticides and autism. *Environmental Health Perspectives, 120*(7), 944–951. https://doi.org/10.1289/EHP.1104553

Shelton, J. F., Geraghty, E. M., Tancredi, D. J., Delwiche, L. D., Schmidt, R. J., Ritz, B., Hansen, R. L., & Hertz-Picciotto, I. (2014). Neurodevelopmental disorders and prenatal residential proximity to agricultural pesticides: The charge study. *Environmental Health Perspectives, 122*(10), 1103–1109. https://doi.org/10.1289/EHP.1307044

Shen, W. Z., Luo, Z. B., Zheng, D. R., & Yew, D. T. (1989). Immunohistochemical studies on the development of 5-HT (serotonin) neurons in the nuclei of the reticular formations of human fetuses. *Pediatric Neuroscience, 15*(6), 291–295. https://doi.org/10.1159/000120485

Shokrzadeh, M., & Saravi, S. S. S. (2011). Pesticides in agricultural products: Analysis, reduction, prevention. In *Pesticides – Formulations, effects, fate* (pp. 225–242). InTech. www.intechopen.com

Shrader-Frechette, K., & ChoGlueck, C. (2017). Pesticides, neurodevelopmental disagreement, and Bradford Hill's guidelines. *Accountability in Research, 24*(1), 30–42. https://doi.org/10.1080/08989621.2016.1203786

Shuid, A. N., Jayusman, P. A., Shuid, N., Ismail, J., Nor, N. K., & Mohamed, I. N. (2021). Association between viral infections and risk of autistic disorder: An overview. *International Journal of Environmental Research and Public Health, 18*(6), 1–17. https://doi.org/10.3390/IJERPH18062817

Simchon-Tenenbaum, Y., Weizman, A., & Rehavi, M. (2015). Alterations in brain neurotrophic and glial factors following early age chronic methylphenidate and cocaine administration. *Behavioural Brain Research, 282*, 125–132. https://doi.org/10.1016/j.bbr.2014.12.058

Singha, S. P., Memon, S., Bano, U., Isaac, A. D., & Shahani, M. Y. (2022). Evaluation of p21 expression and related autism-like behavior in bisphenol-A exposed offspring of Wistar albino rats. *Birth Defects Research, 114*, 536–550. https://doi.org/10.1002/BDR2.2022

Skalny, A. V., Simashkova, N. V., Klyushnik, T. P., Grabeklis, A. R., Bjørklund, G., Skalnaya, M. G., Nikonorov, A. A., & Tinkov, A. A. (2017). Hair toxic and essential trace elements in children with autism spectrum disorder. *Metabolic Brain Disease, 32*(1), 195–202. https://doi.org/10.1007/S11011-016-9899-6

Skalny, A. V., Simashkova, N. V., Skalnaya, M. G., Klyushnik, T. P., Chernova, L. N., & Tinkov, A. A. (2018). Mercury and autism spectrum disorders. *Zhurnal Nevrologii i Psikhiatrii Imeni S.S. Korsakova, 118*(5), 75–79. https://doi.org/10.17116/JNEVRO20181185275

Staal, W. G., De Krom, M., & De Jonge, M. V. (2012). Brief report: The dopamine-3-receptor gene (DRD3) is associated with specific repetitive behavior in autism spectrum disorder (ASD). *Journal of Autism and Developmental Disorders, 42*(5), 885–888. https://doi.org/10.1007/S10803-011-1312-Z

Steegers-Theunissen, R. P., Obermann-Borst, S. A., Kremer, D., Lindemans, J., Siebel, C., Steegers, E. A., Slagboom, P. E., & Heijman, B. T. (2009). Periconceptional maternal folic acid use of 400 µg per day is related to increased methylation of the IGF2 gene in the very young child. *PLoS One, 4*(11), e7845. https://doi.org/10.1371/JOURNAL.PONE.0007845

Steenweg-de Graaff, J., Ghassabian, A., Jaddoe, V. W. V., Tiemeier, H., & Roza, S. J. (2015). Folate concentrations during pregnancy and autistic traits in the offspring. The Generation R Study. *European Journal of Public Health, 25*(3), 431–433. https://doi.org/10.1093/EURPUB/CKU126

Stein, T. P., Schluter, M. D., Steer, R. A., Guo, L., & Ming, X. (2015). Bisphenol A exposure in children with autism spectrum disorders. *Autism Research, 8*(3), 272–283. https://doi.org/10.1002/AUR.1444

Stephenson, J., Heslehurst, N., Hall, J., Schoenaker, D. A. J. M., Hutchinson, J., Cade, J. E., Poston, L., Barrett, G., Crozier, S. R., Barker, M., Kumaran, K., Yajnik, C. S., Baird, J., & Mishra, G. D. (2018). Before the beginning: Nutrition and lifestyle in the preconception period and its importance for future health. *Lancet, 391*(10132), 1830–1841. https://doi.org/10.1016/S0140-6736(18)30311-8

Strömland, K., Nordin, V., Miller, M., Akerström, B., & Gillberg, C. (1994). Autism in thalidomide embryopathy: A population study. *Developmental Medicine and Child Neurology, 36*(4), 351–356. https://doi.org/10.1111/J.1469-8749.1994.TB11856.X

Suades-González, E., Gascon, M., Guxens, M., & Sunyer, J. (2015). Air pollution and neuropsychological development: A review of the latest evidence. *Endocrinology, 156*(10), 3473–3482. https://doi.org/10.1210/EN.2015-1403

Sui, L., & Chen, M. (2012). Prenatal exposure to valproic acid enhances synaptic plasticity in the medial prefrontal cortex and fear memories. *Brain Research Bulletin, 87*(6), 556–563. https://doi.org/10.1016/j.brainresbull.2012.01.011

Sun, Y., Nakashima, M. N., Takahashi, M., Kuroda, N., & Nakashima, K. (2002). Determination of bisphenol A in rat brain by microdialysis and column switching high-performance liquid chromatrography with fluorescence detection. *Biomedical Chromatography, 16*(5), 319–326. https://doi.org/10.1002/BMC.161

Sundström, E., Kölare, S., Souverbic, F., Samuelsson, E. B., Pschera, H., Lunell, N. O., & Seiger, Å. (1993). Neurochemical differentiation of human bulbospinal monoaminergic neurons during the first trimester. *Brain Research. Developmental Brain Research, 75*(1), 1–12. https://doi.org/10.1016/0165-3806(93)90059-J

Tabatadze, T. et al. (2015). Hair heavy metal and essential trace element concentration in children with autism spectrum disorder. *Georgian Medical News, (248)*, 77–82.

Takuma, K., Hara, Y., Kataoka, S., Kawanai, T., Maeda, Y., Watanabe, R., Takano, E., HayataTakano, A., Hashimoto, H., Ago, Y., & Matsuda, T. (2014). Chronic treatment with valproic acid or sodium butyrate attenuates novel object recognition deficits and hippocampal dendritic spine loss in a mouse model of autism. *Pharmacology, Biochemistry, and Behavior, 126*, 43–49. https://doi.org/10.1016/j.pbb.2014.08.013

Tanaka, M., Sato, A., Kasai, S., Hagino, Y., Kotajima-Murakami, H., Kashii, H., Takamatsu, Y., Nishito, Y., Inagaki, M., Mizuguchi, M., Hall, F. S., Uhl, G. R., Murphy, D., Sora, I., & Ikeda, K. (2018). Brain hyperserotonemia causes autism-relevant social deficits in mice. *Molecular Autism, 9*(1), 60. https://doi.org/10.1186/s13229-018-0243-3

Tau, G. Z., & Peterson, B. S. (2010). Normal development of brain circuits. *Neuropsychopharmacology, 35*(1), 147–168. https://doi.org/10.1038/NPP.2009.115

Thompson, M. R., & Boekelheide, K. (2013). Multiple environmental chemical exposures to lead, mercury and polychlorinated biphenyls among childbearing-aged women (NHANES 1999-2004): Body burden and risk factors. *Environmental Research, 121*, 23–30. https://doi.org/10.1016/J.ENVRES.2012.10.005

Thongkorn, S., Kanlayaprasit, S., Jindatip, D., Tencomnao, T., Hu, V. W., & Sarachana, T. (2019). Sex differences in the effects of prenatal bisphenol a exposure on genes associated with autism spectrum disorder in the hippocampus. *Scientific Reports, 9*(1), 3038. https://doi.org/10.1038/S41598-019-39386-W

Thongkorn, S., Kanlayaprasit, S., Panjabud, P., Saeliw, T., Jantheang, T., Kasitipradit, K., Sarobol, S., Jindatip, D., Hu, V. W., Tencomnao, T., Kikkawa, T., Sato, T., Osumi, N., & Sarachana, T. (2021). Sex differences in the effects of prenatal bisphenol A exposure on autism-related genes and their relationships with the hippocampus functions. *Scientific Reports, 11*(1), 1241. https://doi.org/10.1038/s41598-020-80390-2

Tirouvanziam, R., Obukhanych, T. V., Laval, J., Aronov, P. A., Libove, R., Banerjee, A. G., Parker, K. J., O'Hara, R., Herzenberg, L. A., Herzenberg, L. A., & Hardan, A. Y. (2012). Distinct plasma profile of polar neutral amino acids, leucine, and glutamate in children with autism spectrum disorders. *Journal of Autism and Developmental Disorders, 42*(5), 827–836. https://doi.org/10.1007/S10803-011-1314-X

Topper, V. Y., Reilly, M. P., Wagner, L. M., Thompson, L. M., Gillette, R., Crews, D., & Gore, A. C. (2019). Social and neuromolecular phenotypes are programmed by prenatal exposures to endocrine-disrupting chemicals. *Molecular and Cellular Endocrinology, 479*, 133–146. https://doi.org/10.1016/J.MCE.2018.09.010

Ulbrich, B., & Stahlmann, R. (2004). Developmental toxicity of polychlorinated biphenyls (PCBs): A systematic review of experimental data. *Archives of Toxicology, 78*(5), 252–268. https://doi.org/10.1007/S00204-003-0519-Y

Vandenberg, L. N., Hauser, R., Marcus, M., Olea, N., & Welshons, W. V. (2007). Human exposure to bisphenol A (BPA). *Reproductive Toxicology, 24*(2), 139–177. https://doi.org/10.1016/J.REPROTOX.2007.07.010

Veenstra-VanderWeele, J., Muller, C. L., Iwamoto, H., Sauer, J. E., Owens, W. A., Shah, C. R., Cohen, J., Mannangatti, P., Jessen, T., Thompson, B. J., Ye, R., Kerr, T. M., Carneiro, A. M., Crawley, J. N., Sanders-Bush, E., McMahon, D. G., Ramamoorthy, S., Daws, L. C., Sutcliffe, J. S., & Blakely, R. D. (2012). Autism gene variant causes hyperserotonemia, serotonin receptor hypersensitivity, social impairment and repetitive behavior. *Proceedings of the National Academy of Sciences of the United States of America, 109*(14), 5469–5474. https://doi.org/10.1073/pnas.1112345109

Vinkhuyzen, A. A. E., Eyles, D. W., Burne, T. H. J., Blanken, L. M. E., Kruithof, C. J., Verhulst, F., Jaddoe, V. W., Tiemeier, H., & McGrath, J. J. (2018). Gestational vitamin D deficiency and autism-related traits: The Generation R Study. *Molecular Psychiatry, 23*(2), 240–246. https://doi.org/10.1038/MP.2016.213

Vitor-Vieira, F., Vilela, F. C., & Giusti-Paiva, A. (2021). Hyperactivation of the amygdala correlates with impaired social play behavior of prepubertal male rats in a maternal immune activation model. *Behavioural Brain Research, 414*, 113503. https://doi.org/10.1016/j.bbr.2021.113503

Vojdani, A. et al. (2003). Infections, toxic chemicals and dietary peptides binding to lymphocyte receptors and tissue enzymes are major instigators of autoimmunity in autism. *International Journal of Immunopathology and Pharmacology, 16*(3), 189–199. https://doi.org/10.1177/039463200301600302

Volk, H. E., Lurmann, F., Penfold, B., Hertz-Picciotto, I., & McConnell, R. (2013). Traffic-related air pollution, particulate matter, and autism. *JAMA Psychiatry, 70*(1), 71–77. https://doi.org/10.1001/JAMAPSYCHIATRY.2013.266

Volkmar, F. R., & Pauls, D. (2003). Autism. *Lancet, 362*(9390), 1133–1141. https://doi.org/10.1016/S0140-6736(03)14471-6

Von Ehrenstein, O. S., Ling, C., Cui, X., Cockburn, M., Park, A. S., Yu, F., Wu, J., & Ritz, B. (2019). Prenatal and infant exposure to ambient pesticides and autism spectrum disorder in children: Population based case-control study. *BMJ, 364*, 1962. https://doi.org/10.1136/BMJ.L962

Walder, D. J., Laplante, D. P., Sousa-Pires, A., Veru, F., Brunet, A., & King, S. (2014). Prenatal maternal stress predicts autism traits in 61/2 year-old children: Project Ice Storm. *Psychiatry Research, 219*(2), 353–360. https://doi.org/10.1016/j.psychres.2014.04.034

Wang, L., Almeida, L. E. F., Spornick, N. A., Kenyon, N., Kamimura, S., Khaibullina, A., Nouraie, M., & Quezado, Z. M. N. (2015). Modulation of social deficits and repetitive behaviors in a mouse model of autism: The role of the nicotinic cholinergic system. *Psychopharmacology, 232*(23), 4303–4316. https://doi.org/10.1007/S00213-015-4058-Z

Wang, M., Li, K., Zhao, D., & Li, L. (2017). The association between maternal use of folic acid supplements during pregnancy and risk of autism spectrum disorders in children: A metaanalysis. *Molecular Autism, 8*(1), 51. https://doi.org/10.1186/S13229-017-0170-8

Wang, Z., Ding, R., & Wang, J. (2021). The association between vitamin d status and autism spectrum disorder (ASD): A systematic review and meta-analysis. *Nutrients, 13*(1), 1–14. https://doi.org/10.3390/NU13010086

Watts, T. J. (2008). The pathogenesis of autism. *Clinical Medicine. Pathology, 1*, 99–103. PMC3160002.

Welch, C., & Mulligan, K. (2022). Does bisphenol a confer risk of neurodevelopmental disorders? What we have learned from developmental neurotoxicity studies in animal models. *International Journal of Molecular Sciences, 23*(5), 2894. https://doi.org/10.3390/IJMS23052894

Whitaker-Azmitia, P. M. (2005). Behavioral and cellular consequences of increasing serotonergic activity during brain development: A role in autism? *International Journal of Developmental Neuroscience, 23*(1), 75–83. https://doi.org/10.1016/j.ijdevneu.2004.07.022

Winneke, G. (2011). Developmental aspects of environmental neurotoxicology: Lessons from lead and polychlorinated biphenyls. *Journal of the Neurological Sciences, 308*(1–2), 9–15. https://doi.org/10.1016/J.JNS.2011.05.020

Wood, A. G., Nadebaum, C., Anderson, V., Reutens, D., Barton, S., O'Brien, T. J., & Vajda, F. (2015). Prospective assessment of autism traits in children exposed to antiepileptic drugs during pregnancy. *Epilepsia, 56*(7), 1047–1055. https://doi.org/10.1111/EPI.13007

Wu, H. et al. (2017). Fingolimod (FTY720) attenuates social deficits, learning and memory impairments, neuronal loss and neuroinflammation in the rat model of autism. *Life Sciences, 173*, 43–54. https://doi.org/10.1016/j.lfs.2017.01.012

Wu, H. et al. (2019). Beyond a neurotransmitter: The role of serotonin in inflammation and immunity. *Pharmacological Research, 140*, 100–114. https://doi.org/10.1016/j.phrs.2018.06.015

Xu, X. H., Zhang, J., Wang, Y. M., Ye, Y. P., & Luo, Q. Q. (2010). Perinatal exposure to bisphenol-A impairs learning-memory by concomitant down-regulation of N-methyl-daspartate receptors of hippocampus in male offspring mice. *Hormones and Behavior, 58*(2), 326–333. https://doi.org/10.1016/J.YHBEH.2010.02.012

Xu, X.-J. et al. (2015). Prenatal hyperandrogenic environment induced autistic-like behavior in rat offspring. *Physiology & Behavior, 138*, 13–20. https://doi.org/10.1016/j.physbeh.2014.09.014

Yang, C.-J. J., Tan, H.-P. P., & Du, Y.-J. J. (2014). The developmental disruptions of serotonin signaling may involved in autism during early brain development. *Neuroscience, 267*, 1–10. https://doi.org/10.1016/j.neuroscience.2014.02.021

Yassa, H. A. (2014). Autism: A form of lead and mercury toxicity. *Environmental Toxicology and Pharmacology, 38*(3), 1016–1024. https://doi.org/10.1016/J.ETAP.2014.10.005

Yasuda, H., Kobayashi, M., Yasuda, Y., & Tsutsui, T. (2013). Estimation of autistic children by metallomics analysis. *Scientific Reports, 3*, 1199. https://doi.org/10.1038/SREP01199

Yektaş, Ç., Alpay, M., & Tufan, A. E. (2019). Comparison of serum B12, folate and homocysteine concentrations in children with autism spectrum disorder or attention deficit hyperactivity disorder and healthy controls. *Neuropsychiatric Disease and Treatment, 15*, 2213–2219. https://doi.org/10.2147/NDT.S212361

Yoo, H. J. et al. (2014). Polymorphisms of BDNF gene and autism spectrum disorders: family based association study with korean trios. *Psychiatry Investigation, 11*(3), 319–324. https://doi.org/10.4306/pi.2014.11.3.319

Zamberletti, E., Gabaglio, M., Woolley-Roberts, M., Bingham, S., Rubino, T., & Parolaro, D. (2019). Cannabidivarin treatment ameliorates autism-like behaviors and restores hippocampal endocannabinoid system and glia alterations induced by prenatal valproic acid exposure in rats. *Frontiers in Cellular Neuroscience, 13*, 367. https://doi.org/10.3389/fncel.2019.00367

Zhang, H., Wang, Z., Meng, L., Kuang, H., Liu, J., Lv, X., Pang, Q., & Fan, R. (2020). Maternal exposure to environmental bisphenol A impairs the neurons in hippocampus across generations. *Toxicology, 432*, 152393. https://doi.org/10.1016/J.TOX.2020.152393

Zangrossi, H., & Graeff, F. G. (2014). Serotonin in anxiety and panic: contributions of the elevated T-maze. *Neuroscience and Biobehavioral Reviews, 46*(3), 397–406. https://doi.org/10.1016/j.neubiorev.2014.03.007

Zlatnik, M. G. (2016). Endocrine-disrupting chemicals and reproductive health. *Journal of Midwifery & Women's Health, 61*(4), 442–455. https://doi.org/10.1111/JMWH.12500

Chapter 4
Animal Models of ASD

Bruna Lotufo-Denucci

The physiological and anatomical similarities between humans and animals, especially mammals, are useful for studies of the central nervous system, and animal models have often been used in neuroscience research, with rodents being the most frequent ones. Several studies using rodents, such as rats and mice, showed clinical characteristics present in individuals diagnosed with autism. Autism Spectrum Disorder (ASD) is defined as a neurodevelopmental disorder (DSM-5; APA, 2013), and though the pathophysiology of ASD remains unclear, the hypothesis in the etiology suggests a combination of genetic predisposition and environmental factors. The development of animal models that reflect the pathological, behavioral, and genetic aspects is extremely important for the understanding of ASD. Thus, the use of animal models for investigating mechanisms linked to ASD may contribute to understanding the neurobiological bases of this disorder. There are different types of animal models for ASD: genetic and non-genetic models. The genetic animal models are induced by genetic manipulations and mimic well-defined human diseases (e.g., Fragile X). The non-genetic models are induced by the environment, by exposure of pregnant animals to infection or inflammation (e.g., rubella, maternal and fetal inflammation) or chemical (e.g., valproic acid) and show ASD-like behavior.

An animal model should satisfy the criteria for model validation that display three types of validity: predictive, face, and construct validity. The predictive validity requires that induced animal behavior is expected to be reversed or prevented with treatment using effective drugs in humans for the condition. The face validity assesses how much the model is capable of reproducing symptoms and/or behaviors of the disease. The construct validity evaluates how much the model reproduces the biological mechanisms of the disease. Thus, animal models of ASD make it possible to understand the behavior, such as the core symptoms, potential therapeutics for

B. Lotufo-Denucci (✉)
Laboratory of Neurophysiology, University of State of Rio de Janeiro, Rio de Janeiro, Brazil

A. El Idrissi, D. McCloskey (eds.), *Neurobiology of Autism Spectrum Disorders*,
https://doi.org/10.1007/978-3-031-42383-3_4

ASD, and the neurobiology of autism. This chapter will provide an overview of animal models of autism, including genetic and non-genetic models, as well as some findings from the literature.

4.1 Non-genetic Animal Models

Among the different environmental causes that induce ASD in animal models, the most common model is induced by Antiepileptic drugs (AEDs) in rodents. These drugs often are used to control seizures and as mood stabilizers. Among the several AEDs, valproic acid and carbamazepine are important to the study of ASD. Valproic acid and carbamazepine are teratogenic and birth defects in humans have been associated with maternal use of AEDs in pregnancy (Mutlu-Albayrak et al., 2017). They can lead to the Fetal Anticonvulsant Syndrome, which is described as a group of malformations observed in individuals exposed to AEDs during the prenatal period. The use of carbamazepine during pregnancy increases the risk of spina bifida (Jentink et al., 2010), and the use of valproic acid has been associated with a greater risk of major congenital malformations such as neural tube defects (NTDs), cleft lip and palate, limb defects, cardiovascular abnormalities, genitourinary defects, developmental delay, and autism (Ornoy, 2009). The use of VPA in pregnancy is related to a higher prevalence of congenital malformations compared to other antiepileptic drugs (Tomson et al., 2019).

One of the most validated animal models to study ASD uses prenatal or early postnatal exposure to VPA in rodents. The VPA exposure in rodents results in several behavioral characteristics observed in humans, such as repetitive or stereotypical behaviors and deficit in social interaction (Schneider & Przewlocki, 2005; Schneider et al., 2008; Markram et al., 2008; Roullet et al., 2013; Wagner, 2006). The model is based on clinical findings in which VPA significantly increased the incidence of autism in children of mothers who used this drug during the first trimester. Thus, Rodier et al. (1997) proposed prenatal VPA exposure in rodents as an animal model of autism; he showed that injection of VPA in utero on embryonic day 11.5–12.5 produced impairment to the motor cranial nerve nuclei and a reduced number of neurons.

In this model, several researchers developed a model of exposure to VPA at different stages of the prenatal period, as well as in the first two postnatal weeks, a single VPA exposure during different phases of brain development is capable to induce autism-like behaviors. Some studies have administrated VPA on embryonic day 9, affecting neural migration (Tsujino et al., 2007; Narita et al., 2010; Dufour-Rainfray et al., 2010); others during embryonic day 12, affecting neural differentiation (Stanton et al., 2007; Kim et al., 2011); or on embryonic day 11.5–12.5, the day the neural tube closes (Rodier et al., 1997). Besides that, the exposure on postnatal day 14 demonstrated neurodevelopmental deficits similar to human ASD (Wagner, 2006). In addition to the difference in the period that the drug is administrated, the dosage of VPA used for inducing autism-related traits also varies across studies.

There is strong evidence of the association between pregnant women exposed to infections or inflammation and high rates of ASD in children. Immune dysregulation has been implicated as a mechanism in ASD pathogenesis and the exposure to maternal immune activation (MIA) during the prenatal period is suggested as an environmental risk factor for ASD. In a literature review, Wong and Hoeffer (2018) proposed a role for interleukin- 17A (IL-17A) in the development of ASD. Studies showed elevated IL-17A levels in the blood correlate with ASD individuals (Wong & Hoeffer, 2018). In a murine model, Choi et al. (2016) found that IL-17A activates MIA in the fetal brain and induces an abnormal cortical ASD-like phenotype.

Additionally, viral infection such as influenza virus and Borna disease results in behavioral changes, especially in social interaction, and stereotypic behavior with a decrease in the exploratory activity (Patterson, 2009; Lancaster et al., 2007). The administration of immunogens such as lipopolysaccharide-induced inflammation and polynosinic-polycyclidyl acid during pregnancy induces ASD-like behavior, such as repetitive behavior (Xuan & Hampson, 2014). Therefore, exposure to infections and/or inflammation during pregnancy is an important model of ASD.

Behavioral assays have also been widely used to evaluate behavior deficits present in ASD patients, especially the three core areas affected in the disorder: socialization, non-social patterns of behavior, and communication. The social preference test is an example of sociability behavior assessment, in which the time spent with an unknown animal in comparison to a familiar one is evaluated. Among non-social patterns of behavior, repetitive behavior is widely used; marble-burying has been used to assess repetitive behavior in rodents. Finally, the communication may be evaluated by ultrasonic vocalization; ASD animals presented reduced levels of ultrasonic vocalization or non-usual patterns of acoustic communication, as well as an altered pattern of response.

This chapter provides a summary of behavioral studies in rodents that include deficits in social interaction and communication, repetitive behavior, lower sensitivity to pain, hyperactivity, anxiety, and memory disabilities. Anatomical and cellular changes have also been reported, such as reduced size of cerebellar hemispheres, reduced number of Purkinje cells in the cerebellum, damage to cranial nerve nuclei, enhanced synaptic plasticity in regions like the prefrontal cortex and amygdala, alterations in neurotransmitter systems (e.g., monoamines and amino acid neurotransmission), and reduced cortical expression of brain-derived neurotrophic factor (BDNF) (for review, see Ergaz et al., 2016).

4.2 Genetic Animal Models

Research with genetic animal models is widely used to understand the etiology of the disease as well as the pathophysiology and the investigation of therapies. Studies have shown several risk genes involved in ASD (Parikshak et al., 2013; Coe et al., 2014; Pinto et al., 2014). It is believed that up to 1000 genes are implicated in susceptibility to ASD (Geschwind, 2011; Murdoch & State, 2013). These models include genetic mutations in code for proteins, copy number variants associated with chromosomal loci, and single-gene mutations associated with other neurodevelopmental disorders, such as Fragile X Syndrome and Rett Syndrome (for review, see Kazdoba et al., 2016). Genetic animal models present behaviors similar to those observed in ASD individuals. Thus, these models are an important tool for studies on ASD.

An example is knockout mice that lack the FMRP translational regulator 1 (FMR1). The FMR1 is a gene that codes a protein named Fragile X mental retardation protein (FMRP), a very important protein for normal cognitive development. Mutations in that gene may lead to autism, cognitive deficits, and Fragile X syndrome, a neurodevelopmental disorder linked to chromosome X and associated with ASD. The FMRP activity is very important for appropriate neuronal development and some brain functions (e.g., synaptic connectivity, plasticity, and dendritic architecture) (Comery et al., 1997; Cruz-Martin et al., 2010; He & Portera-Cailliau, 2013; Schutt et al., 2009). In a recent review, Fyke and Velinov (2021) highlighted the connection between FMR1 mutations and the etiology of ASD, like presynaptic hypotheses, synaptic dysfunction, and the endocannabinoid system (for review see Fyke & Velinov, 2021).

Concerning behavioral characteristics, Fragile X syndrome patients present attention deficit hyperactivity disorder, sensory processing deficits, anxiety, and intellectual disability. Animal studies have shown autistic-like behavior in this model, in a review, Kazdoba et al. (2016), summarized behavioral assays relevant to the core features of autism in a translational mouse model of ASD. Some behavioral characteristics were observed in FMR1 animals, such as low levels of social play behavior, and higher levels of repetitive block chewing. In addition to the core symptoms, FMR1 animals display sensory symptoms like increased response to sensory stimuli. Attention deficits and hyperactivity have also been observed in FMR1 animals with increased exploratory locomotion in the open field test and alterations in sleep-wake cycle stability. Another important symptom is a seizure disorder that affects at least 20% of ASD individuals. Some genetic models show increased seizure susceptibility, including in FMR1 animals (for review see, Kazdoba et al., 2016). Other traits include abnormal dendritic spine morphology, increased phosphorylation of translational control molecules, exaggerated protein synthesis in the hippocampus, and enhanced metabotropic glutamate receptor-dependent long-term depression (for review, see Ergaz et al., 2016).

Another animal model commonly used for the study of ASD is an inbred mouse strain called BTBR T + tf/j (BTBR), used as a model of the core symptoms in autism. The association between BTBR mice and ASD was found when researchers studied autistic-like behavior in 10 different inbred strains (Moy et al., 2007). BTBR animals have mutations in the gene encoding for Methyl-CpG-binding protein-2 (MECP2), a critical regulator for neural development. MECP2 is widely expressed and most abundant in neurons of the mature nervous system, and it is involved with the development of typical Rett syndrome and autistic-like behaviors (Samaco et al., 2013). Animal studies using the BTBR model presented interesting findings, including differences between knockout for MECP2 and MECP2+/−; knockout female animals of the MECP2 have a short life span while males present a total loss of function; and autistic traits varied depending on the background strain in the female MECP2+/−. At last, it was suggested that heterozygote strain may be used in future studies of ASD (Samaco et al., 2013).

These animals displayed social communication impairment and repetitive movements. Furthermore, the social deficits may be supported by high anxiety-like behavior or motor impairments (McFarlane et al., 2008). In addition, there are reports of decreased anxiety in the elevated plus-maze and decreased sociability in the 3-chamber social test (Moy et al., 2007). According to Moy et al. (2007), BTBR is a very interesting strain for ASD studies because it displays a low social approach and resistance to change in routine, these behaviors are in accordance with an autistic-like phenotype. This model of ASD is validated to the core symptoms and therefore an important tool for studying ASD.

Animal models of ASD are not restricted to rodents, other models include zebrafish and *Caenorhabditis elegans*. Zebrafish has been widely used as an animal model in scientific research. The sequence of the zebrafish genome showed that approximately 71% of human genes have at least one zebrafish orthologue (Howe et al., 2013). Cynomolgus monkeys are very closer to humans regarding cognitive ability, social behavior and brain structure and function. Thus, it is very interesting to use as a non-primate model for studying brain disorders. The genetic non-human primate ASD model might better approximate the behavioral phenotype compared to rodent models. The SHANK3 encodes proteins involved with excitatory synapses. And the mutation of SHANK3 is related with 1% of idiopathic forms of ASD. Cynomolgus macaques with mutation in the SHANK3 showed reduced levels of SHANK protein and exhibited sleep disturbances, motor deficits and increased repetitive behaviors, social and learning impairments (Zhou et al., 2016).

Genetic manipulations of a particular gene target are widely used in the zebrafish ASD model. In a recent review, Pensado-López et al. (2020) summarized several ASD-candidate genes and the phenotype observed in ASD-associated gene models. Some examples of the gene-associated phenotype are neuronal alterations, social behavior and motor alterations, hyperactivity, anxiety, learning difficulties, memory alterations in social and non-social recognition, and alterations in neurotransmitter systems, such as the serotonin signaling pathway.

Regarding behavior, sociability is an important component of zebrafish behavior. To assess sociability, social interactions can be evaluated through the presence of some specific behaviors, such as approaching, circling, mouth opening, biting, and chasing. Additionally, presence of repetitive patterns of locomotor activity has been used as non-social pattern of behavior (Pensado-López et al., 2020).

Another model used to understand the function of ASD associated genes is the *C. elegans*. *C. elegans* is a non-parasitic nematode used for investigating the development of the nervous system. Compared to another ASD model, the c. elegans presented some advantages like genetic amenability, small size, short life, and easy handling (Scmeisser & Parker, 2018).

Studies in humans demonstrated that mutations in genes such as neuroligin (NLGN1) and neurexin (NRXN1) have been associated with ASD (Feng et al.,

2006; Glessner et al., 2009; Sudhof, 2008; Yan et al., 2008). Neuroligin is a cell adhesion protein present on the postsynaptic membrane that mediates the formation and the maintenance of synapses in the neurons. The neurexins are cell adhesion molecules, a protein that mediates cell-cell interactions in the nervous system. These play an important role in synapse formation and maintenance. Mutations in these genes are involved with alterations in synapses function and development, thus affecting cognition.

Investigation of *C. elegans* found orthologue genes named *nlg-1* and *nrx-1* and mutations in these genes caused phenotypes related to ASD. Among these phenotypes we can mention the absence of thermal responses, deficits in touch response, osmotic avoidance, and impairment in dopaminergic and serotoninergic systems (Calahorro & Ruiz-Rubio, 2012; Duerden et al., 2015; Hedgecock & Russel, 1975; Izquierdo et al., 2013). Mutations in other genes like *shn-1, sel-2, sax-7/lad-1* and *lad-2* were observed in a study with *C. elegans* (for review, see Scmeisser & Parker, 2018). Genes related to ASD were investigated using SFARI Gene database to identify orthologous genes in *C. elegans* model (Rawsthorne et al., 2021). The study selected genes involved with synapses, neurons, cell signaling, epigenetic modification, and metabolism of phospholipids. They observed deficits in social behavior and habituation related to mutated synaptic genes (Rawsthorne et al., 2021).

4.3 Conclusion

Using as keyword "Animal model of ASD" in Pubmed, we found more than 1300 results only in the 2000s. If we search for "Autism Spectrum Disorder", during the same period, more than 43,000 scientific articles were found. This large number of scientific articles indicates the importance of the study of autism, whether the objective is the development of pharmacological therapy or the etiology of the disease. It is important to remember that a good model must satisfy the validation criteria. We can conclude that the use of animal models of ASD is important to investigate the role of genetic mutations and environmental factors in the etiology of this spectrum.

References

American Psychiatric Association. (2013). *Diagnostic and statistical manual of mental disorders (DSM-5R)*. American Psychiatric Pub.

Calahorro, F., & Ruiz-Rubio, M. (2012). Functional phenotypic rescue of *Caenorhabditis elegans* neuroligin-deficient mutants by the human and rat NLGN1 genes. *PLoS One, 7*, e39277.

Choi, G. B., Yeong, S. Y., Wong, H., Kim, S., Kim, H., Kim, S. V., Hoeffer, C. A., Littman, D. R., & Huh, J. R. (2016). The maternal interleukin-17a pathway in mice promotes autism-like phenotypes in offspring. *Science, 351*(6276), 933–939.

Coe, B. P., Witherspoon, K., Rosenfeld, J. A., van Bon, B. W., Vulto-van Silfhout, A. T., Bosco, P., Friend, K. L., Baker, C., Buono, S., Vissers, L. E., Schuurs-Hoeijmakers, J. H., Hoischen, A.,

Pfundt, R., Krumm, N., Carvill, G. L., Li, D., Amaral, D., Brown, N., Lockhart, P. J., Scheffer, I. E., Alberti, A., Shaw, M., Pettinato, R., Tervo, R., de Leeuw, N., Reijnders, M. R., Torchia, B. S., Peeters, H., O'Roak, B. J., Fichera, M., Hehir-Kwa, J. Y., Shendure, J., Mefford, H. C., Haan, E., Gecz, J., de Vries, B. B., Romano, C., & Eichler, E. E. (2014). Refining analyses of copy number variation identifies specific genes associated with developmental delay. *Nature Genetics, 46*, 1063–1071.

Comery, T. A., Harris, J. B., Willems, P. J., Oostra, B. A., Irwin, S. A., Weiler, I. J., & Greenough, W. T. (1997). Abnormal dendritic spines in fragile X knockout mice: Maturation and pruning deficits. *Proceedings of the National Academy of Sciences of the United States of America, 94*, 5401–5404.

Cruz-Martin, A., Crespo, M., & Portera-Cailliau, C. (2010). Delayed stabilization of dendritic spines in fragile X mice. *The Journal of Neuroscience, 30*, 7793–7803.

Duerden, E. G., Taylor, M. J., Lee, M., McGrath, P. A., Davis, K. D., & Roberts, S. W. (2015). Decreased sensitivity to thermal stimuli in adolescents with autism spectrum disorder: Relation to symptomatology and cognitive ability. *The Journal of Pain, 16*, 463–471.

Dufour-Rainfray, D., Vourc'h, P., Le Guisquet, A. M., Garreau, L., Ternant, D., Bodard, S., et al. (2010). Behavior and serotonergic disorders in rats exposed prenatally to valproate: A model for autism. *Neuroscience Letters, 470*, 55–59.

Ergaz, Z., Weinstein-Fudim, L., & Ornoy, A. (2016). Genetic and non-genetic animal models for autism spectrum disorders (ASD). *Reproductive Toxicology, 64*, 116–140.

Feng, J., Schroer, R., Yan, J., Song, W., Yang, C., Bockholt, A., Cook, E. H., Jr., Skinner, C., Schwartz, C. E., & Sommer, S. S. (2006). High frequency of neurexin 1beta signal peptide structural variants in patients with autism. *Neuroscience Letters, 409*, 10–13.

Fyke, W., & Velinov, M. (2021). FMR1 and autism, an intriguing connection revisited. *Genes, 12*, 1218. https://doi.org/10.3390/genes12081218

Geschwind, D. H. (2011). Genetics of autism spectrum disorders. *Trends in Cognitive Sciences, 15*, 409–416.

Glessner, J. T., Wnag, K., Cai, G., Korvatska, O., Kim, C. E., Wood, S., et al. (2009). Autism genome-wide copy number variation reveals ubiquitin and neuronal genes. *Nature, 459*, 569–573.

He, C. X., & Portera-Cailliau, C. (2013). The trouble with spines in fragile X syndrome: Density, maturity and plasticity. *Neuroscience, 251*, 120–128.

Hedgecock, E. M., & Russel, R. L. (1975). Normal and mutant thermotaxis in the nematode *Caenorhabditis elegans*. *Proceedings of the National Academy of Sciences of the United States of America, 72*, 4061–4065.

Howe, K., Clark, M. D., Torroja, C. F., Torrance, J., Berthelot, C., Muffato, M., Collins, J. E., Humphray, S., McLaren, K., Matthews, L., et al. (2013). The zebrafish reference genome sequence and its relationship to the human genome. *Nature, 496*, 498–503.

Izquierdo, P. G., Calahorro, F., & Ruiz-Rubio, M. (2013). Neuroligin modulates the locomotory dopaminergic and serotonergic neuronal pathways of *C. elegans*. *Neurogenetics, 14*, 233–242.

Jentink, J., Dolk, H., Loane, M. A., et al. (2010). Intrauterine exposure to carbamazepine and specific congenital malformations: Systematic review and case-control study. *BMJ, 341*, c6581.

Kazdoba, T. M., Leach, P. T., Yang, M., Silverman, J. L., Solomon, M., & Crawley, J. N. (2016). Translational mouse models of autism: Advancing toward pharmacological therapeutics. *Current Topics in Behavioral Neuroscience, 28*, 1–52.

Kim, K. C., Kim, P., Go, H. S., Choi, C. S., Yang, S. I., Cheong, J. H., et al. (2011). The critical period of valproate exposure to induce autistic symptoms in Sprague-Dawley rats. *Toxicology Letters, 201*, 137–142.

Lancaster, K., Dietz, D. M., Moran, T. H., & Pletnikov, M. V. (2007). Abnormal social behaviors in young and adult rats neonatally infected with Borna disease virus. *Behavioural Brain Research, 176*, 141–148.

Markram, K., Rinalde, T., Mendola, D. L., Sandi, C., & Markram, H. (2008). Abnormal fear conditioning and amygdala processing in an animal model of autism. *Neuropsychopharmacology, 33*, 901–912. https://doi.org/10.1038/sj.npp.1301453

McFarlane, H. G., Yang, M., Phoenix, J. L., Bolivar, V. J., & Crawley, J. N. (2008). Autism-like behavioral phenotypes in BTBR T+tf/J mice. *Genes, Brain, and Behavior, 7*(2), 152–163.

Moy, S. S., Nadler, J. J., Young, N. B., Perez, A., Holloway, L. P., Barbaro, R. P., Barbaro, J. R., et al. (2007). Mouse behavioral tasks relevant to autism: Phenotypes of 10 inbred strains. *Behavioural Brain Research, 176*, 4–20.

Murdoch, J. D., & State, M. W. (2013). Recent developments in the genetics of autism spectrum disorders. *Current Opinion in Genetics & Development, 23*, 310–315.

Mutlu-Albayrak, H., Bulut, C., & Çaksen, H. (2017). Fetal valproate syndrome. *Pediatrics and Neonatology, 58*, 158–164.

Narita, M., Oyabu, A., Imura, Y., Kamada, N., Yokoyama, T., Tano, K., et al. (2010). Nonexploratory movement and behavioral alterations in a thalidomide or valproic acid-induced autism model rat. *Neuroscience Research, 66*, 2–6.

Ornoy, A. (2009). Valproic acid in pregnancy: How much are we endangering the embryo and fetus? *Reproductive Toxicology, 28*, 1–10.

Parikshak, N. N., Luo, R., Zhang, A., Won, H., Lowe, J. K., Chandran, V., Horvath, S., & Geschwind, D. H. (2013). Integrative functional genomic analyses implicate specific molecular pathways and circuits in autism. *Cell, 155*, 1008–1021.

Patterson, P. H. (2009). Immune involvement in schizophrenia and autism: Etiology, pathology and animal models. *Behavioural Brain Research, 204*, 313–321.

Pensado-López, A., Veiga-Rúa, S., Carracedo, A., Allegue, C., & Sánchez, L. (2020). Experimental models to study autism spectrum disorders: hiPSCs, rodents and Zebrafish. *Genes (Basel), 11*, 1376. https://doi.org/10.3390/genes11111376

Pinto, D., Delaby, E., Merico, D., Barbosa, M., Merikangas, A., Klei, L., et al. (2014). Convergence of genes and cellular pathways dysregulated in autism spectrum disorders. *American Journal of Human Genetics, 94*, 677–694.

Rawsthorne, H., Calahorro, F., Holden-Dye, L., O'Connor, V., & Dillon, J. (2021). Investigating autism associated genes in C. elegans reveals candidates with a role in social behaviour. *PLoS One, 16*(5), e0243121. https://doi.org/10.1371/journal.pone.0243121

Rodier, P. M., Ingram, J. L., Tisdale, B., & Croog, V. J. (1997). Linking etiologies in humans and animal models: Studies on autism. *Reproductive Toxicology, 11*, 417–422.

Roullet, F. I., Lai, J. K., & Foster, J. A. (2013). In utero exposure to valproic acid and autism—A current review of clinical and animal studies. *Neurotoxicology and Teratology, 36*, 47–56. https://doi.org/10.1016/j.ntt.2013.01.004

Samaco, R. C., McGraw, C. M., Ward, C. S., Sun, Y., Neul, J. L., & Zoghbi, H. Y. (2013). Female Mecp2(+/−) mice display robust behavioral deficits on two different genetic backgrounds providing a framework for pre-clinical studies. *Human Molecular Genetics, 22*, 96–109.

Schneider, T., & Przewlocki, R. (2005). Behavioral alterations in rats prenatally exposed to valproic acid: Animal model of autism. *Neuropsychopharmacology, 30*, 80–89. https://doi.org/10.1038/sj.npp.1300518

Schneider, T., Roman, A., Basta-Kim, A., Kubera, M., Budziszewska, B., Schneider, K., & Przewlocki, R. (2008). Gender-specific behavioral and immunological alterations in an animal model of autism induced by prenatal exposure to valproic acid. *Psychoneuroendocrinology, 33*, 728–740. https://doi.org/10.1016/j.psyneuen.2008.02.011

Schutt, J., Falley, K., Richter, D., Kreienkamp, H. J., & Kindler, S. (2009). Fragile X mental retardation protein regulates the levels of scaffold proteins and glutamate receptors in postsynaptic densities. *The Journal of Biological Chemistry, 284*, 25479–25487.

Scmeisser, K., & Parker, J. A. (2018). Worms on the spectrum – C. elegans models in autism research. *Experimental Neurology, 299*, 199–206. https://doi.org/10.1016/j.expneurol.2017.04.007

Stanton, M. E., Peloso, E., Brown, K. L., & Rodier, P. (2007). Discrimination learning and reversal of the conditioned eyeblink reflex in a rodent model of autism. *Behavioural Brain Research, 176*, 133–140.

Sudhof, T. C. (2008). Neuroligins and neurexins link synaptic function to cognitive disease. *Nature, 455*, 903–911.

Tomson, T., Battino, D., & Perucca, E. (2019). Teratogenicity of antiepileptic drugs. *Current Opinion in Neurology, 32*(2), 246–252. https://doi.org/10.1097/WCO.0000000000000659

Tsujino, N., Nakatani, Y., Seki, Y., Nakasato, A., Nakamura, M., Sugawara, M., et al. (2007). Abnormality of circadian rhythm accompanied by an increase in frontal cortex serotonin in animal model of autism. *Neuroscience Research, 57*, 289–295.

Wagner, G. C., Reuhl, K. R., Cheh, M., McRae, P., & Halladay, A. K. (2006). A new neurobehavioral model of autism in mice: pre- and postnatal exposure to sodium valproate. *J Autism Dev Disord, 36*, 779–793.

Wong, H., & Hoeffer, C. (2018). Maternal IL-17A in autism. *Experimental Neurology, 299*(Pt A), 228–240.

Xuan, I. C., & Hampson, D. R. (2014). Gender-dependent effects of maternal immune activation on the behavior of mouse offspring. *PLoS One, 9*, e104433.

Yan, J., Noltner, K., Feng, J., Li, W., Schroer, R., Skinner, C., Zeng, W., Schwartz, C. E., & Sommer, S. S. (2008). Neurexin 1alpha structural variants associated with autism. *Neuroscience Letters, 438*, 368–370.

Zhou, Y., Kaiser, T., Monteiro, P., Zhang, X., Van der Goes, M. S., Wang, D., Barak, B., Zeng, M., Li, C., Lu, C., Wells, M., Amaya, A., Nguyen, S., Lewis, M., Sanjana, N., Zhou, Y., Zhang, M., Zhang, F., Fu, Z., & Feng, G. (2016). Mice with shank3 mutations associated with ASD and schizophrenia display both shared and distinct defects. *Neuron, 89*(1), 147–162.

Chapter 5
Mitochondrial Dysfunction in Autism Spectrum Disorders

Thiago Nunes, Alexandra Latini, and Joana M. Gaspar

Highlights

- Neuroinflammation is involved in the pathophysiology of ASD
- Mitochondrial dysfunction is involved in the pathophysiology of ASD
- Increased biomarkers of mitochondrial dysfunction are found in ASD samples
- ASD individuals have reduced activities of electron chain transport complexes
- Oxidative Stress is increased in the ASD brain

5.1 Introduction

Autism spectrum disorder (ASD) is a group of neurodevelopmental disorders characterized by deficits in communication and social interaction, restricted interests and repetitive behaviors. ASD symptoms start to manifest at the beginning of childhood, and diagnosis is completed according to the fifth edition of the Diagnostic and

T. Nunes · A. Latini
Graduate Course in Biochemistry, Department of Biochemistry, Center for Biological Sciences, Federal University of Santa Catarina, Florianopolis, SC, Brazil

Laboratory of Bioenergetics and Oxidative Stress (LABOX), Department of Biochemistry, Center for Biological Sciences, Federal University of Santa Catarina, Florianopolis, SC, Brazil

J. M. Gaspar (✉)
Graduate Course in Biochemistry, Department of Biochemistry, Center for Biological Sciences, Federal University of Santa Catarina, Florianopolis, SC, Brazil

Laboratory of Neuroimmune-Metabolism, Federal University of Santa Catarina, Brazil, Florianópolis, Brazill
e-mail: joana.gaspar@ufsc.br

A. El Idrissi, D. McCloskey (eds.), *Neurobiology of Autism Spectrum Disorders*,
https://doi.org/10.1007/978-3-031-42383-3_5

Statistical Manual of Mental Disorders (DSM-5). The symptoms of ASD range from mild to severe (American Psychiatric Association, 2014). The world prevalence of ASD is estimated in about 1.5% (Lyall et al., 2017a), with a higher incidence in males compared with females (Maenner et al., 2020). In the USA, in 2020, the prevalence of ASD was 1 in 54 children, showing a significantly increase in the last decade (Maenner et al., 2020). In Europe the ASD prevalence changes across different countries. Spain has a higher prevalence of ASD, 1 in 65 children aged 3–5 years old, and Germany with a lowest prevalence of 1 in 264 (Bougeard et al., 2021).

ASD is associated with comorbidities that includes motor impairments (Pan, 2012; Ament et al., 2015), sleep disorders (MacDuffie et al., 2020), food selectivity (Prosperi et al., 2017; Lucarelli et al., 2017), gastrointestinal problems (Prosperi et al., 2017) and a high prevalence of obesity. An estimated 42% of ASD children are overweight and 21% are obese compared to healthy children that have a prevalence of 26% overweight and 12% obese (Criado et al., 2018; Toscano et al., 2019).

The etiology of ASD is still unclear, but some evidence has pointed that a significant proportion of individuals with autism have concomitant symptoms associated with as mitochondrial dysfunction and abnormalities of energy generation (Mahalaxmi et al., 2021; Frye et al., 2013; Oliveira et al., 2005; Guevara-Campos et al., 2013). This has therefore led to the hypothesis that mitochondrial dysfunction can be the linking factor for explaining the development of ASD symptoms and associated comorbidities. In this review we will highlight pathological mechanisms of mitochondrial dysfunction observed in ASD trying to understand which are the relationships for the development of comorbidities associated with ASD.

5.2 Etiology and Pathophysiology of ASD

ASD symptoms are complex with high clinical, phenotypic and neuroanatomic heterogeneity (Toal et al., 2010). ASD has as primary symptoms repetitive and stereotyped behaviors, restricted interests, repetitive movements, deficits in social communication and interaction, altered non-verbal communication behaviors, and hypersensitivity or hyposensitivity to sensory stimuli (American Psychiatric Association, 2014).

ASD is usually associated with other neurological and psychiatric dysfunctions like epilepsy (Ewen et al., 2019), cognitive deficit, attention deficit and hyperactivity (Godoy et al., 2021; Zablotsky et al., 2020), and anxiety (Postorino et al., 2017). ASD is also associated with food selectivity (specially refusing vegetables) (Beighley et al., 2013; Lucarelli et al., 2017), gastrointestinal dysfunctions (Prosperi et al., 2017), dyslipidemia (Benachenhou et al., 2019; Luo & Eran, 2020), and obesity (Toscano et al., 2019).

The causes of different severity levels of ASD are likely to be very complex. It is well accepted that ASD development results from a complex combination and interaction among genetic, epigenetic, and environmental factors, as well as the dysregulation and overactivation of the immune system (Lai et al., 2014; Ghirardi et al., 2019; Hegarty et al., 2020) (Fig. 5.1). More than 600 genes have been implicated in

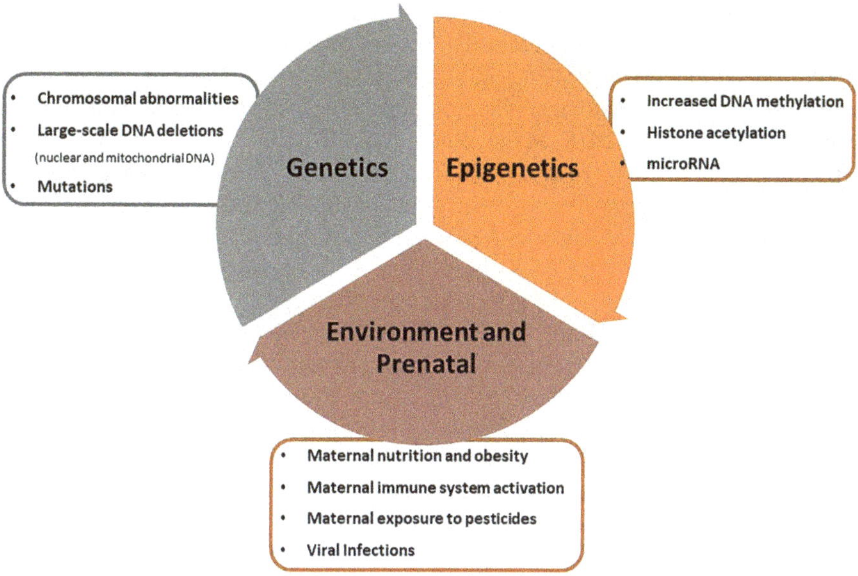

Fig. 5.1 The etiology of ASD. Although the complete etiology and pathogenesis underlying ASD have not yet been clear, most of the evidence identified several risk factors, such as environmental, genetic, and epigenetic factors

the etiology of ASD. Many common and rare variants can contribute additively in the heritability of ASD, with inherited truncated mutations found in excess in subjects with ASD, including genes implicated in the encoding of proteins for synaptic formation, transcriptional regulation, and chromatin-remodeling pathways (For a review of genetics of ASD (Waye & Cheng, 2018)). Also, many genetic changes detected in ASD patients are associated with mitochondrial dysfunction, including chromosomal abnormalities, large-scale deletions, and mutations in both mitochondrial and non-mitochondrial nuclear genes (Rose et al., 2018; Valiente-Pallejà et al., 2018; Brown & Rais, 2015).

Epigenetic mechanisms modulate the chromatin conformation and regulate the expression of genes without alterations in the DNA sequence. Genome-scale studies revealed exaggerated DNA methylation in the post-mortem brain tissue (temporal and prefrontal cortex, and cerebellum) of ASD individuals (Ladd-Acosta et al., 2014; Ellis et al., 2017). Difference in DNA methylation was detected for the genes expressed in the brain and encoding protein that interacts with those involved with the development of ASD (Loke et al., 2015; Vogel Ciernia & Lasalle, 2016). Histone acetylation was observed greater at >5000 cis-regulatory elements in ASD samples in prefrontal and temporal cortex (Sun et al., 2016).

Among the environmental factors that have association with ASD development are maternal nutrition, obesity and metabolic disorders (Lyall et al., 2013; Sullivan et al., 2012; Eow & Gan, 2021), maternal exposure to environmental pollution and pesticides (Nuttall, 2017; Lyall et al., 2017b), and maternal immune system activation (Brown et al., 2014; Urakubo et al., 2001; Lee et al., 2015).

Individuals with autism commonly present neuroanatomical, neurophysiological and molecular changes in the brain regions involved in social behavior, memory, cognition and emotions (Postema et al., 2019; Toal et al., 2010). Meta-analyses studies suggest neuroanatomical changes both in grey-matter and in white-matter structures (Eissa et al., 2018; Mann et al., 2021; Mei et al., 2020; Varghese et al., 2017; Bourgeron, 2015). In ASD subjects, the medial prefrontal cortex, superior temporal sulcus, temporoparietal junction, amygdala, and fusiform gyrus are hypoactive across tasks in which social perception and cognition are used (Varghese et al., 2017). Magnetic resonance images show that ASD subjects have lower accuracy on response shifting trials, independent of whether those trials also required a shift in cognitive set. ASD subjects also showed reduced activation in frontal, striatal, and parietal regions in a manner predictive of cognitive capacity. Also, the severity of restricted, repetitive behaviors was negatively correlated with activation in anterior cingulate and posterior parietal regions (Shafritz et al., 2008).

At cellular level, *postmortem* studies have have shown persistent neuroinflammation signs, and a reduction in neuronal number in the amygdala, fusiform gyrus, and cerebellum. Prefrontal cortical areas have increased markers of neuronal overgrowth and an increase in the neuronal dendritic spine in young children diagnosed with ASD compared to controls (Hutsler & Zhang, 2010). In the amygdala, neuronal cells are reduced in the lateral nucleus (Schumann & Amaral, 2006). ASD subjects also present a significant increase in the neuropil in the frontal cortex and the anterior cingulate gyrus, reduced dendritic branching in the hippocampus, and decreased pyramidal neurons the inferior frontal cortex; reduced number and size of Purkinje cells in the cerebellum (Blatt, 2012). Postmortem brain samples of ASD demonstrated thickening of subependymal cell layer and subependymal nodular dysplasia (indicating active neurogenesis). Heterotopias were observed in hippocampus, subcortical and cerebellar areas, indicating abnormal neuronal migration (Wegiel et al., 2010).

At the molecular level, several mechanisms are proposed to be involved in the pathophysiology of ASD, including changes in synaptic proteins, neurotrophins, neurotransmitters (Amal et al., 2020; Cast et al., 2021; Chana et al., 2015; El-Ansary & Al-Ayadhi, 2014; Fatemi et al., 2009), neuroglia activation and neuroinflammation (Sotgiu et al., 2020; Suzuki et al., 2013; Wei et al., 2011; Wei et al., 2012; Vargas et al., 2005), and mitochondrial dysfunction (Mahalaxmi et al., 2021; Frye et al., 2013; Oliveira et al., 2005; Guevara-Campos et al., 2013).

Neuroinflammation, inflammation and dysregulation of the immune system are clinical features of ASD, and can influence the severity of ASD symptoms (Ashwood et al., 2011; Ashwood & Wakefield, 2006). Elevated pro-inflammatory cytokine expression, such as IL-1β, IL-6, IL-17, and TNF-α, was observed in plasma/serum samples of ASD subjects (Inga Jácome et al., 2016; Masi et al., 2015; Nadeem et al., 2019). The increase in cytokine levels was associated with impaired communication and aberrant behaviors, i.e., correlated with the severity of ASD symptom, and predicted associated comorbidities (Ashwood et al., 2011; Inga Jácome et al., 2016). It was found that ASD individuals have alterations in immune cells such as T cells, B cells, monocytes, natural killer cells, and dendritic cells, as well as alterations in immunoglobulins and increased autoantibodies. ASD individuals have higher expression of the pro- and anti-inflammatory genes (Sciara et al., 2020), and elevated peripheral cytokines and chemokines and associated neuroinflammation (Bjorklund et al., 2016). Immunocytochemical studies identified astrogliosis and microglia activation in postmortem brain samples of ASD subjects, being identified in cerebellum, cerebral cortex, cingulate gyrus, cortex, white matter and hippocampus (Suzuki et al., 2013; Tetreault et al., 2012; Vargas et al., 2005; Rodriguez & Kern, 2011).

Altogether, ASD is a neurodevelopmental condition with a complex interaction of mechanisms involved in the etiology and pathophysiology. The precise influence and interplay of anatomical, genetic, cellular and molecular mechanisms has yet to be elucidated, as well as the mechanisms responsible for the phenotypic heterogeneity and severity observed in patients with autism. The hypothesis of genetic-epigenetic-environmental interactions as a cause of ASD provides a clue to its possible biological underpinnings.

Mitochondria, which are very vulnerable to environmental factors (such as parental age, maternal medication, maternal diseases, maternal nutrition, and bacterial infection), can be the link between environmental-genetic-epigenetic interactions for ASD development. ASD is associated with abnormal mitochondrial metabolism (Fig. 5.1), and pathophysiological abnormalities found in ASD, individuals is associated with mitochondrial dysfunction and clinical aspects of ASD (Fig. 5.2) including neurodevelopmental regression triggered by an inflammatory event, gastrointestinal symptoms, seizures, motor delays, fatigue and lethargy (Rossignol & Frye, 2012; Singh et al., 2020; Varga et al., 2018; Hassan et al., 2021).

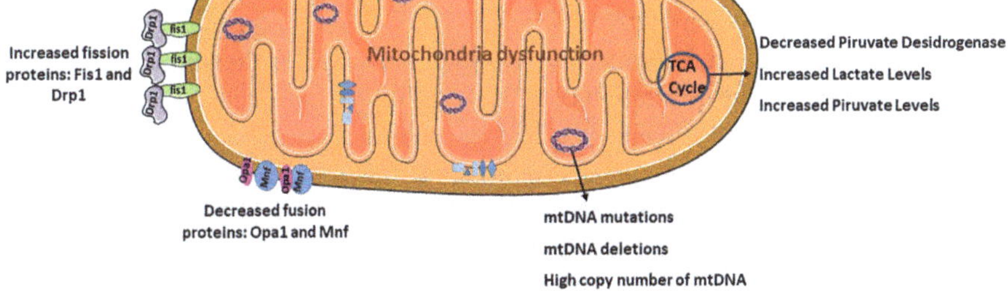

Fig. 5.2 Mitochondria dysfunction in ASD. Alterations of mitochondrial bioenergetics, dynamics, morphology and oxidative stress were reported in autism spectrum disorder

5.3 Mitochondrial Dysfunction in ASD

5.3.1 Mitochondrial Dysfunction in the Brain of ASD Subjects

Mitochondria are essential organelles for homeostasis of many cellular factors, including calcium. They are critical for energy production in the cell as well as apoptosis. Because of these numerous roles within cells, mitochondria are important to maintain cellular and metabolic health. The human brain consumes 20% of all mitochondrial energy. To fulfill these energy requirements, mitochondria supply neurons with energy by generating metabolites via the tricarboxylic acid cycle and ATP through oxidative phosphorylation.

Mitochondrial dysfunction occurs when the mitochondria fail to meet the energy requirements of the cell. Mitochondrial dysfunction occurs as a result of a mutation in a gene that is directly involved in the ATP-generating pathway or to the dysfunction occurring due to other genetic, biochemical or metabolic abnormalities and deficiencies that may impair the mitochondria's functions. Alteration of mitochondrial function has been implicated in several psychiatric and neurological disorders, including ASD.

Meta-analysis has demonstrated that 5% of children with ASD can be diagnosed with mitochondrial disease, which is defined by very low electron transport chain (ETC) complex activity or a clear genetic defect. ASD individuals demonstrate an increase in multiple biomarkers of mitochondrial dysfunction (lactate, alanine-to-lysine ratio, and acyl-carnitine) (Rossignol & Frye, 2012; Frye, 2012; Ghanizadeh et al., 2013). Therefore, understanding mitochondrial dysfunction and its implication in pathological states is of prime importance, especially for ASD etiology and symptoms. Social impairment and altered neurotransmission are frequently associated with mitochondrial dysfunction. In fact, many of the clinical characteristics of ASD are conditions that are commonly associated with mitochondrial disorders. Although mitochondrial function is crucial for brain homeostasis, it remains unknown whether mitochondrial dysfunction contributes to social deficits.

In ASD, the mitochondrial dysfunction hypothesis was based on the presence of lactic acidosis, high urine level of metabolites of the tricarboxylic acid (TCA) cycle, plasma carnitine deficiency, and decreased brain glucose utilization and ATP levels in autistic patients (Lombard, 1998). A reduction in cellular phoshpocreatine levels was found in the prefrontal cortex of individuals with ASD, indicating the high use of this compound for the generation of ATP, suggesting a failure in the generation of energy by mitochondrial respiration. The low levels of phosphocreatine indicate alterations in brain energy and phospholipid metabolism in ASD that correlate with the neuropsychologic and language deficits (Minshew et al., 1993). Indirect biochemical evidence of mitochondrial ETC dysfunction includes increased blood lactate and pyruvate levels, elevated plasma alanine level, AST, ALT and increased urinary levels of TCA cycle intermediates (Weissman et al., 2008), observed in ASD children (Weissman et al., 2008; Chugani et al., 1999).

The metabolite, N-acetyl-aspartate (NAA), a marker of mitochondrial dysfunction, was found decreased in the parietal cortex, cerebellum and anterior cingulate gyrus in individuals with ASD (Ipser et al., 2012; Chugani et al., 1999). A significant decrease in NAA was evident across age categories in the parietal cortex, the cerebellum, and the anterior cingulate cortex. Higher levels of creatine were observed for ASD adults in global grey matter (specifically in temporal lobe) and decreased creatine in the occipital lobe in children with ASD (Ipser et al., 2012). In ASD individuals, lactate doubled compared to healthy controls, detected most frequently within the cingulate gyrus but also present in the subcortical gray matter nuclei, corpus callosum, superior temporal gyrus, and pre- and postcentral gyri (Goh et al., 2014). The activity of the enzyme pyruvate dehydrogenase was also reduced in 35% of individuals with ASD compared to healthy individuals (Gu et al., 2013b). Pyruvate dehydrogenase is an enzyme involved in the conversion of pyruvate to acetyl-CoA. The reduced activity of pyruvate dehydrogenase would result in insufficient removal of pyruvate and lactate for the TCA cycle, causing insufficient ATP generation, supporting the association of mitochondrial dysfunction, abnormal pyruvate and lactate levels (Gu et al., 2013a).

In ASD the most common mitochondrial electron transport chain disorders were deficiencies of complex I and complex III (Weissman et al., 2008). It was observed a decrease in complexes I and IV activities, as well as altered protein levels of mitochondria respiratory chain protein complexes, in postmortem Brodmann area 22 temporal cortex from ASD patients (Tang et al., 2013). Brain tissue from the anterior cingulate gyrus, motor cortex, and thalamus of ASD patients has reduced expression of several ETC genes. Eleven genes of complex I, five genes each of complex III and complex IV, and seven genes of complex V showed brain region-specific reduced expression in autism. ATP5A1 (complex V), ATP5G3 (complex V) and NDUFA5 (complex I) showed consistently reduced expression in all the brain regions of ASD patients. In total, 11 genes from the ETC in the anterior cingulate gyrus, 12 genes in the motor cortex and 19 genes in the thalamus were significantly less expressed (Anitha et al., 2013). In cells of the frontal cortex was observed a reduction in the activities of all mitochondrial complexes (I, II, III, IV and V) of the ETC in ASD individuals compared with healthy individuals (Gu et al., 2013b). The total protein levels of complex II, III and V of the respiratory chain in the frontal cortex of autistic children were significantly reduced compared to healthy age-matched controls (Chauhan et al., 2011).

There was significant aberrant expression in genes between an autistic group compared to healthy control group, in the anterior cingulate cortex, motor cortex, and in the thalamus (Anitha et al., 2012). However, several genes showed brain region-specific expression alterations in ASD patients compared to controls. These genes that showed reduced expression have functions in membrane polarization and potential, mitochondrial transport, small molecule transport, expression of mitochondrial target proteins, outer and inner membrane translocations proteins, mitochondrial fusion and fissions proteins, mitochondrial location and apoptosis. Genes belonging to the SLC25A mitochondrial transporter family and to translocases, showed reduced expression in ASD (Anitha et al., 2012). Mitochondrial membrane mass was higher in ASD temporal cortex, as indicated by higher protein levels of mitochondrial membrane proteins Tom20, Tim23 and porin (Tang et al., 2013). In temporal cortex, levels of the mitochondrial gene transcription factor TFAM or cofactor PGC1α, were not changed in ASD subjects suggesting that mitochondrial biogenesis may not be involved in mitochondrial abnormalities (Tang et al., 2013). Mitochondria undergoes coordinated cycles of fission and fusion, referred as "mitochondrial dynamics", in order to maintain their shape, distribution and size. Their transient and rapid morphological adaptations are crucial for many cellular processes such as immunity, apoptosis and mitochondrial quality control. The mitochondrial fission proteins are increased (Fis1 and Drp1) and levels of the fusion proteins are decreased (Mfn1, Mfn2 and Opa1) in ASD patients, indicating altered mitochondrial dynamics in ASD brain (Tang et al., 2013).

In summary, different brain regions of ASD individuals present changes in the ECT complexes, as well as reduced expression of several proteins involved in mitochondrial function (mitochondrial transport, mitochondrial fusion and fission proteins, and apoptosis) which can compromise ATP production, as well as induced mitochondrial dysfunction.

5.3.2 Alterations in Mitochondrial DNA in ASD Subjects

Mitochondria have their own genome, known as mtDNA, a double-stranded, maternally inherited, circular molecule that contains the genetic information necessary for the synthesis of 13 essential polypeptides in the ETC. The mtDNA, and genetic variants in this molecule can alter the normal energy metabolism in the brain, contributing to the development of neurological disorders, including neurodevelopmental syndromes (Valiente-Pallejà et al., 2018). Most studies have reported that mutations in the mtDNA could be involved in the ASD phenotype (Varga et al., 2018; Gu et al., 2013b; Graf et al., 2000; Napoli et al., 2013).

The first case report involving mtDNA in ASD development was the identification of mtDNA G8363A transfer RNA(Lys) mutation in a boy of 2 years old (Graf et al., 2000). During his second year of life, he lost previously acquired language skills and developed marked hyperactivity with toe-walking, abnormal reciprocal social interaction, stereotyped mannerisms, restricted interests, self-injurious behavior, and seizures. This suggests that certain mitochondrial point mutations could be the basis for autism in some individuals (Graf et al., 2000). Another case report study analyses of five ASD patients have identified the A3243G mtDNA mutation and mtDNA depletion (Pons et al., 2004). Cells with 3243G mtDNAs had reduced mtDNA mRNA levels, rounded mitochondria, small cell size, and manifest induction of glycolytic genes, mitochondrial elongation (Picard et al., 2014). In addition, two Bulgarian ASD individuals were identified with the mutation m.6852G>A that produces an amino acid change of glycine to serine in the MT-CO1 gene and the mutation m.8033A>G (Ile → Val) in the MT-CO2 gene (Avdjieva-Tzavella et al., 2012). Increased mtDNA damage in ASD children was evidenced by twofold higher frequency of mtDNA deletions, 2,4-fold higher number of GC → AT transitions, and higher frequency of G,C,T → A transitions probably originated from oxidative stress. The authors point the hypothesis that reactive oxygen species-mediated damage of mtDNA in children with typical autism (Napoli et al., 2013). Tumor suppressor p53 responds to environmental stressors by regulating bioenergetics and mtDNA copy number maintenance. mtDNA deletions and higher p53 gene copy ratios were more common in children with ASD, suggesting a role for deficient DNA repair capacity. Deletions in mtDNA and altered p53 gene copy ratios seem to result from genetics and/or act in concert with environmental factors (Wong et al., 2016).

Despite that, analysis of ASD for mtDNA mutations has demonstrated no statistical support for an association between mtDNA mutations or polymorphisms and ASD (Álvarez-Iglesias et al., 2011; Hadjixenofontos et al., 2013). The authors conclude that the frequent observation of concomitant mitochondrial dysfunction and ASD could be due to nuclear factors influencing mitochondrion functions or to a more complex interplay between the nucleus and the mtDNA (Álvarez-Iglesias et al., 2011; Hadjixenofontos et al., 2013). A study conducted in peripheral blood cells from children with ASD reported an elevated mtDNA copy number compared with that in healthy children, although no significant correlations between mtDNA copy number and ASD clinical features were found (Chen et al., 2015; Bam et al., 2021).

The mtDNA sequence derived from whole-exome sequencing analyzed in ten multiplex families of genes, have identified two variants in the MT-ND5 gene that were previously identified to impair mitochondrial function (Patowary et al., 2017). Two variants of interest were also identified, within mitochondrial respiratory chain complex, the mtDNA variant in MT-ATP6 and nuclear DNA variant in NDUFS4 (Patowary et al., 2017). The higher mtDNA copy number in autism was also suggested by the increased ratios of mtDNA of three mitochondrial genes ND1, ND4 and Cyt B (that encode for subunits of complexes I and III) to nuclear DNA (Gu et al., 2013b). In another study, mtDNA deletions were confirmed in 16.6% of patients with ASD, and in 90% of those mtDNA deletions were found rare single nucleotide variants in ASD-associated genes (one of those was pathogenic) (Varga et al., 2018). Deletions were observed in ND4 (44%) and Cyt B (33%) in ASD children (Gu et al., 2013b). PGC-1alpha (peroxisome proliferator-activated receptor gamma coactivator-1α) is a master regulator of mitochondrial biogenesis. mtDNA copy number has a significant positive correlation with DNA methylation at the PGC-1α promoter in ASD individuals, which suggest an increase in mitochondrial biogenesis. Dense DNA methylation at promoters is associated with gene repression, whereas unmethylated promoters are mostly associated with gene activation. Furthermore, DNA methylation at PGC-1α promoter and mtDNA copy number correlated significantly with levels of urinary organic acids associated with mitochondrial dysfunction, oxidative stress, and neuroendocrinology (Bam et al., 2021).

In summary, studies have revealed that ASD subjects present changes to mtDNA (higher copy number, mutations and deletions), indicating that there may be mitochondrial dysfunction in children with autism. mtDNA deletions in ASD coexist either with other ASD-associated genetic risk factors or with alterations in genes responsible for intergenomic communication. However further comprehensive genetic investigations of the role of mitochondria genes in ASD are needed.

5.3.3 *Brain Oxidative Stress in ASD*

Oxidative stress was initially defined as a serious imbalance between oxidation and antioxidants, "a disturbance in the pro-oxidant–antioxidant balance in favor of the former, leading to potential damage". Reactive oxygen species (ROS) are produced during the oxidative phosphorylation by the mitochondrial electron transport chain. However, the excessive production and accumulation of ROS can negatively affect macromolecules such as proteins, DNA and lipids, causing cell damage which can eventually lead to cytotoxicity. Cell damage caused by ROS is implicated in the pathophysiology of a number of diseases, including ASD. Several antioxidant defense mechanisms are used by the cells to maintain the level of ROS and regulate the damage they can cause. Cellular antioxidant strategies can be enzymatic (e.g., superoxide dismutase (SOD), catalase, and glutathione peroxidase) and non-enzymatic (e.g., glutathione, lipoic acid, L-arginine, and coenzyme Q10) molecules, all of them being endogenous antioxidants. Beside these, there are several exogenous antioxidant molecules mainly introduced by the diet.

Children with ASD had higher mitochondrial rates of hydrogen peroxide production, and other reactive oxygen species compared with controls (Giulivi et al., 2010; Al-Ayadhi, 2012). Increase in ROS and lipid hydroperoxides (resulting from the oxidation of fatty acids) was found in samples of cerebellum, parietal, occipital, temporal and frontal cortex from ASD patients (Chauhan et al., 2011). The levels of 3-nitrotyrosine (3-NT), a marker of oxidative protein damage, was found to be increased in orbitofrontal cortex, Wernicke's area, cerebellar vermis, cerebellar hemisphere, and pons of the ASD. These particular brain regions are associated with speech processing, sensory and motor coordination, emotional and social behavior, and memory (Sajdel-Sulkowska et al., 2011). Oxidative stress increases in ASD children, resulting from elevated plasma F2-isoprostane and reduced glutathione peroxidase. This effects had significant risk for neuronal autoimmunity in autistic children (Mostafa et al., 2010). In frozen samples from the cerebellum and temporal cortex, the 3-nitrotyrosine and the biomarker for oxidative DNA damage (8-oxo-deoxyguanosine) were found significantly increased in ASD samples. The aconitase activity, a biomarker of mitochondrial superoxide production, were found decreased in ASD samples (Rose et al., 2012).

Carboxyethyl pyrrole (CEP) and iso[4]levuglandin (iso[4]LG)E2-protein adducts, are generated through peroxidation of docosahexaenoate and arachidonate-containing lipids respectively (Teresa et al., 2008). In post-mortem ASD samples, CEP and iso[4]LGE2 protein adducts, and heme-oxigenase-1 were found significant immunoractive in the white matter and into the grey matter. It was also observed that neurofilament heavy chain is a major target of CEP-modification, providing a direct evidence of oxidative injury of proteins in the brain that can provide a cellular basis for ASD neurological abnormalities (Teresa et al., 2008).

Associated with an increase in the ROS, changes in the antioxidant properties of the neuronal cells were also observed. The enzymatic activity of superoxide dismutase (SOD) and superoxide dismutase 2 (SOD2) were reduced in post-mortem ASD samples from the temporal lobe compared to healthy control subjects (Tang et al., 2013). Glutathione (GSH), the major endogenous antioxidant in the cells were analyzed in several studies. In the cerebellum and temporal cortex from ASD subjects the levels of the reduced form of glutathione (GSH) were significantly decreased, with a concomitant increase in the oxidized levels of glutathione (GSSG). However in other brain regions, frontal, pariental and occipital cortices the levels of GSH and GSSG were not changed (Chauhan et al., 2011). GSH and the ratio GSH/GSSG (redox/antioxidant capacity) were significantly decreased in the Brodmann area 22 (Rose et al., 2012). The activities of GSH-related enzymes (glutathione peroxidase (GPx), glutathione-S-transferase and glutathione reductase (GR)) were analyzed in cerebellum from autism and age-matched control subjects. It was observed a significantly decreased in the levels of GPx and glutathione-S-transferase in autism samples (Gu et al., 2013a). Although there was no significant difference in GR activity between autism and control groups, 40% of autistic subjects showed lower GR activity. Glutamate cysteine ligase activity was also significantly reduced % in the autistic subjects (Gu et al., 2013a). A case report study has demonstrated that administration of N-acetylcysteine (800 mg/day, in three divided doses), a precursor for GSH, significantly improved autism symptoms (increased social, and decreased aggressive behaviors) (Ghanizadeh & Derakhshan, 2012). However, in adults with autism the GSH concentrations in the basal ganglia and in the dorsomedial prefrontal cortex, were similar to control individuals, measured in-vivo using proton magnetic resonance spectroscopy ([1H]MRS) (Durieux et al., 2016). This study also observed that no correlation exist between GSH levels and clinical severity in patients (Durieux et al., 2016). Also measured in vivo by magnetic resonance spectroscopy, a trend to decrease in the GSH was observed dorsal anterior cingulate cortex, but no changes were observed in the dorsolateral prefrontal cortex of ASD patients (Endres et al., 2017). Polymorphisms of genes involved in glutathione metabolism are reportedly associated with autistic disorder (Ming et al., 2010).

Oxidative stress can also be responsible for the endoplasmic reticulum (ER) stress activation, which can induce cellular stress and apoptosis. ER stress was detected cerebellum, prefrontal cortex and hippocampus of ASD subjects (Dong et al., 2018). However, changes in the ER stress signals varies in the different brain regions studied (protein kinase R-like endoplasmic reticulum kinase [PERK], activating transcription factor 6 [ATF6], inositol-requiring enzyme 1 alpha [IRE1α]) (Dong et al., 2018). Consistent with the induction of ER stress, apoptosis was found in the three prefrontal cortex, hippocampus and cerebellum (Dong et al., 2018).

These findings indicate a possible role of increased oxidative stress and altered enzymatic antioxidants, both of which may be relevant to the pathophysiology of autism. The decreased antioxidant capacity and increased oxidative stress in the

ASD brain may have functional consequence in terms of a chronic inflammatory response, and oxidative damage in protein and DNA.

5.4 Conclusion

The studies reviewed herein provide support to the hypothesis of the involvement of mitochondrial dysfunction in the development of ASD. Since mitochondrial function was found to be related to ASD symptoms, mitochondria could be a potential target for novel therapeutics. Additionally, identifying those with vulnerable mitochondrial defects could result in prevention of ASD. Changes in bioenergetic characteristics imply a disruption of nutrient homeostasis as a potential cause of the pathological symptoms prevalent in autism. Autism is also associated with deficits in enzymatic antioxidant defense in some brain regions which contributes to oxidative stress and immune dysfunction that may lead to neurodevelopmental abnormalities in ASD. Early detection of oxidative stress and early treatment with antioxidant may result in a better prognosis in the brain before it can induce more irreversible brain damage. Restoration of mitochondrial integrity, redox status, and cognitive behavior defines potential therapeutic targets for neurodevelopmental disorders.

Author Contribution All authors revised and edited the manuscript, approved the final version to be submitted, and take responsibility for the manuscript's integrity and accuracy.

Declaration of Conflicting Interests The author(s) declared no potential conflicts of interest for the research, authorship, and/or publication of this article.

Funding TN was financed by the Coordenação de Aperfeiçoamento de Pessoal de Nível Superior – Brasil (CAPES) – Finance Code 001. JMG was supported by the National Council for Scientific and Technological Development (CNPq) Universal Grant number 432934/2018-9. AL is a CNPq fellow (Grant 312854/2019-6).

References

Al-Ayadhi, L. Y. (2012). Relationship between Sonic hedgehog protein, brain-derived neurotrophic factor and oxidative stress in autism spectrum disorders. *Neurochemical Research, 37*, 394–400.

Álvarez-Iglesias, V., Mosquera-Miguel, A., Cuscó, I., Carracedo, Á., Pérez-Jurado, L. A., & Salas, A. (2011). Reassessing the role of mitochondrial DNA mutations in autism spectrum disorder. *BMC Medical Genetics, 12*, 50.

Amal, H., Barak, B., Bhat, V., Gong, G., & Joughin, B. A. (2020). Shank3 mutation in a mouse model of autism leads to changes in the S-nitroso-proteome and affects key proteins involved in vesicle release and synaptic function. *Molecular Psychiatry, 25*, 1835–1848.

Ament, K., Mejia, A., Buhlman, R., Erklin, S., Caffo, B., Mostofsky, S., & Wodka, E. (2015). Evidence for specificity of motor impairments in catching and balance in children with autism. *Journal of Autism and Developmental Disorders, 45*, 742–751.

American Psychiatric Association. (2014). *Diagnostic and statistical manual of mental disorders*. APA.

Anitha, A., Nakamura, K., Thanseem, I., Yamada, K., Iwayama, Y., Toyota, T., Matsuzaki, H., Miyachi, T., Yamada, S., Tsujii, M., Tsuchiya, K. J., Matsumoto, K., Iwata, Y., Suzuki, K., Ichikawa, H., Sugiyama, T., Yoshikawa, T., & Mori, N. (2012). Brain region-specific altered expression and association of mitochondria-related genes in autism. *Molecular Autism, 3*, 12.

Anitha, A., Nakamura, K., Thanseem, I., Matsuzaki, H., Miyachi, T., Tsujii, M., Iwata, Y., Suzuki, K., Sugiyama, T., & Mori, N. (2013). Downregulation of the expression of mitochondrial electron transport complex genes in autism brains. *Brain Pathology, 23*, 294–302.

Ashwood, P., & Wakefield, A. J. (2006). Immune activation of peripheral blood and mucosal CD3+ lymphocyte cytokine profiles in children with autism and gastrointestinal symptoms. *Journal of Neuroimmunology, 173*, 126–134.

Ashwood, P., Krakowiak, P., Hertz-Picciotto, I., Hansen, R., Pessah, I., & van de Water, J. (2011). Elevated plasma cytokines in autism spectrum disorders provide evidence of immune dysfunction and are associated with impaired behavioral outcome. *Brain, Behavior, and Immunity, 25*, 40–45.

Avdjieva-Tzavella, D., Mihailova, S., Lukanov, C., Naumova, E., Simeonov, E., Tincheva, R., & Toncheva, D. (2012). Mitochondrial DNA mutations in two bulgarian children with autistic spectrum disorders. *Balkan Journal of Medical Genetics: BJMG, 15*, 47–54.

Bam, S., Buchanan, E., Mahony, C., & O'Ryan, C. (2021). DNA methylation of PGC-1α is associated with elevated mtDNA copy number and altered urinary metabolites in autism spectrum disorder. *Frontiers in Cell and Development Biology, 9*, 696428.

Beighley, J. S., Matson, J. L., Rieske, R. D., & Adams, H. L. (2013). Food selectivity in children with and without an autism spectrum disorder: Investigation of diagnosis and age. *Research in Developmental Disabilities, 34*, 3497–3503.

Benachenhou, S., Etcheverry, A., & Galarneau, L. (2019). Implication of hypocholesterolemia in autism spectrum disorder and its associated comorbidities: A retrospective case-control study. *Autism Research, 12*, 1860–1869.

Bjorklund, G., Saad, K., Chirumbolo, S., Kern, J. K., Geier, D. A., Geier, M. R., & Urbina, M. A. (2016). Immune dysfunction and neuroinflammation in autism spectrum disorder. *Acta Neurobiologiae Experimentalis (Wars), 76*, 257–268.

Blatt, G. J. (2012). The neuropathology of autism. *Scientifica, 2012*, 703675.

Bougeard, C., Picarel-Blanchot, F., Schmid, R., Campbell, R., & Buitelaar, J. (2021). Prevalence of autism spectrum disorder and co-morbidities in children and adolescents: A systematic literature review. *Frontiers in Psychiatry, 12*, 744709.

Bourgeron, T. (2015). From the genetic architecture to synaptic plasticity in autism spectrum disorder. *Nature Reviews. Neuroscience, 16*, 551–563.

Brown, B. D., & Rais, T. (2015). Autism in the son of a woman with mitochondrial myopathy and dysautonomia: A case report. *Innovations in Clinical Neuroscience, 12*, 29–32.

Brown, A. S., Sourander, A., Hinkka-Yli-Salomäki, S., Mckeague, I. W., Sundvall, J., & Surcel, H. M. (2014). Elevated maternal C-reactive protein and autism in a national birth cohort. *Molecular Psychiatry, 19*, 259–264.

Cast, T. P., Boesch, D. J., Smyth, K., Shaw, A. E., Ghebrial, M., & Chanda, S. (2021). An autism-associated mutation impairs Neuroligin-4 glycosylation and enhances excitatory synaptic transmission in human neurons. *The Journal of Neuroscience, 41*, 392–407.

Chana, G., Laskaris, L., Pantelis, C., Gillett, P., Testa, R., Zantomio, D., Burrows, E. L., Hannan, A. J., Everall, I. P., & Skafidas, E. (2015). Decreased expression of mGluR5 within the dorsolateral prefrontal cortex in autism and increased microglial number in mGluR5 knockout mice: Pathophysiological and neurobehavioral implications. *Brain, Behavior, and Immunity, 49*, 197–205.

Chauhan, A., Gu, F., Essa, M. M., Wegiel, J., Kaur, K., Brown, W. T., & Chauhan, V. (2011). Brain region-specific deficit in mitochondrial electron transport chain complexes in children with autism. *Journal of Neurochemistry, 117*, 209–220.

Chen, S., Li, Z., He, Y., Zhang, F., Li, H., Liao, Y., Wei, Z., Wan, G., Xiang, X., Hu, M., Xia, K., Chen, X., & Tang, J. (2015). Elevated mitochondrial DNA copy number in peripheral blood cells is associated with childhood autism. *BMC Psychiatry, 15*, 50.

Chugani, D. C., Sundram, B. S., Behen, M., Lee, M. L., & Moore, G. J. (1999). Evidence of altered energy metabolism in autistic children. *Progress in Neuro-Psychopharmacology and Biological Psychiatry, 23*, 635–641.

Criado, K. K., Sharp, W. G., Mccracken, C. E., De Vinck-Baroody, O., Dong, L., Aman, M. G., Mcdougle, C. J., Mccracken, J. T., Eugene Arnold, L., Weitzman, C., Leventhal, J. M., Vitiello, B., & Scahill, L. (2018). Overweight and obese status in children with autism spectrum disorder and disruptive behavior. *Autism, 22*(4), 450–459.

Dong, D., Zielke, H. R., Yeh, D., & Yang, P. (2018). Cellular stress and apoptosis contribute to the pathogenesis of autism spectrum disorder. *Autism Research, 11*, 1076–1090.

Durieux, A. M. S., Horder, J., Mendez, M. A., Egerton, A., Williams, S. C. R., Wilson, C. E., Spain, D., Murphy, C., Robertson, D., Barker, G. J., Murphy, D. G., & Mcalonan, G. M. (2016). Cortical and subcortical glutathione levels in adults with autism spectrum disorder. *Autism Research, 9*, 429–435.

Eissa, N., Al-Houqani, M., Sadeq, A., Ojha, S. K., Sasse, A., & Sadek, B. (2018). Current enlightenment about etiology and pharmacological treatment of autism spectrum disorder. *Frontiers in Neuroscience, 12*, 304.

El-Ansary, A., & Al-Ayadhi, L. (2014). GABAergic/glutamatergic imbalance relative to excessive neuroinflammation in autism spectrum disorders. *Journal of Neuroinflammation, 11*, 189.

Ellis, S. E., Gupta, S., Moes, A., West, A. B., & Arking, D. E. (2017). Exaggerated CpH methylation in the autism-affected brain. *Molecular Autism, 8*, 6.

Endres, D., Tebartz Van Elst, L., Meyer, S. A., Feige, B., Nickel, K., Bubl, A., Riedel, A., Ebert, D., Lange, T., Glauche, V., Biscaldi, M., Philipsen, A., Maier, S. J., & Perlov, E. (2017). Glutathione metabolism in the prefrontal brain of adults with high-functioning autism spectrum disorder: An MRS study. *Molecular Autism, 8*, 10.

Eow, S. Y., & Gan, W. Y. (2021). Parental feeding practices and child-related factors are associated with overweight and obesity in children and adolescents with autism spectrum disorder. *Journal of Autism and Developmental Disorders, 52*(8), 3655–3667.

Ewen, J. B., Marvin, A. R., Law, K., & Lipkin, P. H. (2019). Epilepsy and autism severity: A study of 6,975 children. *Autism Research, 12*, 1251–1259.

Fatemi, S. H., Reutiman, T. J., Folsom, T. D., & Thuras, P. D. (2009). GABA(A) receptor downregulation in brains of subjects with autism. *Journal of Autism and Developmental Disorders, 39*, 223–230.

Frye, R. (2012). Biomarkers of abnormal energy metabolism in children with autism spectrum disorder. *North American Journal of Medical Sciences, 5*, 141–147.

Frye, R. E., Melnyk, S., & Macfabe, D. F. (2013). Unique acyl-carnitine profiles are potential biomarkers for acquired mitochondrial disease in autism spectrum disorder. *Translational Psychiatry, 3*, e220.

Ghanizadeh, A., & Derakhshan, N. (2012). N-acetylcysteine for treatment of autism, a case report. *Journal of Research in Medical Sciences: The Official Journal of Isfahan University of Medical Sciences, 17*, 985–987.

Ghanizadeh, A., Berk, M., Farrashbandi, H., Alavi Shoushtari, A., & Villagonzalo, K.-A. (2013). Targeting the mitochondrial electron transport chain in autism, a systematic review and synthesis of a novel therapeutic approach. *Mitochondrion, 13*, 515–519.

Ghirardi, L., Pettersson, E., Taylor, M. J., Freitag, C. M., Franke, B., Asherson, P., Larsson, H., & Kuja-Halkola, R. (2019). Genetic and environmental contribution to the overlap between ADHD and ASD trait dimensions in young adults: A twin study. *Psychological Medicine, 49*, 1713–1721.

Giulivi, C., Zhang, Y. F., Omanska-Klusek, A., Ross-Inta, C., Wong, S., Hertz-Picciotto, I., Tassone, F., & Pessah, I. N. (2010). Mitochondrial dysfunction in autism. *JAMA, 304*, 2389–2396.

Godoy, P. B. G., Shephard, E., Milosavljevic, B., Johnson, M. H., & Charman, T. (2021). Brief report: Associations between cognitive control processes and traits of autism spectrum disorder (ASD), attention-deficit/hyperactivity disorder (ADHD) and anxiety in children at elevated and typical familial likelihood for ASD. *Journal of Autism and Developmental Disorders, 51*, 3001–3013.

Goh, S., Dong, Z., Zhang, Y., Dimauro, S., & Peterson, B. S. (2014). Mitochondrial dysfunction as a neurobiological subtype of autism spectrum disorder: Evidence from brain imaging. *JAMA Psychiatry, 71*, 665–671.

Graf, W. D., Marin-Garcia, J., Gao, H. G., Pizzo, S., Naviaux, R. K., Markusic, D., Barshop, B. A., Courchesne, E., & Haas, R. H. (2000). Autism associated with the mitochondrial DNA G8363A transfer RNA(Lys) mutation. *Journal of Child Neurology, 15*, 357–361.

Gu, F., Chauhan, V., & Chauhan, A. (2013a). Impaired synthesis and antioxidant defense of glutathione in the cerebellum of autistic subjects: Alterations in the activities and protein expression of glutathione-related enzymes. *Free Radical Biology & Medicine, 65*, 488–496.

Gu, F., Chauhan, V., Kaur, K., Brown, W. T., Lafauci, G., Wegiel, J., & Chauhan, A. (2013b). Alterations in mitochondrial DNA copy number and the activities of electron transport chain complexes and pyruvate dehydrogenase in the frontal cortex from subjects with autism. *Translational Psychiatry, 3*, e299.

Guevara-Campos, J., González-Guevara, L., Puig-Alcaraz, C., & Cauli, O. (2013). Autism spectrum disorders associated to a deficiency of the enzymes of the mitochondrial respiratory chain. *Metabolic Brain Disease, 28*, 605–612.

Hadjixenofontos, A., Schmidt, M. A., Whitehead, P. L., Konidari, I., Hedges, D. J., Wright, H. H., Abramson, R. K., Menon, R., Williams, S. M., Cuccaro, M. L., Haines, J. L., Gilbert, J. R., Pericak-Vance, M. A., Martin, E. R., & Mccauley, J. L. (2013). Evaluating mitochondrial DNA variation in autism spectrum disorders. *Annals of Human Genetics, 77*, 9–21.

Hassan, H., Zakaria, F., & Makpol, S. (2021). A link between mitochondrial dysregulation and idiopathic autism spectrum disorder (ASD): Alterations in mitochondrial respiratory capacity and membrane potential. *Current Issues in Molecular Biology, 43*, 2238–2252.

Hegarty, I. J., Lazzeroni, L. C., Raman, M. M., Hallmayer, J. F., & Cleveland, S. C. (2020). Genetic and environmental influences on corticostriatal circuits in twins with autism. *Journal of Psychiatry & Neuroscience, 45*, 188–197.

Hutsler, J. J., & Zhang, H. (2010). Increased dendritic spine densities on cortical projection neurons in autism spectrum disorders. *Brain Research, 1309*, 83–94.

Inga Jácome, M. C., Morales Chacòn, L. M., Vera Cuesta, H., Maragoto Rizo, C., Whilby Santiesteban, M., Ramos Hernandez, L., Noris García, E., González Fraguela, M. E., Fernandez Verdecia, C. I., Vegas Hurtado, Y., Siniscalco, D., Gonçalves, C. A., & Robinson-Agramonte, M. D. L. A. (2016). Peripheral inflammatory markers contributing to comorbidities in autism. *Behavioral Sciences (Basel), 6*, 29. Available: http://europepmc.org/abstract/MED/27983615; https://doi.org/10.3390/bs6040029; https://europepmc.org/articles/PMC5197942; https://europepmc.org/articles/PMC5197942?pdf=render. Accessed 2016/12

Ipser, J. C., Syal, S., Bentley, J., Adnams, C. M., Steyn, B., & Stein, D. J. (2012). 1H-MRS in autism spectrum disorders: A systematic meta-analysis. *Metabolic Brain Disease, 27*, 275–287.

Ladd-Acosta, C., Hansen, K. D., Briem, E., Fallin, M. D., Kaufmann, W. E., & Feinberg, A. P. (2014). Common DNA methylation alterations in multiple brain regions in autism. *Molecular Psychiatry, 19*, 862–871.

Lai, M. C., Lombardo, M. V., & Baron-Cohen, S. (2014). Autism. *Lancet, 383*, 896–910.

Lee, B. K., Magnusson, C., Gardner, R. M., Blomström, Å., Newschaffer, C. J., Burstyn, I., Karlsson, H., & Dalman, C. (2015). Maternal hospitalization with infection during pregnancy and risk of autism spectrum disorders. *Brain, Behavior, and Immunity, 44*, 100–105.

Loke, Y. J., Hannan, A. J., & Craig, J. M. (2015). The role of epigenetic change in autism spectrum disorders. *Frontiers in Neurology, 6*, 107.

Lombard, J. (1998). Autism: A mitochondrial disorder? *Medical Hypotheses, 50*, 497–500.

Lucarelli, J., Pappas, D., Welchons, L., & Augustyn, M. (2017). Autism spectrum disorder and avoidant/restrictive food intake disorder. *Journal of Developmental and Behavioral Pediatrics, 38*, 79–80.

Luo, Y., & Eran, A. (2020). A multidimensional precision medicine approach identifies an autism subtype characterized by dyslipidemia. *Nature Medicine, 26*, 1375–1379.

Lyall, K., Munger, K. L., O'Reilly, É. J., Santangelo, S. L., & Ascherio, A. (2013). Maternal dietary fat intake in association with autism spectrum disorders. *American Journal of Epidemiology, 178*, 209–220.

Lyall, K., Croen, L., Daniels, J., Fallin, M. D., Ladd-Acosta, C., Lee, B. K., Park, B. Y., Snyder, N. W., Schendel, D., Volk, H., Windham, G. C., & Newschaffer, C. (2017a). The changing epidemiology of autism spectrum disorders. *Annual Review of Public Health, 38*, 81–102.

Lyall, K., Croen, L. A., Sjödin, A., Yoshida, C. K., Zerbo, O., Kharrazi, M., & Windham, G. C. (2017b). Polychlorinated biphenyl and organochlorine pesticide concentrations in maternal mid-pregnancy serum samples: Association with autism spectrum disorder and intellectual disability. *Environmental Health Perspectives, 125*, 474–480.

Macduffie, K. E., Shen, M. D., Dager, S. R., Styner, M. A., Kim, S. H., Paterson, S., Pandey, J., St John, T., Elison, J. T., Wolff, J. J., Swanson, M. R., Botteron, K. N., Zwaigenbaum, L., Piven, J., & Estes, A. M. (2020). Sleep onset problems and subcortical development in infants later diagnosed with autism spectrum disorder. *The American Journal of Psychiatry, 177*, 518–525.

Maenner, M. J., Shaw, K. A., Baio, J., Washington, A., Patrick, M., Dirienzo, M., Christensen, D. L., Wiggins, L. D., Pettygrove, S., Andrews, J. G., Lopez, M., Hudson, A., Baroud, T., Schwenk, Y., White, T., Rosenberg, C. R., Lee, L. C., Harrington, R. A., Huston, M., Hewitt, A., Esler, A., Hall-Lande, J., Poynter, J. N., Hallas-Muchow, L., Constantino, J. N., Fitzgerald, R. T., Zahorodny, W., Shenouda, J., Daniels, J. L., Warren, Z., Vehorn, A., Salinas, A., Durkin, M. S., & Dietz, P. M. (2020). Prevalence of autism spectrum disorder among children aged 8 years – autism and developmental disabilities monitoring network, 11 sites, United States, 2016. *MMWR Surveillance Summaries, 69*, 1–12.

Mahalaxmi, I., Subramaniam, M. D., Gopalakrishnan, A. V., & Vellingiri, B. (2021). Dysfunction in mitochondrial electron transport chain complex I, pyruvate dehydrogenase activity, and mutations in ND1 and ND4 gene in autism spectrum disorder subjects from Tamil Nadu population, India. *Molecular Neurobiology, 58*, 5303–5311.

Mann, C., Schäfer, T., & Bletsch, A. (2021). Examining volumetric gradients based on the frustum surface ratio in the brain in autism spectrum disorder. *Human Brain Mapping, 42*, 953–966.

Masi, A., Quintana, D. S., Glozier, N., Lloyd, A. R., Hickie, I. B., & Guastella, A. J. (2015). Cytokine aberrations in autism spectrum disorder: A systematic review and meta-analysis. *Molecular Psychiatry, 20*, 440–446.

Mei, T., Llera, A., Floris, D. L., Forde, N. J., Tillmann, J., Durston, S., Moessnang, C., Banaschewski, T., Holt, R. J., Baron-Cohen, S., Rausch, A., Loth, E., Dell'Acqua, F., Charman, T., Murphy, D. G. M., Ecker, C., Beckmann, C. F., & Buitelaar, J. K. (2020). Gray matter covariations and core symptoms of autism: The EU-AIMS Longitudinal European Autism Project. *Molecular Autism, 11*, 86.

Ming, X., Johnson, W. G., Stenroos, E. S., Mars, A., Lambert, G. H., & Buyske, S. (2010). Genetic variant of glutathione peroxidase 1 in autism. *Brain and Development, 32*, 105–109.

Minshew, N. J., Goldstein, G., Dombrowski, S. M., Panchalingam, K., & Pettegrew, J. W. (1993). A preliminary 31P MRS study of autism: Evidence for undersynthesis and increased degradation of brain membranes. *Biological Psychiatry, 33*, 762–773.

Mostafa, G. A., El-Hadidi, E. S., Hewedi, D. H., & Abdou, M. M. (2010). Oxidative stress in Egyptian children with autism: Relation to autoimmunity. *Journal of Neuroimmunology, 219*, 114–118.

Nadeem, A., Ahmad, S. F., Attia, S. M., Al-Ayadhi, L. Y., Bakheet, S. A., & Al-Harbi, N. O. (2019). Oxidative and inflammatory mediators are upregulated in neutrophils of autistic children: Role of IL-17A receptor signaling. *Progress in Neuro-Psychopharmacology and Biological Psychiatry, 90*, 204–211.

Napoli, E., Wong, S., & Giulivi, C. (2013). Evidence of reactive oxygen species-mediated damage to mitochondrial DNA in children with typical autism. *Molecular Autism, 4*, 2.

Nuttall, J. R. (2017). The plausibility of maternal toxicant exposure and nutritional status as contributing factors to the risk of autism spectrum disorders. *Nutritional Neuroscience, 20*, 209–218.

Oliveira, G., Diogo, L., Grazina, M., Garcia, P., Ataíde, A., Marques, C., Miguel, T., Borges, L., Vicente, A. M., & Oliveira, C. R. (2005). Mitochondrial dysfunction in autism spectrum disorders: A population-based study. *Developmental Medicine and Child Neurology, 47*, 185–189.

Pan, C.-Y. (2012). Motor proficiency and physical fitness in adolescent males with and without autism spectrum disorders. *Autism, 18*, 156–165.

Patowary, A., Nesbitt, R., Archer, M., Bernier, R., & Brkanac, Z. (2017). Next generation sequencing mitochondrial DNA analysis in autism spectrum disorder. *Autism Research, 10*, 1338–1343.

Picard, M., Zhang, J., Hancock, S., Derbeneva, O., Golhar, R., Golik, P., O'Hearn, S., Levy, S., Potluri, P., Lvova, M., Davila, A., Lin, C. S., Perin, J. C., Rappaport, E. F., Hakonarson, H., Trounce, I. A., Procaccio, V., & Wallace, D. C. (2014). Progressive increase in mtDNA 3243A>G heteroplasmy causes abrupt transcriptional reprogramming. *Proceedings of the National Academy of Sciences of the United States of America, 111*, E4033–42.

Pons, R., Andreu, A. L., Checcarelli, N., Vilà, M. R., Engelstad, K., Sue, C. M., Shungu, D., Haggerty, R., De Vivo, D. C., & Dimauro, S. (2004). Mitochondrial DNA abnormalities and autistic spectrum disorders. *The Journal of Pediatrics, 144*, 81–85.

Postema, M. C., Van Rooij, D., Anagnostou, E., Arango, C., & Auzias, G. (2019). Altered structural brain asymmetry in autism spectrum disorder in a study of 54 datasets. *Nature Communications, 10*, 4958.

Postorino, V., Kerns, C. M., Vivanti, G., Bradshaw, J., Siracusano, M., & Mazzone, L. (2017). Anxiety disorders and obsessive-compulsive disorder in individuals with autism spectrum disorder. *Current Psychiatry Reports, 19*, 92.

Prosperi, M., Santocchi, E., Balboni, G., Narzisi, A., Bozza, M., Fulceri, F., Apicella, F., Igliozzi, R., Cosenza, A., Tancredi, R., Calderoni, S., & Muratori, F. (2017). Behavioral phenotype of ASD preschoolers with gastrointestinal symptoms or food selectivity. *Journal of Autism and Developmental Disorders, 47*, 3574–3588.

Rodriguez, J. I., & Kern, J. K. (2011). Evidence of microglial activation in autism and its possible role in brain underconnectivity. *Neuron Glia Biology, 7*, 205–213.

Rose, S., Melnyk, S., Pavliv, O., Bai, S., Nick, T. G., Frye, R. E., & James, S. J. (2012). Evidence of oxidative damage and inflammation associated with low glutathione redox status in the autism brain. *Translational Psychiatry, 2*, e134.

Rose, S., Niyazov, D. M., Rossignol, D. A., Goldenthal, M., Kahler, S. G., & Frye, R. E. (2018). Clinical and molecular characteristics of mitochondrial dysfunction in autism spectrum disorder. *Molecular Diagnosis & Therapy, 22*, 571–593.

Rossignol, D. A., & Frye, R. E. (2012). Mitochondrial dysfunction in autism spectrum disorders: A systematic review and meta-analysis. *Molecular Psychiatry, 17*, 290–314.

Sajdel-Sulkowska, E. M., Xu, M., Mcginnis, W., & Koibuchi, N. (2011). Brain region-specific changes in oxidative stress and neurotrophin levels in autism spectrum disorders (ASD). *Cerebellum, 10*, 43–48.

Schumann, C. M., & Amaral, D. G. (2006). Stereological analysis of amygdala neuron number in autism. *The Journal of Neuroscience, 26*, 7674–7679.

Sciara, A. N., Beasley, B., Crawford, J. D., Anderson, E. P., Carrasco, T., Zheng, S., Ordway, G. A., & Chandley, M. J. (2020). Neuroinflammatory gene expression alterations in anterior cingulate cortical white and gray matter of males with autism spectrum disorder. *Autism Research, 13*, 870–884.

Shafritz, K. M., Dichter, G. S., Baranek, G. T., & Belger, A. (2008). The neural circuitry mediating shifts in behavioral response and cognitive set in autism. *Biological Psychiatry, 63*, 974–980.

Singh, K., Singh, I. N., Diggins, E., Connors, S. L., Karim, M. A., Lee, D., Zimmerman, A. W., & Frye, R. E. (2020). Developmental regression and mitochondrial function in children with autism. *Annals of Clinical Translational Neurology, 7*, 683–694.

Sotgiu, S., Manca, S., Gagliano, A., Minutolo, A., Melis, M. C., Pisuttu, G., Scoppola, C., Bolognesi, E., Clerici, M., Guerini, F. R., & Carta, A. (2020). Immune regulation of neurodevelopment at the mother-foetus interface: The case of autism. *Clinical & Translational Immunology, 9*, e1211.

Sullivan, E. L., Nousen, E. K., Chamlou, K. A., & Grove, K. L. (2012). The impact of maternal high-fat diet consumption on neural development and behavior of offspring. *International Journal of Obesity Supplements, 2*, S7–S13.

Sun, W., Poschmann, J., Cruz-Herrera del Rosario, R., Parikshak, N. N., Hajan, H. S., Kumar, V., Ramasamy, R., Belgard, T. G., Elanggovan, B., Wong, C. C. Y., Mill, J., Geschwind, D. H., & Prabhakar, S. (2016). Histone acetylome-wide association study of autism spectrum disorder. *Cell, 167*, 1385–1397.e11.

Suzuki, K., Sugihara, G., Ouchi, Y., Nakamura, K., Futatsubashi, M., Takebayashi, K., Yoshihara, Y., Omata, K., Matsumoto, K., Tsuchiya, K. J., Iwata, Y., Tsujii, M., Sugiyama, T., & Mori, N. (2013). Microglial activation in young adults with autism spectrum disorder. *JAMA Psychiatry, 70*, 49–58.

Tang, G., Gutierrez Rios, P., Kuo, S. H., Akman, H. O., Rosoklija, G., Tanji, K., Dwork, A., Schon, E. A., Dimauro, S., Goldman, J., & Sulzer, D. (2013). Mitochondrial abnormalities in temporal lobe of autistic brain. *Neurobiology of Disease, 54*, 349–361.

Teresa, A. E., Sandra, L. S., Liang, L., Xiaoming, F., Zeneng, W., Woody, R. M., Evelyn, F., Rudy, J. C., Stanley, L. H., William, J. W., Allen, T. L., Robert, G. S., Mark, A. S., George, P., & Xiongwei, Z. (2008). The autistic phenotype exhibits a remarkably localized modification of brain protein by products of free radical-induced lipid oxidation. *American Journal of Biochemistry and Biotechnology, 4*, 61–72.

Tetreault, N. A., Hakeem, A. Y., Jiang, S., Williams, B. A., Allman, E., Wold, B. J., & Allman, J. M. (2012). Microglia in the cerebral cortex in autism. *Journal of Autism and Developmental Disorders, 42*, 2569–2584.

Toal, F., Daly, E. M., Page, L., Deeley, Q., Hallahan, B., Bloemen, O., Cutter, W. J., Brammer, M. J., Curran, S., Robertson, D., Murphy, C., Murphy, K. C., & Murphy, D. G. (2010). Clinical and anatomical heterogeneity in autistic spectrum disorder: A structural MRI study. *Psychological Medicine, 40*, 1171–1181.

Toscano, C. V. A., Ferreira, J. P., Gaspar, J. M., & Carvalho, H. M. (2019). Growth and weight status of Brazilian children with autism spectrum disorders: A mixed longitudinal study. *Jornal de Pediatria, 95*, 705–712.

Urakubo, A., Jarskog, L. F., Lieberman, J. A., & Gilmore, J. H. (2001). Prenatal exposure to maternal infection alters cytokine expression in the placenta, amniotic fluid, and fetal brain. *Schizophrenia Research, 47*, 27–36.

Valiente-Pallejà, A., Torrell, H., Muntané, G., Cortés, M. J., Martínez-Leal, R., Abasolo, N., Alonso, Y., Vilella, E., & Martorell, L. (2018). Genetic and clinical evidence of mitochondrial dysfunction in autism spectrum disorder and intellectual disability. *Human Molecular Genetics, 27*, 891–900.

Varga, N., Pentelényi, K., Balicza, P., Gézsi, A., Reményi, V., Hársfalvi, V., Bencsik, R., Illés, A., Prekop, C., & Molnár, M. J. (2018). Mitochondrial dysfunction and autism: Comprehensive genetic analyses of children with autism and mtDNA deletion. *Behavioral and Brain Functions, 14*, 4.

Vargas, D. L., Nascimbene, C., Krishnan, C., Zimmerman, A. W., & Pardo, C. A. (2005). Neuroglial activation and neuroinflammation in the brain of patients with autism. *Annals of Neurology, 57*, 67–81.

Varghese, M., Keshav, N., Jacot-Descombes, S., Warda, T., Wicinski, B., Dickstein, D. L., Harony-Nicolas, H., De Rubeis, S., Drapeau, E., Buxbaum, J. D., & Hof, P. R. (2017). Autism spectrum disorder: Neuropathology and animal models. *Acta Neuropathologica, 134*, 537–566.

Vogel Ciernia, A., & Lasalle, J. (2016). The landscape of DNA methylation amid a perfect storm of autism aetiologies. *Nature Reviews. Neuroscience, 17*, 411–423.

Waye, M. M. Y., & Cheng, H. Y. (2018). Genetics and epigenetics of autism: A review. *Psychiatry and Clinical Neurosciences, 72*, 228–244.

Wegiel, J., Kuchna, I., Nowicki, K., Imaki, H., Wegiel, J., Marchi, E., Ma, S. Y., Chauhan, A., Chauhan, V., Bobrowicz, T. W., De Leon, M., Louis, L. A., Cohen, I. L., London, E., Brown, W. T., & Wisniewski, T. (2010). The neuropathology of autism: Defects of neurogenesis and neuronal migration, and dysplastic changes. *Acta Neuropathologica, 119*, 755–770.

Wei, H., Zou, H., Sheikh, A. M., Malik, M., Dobkin, C., Brown, W. T., & Li, X. (2011). IL-6 is increased in the cerebellum of autistic brain and alters neural cell adhesion, migration and synaptic formation. *Journal of Neuroinflammation, 8*, 52.

Wei, H., Mori, S., Hua, K., & Li, X. (2012). Alteration of brain volume in IL-6 overexpressing mice related to autism. *International Journal of Developmental Neuroscience, 30*, 554–559.

Weissman, J. R., Kelley, R. I., Bauman, M. L., Cohen, B. H., Murray, K. F., Mitchell, R. L., Kern, R. L., & Natowicz, M. R. (2008). Mitochondrial disease in autism spectrum disorder patients: A cohort analysis. *PLoS One, 3*, e3815.

Wong, S., Napoli, E., Krakowiak, P., Tassone, F., Hertz-Picciotto, I., & Giulivi, C. (2016). Role of p53, mitochondrial DNA deletions, and paternal age in autism: A case-control study. *Pediatrics, 137*, e20151888.

Zablotsky, B., Bramlett, M. D., & Blumberg, S. J. (2020). The co-occurrence of autism spectrum disorder in children with ADHD. *Journal of Attention Disorders, 24*, 94–103.

Chapter 6
The Usability of Mouse Models to Study the Neural Circuity in Autism Spectrum Disorder: Regulatory Mechanisms of Core Behavioral Symptoms

Hiroyuki Arakawa and Yuki Higuchi

6.1 Current Trends in ASD Studies Using Animal Models

Animal models can provide coherent investigations into the neurobiological characteristics of human psychiatric diseases by looking at cells, neural circuits, and the pathophysiology (Mapelli et al., 2022; Silverman et al., 2022). When using animal models in translational studies, the animals used for research need not have the disease they are modeling or even reproduce all aspects of the disease, but may instead mimic certain physiological or behavioral states that are characteristic of the disease. Human ASD is defined by two discrete behavioral symptoms, including persistent deficits in social and communicative interactions and abnormal and fixed interests and repetitive behavior (DSM-5TR, 2022). A representative neural and genetic mechanism underlying these behavioral characteristics would be the target responsible for causing the disease; however, multiple malconstructed systems form during development that produce the multilateral characteristics of ASD (Arakawa, 2020). Therefore, we could take advantage of animal 'models' to dissect a particular piece of ASD-relevant mechanism and mimic an aspect of human symptoms, including neurobiological conditions affecting cells, circuits, and behaviors. In particular, rodent models have allowed us to delve into the exhaustive mechanistic depth of neural circuits in genetically manipulated highly social species to investigate the complex neural organization underlying ASD-related behaviors (Silverman et al., 2022).

Recent trends in ASD research using rodent models have shifted the focus of investigation from exploring genomic factors with transgenic manipulations relevant to the embryologic perspective of human ASD to neural circuits regulating

H. Arakawa (✉) · Y. Higuchi
Department of Systems Physiology, Graduate School of Medicine, University of the Ryukyus, Okinawa, Japan
e-mail: harakawa@med.u-ryukyu.ac.jp

A. El Idrissi, D. McCloskey (eds.), *Neurobiology of Autism Spectrum Disorders*,
https://doi.org/10.1007/978-3-031-42383-3_6

concomitant behavioral symptoms of ASD. The genetic basis of ASD has been documented since early studies on human ASD (Klauck, 2006; Geschwind, 2011). Numerous clinical studies have indicated that the risk factor for developing ASD is ascribed to a genetic variability of approximately 40%, while the remaining 60% is due to epigenetic (i.e., gene x environmental) factors (Gardener et al., 2011; Guinchat et al., 2012; Maramara et al., 2014). Most of the known genetic alterations that contribute to increased ASD risk affect the formation and maintenance of synapses/neurons and their sustainer cells during the neural development (Mahfouz et al., 2015; Bonsi et al., 2022). Accordingly, some distinct human genetic diseases that are caused by a specific set of (or single) mutations occur in conjunction with behavioral symptoms characterized as ASD. Diseases caused by identified mutations include the fragile X mental retardation syndrome (Fmr1), Rett syndrome (Mecp2), tuberous sclerosis disease (TSC1/TSC2), and Angelman syndrome (UBE3A) (Provenzano et al., 2012; Arakawa, 2020; Crawley, 2022), all of which are related to mutations that cause deficits in neuronal formation and maintenance. Although the behavioral symptoms of these genetic diseases are encompassed within the spectrum range of ASD, the manipulation of target genes relevant to these diseases in animal models could reveal only a linkage between neuronal structures/mechanisms mediated by a target gene and the underlying behavioral phenotypes reminiscent of human ASD. Similarly, several research groups have developed mouse models of ASD, driven by a search for candidate genes/chromosomes relevant to human ASD, by using whole-genome sequences from ASD patients (Chaste & Leboyer, 2012). These studies have raised a substantial list of abnormalities in the brain, as well as behavioral modifications relevant to target mutations, thus providing valuable insights into neurophysiological mechanisms (Grove et al., 2019). At the same time, the miscellaneous lists of linkages between mutations and behaviors have made it difficult to identify the primal neural mechanisms underlying abnormalities in ASD.

Most candidate mutations produce a substantial set of malformations in neuronal systems during perinatal development, whereas ASD-like symptoms are regulated by multiple concomitant neural circuits that are malformed through development (Fig. 6.1). Therefore, a major challenge of the genetic approach to ASD pathology is distinguishing the effects of mutations on neuronal development from the effects of temporal behavioral regulations pertaining to ASD symptoms. Although many

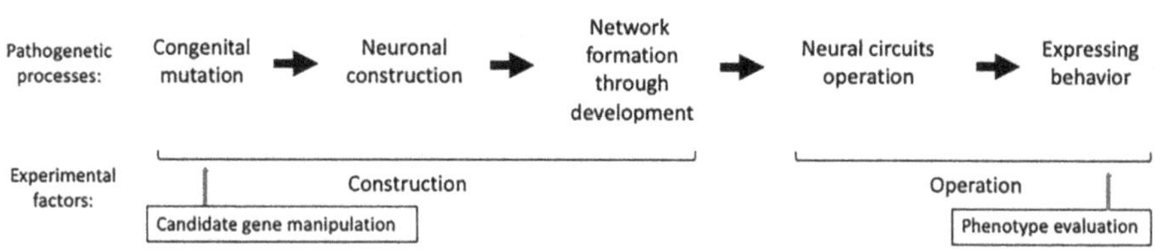

Fig. 6.1 Framework illustrating pathogenetic processes through development and genetic approach on ASD studies using animal models. Candidate gene manipulation affects construction processes of neuronal circuitry during development. The formed (matured) neuronal circuits operate to express a particular behavior that is evaluated as ASD-related phenotypes

candidate genes for concurrently regulating ASD pathology have been nominated, investigations on the function of these genes tend to focus on the construction processes of neural circuits associated with ASD symptom. Thus, these gene candidates are far from unraveling the neural mechanisms underlying particular ASD-like behaviors.

6.2 Assessment of Atypical Social Behavior in Animal Models

ASD is characterized by the presence of behavioral abnormalities, including social deficits, verbal and intellectual disabilities, insistence on sameness, sensory abnormalities, and increased stereotypical behavioral patterns (Baron-Cohen et al., 2005; Geschwind, 2011; Jiujias et al., 2017). The components of these behavioral features in ASD mutually influence one another for adequate implementation of the listed atypical behaviors (Bourgeron, 2015; Orefice et al., 2016). For example, sensory abnormalities disrupt appropriate recognition of social stimuli, both of which deteriorate the symptoms of social deficits. Therefore, atypical behaviors may be analyzed by dissecting the assembled processes of whole behavior into circuit-based micro-behavior units.

The use of rodent models to perform these dissecting processes of ASD-related behaviors must account for species-specific/typical differences in the target of interest. The validity of rodent models suitable for ASD symptoms is primarily based on behavioral characteristics having a compelling analogy to human behavioral symptoms. However, the behavior of animals is distinct from that of humans (Arakawa & Iguchi, 2018). For example, mice, as a representative rodent model, are nocturnal and use olfaction and tactility as primary senses, that differs in humans (Brennan & Zufall, 2006). Likewise, mice and rats are highly social animals, living substantial parts of their lives in groups in which they use complex ways to communicate with one another to form flexible but secure social relationships, including social hierarchies and cooperative relationships among close relatives (Sluyter & Van Oortmerssen, 2000; Blanchard et al., 2013). Such complex group living requires multimodal processing of social information in appropriate ways, including an understanding of the social rules and familiarity within certain groups (Hurst & Beynon, 2004; Arakawa, 2018), and adequate expressions of behavioral responses in a situation-dependent manner. This sequential process must be dynamic and flexible as the social context continuously changes and is updated with new information (Fig. 6.2). We must understand the primary systems and mechanisms that regulate the behaviors of mice in a particular test setting based on the ethological perspectives of mouse behavior and identify the factors underlying ASD-like behaviors for suitable ASD animal models.

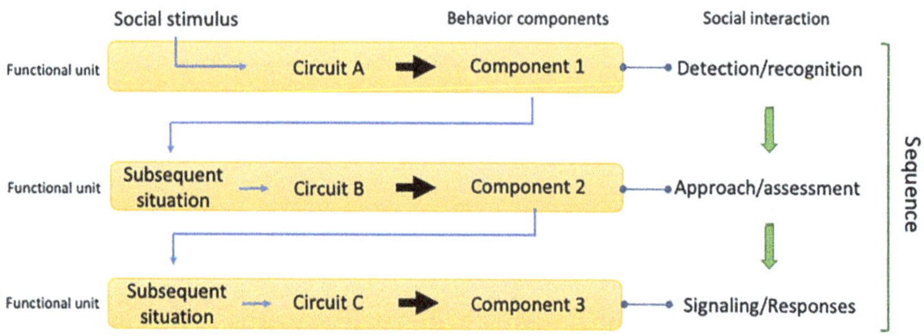

Fig. 6.2 A schematic for neural circuits-based sequential behavior processes. Whole sequence of social behavior consists of several behavior components, such as detection/recognition, approach/assessment, and signaling/responses. These components are coupled with a responsible neural circuit, forming functional units. The functional units are linked one another with stimulus/situation that are affected by activities of other functional units

6.3 Neural Circuits Responsible for the Regulation of ASD-Like Behavior

Here, we take steps toward a circuit-level understanding of the neural basis of complex behaviors relevant to ASD symptoms. As humans are dependent on different modalities while making decisions regarding social interactions, mice must integrate external sensory cues mainly via olfaction and tactility with internal information processes that involve recognition and assessment of and response (signaling) to a confronted social situation, each of which is coupled with representative neural circuits. Therefore, preclinical rodent models for the study of social behavior pertaining to ASD must undergo a comprehensive analysis to assess the variables involved in social recognition (via initial olfactory detection), assessment of potential social partners (via investigatory contacts), and subsequent responses (via social signaling) in accordance with discrete activities of responsible neural circuits as a functional unit during a bout of social engagement. A proper understanding of the functions of each unit leads to a reliable interpretation of behaviors that have substantial implications for human health.

Figure 6.2 illustrates the sequential process of the circuit-based regulation of behavior. Atypical social behavior, represented as ASD symptom includes multiple processes of information (whereby formation of the brain circuit takes place) and otherwise, disruptions that are regulated by a functional unit of a representative neural circuit. These circuit-level interfaces for behavioral regulation are crucial for determining how individual animals organize themselves within broad social frameworks and unraveling the neural mechanisms whereby the neural coding of social information is disrupted under ASD-like atypical behavior that broadly disrupts social function.

Accordingly, social recognition is essential for identifying potential social stimuli, and successful group living. Sensory information is processed mainly via the olfactory and tactile neural pathways based on signal transmission through basal ganglia and forebrain regions (Brennan & Kendrick, 2006). Activation of the recognition circuits leads to a subsequent assessment process, as shown by the approaching and sniffing behavior toward social stimuli (Arakawa & Iguchi, 2018). The expression of investigatory assessment also facilitates consequent behavioral responses that determine temporal social relationships, i.e., accepting or rejecting the social stimuli.

Huddling with familiar partners is common among most mammals to maintain familiar (e.g., friendly) relationships, representing acceptance of their counterparts (Alberts, 2007; Morrison, 2016). Otherwise, animals exhibit a signal or reaction, indicating rejection of social stimuli. Withdrawal and directing away from the social stimuli as avoidance or, if it is difficult to express, defensive attack in aggression, act as representative behaviors for rejection (Blanchard et al., 2011). In this context, male rodents use the deposition of scent marks to counter potential competitors and on their own territorial boundaries to advertise their rejection signals and show ownership (Hurst & Beynon, 2004; Arakawa & Higuchi, 2022). These sequential behaviors during social interactions consist of multiple sub-components of behavioral processes along with responsible circuit activities that involve specific cells, projections, and targets, as well as neurotransmitters, as functional units. In the following section, we discuss two peptide hormones, oxytocin (OXT) and arginine vasopressin (AVP), which play a pivotal role in the regulation of these social functional units (Lukas & Neumann, 2013; Arakawa, 2021a).

6.4 Peptide Hormones Regulating Complex Social Behavior

Several clinical studies have indicated that OXT and AVP are involved in the regulation of social behavior, along with playing a role in the pathophysiology of ASD (Carter, 2017; Crespi et al., 2022). These two neuropeptides consist of a unique biological cassette that has regulated the physiology and behavior of mammals and their ancestors (Manning et al., 2012; van Kesteren et al., 1992). The receptors of these neuropeptides belong to the G protein-coupled receptor superfamily and form a small receptor subfamily with three structurally related AVP receptors (AVPR) (1a receptor [V1aR], 1b receptor [V1bR], and type 2 receptor [V2R]) and its homolog, OXT receptor (OTR) (Birnbaumer, 2000; Song & Albers, 2018). AVP and OXT are closely related by homology, as they differ structurally by only two amino acids (Zingg & Laporte, 2003). The ligand-binding pockets in the AVPR subfamily and OTR are highly conserved and have high binding affinity to each other (e.g., AVP to V1aR affinity value: 1.11 nM, AVP to V1bR affinity value: 0.43 nM, and AVP to OTR affinity value: 0.87 nM) (Busnelli et al., 2013; Manning et al., 2012). These subtype-selective analogs are crucial for examining the role of different receptor subtypes, as they endogenously bind either receptor with similar affinity. Instead of high endogenous affinities, AVP and OXT are either expressed in different segregated populations of neurons or synthesized via distinguishable projection pathways throughout the central and peripheral nervous systems (Jurek & Neumann, 2018; Rae et al., 2021).

OXT is discretely synthesized in the hypothalamic regions, including the paraventricular nucleus (PVN) and supraoptic nucleus of the hypothalamus (SON), and sparsely synthesized in the accessory nuclei of the hypothalamus (Swanson & Sawchenko, 1983; Landgraf & Neumann, 2004). As OTRs are expressed throughout the brain, OXT neurons send innervations at various distances from the hypothalamus via long-range axonal projections and dendritic release (Knobloch et al., 2012; Tang et al., 2020). High-density axonal projections are present in the basal ganglia area (including the nucleus accumbens [NAc], lateral septal nucleus [LS], bed nucleus of the stria terminalis [BnST]), amygdala complex (including the medial amygdala [MeA] and central amygdala), paraventricular thalamic nucleus, and midbrain areas (including the raphe nucleus and periaqueductal gray) (Mitre et al., 2016; Grinevich et al., 2016; Jurek & Neumann, 2018), with only minor variations across mammals (Lefevre et al., 2021).

In contrast to OXT, conspicuous species differences as well as subtype specificities exist in AVP receptor localization. V1aRs are distributed throughout the central and peripheral regions (Henry & Hohmann, 2012; Day et al., 2020). V1bRs are localized in restricted regions in rodents, including corticotrophs of the anterior pituitary and the hippocampus, while V2Rs are predominantly located in the kidney and other peripheral tissues (Dubois-Dauphin et al., 1996; Young et al., 2006). Therefore, V1aR has been identified as the main receptor responsible for various central processes of AVP (Tickerhoof & Smith, 2017). In the mouse brain, AVP is made in various nuclei in the hypothalamic regions including the PVN, SON, and accessory nuclei, and extended amygdala regions (including BnST and MeA) (Rood & De Vries, 2011; Rood et al., 2013). AVP-expressing cells in the later regions are known to be steroid sensitive (De Vries, 2008).

An increasing number of studies in humans using intranasal administration of OXT have indicated its role in enhancing positive social behaviors such as trust, affiliation, and empathy (Kosfeld et al., 2005; Bartz et al., 2011; Keech et al., 2018) and ameliorating symptoms of social dysfunction in various psychiatric conditions, including ASD, schizophrenia, psychopathy, and social anxiety (Hollander et al., 2003, 2007; Bartz et al., 2011; Ma et al., 2016). Although a large cohort study recently reported that intranasal (i.e., whole-brain) administration of OXT may fail to alleviate ASD symptoms (Sikich et al., 2021), it is well established that OXT treatment facilitates prosocial attention and decreases social vigilance (Le et al., 2021; Ebitz et al., 2013). Therefore, OXT acts to specifically mediate neural circuits that are not involved in the regulation of whole social processes, but rather discrete social components, such as social attention (Neumann & Slattery, 2016; Arakawa, 2021a).

AVP levels, but not OXT levels, are significantly lower in the cerebrospinal fluid of children with ASD (Oztan et al., 2018, 2020), and in primate models of autism (Parker et al., 2018). Similarly, plasma AVP levels are correlated with the quality of social relationships, particularly in men (Taylor et al., 2010; Gouin et al., 2012). AVP administration can alleviate behavioral symptoms of ASD in transgenic mouse

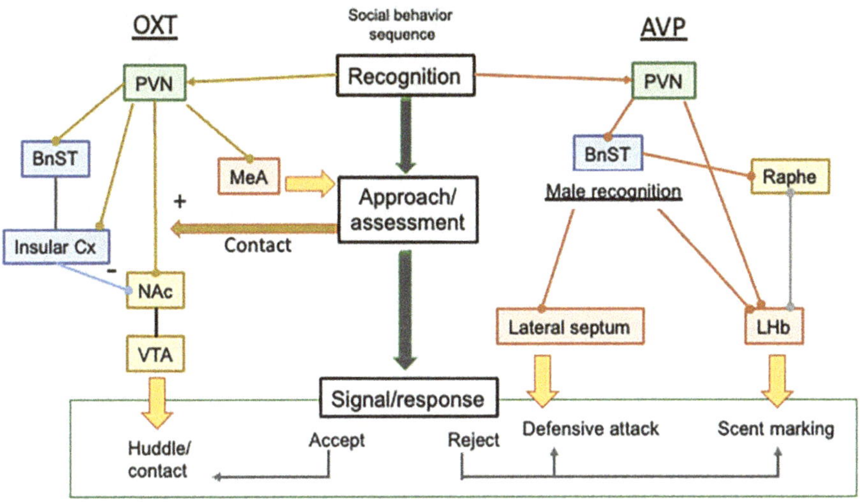

Fig. 6.3 A schematic diagram for neural circuit-based processing of sequential behavior mediated by two neuropeptides, oxytocin (OXT) and vasopressin (AVP). Upon the detection of potential social stimuli, PVN-based OXT projections to the MeA facilitate approach behavior, while those to the BnST-insular Cx circuit inhibit social contact. Activation of the NAc-VTA circuit facilitates huddling contact with encountered familiar conspecifics. AVP system is also activated when confronted with potential social stimuli. The BnST-lateral septum circuit may be involved in the expression of rejection signal as defensive attack, whereas the BnST-LHb circuit coordinated with Raphe plays a role in exhibiting scent marking behavior, as another rejection signal. Abbreviations: BnST (bed nucleus of the stria terminalis), Insular Cx (insular cortex), LHb (lateral habenula), NAc (nucleus accumbens), MeA (medial amygdala), PVN (paraventricular nucleus of the hypothalamus), Raphe (raphe nucleus), and VTA (ventral tegmental area)

models (Borie et al., 2021; Wu et al., 2021) and human children (Hendaus et al., 2019). Although these findings have implicated a consecutive linkage between these peptides and behavioral symptoms of ASD, intranasal treatment with these peptides stimulates their reciprocal ligand receptors or circuits across the board area of the brain. The circuit-based regulation of these peptides should be clarified, in particular social processes relevant to ASD symptoms (Fig. 6.3).

6.5 OXT-Mediated Neural Circuits

Several studies using rodent models have demonstrated that OXT neurons in the PVN could be the key to mediating approach-contact patterns toward social stimuli (Ferguson et al., 2001; Young & Barrett, 2015) and ASD-like phenotypes in an ASD mouse model, for example, Cntnap2 KO mice (Peñagarikano et al., 2015) and Shank3 KO mice (Harony-Nicolas et al., 2017). Magnocellular and parvocellular OXT neurons in the PVN send massive projections to several brain regions, which relay the systematic regulation of behavioral components that comprise sequential social behaviors (Jurek & Neumann, 2018; Lefevre et al., 2021). Among these, the OXT projection to the MeA plays a crucial role in the expression of social approach (Arakawa et al., 2010, 2015; Yao et al., 2017). While systemic OTR blockage does not influence social approaches in male mice (Haskal de la Zerda et al., 2020), microinfusion of an OTR antagonist into the MeA blocks the activation of social approach in response to social stimuli (Ferguson et al., 2001; Arakawa et al., 2010). A cell type-specific analysis illustrated that magnocellular OXT neurons in the PVN modulate social approach without physical contact (i.e., olfactory investigation), which likely switches to control by parvocellular OXT neurons when rats can physically contact with confronted social stimuli (Tang et al., 2020). Signals from parvocellular OXT neurons activated by social contact are sent to the NAc, which is linked to the dopamine (DA)-mediated reward circuit (Dölen & Malenka, 2014).

Social interaction with familiar conspecifics has reward properties that motivate animals to engage in physical contact involving familiar social stimuli, such as huddling, and maintain close relationships with partners (Alberts, 2007; Young & Barrett, 2015). As shown by cuddling in humans, a strong preference for huddling in rodent models indicates an analog of social attachment via bodily contact with familiar partners (Morrison, 2016; Arakawa, 2020). Thus, there is a direct relationship between the reward value of social contacts and the frequency of seeking such interactions (Borland et al., 2018; Netser et al., 2020). OXT acts on the DA circuit in the NAc, which receives projections from the VTA (Buffington et al., 2016; Song et al., 2016) and plays a role in the regulation of signaling responses, such as social acceptance represented by huddling (Dölen et al., 2013; Gunaydin et al., 2014; Hung et al., 2017; Borland et al., 2018).

The DA circuits of the VTA-NAc are stimulated by OXT from the PVN and maintain prosocial contacts, while these circuits also receive inputs from other OXT projections into the insular cortex (Lefevre et al., 2021). The OXT circuit through the insular cortex acts via OTR-expressing (i.e., GABAergic) neurons (Gamal-Eltrabily et al., 2020) projecting to the NAc as mediators of stress-related defensive behavior and affiliative contact (Steinman et al., 2019). Accordingly, inactivation of the posterior insular cortex reduces the freezing response of rats to predatory odors (Rodríguez et al., 2019), and inactivation of the anterior insular cortex attenuates the acquisition of contextual fear response in rats (Alves et al., 2013). Selective blockage of the OTR in the posterior insular cortex reduces the social approach in rats (Rogers-Carter et al., 2019). Furthermore, the BnST to insular cortex circuit mediates active defense (e.g., struggling-avoidance) responses in mice (Luchsinger et al., 2021). Under stressful defensive conditions, OTR-expressing neurons in the BnST are activated to mediate social vigilance, thus inducing social avoidance in several rodent models (Duque-Wilckens et al., 2018; Nasanbuyan et al., 2018). Therefore, the BnST-insular cortex circuit projecting to the NAc likely inhibits affiliative contacts with a potential social partner. These circuits would allow animals to regulate flexible approach-contact responses depending on the social reward characteristics of their opponents (Steinman et al., 2019; Williams et al., 2020).

6.6 AVP-Mediated Neural Circuits

Studies using AVP-receptor knockout mice have indicated that the AVP system plays a marginal role in either the regulation of social recognition or the investigatory approach (Lukas & Neumann, 2013). V1aR KO mice showed normal recognition of social odor and a sufficient level of approach behavior toward social stimuli (Bielsky et al., 2004). Similarly, mice with the deletion of V1bRs exhibited no differences in social approach in a three-chamber sociability test compared with heterozygous controls (Yang et al., 2007). Given that AVP-expressing cells in the PVN and other regions send projections to several target brain areas, the function of AVP must be elucidated at the micro-behavior circuit level. For example, optogenetic activation of AVP neurons in the PVN induces robust self-grooming behavior in male mice (Islam et al., 2022), as this excessive grooming is considered a type of persistent repetitive behavior resembling the stereotypical symptoms of human ASD (Zhou et al., 2016; Arakawa, 2021b; Higuchi & Arakawa, 2022). However, genetic deletion of AVP-expressing cells in the PVN has no effect on the investigatory approach in male mice, whereas it increased that in female mice (Rigney et al., 2021). Sex differences in AVP function have been documented as sex dimorphic steroid hormones predominantly control the activity of AVP-expressing cells in selected brain regions (De Vries & Panzica, 2006; De Vries, 2008).

AVP neurons in the PVN send a projection to the BnST, in which other male-dominant AVP-expressing cells are involved in controlling defensive behavior and social investigation (Dantzer et al., 1988; Koob et al., 1989; Endres & Fendt, 2008). Genetic deletion of AVP-expressing cells in the BnST reduced the investigatory approach between males but not in females (Rigney et al., 2019). As the deletion of AVP-expressing cells in the BnST has little impact on locomotor activity or olfactory recognition, this circuit is involved in facilitating male-oriented social approach behavior. AVP-expressing cells in the BnST are interconnected with the extra-hypothalamic regions, including the LS and LHb (De Vries & Panzica, 2006; Rood & De Vries, 2011). Overexpression of V1aRs (Landgraf et al., 2003) or AVP injections (Koolhaas et al., 1998) in the LS facilitates male-to-male aggressive interactions in rats (Dantzer et al., 1988; Koob et al., 1989; Le Moal et al., 1987; Engelmann, 2008), and microinfusion of a V1aR antagonist into the LS reduces territorial aggression toward intruders in male rats (Veenema et al., 2010). Therefore, AVP in the LS plays a marginal role in the regulation of social approach, but plays a pivotal role in the expression of signaling responses, implying a rejection of social stimuli (i.e., defensive attack).

The signaling response during social interaction determines temporal relationships with the confronted counterparts informing their evaluation (e.g., acceptance or rejection). Huddling or maintaining peer contact represents a type of signaling response indicating the acceptance of social stimuli, which is mediated by a specific OXT circuit through the VTA-NAc pathway. On the other hand, those indicating rejection of social stimuli are expressed as defensive attacks on social stimuli or deposition of territorial scent marks. Scent marking is a male-dominant signal that advertises rejection to male competitors and acceptance (attraction) to potential reproductive partners (Brennan & Kendrick, 2006; Arakawa et al., 2008). In this context, genetic ablation of AVP-expressing cells in the PVN of male mice did not alter the production of scent markings to male counterparts (Rigney et al., 2021). In contrast, deletion of AVP-expressing cells in the BnST of male mice enhanced scent marking in response to a female mouse but not a male mouse (Rigney et al., 2019). AVP neurons in the BnST may manipulate signaling responses that differ depending on the sex of opponents. Interestingly, the thalamic LHb, which receives AVP projections from both the BnST and PVN (Rood & De Vries, 2011; Hernández et al., 2015), is clearly involved in the regulation of scent marking. V1aR blockage in the LHb of male mice suppressed scent marking to unfamiliar males without concomitant changes in locomotor activity (Rigney et al., 2020). These circuit studies demonstrated that a regulatory function of the BnST to the LHb circuit is associated with male-oriented defensiveness expressed as a scent marking towards social competitors.

Lastly, recent findings indicate that raphe nuclei that receive AVP projections from gonadal- sensitive AVP-expressing cells, mainly in the BnST (Rood et al., 2013), are also involved in the regulation of scent marking and rejection signals in male mice (Rigney et al., 2020). One of the V1aR expressing neurons in the raphe is glutamatergic (Rood & Beck, 2014) projecting to the MeA, which facilitates aggression in male mice (Nordman & Li, 2020) and those neurons which co-express either glutamine, GABA, or serotonin are activated by the presence of female stimuli (Patel et al., 2022). Taken together, the BnST-raphe circuits are involved in the regulation of signaling processes in an opponent-sex dependent manner. These circuit-based analyses of neuronal function in male mice illustrate how the AVP signal in the BnST determines the sex-directed signaling responses. The projections from the BnST to the LS activate defensive attack, those to the LHb facilitate scent marking, and those to the raphe nucleus sustain sex-dependent processes, all of which are in coordination with the expression of signaling responses.

6.7 Concluding Remarks

While a large portion of the literature describes linkages between target genes or mechanisms associated with ASD pathology and atypical behaviors resembling human ASD symptoms, little is known about the circuit-based unit mediating a particular behavioral component. Recent technical advantages of neuronal circuit-based analysis (Ozawa & Arakawa, 2021) allow us to dissect the whole behavior sequence into a functional unit of a micro behavior component that is coupled with a certain neural circuit. Thus, our strategy provides a better understanding of the molecular circuit mechanisms underlying the regulation of atypical behavior associated with human ASD.

The neuropeptides, OXT and AVP are tightly involved in the regulation of ASD-like behaviors. OXT circuits are coordinated with several projecting regions to control the behavioral sequence, from a cue-induced social approach to the maintenance of social contacts. AVP circuits play a significant role in the regulation of signaling responses, including defensive attack reactions and deposition of scent marking rejection signals. Although more investigations on neural circuits and transmitters should be conducted, this strategy of dissecting ASD pathological symptoms into micro functional units allows us to deepen our understanding of how certain atypical behaviors are regulated at the neural circuitry level. This insight is crucial for developing therapeutic strategies to prevent or ameliorate the temporal behavioral symptoms of ASD using chemo- or opto-genetic neural manipulation as a gene therapy.

Acknowledgments The authors declare no conflict of interest. This work was supported by JSPS KAKENHI Grant Number JP19K24681 (HA) and JSPS KAKENHI Grant Number JP20K16232 (YH).

References

Alberts, J. R. (2007). Huddling by rat pups: Ontogeny of individual and group behavior. *Developmental Psychobiology, 49*, 22–32.

Alves, F. H. F., Gomes, F. V., Reis, D. G., Crestani, C. C., Corrêa, F. M. A., Guimarães, F. S., et al. (2013). Involvement of the insular cortex in the consolidation and expression of contextual fear conditioning. *The European Journal of Neuroscience, 38*, 2300–2307.

American Psychiatric Association (2022). Diagnostic and statistical manual of mental disorders, Fifth Edition, Text rev. ISBN 978-0-89042-576-3. https://doi.org/10.1176/appi.books.9780890425787

Arakawa, H. (2018). Ethological approach to social isolation effects in behavioral studies of laboratory rodents. *Behavioural Brain Research, 341*, 98–108.

Arakawa, H. (2020). From multisensory assessment to functional interpretation of social behavioral phenotype in transgenic mouse models for autism Spectrum disorders. *Frontiers in Psychiatry, 11*, 592408.

Arakawa, H. (2021a). Dynamic regulation of oxytocin neuronal circuits in the sequential processes of prosocial behavior in rodent models. *Current Research in Neurobiology, 2*, 100011.

Arakawa, H. (2021b). Implication of the social function of excessive self-grooming behavior in BTBR T+ltpr3tf/J mice as an idiopathic model of autism. *Physiology & Behavior, 237*, 113432.

Arakawa, H., & Higuchi, Y. (2022). Exocrine scent marking: Coordinative role of arginine vasopressin in the systemic regulation of social signaling behaviors. *Neuroscience and Biobehavioral Reviews, 136*, 104597.

Arakawa, H., & Iguchi, Y. (2018). Ethological and multi-behavioral analysis of learning and memory performance in laboratory rodent models. *Neuroscience Research, 135*, 1–12.

Arakawa, H., Blanchard, D. C., Arakawa, K., Dunlap, C., & Blanchard, R. J. (2008). Scent marking behavior as an odorant communication in mice. *Neuroscience and Biobehavioral Reviews, 32*, 1236–1248.

Arakawa, H., Arakawa, K., & Deak, T. (2010). Oxytocin and vasopressin in the medial amygdala differentially modulate approach and avoidance behavior toward illness-related social odor. *Neuroscience, 171*, 1141–1151.

Arakawa, H., Blanchard, D. C., & Blanchard, R. J. (2015). Central oxytocin regulates social familiarity and scent marking behavior that involves amicable odor signals between male mice. *Physiology & Behavior, 146*, 36–46.

Baron-Cohen, S., Knickmeyer, R. C., & Belmonte, M. K. (2005). Sex differences in the brain: Implications for explaining autism. *Science, 310*, 819–823.

Bartz, J., Simeon, D., Hamilton, H., Kim, S., Crystal, S., Braun, A., et al. (2011). Oxytocin can hinder trust and cooperation in borderline personality disorder. *Social Cognitive and Affective Neuroscience, 6*, 556–563.

Bielsky, I. F., Hu, S.-B., Szegda, K. L., Westphal, H., & Young, L. J. (2004). Profound impairment in social recognition and reduction in anxiety-like behavior in vasopressin V1a receptor knockout mice. *Neuropsychopharmacology, 29*, 483–493.

Birnbaumer, M. (2000). Vasopressin receptors. *Trends in Endocrinology and Metabolism, 11*, 406–410.

Blanchard, D. C., Griebel, G., Pobbe, R., & Blanchard, R. J. (2011). Risk assessment as an evolved threat detection and analysis process. *Neuroscience and Biobehavioral Reviews, 35*, 991–998.

Blanchard, D. C., Crawley, J. N., Arakawa, H., & Blanchard, R. J. (2013). Social behaviors in wild and laboratory mice with a special emphasis on the C57BL/6J inbred strain. In W. E. Crusio (Ed.), *Behavioral genetics of the mouse* (Vol. xiii, pp. 205–217). Cambridge University Press.

Bonsi, P., De Jaco, A., Fasano, L., & Gubellini, P. (2022). Postsynaptic autism spectrum disorder genes and synaptic dysfunction. *Neurobiology of Disease, 162*, 105564.

Borie, A. M., Dromard, Y., Guillon, G., Olma, A., Manning, M., Muscatelli, F., et al. (2021). Correction of vasopressin deficit in the lateral septum ameliorates social deficits of mouse autism model. *The Journal of Clinical Investigation, 131*(2), e144450.

Borland, J. M., Grantham, K. N., Aiani, L. M., Frantz, K. J., & Albers, H. E. (2018). Role of oxytocin in the ventral tegmental area in social reinforcement. *Psychoneuroendocrinology, 95*, 128–137.

Bourgeron, T. (2015). From the genetic architecture to synaptic plasticity in autism spectrum disorder. *Nature Reviews. Neuroscience, 16*, 551–563.

Brennan, P. A., & Kendrick, K. M. (2006). Mammalian social odours: Attraction and individual recognition. *Philosophical Transactions of the Royal Society of London. Series B, Biological Sciences, 361*, 2061–2078.

Brennan, P. A., & Zufall, F. (2006). Pheromonal communication in vertebrates. *Nature, 444*(7117), 308–315. https://doi.org/10.1038/nature05404

Buffington, S. A., Di Prisco, G. V., Auchtung, T. A., Ajami, N. J., Petrosino, J. F., & Costa-Mattioli, M. (2016). Microbial reconstitution reverses maternal diet-induced social and synaptic deficits in offspring. *Cell, 165*, 1762–1775.

Busnelli, M., Bulgheroni, E., Manning, M., Kleinau, G., & Chini, B. (2013). Selective and potent agonists and antagonists for investigating the role of mouse oxytocin receptors. *The Journal of Pharmacology and Experimental Therapeutics, 346*, 318–327.

Carter, C. S. (2017). The role of oxytocin and vasopressin in attachment. *Psychodynamic Psychiatry, 45*, 499–517.

Chaste, P., & Leboyer, M. (2012). Autism risk factors: Genes, environment, and gene-environment interactions. *Dialogues in Clinical Neuroscience, 14*, 281–292.

Crawley, J. N. (2022). Translational animal models of autism and neurodevelopmental disorders. *Dialogues in Clinical Neuroscience, 14*, 293–305.

Crespi, B., Procyshyn, T., & Mokkonen, M. (2022). Natura non facit Saltus: The adaptive significance of arginine vasopressin in human affect, cognition, and behavior. *Frontiers in Behavioral Neuroscience, 16*, 814230.

Dantzer, R., Koob, G. F., Bluthé, R. M., & Le Moal, M. (1988). Septal vasopressin modulates social memory in male rats. *Brain Research, 457*, 143–147.

Day, K. R., Coleman, A., Greenwood, M. A., & Hammock, E. A. D. (2020). AVPR1A distribution in the whole C57BL/6J mouse neonate. *Scientific Reports, 10*, 14512.

De Vries, G. J. (2008). Sex differences in vasopressin and oxytocin innervation of the brain. In I. D. Neumann & R. Landgraf (Eds.), *Progress in brain research* (pp. 17–27). Elsevier.

De Vries, G. J., & Panzica, G. C. (2006). Sexual differentiation of central vasopressin and vasotocin systems in vertebrates: Different mechanisms, similar endpoints. *Neuroscience, 138*, 947–955.

Dölen, G., & Malenka, R. C. (2014). The emerging role of nucleus accumbens oxytocin in social cognition. *Biological Psychiatry, 76*, 354–355.

Dölen, G., Darvishzadeh, A., Huang, K. W., & Malenka, R. C. (2013). Social reward requires coordinated activity of nucleus accumbens oxytocin and serotonin. *Nature, 501*, 179–184.

Dubois-Dauphin, M., Barberis, C., & de Bilbao, F. (1996). Vasopressin receptors in the mouse (Mus musculus) brain: Sex-related expression in the medial preoptic area and hypothalamus. *Brain Research, 743*, 32–39.

Duque-Wilckens, N., Steinman, M. Q., Busnelli, M., Chini, B., Yokoyama, S., Pham, M., et al. (2018). Oxytocin receptors in the anteromedial bed nucleus of the stria terminalis promote stress-induced social avoidance in female California mice. *Biological Psychiatry, 83*, 203–213.

Ebitz, R. B., Watson, K. K., & Platt, M. L. (2013). Oxytocin blunts social vigilance in the rhesus macaque. *Proceedings of the National Academy of Sciences of the United States of America, 110*, 11630–11635.

Endres, T., & Fendt, M. (2008). Inactivation of the lateral septum blocks fox odor-induced fear behavior. *Neuroreport, 19*, 667–670.

Engelmann, M. (2008). Vasopressin in the septum: Not important versus causally involved in learning and memory — two faces of the same coin? In *Advances in vasopressin and oxytocin — from genes to behaviour to disease* (pp. 389–395). Elsevier.

Ferguson, J. N., Aldag, J. M., Insel, T. R., & Young, L. J. (2001). Oxytocin in the medial amygdala is essential for social recognition in the mouse. *The Journal of Neuroscience, 21*, 8278–8285.

Gamal-Eltrabily, M., de Los, E., Monteros-Zúñiga, A., Manzano-García, A., Martínez-Lorenzana, G., Condés-Lara, M., & González-Hernández, A. (2020). The rostral agranular insular cortex, a new site of oxytocin to induce antinociception. *The Journal of Neuroscience, 40*, 5669–5680.

Gardener, H., Spiegelman, D., & Buka, S. L. (2011). Perinatal and neonatal risk factors for autism: A comprehensive meta-analysis. *Pediatrics, 128*, 344–355.

Geschwind, D. H. (2011). Genetics of autism spectrum disorders. *Trends in Cognitive Sciences, 15*, 409–416.

Gouin, J.-P., Carter, C. S., Pournajafi-Nazarloo, H., Malarkey, W. B., Loving, T. J., Stowell, J., et al. (2012). Plasma vasopressin and interpersonal functioning. *Biological Psychology, 91*, 270–274.

Grinevich, V., Knobloch-Bollmann, H. S., Eliava, M., Busnelli, M., & Chini, B. (2016). Assembling the puzzle: Pathways of oxytocin signaling in the brain. *Biological Psychiatry, 79*, 155–164.

Grove, J., Ripke, S., Als, T. D., Mattheisen, M., Walters, R. K., Won, H., et al. (2019). Identification of common genetic risk variants for autism spectrum disorder. *Nature Genetics, 51*, 431–444.

Guinchat, V., Thorsen, P., Laurent, C., Cans, C., Bodeau, N., & Cohen, D. (2012). Pre-, peri- and neonatal risk factors for autism. *Acta Obstetricia et Gynecologica Scandinavica, 91*, 287–300.

Gunaydin, L. A., Grosenick, L., Finkelstein, J. C., Kauvar, I. V., Fenno, L. E., Adhikari, A., et al. (2014). Natural neural projection dynamics underlying social behavior. *Cell, 157*, 1535–1551.

Harony-Nicolas, H., Kay, M., Hoffmann, J., du Klein, M. E., Bozdagi-Gunal, O., Riad, M., et al. (2017). Oxytocin improves behavioral and electrophysiological deficits in a novel Shank3-deficient rat. *eLife, 6*, e18904.

Haskal de la Zerda, S., Netser, S., Magalnik, H., & Wagner, S. (2020). Impaired sex preference, but not social and social novelty preferences, following systemic blockade of oxytocin receptors in adult male mice. *Psychoneuroendocrinology, 116*, 104676.

Hendaus, M. A., Jomha, F. A., & Alhammadi, A. H. (2019). Vasopressin in the amelioration of social functioning in autism spectrum disorder. *Journal of Clinical Medical Research, 8*(7), 1061.

Henry, A. M., & Hohmann, J. G. (2012). High-resolution gene expression atlases for adult and developing mouse brain and spinal cord. *Mammalian Genome, 23*, 539–549.

Hernández, V. S., Vázquez-Juárez, E., Márquez, M. M., Jáuregui-Huerta, F., Barrio, R. A., & Zhang, L. (2015). Extra-neurohypophyseal axonal projections from individual vasopressin-containing magnocellular neurons in rat hypothalamus. *Frontiers in Neuroanatomy, 9*, 130.

Higuchi, Y., & Arakawa, H. (2022). Contrasting central and systemic effects of arginine-vasopressin on urinary marking behavior as a social signal in male mice. *Hormones and Behavior, 141*, 105128.
Hollander, E., Novotny, S., Hanratty, M., Yaffe, R., DeCaria, C. M., Aronowitz, B. R., et al. (2003). Oxytocin infusion reduces repetitive behaviors in adults with autistic and Asperger's disorders. *Neuropsychopharmacology, 28*, 193–198.
Hollander, E., Bartz, J., Chaplin, W., Phillips, A., Sumner, J., Soorya, L., et al. (2007). Oxytocin increases retention of social cognition in autism. *Biological Psychiatry, 61*, 498–503.
Hung, L. W., Neuner, S., Polepalli, J. S., Beier, K. T., Wright, M., Walsh, J. J., et al. (2017). Gating of social reward by oxytocin in the ventral tegmental area. *Science, 357*, 1406–1411.
Hurst, J. L., & Beynon, R. J. (2004). Scent wars: The chemobiology of competitive signalling in mice. *BioEssays, 26*, 1288–1298.
Islam, M. T., Maejima, T., Matsui, A., & Mieda, M. (2022). Paraventricular hypothalamic vasopressin neurons induce self-grooming in mice. *Molecular Brain, 15*, 47.
Jiujias, M., Kelley, E., & Hall, L. (2017). Restricted, repetitive behaviors in autism spectrum disorder and obsessive-compulsive disorder: A comparative review. *Child Psychiatry and Human Development, 48*, 944–959.
Jurek, B., & Neumann, I. D. (2018). The oxytocin receptor: From intracellular signaling to behavior. *Physiological Reviews, 98*, 1805–1908.
Keech, B., Crowe, S., & Hocking, D. R. (2018). Intranasal oxytocin, social cognition and neurodevelopmental disorders: A meta-analysis. *Psychoneuroendocrinology, 87*, 9–19.
Klauck, S. M. (2006). Genetics of autism spectrum disorder. *European Journal of Human Genetics, 14*, 714–720.
Knobloch, H. S., Charlet, A., Hoffmann, L. C., Eliava, M., Khrulev, S., Cetin, A. H., et al. (2012). Evoked axonal oxytocin release in the central amygdala attenuates fear response. *Neuron, 73*, 553–566.
Koob, G. F., Lebrun, C., Bluthé, R. M., Dantzer, R., & Le Moal, M. (1989). Role of neuropeptides in learning versus performance: Focus on vasopressin. *Brain Research Bulletin, 23*, 359–364.
Koolhaas, J. M., Everts, H., de Ruiter, A. J., de Boer, S. F., & Bohus, B. (1998). Coping with stress in rats and mice: Differential peptidergic modulation of the amygdala-lateral septum complex. *Progress in Brain Research, 119*, 437–448.
Kosfeld, M., Heinrichs, M., Zak, P. J., Fischbacher, U., & Fehr, E. (2005). Oxytocin increases trust in humans. *Nature, 435*, 673–676.
Landgraf, R., & Neumann, I. D. (2004). Vasopressin and oxytocin release within the brain: A dynamic concept of multiple and variable modes of neuropeptide communication. *Frontiers in Neuroendocrinology, 25*, 150–176.
Landgraf, R., Frank, E., Aldag, J. M., Neumann, I. D., Sharer, C. A., Ren, X., et al. (2003). Viral vector-mediated gene transfer of the vole V1a vasopressin receptor in the rat septum: Improved social discrimination and active social behaviour. *The European Journal of Neuroscience, 18*, 403–411.
Le Moal, M., Dantzer, R., Michaud, B., & Koob, G. F. (1987). Centrally injected arginine vasopressin (AVP) facilitates social memory in rats. *Neuroscience Letters, 77*, 353–359.
Le, J., Zhao, W., Kou, J., Fu, M., Zhang, Y., Becker, B., et al. (2021). Oxytocin facilitates socially directed attention. *Psychophysiology, 58*, e13852.
Lefevre, A., Benusiglio, D., Tang, Y., Krabichler, Q., Charlet, A., & Grinevich, V. (2021). Oxytocinergic feedback circuitries: An anatomical basis for neuromodulation of social behaviors. *Frontiers in Neural Circuits, 15*, 688234.
Luchsinger, J. R., Fetterly, T. L., Williford, K. M., Salimando, G. J., Doyle, M. A., Maldonado, J., et al. (2021). Delineation of an insula-BNST circuit engaged by struggling behavior that regulates avoidance in mice. *Nature Communications, 12*, 3561.
Lukas, M., & Neumann, I. D. (2013). Oxytocin and vasopressin in rodent behaviors related to social dysfunctions in autism spectrum disorders. *Behavioural Brain Research, 251*, 85–94.

Ma, Y., Shamay-Tsoory, S., Han, S., & Zink, C. F. (2016). Oxytocin and social adaptation: Insights from neuroimaging studies of healthy and clinical populations. *Trends in Cognitive Sciences, 20*, 133–145.

Mahfouz, A., Ziats, M. N., Rennert, O. M., Lelieveldt, B. P. F., & Reinders, M. J. T. (2015). Shared pathways among autism candidate genes determined by co-expression network analysis of the developing human brain transcriptome. *Journal of Molecular Neuroscience, 57*, 580–594.

Manning, M., Misicka, A., Olma, A., Bankowski, K., Stoev, S., Chini, B., et al. (2012). Oxytocin and vasopressin agonists and antagonists as research tools and potential therapeutics. *Journal of Neuroendocrinology, 24*, 609–628.

Mapelli, L., Soda, T., D'Angelo, E., & Prestori, F. (2022). The cerebellar involvement in autism spectrum disorders: From the social brain to mouse models. *International Journal of Molecular Sciences, 23*, 3894.

Maramara, L. A., He, W., & Ming, X. (2014). Pre- and perinatal risk factors for autism spectrum disorder in a New Jersey cohort. *Journal of Child Neurology, 29*, 1645–1651.

Mitre, M., Marlin, B. J., Schiavo, J. K., Morina, E., Norden, S. E., Hackett, T. A., et al. (2016). A distributed network for social cognition enriched for oxytocin receptors. *The Journal of Neuroscience, 36*, 2517–2535.

Morrison, I. (2016). Keep calm and cuddle on: Social touch as a stress buffer. *Adaptive Human Behavior and Physiology, 2*, 344–362.

Nasanbuyan, N., Yoshida, M., Takayanagi, Y., Inutsuka, A., Nishimori, K., Yamanaka, A., et al. (2018). Oxytocin-oxytocin receptor systems facilitate social defeat posture in male mice. *Endocrinology, 159*, 763–775.

Netser, S., Meyer, A., Magalnik, H., Zylbertal, A., de la Zerda, S. H., Briller, M., et al. (2020). Distinct dynamics of social motivation drive differential social behavior in laboratory rat and mouse strains. *Nature Communications, 11*, 5908.

Neumann, I. D., & Slattery, D. A. (2016). Oxytocin in general anxiety and social fear: A translational approach. *Biological Psychiatry, 79*, 213–221.

Nordman, J., & Li, Z. (2020). The dorsal raphe regulates the duration of attack through the medial orbitofrontal cortex and medial amygdala. *eNeuro, 7*, 0331–0320.

Orefice, L. L., Zimmerman, A. L., Chirila, A. M., Sleboda, S. J., Head, J. P., & Ginty, D. D. (2016). Peripheral mechanosensory neuron dysfunction underlies tactile and behavioral deficits in mouse models of ASDs. *Cell, 166*, 299–313.

Ozawa, A., & Arakawa, H. (2021). Chemogenetics drives paradigm change in the investigation of behavioral circuits and neural mechanisms underlying drug action. *Behavioural Brain Research, 406*, 113234.

Oztan, O., Garner, J. P., Partap, S., Sherr, E. H., Hardan, A. Y., Farmer, C., et al. (2018). Cerebrospinal fluid vasopressin and symptom severity in children with autism. *Annals of Neurology, 84*, 611–615.

Oztan, O., Garner, J. P., Constantino, J. N., & Parker, K. J. (2020). Neonatal CSF vasopressin concentration predicts later medical record diagnoses of autism spectrum disorder. *Proceedings of the National Academy of Sciences of the United States of America, 117*, 10609–10613.

Parker, K. J., Garner, J. P., Oztan, O., Tarara, E. R., Li, J., Sclafani, V., et al. (2018). Arginine vasopressin in cerebrospinal fluid is a marker of sociality in nonhuman primates. *Science Translational Medicine, 10*, 439.

Patel, T. N., Caiola, H. O., Mallari, O. G., Blandino, K. L., Goldenthal, A. R., Dymecki, S. M., et al. (2022). Social interactions increase activation of vasopressin-responsive neurons in the dorsal raphe. *Neuroscience, 495*, 25–46.

Peñagarikano, O., Lázaro, M. T., Lu, X.-H., Gordon, A., Dong, H., Lam, H. A., et al. (2015). Exogenous and evoked oxytocin restores social behavior in the Cntnap2 mouse model of autism. *Science Translational Medicine, 7*, 271ra8.

Provenzano, G., Zunino, G., Genovesi, S., Sgadó, P., & Bozzi, Y. (2012). Mutant mouse models of autism spectrum disorders. *Disease Markers, 33*, 225–239.

Rae, M., Lemos Duarte, M., Gomes, I., Camarini, R., & Devi, L. A. (2021). Oxytocin and vasopressin: Signalling, behavioural modulation and potential therapeutic effects. *British Journal of Pharmacology*. https://doi.org/10.1111/bph.15481

Rigney, N., Whylings, J., Mieda, M., de Vries, G., & Petrulis, A. (2019). Sexually dimorphic vasopressin cells modulate social investigation and communication in sex-specific ways. *eNeuro, 6*, 0415–0418.

Rigney, N., Beaumont, R., & Petrulis, A. (2020). Sex differences in vasopressin 1a receptor regulation of social communication within the lateral habenula and dorsal raphe of mice. *Hormones and Behavior, 121*, 104715.

Rigney, N., Whylings, J., de Vries, G. J., & Petrulis, A. (2021). Sex differences in the control of social investigation and anxiety by vasopressin cells of the paraventricular nucleus of the hypothalamus. *Neuroendocrinology, 111*, 521–535.

Rodríguez, M., Ceric, F., Murgas, P., Harland, B., Torrealba, F., & Contreras, M. (2019). Interoceptive insular cortex mediates both innate fear and contextual threat conditioning to predator odor. *Frontiers in Behavioral Neuroscience, 13*, 283.

Rogers-Carter, M. M., Djerdjaj, A., Gribbons, K. B., Varela, J. A., & Christianson, J. P. (2019). Insular cortex projections to nucleus accumbens core mediate social approach to stressed juvenile rats. *The Journal of Neuroscience, 39*, 8717–8729.

Rood, B. D., & Beck, S. G. (2014). Vasopressin indirectly excites dorsal raphe serotonin neurons through activation of the vasopressin1A receptor. *Neuroscience, 260*, 205–216.

Rood, B. D., & De Vries, G. J. (2011). Vasopressin innervation of the mouse (Mus musculus) brain and spinal cord. *The Journal of Comparative Neurology, 519*, 2434–2474.

Rood, B. D., Stott, R. T., You, S., Smith, C. J. W., Woodbury, M. E., & De Vries, G. J. (2013). Site of origin of and sex differences in the vasopressin innervation of the mouse (Mus musculus) brain. *The Journal of Comparative Neurology, 521*, 2321–2358.

Sikich, L., Kolevzon, A., King, B. H., McDougle, C. J., Sanders, K. B., Kim, S.-J., et al. (2021). Intranasal oxytocin in children and adolescents with autism spectrum disorder. *The New England Journal of Medicine, 385*, 1462–1473.

Silverman, J. L., Thurm, A., Ethridge, S. B., Soller, M. M., Petkova, S. P., Abel, T., et al. (2022). Reconsidering animal models used to study autism spectrum disorder: Current state and optimizing future. *Genes, Brain, and Behavior, 21*, e12803.

Sluyter, F., & Van Oortmerssen, G. A. (2000). A mouse is not just a mouse. *Animal Welfare, 9*, 193–205.

Song, Z., & Albers, H. E. (2018). Cross-talk among oxytocin and arginine-vasopressin receptors: Relevance for basic and clinical studies of the brain and periphery. *Frontiers in Neuroendocrinology, 51*, 14–24.

Song, Z., Borland, J. M., Larkin, T. E., O'Malley, M., & Albers, H. E. (2016). Activation of oxytocin receptors, but not arginine-vasopressin V1a receptors, in the ventral tegmental area of male Syrian hamsters is essential for the reward-like properties of social interactions. *Psychoneuroendocrinology, 74*, 164–172.

Steinman, M. Q., Duque-Wilckens, N., & Trainor, B. C. (2019). Complementary neural circuits for divergent effects of oxytocin: Social approach versus social anxiety. *Biological Psychiatry, 85*, 792–801.

Swanson, L. W., & Sawchenko, P. E. (1983). Hypothalamic integration: Organization of the paraventricular and supraoptic nuclei. *Annual Review of Neuroscience, 6*, 269–324.

Tang, Y., Benusiglio, D., Lefevre, A., Hilfiger, L., Althammer, F., Bludau, A., et al. (2020). Social touch promotes interfemale communication via activation of parvocellular oxytocin neurons. *Nature Neuroscience, 23*, 1125–1137.

Taylor, S. E., Saphire-Bernstein, S., & Seeman, T. E. (2010). Are plasma oxytocin in women and plasma vasopressin in men biomarkers of distressed pair-bond relationships? *Psychological Science, 21*, 3–7.

Tickerhoof, M. C., & Smith, A. S. (2017). Vasopressinergic neurocircuitry regulating social attachment in a monogamous species. *Frontiers in Endocrinology, 8*, 265.

van Kesteren, R. E., Smit, A. B., Dirks, R. W., de With, N. D., Geraerts, W. P., & Joosse, J. (1992). Evolution of the vasopressin/oxytocin superfamily: Characterization of a cDNA encoding a vasopressin-related precursor, preproconopressin, from the mollusc Lymnaea stagnalis. *Proceedings of the National Academy of Sciences of the United States of America, 89*, 4593–4597.

Veenema, A. H., Beiderbeck, D. I., Lukas, M., & Neumann, I. D. (2010). Distinct correlations of vasopressin release within the lateral septum and the bed nucleus of the stria terminalis with the display of intermale aggression. *Hormones and Behavior, 58*, 273–281.

Williams, A. V., Duque-Wilckens, N., Ramos-Maciel, S., Campi, K. L., Bhela, S. K., Xu, C. K., et al. (2020). Social approach and social vigilance are differentially regulated by oxytocin receptors in the nucleus accumbens. *Neuropsychopharmacology, 45*, 1423–1430.

Wu, J., Dai, Y.-C., Lan, X.-Y., Zhang, H.-F., Bai, S.-Z., Hu, Y., et al. (2021). Postnatal AVP treatments prevent social deficit in adolescence of valproic acid-induced rat autism model. *Peptides, 137*, 170493.

Yang, M., Scattoni, M. L., Zhodzishsky, V., Chen, T., Caldwell, H., Young, W. S., et al. (2007). Social approach behaviors are similar on conventional versus reverse lighting cycles, and in replications across cohorts, in BTBR T+ tf/J, C57BL/6J, and vasopressin receptor 1B mutant mice. *Frontiers in Behavioral Neuroscience, 1*, 1.

Yao, S., Bergan, J., Lanjuin, A., & Dulac, C. (2017). Oxytocin signaling in the medial amygdala is required for sex discrimination of social cues. *eLife, 6*, e31373.

Young, L. J., & Barrett, C. E. (2015). Neuroscience. Can oxytocin treat autism? *Science, 347*, 825–826.

Young, W. S., Li, J., Wersinger, S. R., & Palkovits, M. (2006). The vasopressin 1b receptor is prominent in the hippocampal area CA2 where it is unaffected by restraint stress or adrenalectomy. *Neuroscience, 143*, 1031–1039.

Zhou, Y., Kaiser, T., Monteiro, P., Zhang, X., Van der Goes, M. S., Wang, D., et al. (2016). Mice with Shank3 mutations associated with ASD and schizophrenia display both shared and distinct defects. *Neuron, 89*, 147–162.

Zingg, H. H., & Laporte, S. A. (2003). The oxytocin receptor. *Trends in Endocrinology and Metabolism, 14*, 222–227.

Chapter 7
Seizures in Mouse Models of Autism

Alison J. Sebold, Alyssa Strassburg, Natalia Avery, Darya Ryndych, Violeta B. Foss, Preet Sawhney, and Gonzalo H. Otazu

The DSM V defines Autism Spectrum Disorder (ASD) as (1) persistent deficits in social communication/interaction across the contexts of social-emotional reciprocity, nonverbal communicative behaviors, and developing, maintaining, and understanding relationships, and (2) restricted, repetitive patterns of behavior, interests, or activities, with these symptoms present early in the developmental period, impairing function in a clinically significant manner, and not better explained by another disorder or delay (American Psychiatric Association, 2022).

ASD can present with other conditions, including intellectual disability, Attention-Deficit Hyperactivity Disorder (ADHD), and epilepsy. Diagnosing epilepsy in ASD children is challenging because some features of seizures, including repetitive behaviors, overlap with some issues in children with ASD. A 2013 study with 4705 children found that 44% of children with ASD received a subsequent diagnosis of epilepsy, and 54% of children with epilepsy received a diagnosis of ASD (Jokiranta et al., 2014) Prevalence of epilepsy in ASD seems to be positively correlated with presence and severity of intellectual disability in ASD (Hyman et al., 2020) There appears to be a relationship between age and prevalence as well as incidence of epilepsy; most studies show that older children and adolescents seem to exhibit higher prevalence and incidence of epilepsy than younger children (Lukmanji et al., 2019) No particular seizure category or epilepsy type has been determined as most prevalent in individuals with ASD; partial, generalized, and absence seizures have all been reported in patients with ASD (Spence & Schneider, 2009).

Besides diagnosed epilepsy, ASD often shows paroxysmal electroencephalogram (EEG) abnormalities. A prospective study that included 1014 children with

A. J. Sebold · A. Strassburg · N. Avery · D. Ryndych · V. B. Foss
P. Sawhney · G. H. Otazu (✉)
New York Institute of Technology, College of Osteopathic Medicine, Old Westbury, NY, USA
e-mail: gotazual@nyit.edu

A. El Idrissi, D. McCloskey (eds.), *Neurobiology of Autism Spectrum Disorders*,
https://doi.org/10.1007/978-3-031-42383-3_7

ASD, of 3–20 years of age, found a frequent occurrence of EEG abnormalities in children without an epilepsy diagnosis. The children were followed for 3 years, over which time they had an EEG roughly every 6 months, while asleep (Yasuhara, 2010). Epileptic EEG discharges were observed in 85.8%; 65.6% of those were from frontal lobe. This widespread prevalence of epileptic discharges, in particular during sleep, suggests a possible link between epilepsy and ASD.

7.1 Sex Bias of Presentation of ASD and Epilepsy

ASD has been found to have a strong male predominance, with most studies finding the sex bias to be 3–5:1, M:F (Amiet et al., 2008; Ferri et al., 2018; Giarelli et al., 2010; Santos et al., 2022) In contrast, epilepsy has been found to have no sex bias (Fiest et al., 2017; Savic, 2014). Interestingly, when ASD and epilepsy present together, the sex bias becomes significantly less male predominant. A review of 74 studies with 283,549 total patients found that females had a higher prevalence of epilepsy in ASD (Lukmanji et al., 2019). Another study of 671,103 patients, 16,126 of which were autistic, found no significant difference in the sex bias of ASD and epilepsy presenting as comorbid conditions, but did find that female autistic patients had a significantly higher odds ratio of comorbid conditions as compared to the non-autistic population, suggesting that the presence of a comorbid condition makes the diagnosis of ASD more likely for a female patient (Supekar et al., 2017).

Another study of 4790 patients with a diagnosis of ASD stratified prevalence of comorbid conditions in ASD by sex and age. Autistic males and females were found to have a significantly higher prevalence of comorbid conditions as compared to the general population, including epilepsy. Epilepsy was further found to have two peaks in prevalence, the first in early childhood and second in adolescence. Epilepsy had a significantly higher prevalence in female autistic patients as opposed to male autistic patients. Interestingly, when data is further stratified by age, epilepsy prevalence in age groups under 35 years old was highest in autistic females, but in patients above 35 years old, prevalence was highest in autistic males. This suggests that epilepsy may be less likely to remit and harder to treat in older male patients with ASD (Rødgaard et al., 2021).

Lastly, a meta-analysis of 14 studies with a total of 1503 patients also found that autistic females had a higher relative risk of epilepsy as opposed to autistic males (Amiet et al., 2008). This review went on to find an overall ratio of 2:1 M:F sex bias of epilepsy and ASD comorbid, suggesting a slight male predominance of the two conditions when presenting comorbidly (Amiet et al., 2008).

Many theories have been proposed as to why ASD appears more frequently in men, including the female protective effect, the extreme male brain theory, and the underdiagnosis of females. It is not clear the biological mechanism on how an epilepsy comorbidity might rebalance the sex ratio in ASD. An alternative possibility is that the only the more severe cases of ASD are diagnosed in females and those cases are more likely to be comorbid with epilepsy because the presence of epilepsy might facilitate the diagnosis of ASD in females (Rødgaard et al., 2021).

7.2 Early Epileptic Activity and ASD

What is the origin of the clinical association between epilepsy, or abnormal EEG, and ASD? One hypothesis is that the early abnormal electrical activity in epilepsy contributes to the development of ASD. If that were the case, we would expect that early epileptic activity would precede an ASD diagnosis. In a study of 84 children with seizure onset between 28 days and 12 months of age, and excluding those with infantile spasms/febrile seizures, 31 children were investigated for ASD, six (7.1%) children reached the criteria for ASD diagnosis (Saemundsen et al. 2007a). Another study, focused on infantile spasms, found that of 20 children with infantile spasms, 6 had ASD, greater than the prevalence of ASD in the general population (Saemundsen et al., 2007b), although a second study found that while infantile spasms predicted high risk for ASD, this was largely explained by the association of ASD, and the symptomatic origin of seizures and infantile spasms were not an independent risk factor for ASD (Saemundsen et al., 2008).

Another study, looking at 519 people with epilepsy, found that 79 (15.2%) had ASD, 62 of whom had idiopathic ASD, and 17 with secondary ASD (Matsuo et al., 2010). Retrospective analysis of the epilepsy patients with idiopathic ASD showed early onset of abnormal brain activity. 32% of patients had a past history of febrile seizure, 9 had a history of more than three febrile seizures, and 2 patients had a history of benign infantile convulsions (Matsuo et al., 2010). In this sample, 85% of the people studied had seizure onset before age 10 (mode age of onset = 4 years) (Matsuo et al., 2010). In 29 cases (46.8%) ASD was detected after onset of epilepsy, and in 8 of these patients ASD diagnosis was more than 5 years after epilepsy onset, although symptoms of ASD were present prior to seizure onset in these patients (Matsuo et al., 2010). Epileptoid activity seems to precede the onset of ASD diagnosis in some cases. However, the relationship is not straightforward.

Current pharmacologic management of seizures in children with ASD is similar to that of non-ASD patients. Seizure and/or epilepsy treatment in individuals with ASD is solely indicated if the patient is exhibiting clinical epileptic activity. Abnormal EEG in itself is not an indication for the initiation of antiseizure drug treatment (Canitano, 2007) The risk versus benefit and potential medication side effects have to be considered due to the significant side effect profile of most anti-epileptic drugs (AEDs).

Given the potential connection between EEG abnormalities and ASD, clinical trials have directly tested whether AED could improve the ASD symptoms. A meta-analysis of 8 randomized controlled trials (4 on valproate, 2 on levetiracetam, 1 on lamotrigine, 1 on topiramate) with 310 ASD individuals evaluated the effect on core and associated symptoms of ASD including irritability and aggression. However, there was no significant difference noted (Limbu et al., 2022), suggesting a more complex relationship between ASD and epilepsy.

7.3 Imbalance of Excitation and Inhibition as a Common Etiology for ASD and Epilepsy

An alternative hypothesis is that the abnormal EEG associated with epilepsy and ASD are both produced by a common cause. A very influential idea (Rubenstein & Merzenich, 2003) is that both ASD and epilepsy might be caused by an imbalance in neuronal excitation and inhibition leading to a hyperexcitable state. Cortical balance of excitation and inhibition is controlled by the relative activity of glutamatergic and GABAergic neurons, respectively, which are further regulated by extrinsic excitatory and/or inhibitory inputs from subcortical regions, glial cells, vasculature, and CSF system. An imbalance (whether it be due to increased excitation or decreased inhibition) leading to hyperexcitable state can be caused by defects in production, migration, activity, or susceptibility to modulation, of any one of these components. A pathognomonic characteristic of autistic individuals is a hypersensitivity to auditory, visual, and/or tactile stimulus due to poor differentiation during critical periods of development and subsequently poor maturation of associated neural pathways. This poor differentiation leads to a higher susceptibility to positive coupling of neuronal populations throughout the cortex, and has detrimental effects on the development of closely intertwined higher-order pathways such as learning, memory, and emotion, all of which are also impaired in ASD.

However, recent work has put the idea of a causative role of the imbalance of excitation/inhibition in ASD into question, seeing it as a compensatory mechanism instead (Antoine et al., 2019). When synaptic conductance was measured in four common mouse models of ASD, all exhibited decreased inhibition (and mildly decreased excitation) leading to an elevated E-I conductance ratio and theoretically favoring a hyperexcitable state. However, quantification of peak synaptic depolarization appeared normal. This suggests that the observed increase in E-I conduction ratio occurs to compensate for a decrease in overall synaptic conductance and subsequently maintain firing rates. The authors conclude that in contrast to the current theory, alteration in synaptic conductance is a homeostatic mechanism employed to stabilize defects in cortical firing rather than another manifestation of impaired information processing present in ASD.

7.4 Seizures in Monogenetic Mouse Models of ASD

There remain open questions regarding the relations between epileptic activity and ASD. Although clinical observations have pointed at a correlation between these two different conditions, a circuit level explanation is lacking which hampers the development of effective targeted therapies. Intriguing observations like the loss of the protective effect in females and the effect of epileptic activity in the development and onset of ASD require animal models that might point to circuit disfunction. For this chapter, we have focused on currently available (year 2022) mouse lines from Jackson laboratory all of which are part of the Simons foundation list of mouse models of ASD (https://www.sfari.org/resource/mouse-models). We have done an extensive literature review of the research done with each of these models. We specifically focused on the presence of seizures in an attempt to find which models are the most appropriate for the study of the comorbidity of epilepsy and ASD. We filtered through all the citations of each originating paper, for each mouse line. With the use of Google Scholar, we searched for the key terms "seizure" and "epilepsy". This was done with research in both mice and humans to establish the face validity of the presence of seizures in the given genetic variations of the mouse models.

The Simons Foundation list include mouse models with mutations in 16p11.2, ARID1B, CACNA1C, CNTNAP2, CUL3, DYRK1A, GTF2I, RAI1, SHANK3, SYNGAP1, and UBE3A. Each model represents a genetic variation which has been associated with the development of ASD or characteristics similar to those seen in patients with ASD. The diseases which develop from these models, including Timothy syndrome (CACNA1C) and Phelan-McDermid syndrome (SHANK3), are considered syndromic forms of ASD: a disorder with a clinically defined pattern of somatic abnormalities and a neurobehavioral phenotype that may include ASD (Fernandez & Scherer, 2017). Syndromic forms of ASD represent only a small percentage of the entire ASD spectrum, however they provide an opportunity to learn about the neurobiology of ASD-related phenotypes as well as a means to test possible therapeutic interventions. The mouse models are considered to have construct validity because they carry mutations which the current genetic literature has shown to be present, to varying degrees, in the human disorder. They also exhibit face validity because the mice present with physical and behavioral characteristics, such as reduced social interaction, repetitive behavior, and sensitivity to sensory stimuli, that are seen in the human disorder as well. However, there are and will always be limitations to representing a complex disorder such as ASD, which displays immense genetic and clinical diversity, with only a single-gene mutation model. Table 7.1 lists all the mouse models discussed in this chapter, as well as links to the corresponding strain on Jax.org.

7.4.1 16p11.2

The deletion and duplication of 16p11.2 has been associated with ASD and other neuropsychiatric disorders. Moreover, deletions are associated with macrocephaly and sensorimotor deficits in rats with lateral hypothalamic and nigrostriatal lesions, while duplications are associated with microcephaly in patients (Horev et al., 2011). There are several similarities between ASD features in humans with 16p11.2 and its mouse models. The mice exhibit hyperactivity which can correlate ADHD observed in 16p11.2 patients, as well as deficits in smooth motor movements which can correlate to motor delay prevalent in 12p11.2 patients (Portmann et al., 2014). In human subjects, deletion at various regions (550 kb and 593 kb) on 16p11.2 have been

Table 7.1 Mouse models and links for purchase (Alphabetical order)

Mouse Model	Number of citations in google scholar (as June 2022)	Link to strain page in JAX
16p11.2 deletion (0.39 megabases on mouse chromosome 7)	287	https://www.jax.org/strain/013128
16p11.2 deletion (synthenic 440 kilobase-pair region on mouse chromosome 7F3)	174	https://www.jax.org/strain/025100
16p11.2 duplication	(same originating article as 16p11.2 deletion (0.39 megabases on mouse chromosome 7))	https://www.jax.org/strain/013129 and https://www.jax.org/strain/016915
ARID1B floxed	62	https://www.jax.org/strain/032061
CACNA1C (TS2-neo) mutation	212 total (two articles)	https://www.jax.org/strain/019547
CNTNAP2 (CASPR2) deletion	500	https://www.jax.org/strain/017482
CUL3 floxed	64	https://www.jax.org/strain/028349
DYRK1A floxed	58	https://www.jax.org/strain/027801
GTF2I duplication	89	https://www.jax.org/strain/027792
RAI1 floxed	44	https://www.jax.org/strain/029103
RAI1 tagged	(same originating article as RAI1 floxed)	https://www.jax.org/strain/029101
SHANK3 deletion (ankyrin repeat domains (exons 4–9))	1216	https://www.jax.org/strain/017688
SHANK 3 deletion (PDZ domain)	(same originating article as SHANK3 deletion (ankyrin repeat domains (exons 4–9)))	
SYNGAP1 conditional deletion	294	https://www.jax.org/strain/029303
SYNGAP1 conditional rescue	(same originating article as SYNGAP1 conditional deletion)	
UBE3A variant 2, 3 overexpression	236	https://www.jax.org/strain/019730

linked with increased risk for ASD, developmental delay, seizures, early-onset obesity, increased body mass index, and birth defects (Shinawi et al., 2010). In addition, the microdeletions have also been linked to neuroblastomas which, although rarely, can be associated with seizures. This pleiotropic effect on neurodevelopmental phenotypes can be dysregulation of developmental programs involved in both central and peripheral nervous system development (Egolf et al., 2019).

One study found 5 of 16 patients with the deletion presented with seizures, versus three of 10 patients with a duplication variant. In addition, one patient with deletion was identified with non-convulsive epilepsy (Shinawi et al., 2010). Another cohort of individuals with developmental delays and 16p11.2 microdeletions and microduplications was identified to have two out of 18 co-occurring cases of seizures(Rosenfeld et al., 2010). Onset of epileptic activity occurs during the first year of life. The seizures were noted to be easily controlled with antiepileptics and decreased in severity during childhood (Shinawi et al., 2010).

Mouse models with deletion of 16p11.2 were noted to die more often after birth compared to their duplication counterparts, indicating that 16p11.2 loss may compromise survivability (Horev et al., 2011). While the association of seizures in patients with ASD and 16p11.2 variants has been noted, the studies on the mouse model are still lacking in research to determine this correlation. Neither provoked nor spontaneous seizure activity has been described in this mouse line.

7.4.2 *ARID1B*

The ARID1B gene encodes for one of the largest subunits of several different SWI/SNF protein complexes (Hoyer et al., 2012). The SWI/SNF protein complexes help regulate gene activity via chromatin remodeling, and mutations of the ARID1B subunit have been found to be causative agents of Coffin-Siris syndrome (CSS) and moderate to severe intellectual disability (Hoyer et al., 2012; van der Sluijs et al., 2019). The most significant differences differentiating the two diagnoses is the prevalence of dysmorphic features, found more commonly in CSS (van der Sluijs et al., 2019). Patients with CSS have hypoplasia of the fifth digit, dysmorphic facial features, feeding problems, hirsutism, and hypotonia (Iwamoto et al., 2021). Additionally, roughly 50% of patients with CSS have been found to have seizures and about 10% have autistic features (Iwamoto et al., 2021).

In mouse models, $ARID1B^{+/-}$ mice have been found to have mild deficits in areas relevant to some of the social deficits seen in ASD, such as deficits in direct reciprocal social interactions and object recognition (Ellegood et al., 2021). In addition, $ARID1B^{+/-}$ mice are found to have an increased susceptibility to seizures when induced with pentylenetetrazol (PTZ) (Ellegood et al., 2021). When induced with PTZ, $ARID1B^{+/-}$ mice had quicker onset of tonic-clonic extension, loss of righting, and death than $ARID1B^{+/+}$ controls (Ellegood et al., 2021). Interestingly, another study found that administration of a typical GABA allosteric modulator, clonazepam, was able to rescue the mild social deficits and anxiety of ARID1B$^{+/-}$ mice, suggesting that manipulation of GABA modulation pharmacologically may be able to rescue some of the social and/or cognitive deficits seen in ARID1B mutations (Jung et al., 2017). Of importance to note, of the studies cited here, no mice were found to have spontaneous seizures and seizures were not assessed by electroencephalogram.

7.4.3 CACNA1C

CACNA1C encodes voltage-gated L-type Ca2+ channel (Cav1.2). Mutations in this gene result in Timothy Syndrome (TS). Although the primary organ involved in TS in humans is the heart, nervous system involvement has also been observed, including seizures co-occurring with this syndromic form of ASD (Bett et al., 2012). It is suggested that since Cav1.2 plays a crucial role in electrical coupling and long-term potentiation in the nervous system, alteration in its function may lead to impairment in the development of neurons (Kessi et al., 2021; Krey et al., 2013). Additional studies have found that CACNA1C is responsible for the survival of neurons in the hippocampal dentate gyrus (Temme et al., 2016). While there have been reports of cases of epilepsy occurring in TS in humans, definitive correlation has not been established (Bozarth et al., 2018).

Seizures have not been specifically studied in mouse models of TS. TS mice do present with gross brain abnormalities, repetitive behaviors, altered ultrasonic vocalization, and altered social behavior in mice (Bader et al., 2011). Heterozygous TS-neo male mouse models have shown that the gain-of-function point mutation involved in the more severe form of the Timothy syndrome results in a marked increase in resting calcium influx, which ultimately leads to abnormal burst firing (Calorio et al., 2019).

7.4.4 CNTNAP2

Contactin- associated protein-like 2 (Caspr2) is a neurexin superfamily protein (Poliak et al., 1999) that binds to TAG-1 and is essential in the positioning of K+ channels at the axonal membrane of myelinated axons at the juxtaparanodal region (Poliak et al., 2003). Caspr2 has an important in the organization of ion channels; disruption of Caspr2 results in marked reduction of K+ channel accumulation at the juxtaparanodes of axons of the peripheral and central nervous systems, with redistribution of the K+ channels along the intranodal region (Poliak et al., 2003).

Homozygous mutations of Cntnap2, the gene encoding Caspr2, has been identified as the cause of cortical dysplasia, focal epilepsy (CDFE), evidence that Caspr2 affects brain development. Of the 9 children studied (ages 2–9), all had seizures, specifically complex partial (100%), secondary generalized (44%), simple partial (33%) and status epilepticus (33%) in etiology. Age of seizure onset ranged from 14 to 20 months with a median of 16 months; after seizure onset, there was deterioration in learning ability and social behavior (Strauss et al., 2006). 67% of these children also had a pervasive developmental delay or ASD (Strauss et al., 2006). Other studies of individuals with homozygous Cntnap2 mutations (Rodenas-Cuadrado et al., 2016) also found similar and slightly older ages of seizure onset (18 months and 36 months in two siblings), and similar to Strauss et. al's study, both siblings experience complex partial seizures. Epileptic activity precedes the onset of the ASD features in CDFE, making it a good candidate to study the contribution of epileptic activity to the development of ASD features.

Cntnap2$^{-/-}$ mice have been found to reflect deficits in humans with CDFE. One study of a Cntnap2$^{-/-}$ mouse modes by Peñagarikano et al., found a reduction in GABAergic interneurons, defects in migration of cortical projection neurons, reduced cortical-neuronal synchrony, hyperactivity, and spontaneous seizures induced by mild stressors during routine handling in mice over 6 months of age (Peñagarikano et al., 2011); behavioral, neuropathological, physiological, and behavioral characterization of the Cntnap2 mutants was performed prior to seizure onset. Reactive astrocytosis was observed throughout the hippocampus after seizure onset, but not before (Peñagarikano et al., 2011). Another study of male Cntnap2$^{-/-}$ mice also found that there was no spontaneous onset of seizures in mice of 12–16 weeks of age, consistent with the findings of Peñagarikano et al., that seizures started after 6 months of age (Liska et al., 2018). Of note, the heterozygous animals in this study did not display behavioral or neuropathological abnormalities or epileptic seizures present in the homozygotes (Peñagarikano et al., 2011), which as the author notes, is consistent with the recessive nature of CDFE in humans. In EEG studies of Cntnap 2$^{-/-}$ male mice, one study found absence seizure-like events, hypoactivity, and wake fragmentation (Thomas et al., 2017). No spontaneous overt seizures were noted in mice 1.5–8 months of age, however there was an increase in frequency of seizure-like spiking events during 24-hour EEG recordings done at 4 and 8 months (Thomas et al., 2017); on average these events were common in the transition between dark/light, and about 1–2 occurred per hour. In contrast to the human homozygous mutations of Cntnap2, where the epileptic activity precedes the onset of ASD, in these mice the epileptic activity appeared after the behavioral features of ASD were present.

7.4.5 CUL3

The Cul3 gene encodes for the cullin-3 protein which plays a role in the ubiquitin-proteasome system, aiding in the disposal of unwanted proteins in the cell (Ohta et al., 2013). Mutations of Cul3 have been found in ASD and Gordon's hypertension syndrome, also known as pseudohypoaldosteronism type 2 (Amar et al., 2021; Ohta et al., 2013). In reports of human Cul3 mutations, seizures have been documented with an unknown prevalence rate (Iwafuchi et al., 2021; Nakashima et al., 2020).

Of importance to note, homozygous mutations of Cul3 is embryonically lethal in mice, and as such, only heterozygous or floxed Cul3 mice are studied (Amar et al., 2021; McEvoy et al., 2007). Of the review of the 62 studies that cited the Jackson Laboratory's Cul3 floxed mice, no studies mentioned the prevalence of seizures and/or epilepsy in their mice.

7.4.6 *DYRK1A*

DYRK1A gene encodes the dual-specificity tyrosine phosphorylation-regulated kinase 1A. In mice, a frame-shift mutation results in a protein truncation and elimination of its kinase activity site (Raveau et al., 2018). DYRK1A plays a role in controlling brain growth through neuronal proliferation and neurogenesis (Courcet et al., 2012). Mutations affecting DYRK1A gene are frequently associated with neurodevelopmental disorders, such as ASD associated with microcephaly, febrile seizures, cognitive flexibility deficits and severe speech acquisition delay. It is believed that both excitatory and inhibitory imbalance contributes to the seizure activity, ASD behaviors and cognitive deficits associated with this mutation (Arranz et al., 2019).

In humans, DYRK1A mutation is characterized by a speech acquisition delay, which can correlate to the observed communication deficit in juvenile DYRK1A male mice. The 6 day old pups produced fewer and simpler vocalizations pointing to an early onset of autistic traits. A social behavior study with adult DYRK1A mice model showed a decreased sociability and escaped confrontations. However, no sign of repetitive behavior was noted (Raveau et al., 2018). In contrast, another study found that DYRK1A mice spend significantly longer on repetitive self-grooming, therefore characterizing the model with repetitive and stereotyped behaviors (Arranz et al., 2019).

Patients with DYRK1A mutation are more likely to develop febrile seizures in infancy that can evolve to other types of epileptic seizures later in life. An index case reported epileptic seizure development at 1 year of age and requiring valproate treatment. The presence of the disease was characterized by atonic and myoclonic seizures. EEG did not reveal a specific pattern other than theta slow waves bursts (Courcet et al., 2012). Another study found 9/14 individuals with DYRK1A mutation to have reported febrile seizures at 6–18 months which then became generalized tonic-clonic (Ji et al., 2015). However, the role of temperature elevation in DYRK1A mice revealed susceptibility to hyperthermia-induced seizures (Raveau et al., 2018). The epileptic activity was also evaluated with vEGG and found spontaneous myoclonic jerks corresponding to interictal epileptiform activity. EGG showed isolated spikes, polyspikes and spike-wave discharges. Four out of eight DYRK1A mice also showed spontaneous generalized tonic-clonic seizures (Arranz et al., 2019). Of note, DYRK1A mice are sensitive to convulsive agent pentylenetetrazol and exhibit anxiety-like behavior (Arranz et al., 2019).

The DYRK1A mouse model is useful for further investigations of the features and mechanism of ASD as well as seizures. Temperature elevation is a natural way to trigger seizures, which might simplify electrophysiological measurements of seizure activity.

7.4.7 *GTF2I*

The GTF2I gene encodes for 2 proteins: TGFII-I, a generalized transcription factor (Barak et al., 2019), and BAP-135, a protein phosphorylated by Bruton's tyrosine kinase in b-cells (Pérez Jurado et al., 1998). Duplication of the region 7q11.23 containing GTF2I is known to cause significant anxiety disorders (Mervis et al., 2012) and ASD (Mervis et al., 2021). Humans with 7q11.23 duplication syndrome often have neurological disorders, including seizures. In a study of 53 patients, 10 had seizure disorders, making the prevalence of seizures in 7q11.23 duplication roughly 19% (Morris et al., 2015). However, of the 89 articles that cited the Jackson Laboratory's GTF2I duplication mice, no studies mentioned the prevalence of seizures and/or epilepsy in their mice.

7.4.8 *RAI1*

The RAI1 gene encodes for the Retinoic Acid Induced 1 protein, which is on the nucleus of many cells throughout the body with transcription factor activity (Carmona-Mora et al., 2010; Huang et al., 2016). Mutation of the RAI1 gene has been linked to Smith-Magenis Syndrome (SMS), which often presents with ASD, dysmorphic features, and other neurobehavioral alterations (Bi et al., 2007; Carmona-Mora et al., 2010; Elsea & Girirajan, 2008), with seizures reported in up to 30% of patients with SMS (Elsea & Girirajan, 2008). Mouse models with mutations in RAI1 tested at approximately 3 months of age are documented to have neurobehavioral deficits characteristic of SMS including hypoactivity, cognitive deficits, and autistic features. One-third of the homozygous $Rai^{-/-}$ mice developed overt seizures after 3 months of age. In contrast, only about 2% of $Rai^{+/-}$ mice showed overt seizures after 4 months of age (Bi et al., 2007). The mice were monitored during periods of wake and sleep. Their EEG showed generalized spikes, multiple spikes and sharp waves. Mice also presented with clonic-tonic-clonic seizures that lasted for 30 s, accompanied by EEG abnormalities. The EEG abnormalities and the ASD features make the Rai−/− mice an interesting model to study the relationship between epilepsy and ASD. Whether the onset of the seizures precede the behavioral onset of ASD-like traits remains to be explored.

7.4.9 *SHANK3*

The Shank3 gene codes for a protein present on the post-synaptic membrane of glutaminergic neurons. When mutated, it has been implicated as the most common causative agent of Phelan-McDermid Syndrome (PMS) (Monteiro & Feng, 2017; Phelan et al., 2022). Shank3-related PMS presents with variable phenotypes including, intellectual disability, delayed/absent speech, ASD/autistic-like features, and seizures/epilepsy (Boccuto et al., 2022; Phelan et al., 2022).

Studies of persons with PMS have found seizures of variable forms and complexities, with one of the most common types being absence seizures. Holder and Quach (2016) found that that of 11 of their 24 patients with Shank3 mutations had at least one lifetime seizure, with 10 (90%) having absence seizures(Holder & Quach, 2016). Another study of 16 patients found that 75% of their cohort had abnormal epileptiform discharges when recorded via video-electroencephalogram, of which 7 of these patients already had diagnosed epilepsy (Khan et al., 2018). Despite being the most widely used mouse model, none of the studies that utilized the Shank3 deletion mouse model found spontaneous seizures (Dhamne et al., 2017; Kabitzke et al., 2018). However, the detection of spontaneous seizures in these mice, in particular if they are absence type, might require electroencephalograms (EEGs). Furthermore, absence seizures in mice are very difficult to detect and score (Marshall et al., 2021), making it possible that seizures of this nature have been missed in this mouse population.

7.4.10 SYNGAP1

Synaptic GTPase activating protein 1, encoded by the gene SYNGAP1, is part of the NMDA receptor complex found on postsynaptic excitatory glutamatergic neurons. It acts as a negative regulator of small GTPases, such as Ras and Rap, to regulate AMPA receptor activity. This function is quintessential for proper synaptic development, structure, function, and plasticity (Agarwal et al., 2019).

Mutations within the SYNGAP1 gene, all of which cause loss-of-function haploinsufficiency, are commonly associated with a phenotype referred to as Mental Retardation Type 5 (MDR5). MDR5 presents with moderate-severe intellectual disability (ID) including language and motor delays (~100% of patients), ASD(~50%), severe behavioral problems such as aggression, hyperexcitability, opposition, tantrums, and self-injury (~75%), sleeping difficulties (~60%), and eating difficulties (Agarwal et al., 2019). SYNGAP1 has thus become a common target of genetic manipulation in mice for the study of ASD.

Over 80% of patients with SYNGAP1 mutations also present with generalized epilepsy. Developmental and epileptic encephalopathy (DEE) or developmental encephalopathy with or without epilepsy may also arise (Agarwal et al., 2019). The types of seizures that present amongst patients with SYNGAP1 mutations are highly variable and most patients experience more than one type. Atypical absence seizures have been shown to be the most common (Jimenez-Gomez et al., 2019; Lo Barco et al., 2021), but typical absence, generalized tonic-clonic, atonic, drop attack, eyelid myoclonia (EM), and myoclonic jerks have also been observed (Jimenez-Gomez et al., 2019; lo Barco et al., 2021; Mignot et al., 2016; Parker et al., 2015; Vlaskamp et al., 2019; von Stülpnagel et al., 2015). Age of seizure onset is also highly variable, ranging from as early as 2.5 months (von Stülpnagel et al., 2015) to 7 years old (Vlaskamp et al., 2019) with an average onset at ~35–38 months (Jimenez-Gomez et al., 2019; Mignot et al., 2016). Seizures typically present after development of intellectual delay (Mignot et al., 2016; Vlaskamp et al., 2019), however current data on this temporal relationship is limited.

Given the dual presentation of ASD and epilepsy with SYNGAP1 mutations in humans, it has also become a useful mouse model to study the relationship between ASD and epilepsy. Of the 294 articles that cited the Jackson Laboratory's SYNGAP1 conditional deletion and conditional rescue mice, two studies from the same group mentioned the presence of recurrent spontaneous seizures in both juvenile (P7) and adult (P60, P120) SYNGAP1$^{+/-}$ mice during continuous vEEG (Sullivan et al., n.d., 2020). Although the juvenile mice had a greater number of flexor spasms and jerks as compared to wild-type, >50% of the seizures were associated with a motor event indicating that most seizures were identified by abnormal EEG patterns. All of the adult SYNGAP1$^{+/-}$ mice studied had myoclonic seizures by P60, 80% of which began at transitions from NREM to wake. The myoclonic seizures were on average 34.4 s long with 3-4 Hz rhythmic spike-wave discharges. At P120, myoclonic and generalized tonic-clonic seizures were observed during NREM, and electrographic seizures were observed during wake.

An EEG/video monitoring study showed widespread high-amplitude cortical discharge in SYNGAP1$^{+/-}$ mice >12 weeks old as compared to wild-type (Ozkan et al., 2014). These epileptiform discharges, which did not coincide with motor events, were intermittent and had a frequency ranging from 1 to 681/h, with an average of 81/h. Voltage imaging, mEPSC, and mIPSC measurements of brain slices from adult SYNGAP1$^{+/-}$ mice done in the same study indicated an increase in amplitude and frequency of excitatory signals and decrease in amplitude of inhibitory signals in mPFC cortical neurons. These findings reflect an elevated excitation/inhibition (E/I) ratio indicative of hyperexcitability. Normal seizure threshold was reestablished after global reversal of SYNGAP1 haploinsufficiency and reversal in only forebrain glutamatergic neurons was sufficient enough to protect from the development of behavioral impairment such as altered anxiety, risk-taking, working memory, and contextual memory (Ozkan et al., 2014). Similarly, the re-expression of SYNGAP protein in adulthood significantly increased latency to seizure induction by flurothyl, eliminated interictal spiking events during wakefulness and sleep on EEG electrode recordings, and normalized elevated activity suppression ratio (ASR) levels (Creson et al., 2019). These results indicate a protective characteristic of SYNGAP, both on seizure threshold as well as against cognitive impairment such as memory deficits. Enhanced expression of the SYNGAPα isoform also protected against seizures and improved behavioral measures such as learning during spatial navigation (Kilinc et al., 2022). The high penetrance of seizures combined with ASD-like behavioral features make this an important mouse model. The early onset of the seizures makes it an attractive model to explore the contribution of early epileptic activity to the development of ASD traits.

7.4.11 UBE3A

The UBE3A gene encodes for E3 ligase E6-associated protein (E6AP): a substrate-specific ubiquitin-protein ligase which attaches activated ubiquitin to targeted proteins, subsequently leading to their proteasome-mediated degradation. UBE3A expression is based on parental imprinting within the chromosome region 15q11-q13: maternal copy is expressed while paternal copy is silenced in neurons (Khatri & Man, 2019).

Mutations within the UBE3A gene, most commonly deletion of the maternal chromosome region 15q11-q13, causes Angelman Syndrome (AS): a neurodevelopmental disorder characterized by severe intellectual disability, speech impairment, a distinct behavioral phenotype (happy demeanor, frequent and easily provoked laughter/smiling), and autistic features (Margolis et al., 2015). Similar phenotypes, presenting as alterations in rotarod performance, marble burying, open field exploration, and nest building, were observed in mice whose expression of UBE3A was genetically reduced. When expression of UBE3A was reinstated at 3 weeks, 6 weeks, and 14 weeks old, behavioral testing performance returned to levels observed in wild-type mice (Silva-Santos et al., 2015). Overexpression of UBE3A, known as Maternal Dup15q Syndrome, is associated with ASD (~50–80% of patients), intellectual disability, hypotonia, and minor dysmorphic features (Lusk et al. 2016). Due to its significant association with ASD when both reduced or overexpressed, the UBE3A gene has become a commonly used mouse model in ASD research.

Both AS and Maternal Dup15q Syndrome also present with frequent seizures and are associated with epilepsy (Margolis et al., 2015; Lusk et al., 2016). It has been observed that patients with Maternal Dup15q Syndrome have enhanced beta power and reduced delta power on EEG as compared to patients diagnosed with nonsyndromic ASD and those considered typically developed (Frohlich et al., 2016).

Of the 236 articles that cited the Jackson Laboratory's UBE3A overexpression mice, no studies mentioned the prevalence of spontaneous seizures and/or epilepsy in their mice. However, many of the studies established relationships between changes in UBE3A expression and neural activity which may be associated with a hyperexcitable state and thus seizures. For example, mice with an overexpression of UBE3A had a faster onset of PTZ-induced seizures as compared to wild-type indicating a reduced seizure threshold (Copping et al., 2017). UBE3A overexpression has been observed to cause dose-dependent downregulation of CBLN1, a glutamatergic synapse organizer associated with impaired sociability, compromising glutamatergic synaptic transmission. Similarly, PTZ-induced seizures reduced expression of CBLN1 (Krishnan et al., 2017). On the other hand, loss of UBE3A expression (Angelman Syndrome) is associated with audiogenic seizures in mice, which showed to be responsive to treatment with antiepileptic drugs but not postnatal reactivation of the gene (Silva-Santos et al., 2015).

7.5 Conclusions

ASD subjects present with epilepsy at a higher rate than the neurotypical population. EEG abnormalities, especially during sleep, seem to be a common factor in ASD subjects. Intriguingly, the presence of epilepsy in ASD subjects seems to dilute the protective effect of female sex, resulting in a more balanced ratio of male/female ratio for ASD and epilepsy. For a certain fraction of cases, early epileptic activity precedes the onset of ASD, suggesting a causative role. However, randomized clinical trials using antiepileptic drugs have not shown an improvement on ASD traits.

We have identified 4 mice models of syndromic forms of ASD (CNTNAP2, DYRK1A, RAI1 floxed, and SYNGAP1) that present with spontaneous seizures. The SYNGAP1 mouse is particularly interesting for studying the contribution of early seizures for the development of ASD traits. In the other models, the seizures appear after the onset of the behavioral deficits, making them interesting models for exploring the hypothesis of a common trait leading to both epilepsy and ASD.

Of note, sex differences in the presence of seizures, or seizure susceptibility, in these mouse models has not been explored. Although these models show clearly observable seizures, other mouse models might still present with seizures. Seizures during sleep might be hard to detect, requiring continuous EEG recording. Absence seizures, in particular, might also be hard to detect without continuous electrophysiological monitoring.

The use of mouse models of ASD that present with seizures might facilitate measurements and manipulation of specific circuit elements which might help elucidate the pathophysiology of ASD and the commonalities that ASD share with epilepsy.

References

Agarwal, M., Johnston, M. V., & Stafstrom, C. E. (2019). SYNGAP1 mutations: Clinical, genetic, and pathophysiological features. *International Journal of Developmental Neuroscience, 78*, 65–76. https://doi.org/10.1016/j.ijdevneu.2019.08.003. Elsevier Ltd.

Amar, M., Pramod, A. B., Yu, N. K., Herrera, V. M., Qiu, L. R., Moran-Losada, P., Zhang, P., Trujillo, C. A., Ellegood, J., Urresti, J., Chau, K., Diedrich, J., Chen, J., Gutierrez, J., Sebat, J., Ramanathan, D., Lerch, J. P., Yates, J. R., Muotri, A. R., & Iakoucheva, L. M. (2021). Autism-linked Cullin3 germline haploinsufficiency impacts cytoskeletal dynamics and cortical neurogenesis through RhoA signaling. *Molecular Psychiatry, 26*(7), 3586–3613. https://doi.org/10.1038/s41380-021-01052-x

American Psychiatric Association. (2022). *Desk reference to the diagnostic criteria from DSM-5-TR*. American Psychiatric Association Publishing.

Amiet, C., Gourfinkel-An, I., Bouzamondo, A., Tordjman, S., Baulac, M., Lechat, P., Mottron, L., & Cohen, D. (2008). Epilepsy in autism is associated with intellectual disability and gender: Evidence from a meta-analysis. *Biological Psychiatry, 64*(7), 577–582. https://doi.org/10.1016/j.biopsych.2008.04.030

Antoine, M. W., Langberg, T., Schnepel, P., & Feldman, D. E. (2019). Increased excitation-inhibition ratio stabilizes synapse and circuit excitability in four autism mouse models. *Neuron, 101*(4), 648–661.e4. https://doi.org/10.1016/j.neuron.2018.12.026

Arranz, J., Balducci, E., Arató, K., Sánchez-Elexpuru, G., Najas, S., Parras, A., Rebollo, E., Pijuan, I., Erb, I., Verde, G., Sahun, I., Barallobre, M. J., Lucas, J. J., Sánchez, M. P., de la Luna, S., & Arbonés, M. L. (2019). Impaired development of neocortical circuits contributes to the neurological alterations in DYRK1A haploinsufficiency syndrome. *Neurobiology of Disease, 127*, 210–222. https://doi.org/10.1016/j.nbd.2019.02.022

Bader, P. L., Faizi, M., Kim, L. H., Owen, S. F., Tadross, M. R., Alfa, R. W., Bett, G. C. L., Tsien, R. W., Rasmusson, R. L., & Shamloo, M. (2011). Mouse model of Timothy syndrome recapitulates triad of autistic traits. *Proceedings of the National Academy of Sciences, 108*(37), 15432–15437. https://doi.org/10.1073/pnas.1112667108

Barak, B., Zhang, Z., Liu, Y., Nir, A., Trangle, S. S., Ennis, M., Levandowski, K. M., Wang, D., Quast, K., Boulting, G. L., Li, Y., Bayarsaihan, D., He, Z., & Feng, G. (2019). Neuronal deletion of Gtf2i, associated with Williams syndrome, causes behavioral and myelin alterations rescuable by a remyelinating drug. *Nature Neuroscience, 22*(5), 700–708. https://doi.org/10.1038/s41593-019-0380-9

Bett, G. C. L., Lis, A., Wersinger, S. R., Baizer, J. S., Duffey, M. E., & Rasmusson, R. L. (2012). A mouse model of Timothy syndrome: A complex autistic disorder resulting from a point mutation in Cav1.2. *American Chinese Journal of Medicine and Science, 5*(3), 135. https://doi.org/10.7156/najms.2012.053135

Bi, W., Yan, J., Shi, X., Yuva-Paylor, L. A., Antalffy, B. A., Goldman, A., Yoo, J. W., Noebels, J. L., Armstrong, D. L., Paylor, R., & Lupski, J. R. (2007). Rai1 deficiency in mice causes learning impairment and motor dysfunction, whereas Rai1 heterozygous mice display minimal behavioral phenotypes. *Human Molecular Genetics, 16*(15), 1802–1813. https://doi.org/10.1093/HMG/DDM128

Boccuto, L., Mitz, A., Abenavoli, L., Sarasua, S. M., Bennett, W., Rogers, C., DuPont, B., & Phelan, K. (2022). Phenotypic variability in Phelan–McDermid syndrome and its putative link to environmental factors. *Genes, 13*(3), 528. https://doi.org/10.3390/genes13030528

Bozarth, X., Dines, J. N., Cong, Q., Mirzaa, G. M., Foss, K., Lawrence Merritt, J., Thies, J., Mefford, H. C., & Novotny, E. (2018). Expanding clinical phenotype in *CACNA1C* related disorders: From neonatal onset severe epileptic encephalopathy to late-onset epilepsy. *American Journal of Medical Genetics Part A, 176*(12), 2733–2739. https://doi.org/10.1002/ajmg.a.40657

Calorio, C., Gavello, D., Guarina, L., Salio, C., Sassoè-Pognetto, M., Riganti, C., Bianchi, F. T., Hofer, N. T., Tuluc, P., Obermair, G. J., Defilippi, P., Balzac, F., Turco, E., Bett, G. C., Rasmusson, R. L., & Carbone, E. (2019). Impaired chromaffin cell excitability and exocytosis in autistic Timothy syndrome TS2-neo mouse rescued by L-type calcium channel blockers. *The Journal of Physiology, 597*(6), 1705–1733. https://doi.org/10.1113/JP277487

Canitano, R. (2007). Epilepsy in autism spectrum disorders. *European Child & Adolescent Psychiatry, 16*(1), 61–66. https://doi.org/10.1007/s00787-006-0563-2

Carmona-Mora, P., Encina, C. A., Canales, C. P., Cao, L., Molina, J., Kairath, P., Young, J. I., & Walz, K. (2010). Functional and cellular characterization of human Retinoic Acid Induced 1 (RAI1) mutations associated with Smith-Magenis syndrome. *BMC Molecular Biology, 11*, 63. https://doi.org/10.1186/1471-2199-11-63

Copping, N. A., Christian, S. G. B., Ritter, D. J., Islam, M. S., Buscher, N., Zolkowska, D., Pride, M. C., Berg, E. L., LaSalle, J. M., Ellegood, J., Lerch, J. P., Reiter, L. T., Silverman, J. L., & Dindot, S. V. (2017). Neuronal overexpression of Ube3a isoform 2 causes behavioral impairments and neuroanatomical pathology relevant to 15q11.2-q13.3 duplication syndrome. *Human Molecular Genetics, 26*(20), 3995–4010. https://doi.org/10.1093/hmg/ddx289

Courcet, J.-B., Faivre, L., Malzac, P., Masurel-Paulet, A., Lopez, E., Callier, P., Lambert, L., Lemesle, M., Thevenon, J., Gigot, N., Duplomb, L., Ragon, C., Marle, N., Mosca-Boidron, A.-L., Huet, F., Philippe, C., Moncla, A., & Thauvin-Robinet, C. (2012). The DYRK1A gene is a cause of syndromic intellectual disability with severe microcephaly and epilepsy. *Journal of Medical Genetics, 49*(12), 731–736. https://doi.org/10.1136/jmedgenet-2012-101251

Creson, T. K., Rojas, C., Hwaun, E., Vaissiere, T., Kilinc, M., Jimenez-Gomez, A., Holder, J. L., Tang, J., Colgin, L. L., Miller, C. A., & Rumbaugh, G. (2019). Re-expression of SynGAP protein in adulthood improves translatable measures of brain function and behavior. https://doi.org/10.7554/eLife.46752.001

Dhamne, S. C., Silverman, J. L., Super, C. E., Lammers, S. H. T., Hameed, M. Q., Modi, M. E., Copping, N. A., Pride, M. C., Smith, D. G., Rotenberg, A., Crawley, J. N., & Sahin, M. (2017). Replicable in vivo physiological and behavioral phenotypes of the Shank3B null mutant mouse model of autism. *Molecular Autism, 8*(1), 26. https://doi.org/10.1186/S13229-017-0142-Z

Egolf, L. E., Vaksman, Z., Lopez, G., Rokita, J. L., Modi, A., Basta, P. V., Hakonarson, H., Olshan, A. F., & Diskin, S. J. (2019). Germline 16p11.2 microdeletion predisposes to neuroblastoma. *American Journal of Human Genetics, 105*(3), 658–668. https://doi.org/10.1016/j.ajhg.2019.07.020

Ellegood, J., Petkova, S. P., Kinman, A., Qiu, L. R., Adhikari, A., Wade, A. A., Fernandes, D., Lindenmaier, Z., Creighton, A., Nutter, L. M. J., Nord, A. S., Silverman, J. L., & Lerch, J. P. (2021). Neuroanatomy and behavior in mice with a haploinsufficiency of AT-rich interactive domain 1B (ARID1B) throughout development. *Molecular Autism, 12*(1), 1–24. https://doi.org/10.1186/S13229-021-00432-Y/FIGURES/8

Elsea, S. H., & Girirajan, S. S. (2008). Smith–Magenis syndrome. *European Journal of Human Genetics, 16*(4), 412–421. https://doi.org/10.1038/sj.ejhg.5202009

Fernandez, B. A., & Scherer, S. W. (2017). *Syndromic autism spectrum disorders: Moving from a clinically defined to a molecularly defined approach*. www.dialogues-cns.org

Ferri, S. L., Abel, T., & Brodkin, E. S. (2018). Sex differences in autism spectrum disorder: A review. *Current Psychiatry Reports, 20*(2), 1–17. https://doi.org/10.1007/S11920-018-0874-2

Fiest, K. M., Sauro, K. M., Wiebe, S., Patten, S. B., Kwon, C. S., Dykeman, J., Pringsheim, T., Lorenzetti, D. L., & Jetté, N. (2017). Prevalence and incidence of epilepsy. *Neurology, 88*(3), 296–303. https://doi.org/10.1212/WNL.0000000000003509

Frohlich, J., Senturk, D., Saravanapandian, V., Golshani, P., Reiter, L. T., Sankar, R., Thibert, R L., DiStefano, C., Huberty, S., Cook, E. H., Jeste, S. S. (2016). A quantitative electrophysiological biomarker of duplication 15q11.2-q13.1 syndrome. *PLOS ONE, 11*(12), e0167179. https://doi.org/10.1371/journal.pone.0167179

Giarelli, E., Wiggins, L. D., Rice, C. E., Levy, S. E., Kirby, R. S., Pinto-Martin, J., & Mandell, D. (2010). Sex differences in the evaluation and diagnosis of autism spectrum disorders among children. *Disability and Health Journal, 3*(2), 107–116. https://doi.org/10.1016/J.DHJO.2009.07.001

Holder, J. L., & Quach, M. M. (2016). The spectrum of epilepsy and electroencephalographic abnormalities due to SHANK3 loss-of-function mutations. *Epilepsia, 57*(10), 1651–1659. https://doi.org/10.1111/EPI.13506

Horev, G., Ellegood, J., Lerch, J. P., Son, Y.-E. E., Muthuswamy, L., Vogel, H., Krieger, A. M., Buja, A., Henkelman, R. M., Wigler, M., & Mills, A. A. (2011). Dosage-dependent phenotypes in models of 16p11.2 lesions found in autism. *Proceedings of the National Academy of Sciences of the United States of America, 108*(41), 17076–17081. https://doi.org/10.1073/pnas.1114042108

Hoyer, J., Ekici, A. B., Endele, S., Popp, B., Zweier, C., Wiesener, A., Wohlleber, E., Dufke, A., Rossier, E., Petsch, C., Zweier, M., Göhring, I., Zink, A. M., Rappold, G., Schröck, E., Wieczorek, D., Riess, O., Engels, H., Rauch, A., & Reis, A. (2012). Haploinsufficiency of ARID1B, a member of the SWI/SNF-A chromatin-remodeling complex, is a frequent cause of intellectual disability. *American Journal of Human Genetics, 90*(3), 565–572. https://doi.org/10.1016/j.ajhg.2012.02.007

Huang, W. H., Guenthner, C. J., Xu, J., Nguyen, T., Schwarz, L. A., Wilkinson, A. W., Gozani, O., Chang, H. Y., Shamloo, M., & Luo, L. (2016). Molecular and neural functions of Rai1, the causal gene for Smith-Magenis syndrome. *Neuron, 92*(2), 392–406. https://doi.org/10.1016/J.NEURON.2016.09.019

Hyman, S. L., Levy, S. E., Myers, S. M., Kuo, D. Z., Apkon, S., Davidson, L. F., Ellerbeck, K. A., Foster, J. E. A., Noritz, G. H., Leppert, M. O., Saunders, B. S., Stille, C., Yin, L., Weitzman, C. C., Childers, D. O., Levine, J. M., Peralta-Carcelen, A. M., Poon, J. K., Smith, P. J., et al.

(2020). Identification, evaluation, and management of children with autism spectrum disorder. *Pediatrics, 145*(1). https://doi.org/10.1542/peds.2019-3447

Iwafuchi, S., Kikuchi, A., Endo, W., Inui, T., Aihara, Y., Satou, K., Kaname, T., & Kure, S. (2021). A novel stop-gain CUL3 mutation in a Japanese patient with autism spectrum disorder. *Brain and Development, 43*(2), 303–307. https://doi.org/10.1016/J.BRAINDEV.2020.09.015

Iwamoto, O., Koga, C., Matsumoto, H., Terasaki, S., Kusukawa, J., & Kameyama, T. (2021). Coffin-Siris syndrome. *Asian Journal of Oral and Maxillofacial Surgery, 15*(3), 205–207. https://doi.org/10.1016/S0915-6992(03)80044-4

Ji, J., Lee, H., Argiropoulos, B., Dorrani, N., Mann, J., Martinez-Agosto, J. A., Gomez-Ospina, N., Gallant, N., Bernstein, J. A., Hudgins, L., Slattery, L., Isidor, B., le Caignec, C., David, A., Obersztyn, E., Wiśniowiecka-Kowalnik, B., Fox, M., Deignan, J. L., Vilain, E., et al. (2015). DYRK1A haploinsufficiency causes a new recognizable syndrome with microcephaly, intellectual disability, speech impairment, and distinct facies. *European Journal of Human Genetics: EJHG, 23*(11), 1473–1481. https://doi.org/10.1038/ejhg.2015.71

Jimenez-Gomez, A., Niu, S., Andujar-Perez, F., McQuade, E. A., Balasa, A., Huss, D., Coorg, R., Quach, M., Vinson, S., Risen, S., & Holder, J. L. (2019). Phenotypic characterization of individuals with SYNGAP1 pathogenic variants reveals a potential correlation between posterior dominant rhythm and developmental progression. *Journal of Neurodevelopmental Disorders, 11*(1), 18. https://doi.org/10.1186/s11689-019-9276-y

Jokiranta, E., Sourander, A., Suominen, A., Timonen-Soivio, L., Brown, A. S., & Sillanpää, M. (2014). Epilepsy among children and adolescents with autism spectrum disorders: A population-based study. *Journal of Autism and Developmental Disorders, 44*(10), 2547–2557. https://doi.org/10.1007/s10803-014-2126-6

Jung, E. M., Moffat, J. J., Liu, J., Dravid, S. M., Gurumurthy, C. B., & Kim, W. Y. (2017). Arid1b haploinsufficiency disrupts cortical interneuron development and mouse behavior. *Nature Neuroscience, 20*(12), 1694–1707. https://doi.org/10.1038/s41593-017-0013-0

Kabitzke, P. A., Brunner, D., He, D., Fazio, P. A., Cox, K., Sutphen, J., Thiede, L., Sabath, E., Hanania, T., Alexandrov, V., Rasmusson, R., Spooren, W., Ghosh, A., Feliciano, P., Biemans, B., Benedetti, M., & Clayton, A. L. (2018). Comprehensive analysis of two Shank3 and the Cacna1c mouse models of autism spectrum disorder. *Genes, Brain and Behavior, 17*(1), 4–22. https://doi.org/10.1111/GBB.12405

Kessi, M., Chen, B., Peng, J., Yan, F., Yang, L., & Yin, F. (2021). Calcium channelopathies and intellectual disability: A systematic review. *Orphanet Journal of Rare Diseases, 16*(1), 219. https://doi.org/10.1186/s13023-021-01850-0

Khan, O. I., Zhou, X., Leon, J., Kessler, R., Gaughan, T., D'Souza, P., Gropman, A., Cohen, N., Rennert, O., Buckley, A., Inati, S., & Thurm, A. (2018). Prospective longitudinal overnight video-EEG evaluation in Phelan–McDermid syndrome. *Epilepsy & Behavior, 80*, 312–320. https://doi.org/10.1016/J.YEBEH.2017.11.034

Khatri, N., & Man, H.-Y. (2019). The Autism and Angelman syndrome protein Ube3A/E6AP: The gene, E3 ligase Ubiquitination Targets and Neurobiological functions. *Frontiers in Molecular Neuroscience, 12*, 109.

Kilinc, M., Arora, V., Creson, T. K., Rojas, C., Le, A. A., Lauterborn, J., Wilkinson, B., Hartel, N., Graham, N., Reich, A., Gou, G., Araki, Y., Bayés, À., Coba, M., Lynch, G., Miller, C. A., & Rumbaugh, G. (2022). Endogenous Syngap1 alpha splice forms promote cognitive function and seizure protection. *eLife, 11*. https://doi.org/10.7554/eLife.75707

Krey, J. F., Paşca, S. P., Shcheglovitov, A., Yazawa, M., Schwemberger, R., Rasmusson, R., & Dolmetsch, R. E. (2013). Timothy syndrome is associated with activity-dependent dendritic retraction in rodent and human neurons. *Nature Neuroscience, 16*(2), 201–209. https://doi.org/10.1038/nn.3307

Krishnan, V., Stoppel, D. C., Nong, Y., Johnson, M. A., Nadler, M. J. S., Ozkaynak, E., Teng, B. L., Nagakura, I., Mohammad, F., Silva, M. A., Peterson, S., Cruz, T. J., Kasper, E. M., Arnaout, R., & Anderson, M. P. (2017). Autism gene Ube3a and seizures impair sociability by repressing VTA Cbln1. *Nature, 543*(7646), 507–512. https://doi.org/10.1038/nature21678

Limbu, B., Deb, S., Roy, M., Lee, R., Roy, A., & Taiwo, O. (2022). Randomised controlled trials of mood stabilisers for people with autism spectrum disorder: Systematic review and meta-analysis. *BJPsych Open, 8*(2), e52. https://doi.org/10.1192/bjo.2022.18

Liska, A., Bertero, A., Gomolka, R., Sabbioni, M., Galbusera, A., Barsotti, N., Panzeri, S., Scattoni, M. L., Pasqualetti, M., & Gozzi, A. (2018). Homozygous loss of autism-risk gene CNTNAP2 results in reduced local and long-range prefrontal functional connectivity. *Cerebral Cortex, 28*(4), 1141–1153. https://doi.org/10.1093/cercor/bhx022

Lo Barco, T., Kaminska, A., Solazzi, R., Cancés, C., Barcia, G., Chemaly, N., Fontana, E., Desguerre, I., Canafoglia, L., Le Camus, C. H., Losito, E., Villard, L., Eisermann, M., Bernardina, B. D., Villeneuve, N., & Nabbout, R. (2021). SYNGAP1-DEE: A visual sensitive epilepsy. *Clinical Neurophysiology, 132*(4), 841–850.

Lusk, L., Vogel-Farley, V., DiStefano, C, Jeste S. Maternal 15q Duplication Syndrome. 2016 Jun 16 [updated 2021 Jul 15]. In: Adam MP, Feldman J, Mirzaa GM, Pagon RA, Wallace SE, Bean LJH, Gripp KW, Amemiya A, editors. GeneReviews® [Internet]. Seattle (WA), University of Washington, Seattle; 1993–2023. PMID: 27308687.

Lukmanji, S., Manji, S. A., Kadhim, S., Sauro, K. M., Wirrell, E. C., Kwon, C. S., & Jetté, N. (2019). The co-occurrence of epilepsy and autism: A systematic review. *Epilepsy & Behavior, 98*, 238–248. https://doi.org/10.1016/J.YEBEH.2019.07.037

Margolis, S. S. Sell, G. L., Zbinden, M. A., Bird, L. M. (2015). Angelman Syndrome Neurotherapeutics, 12(3), 641–650. https://doi.org/10.1007/s13311-015-0361-y

Marshall, G. F., Gonzalez-Sulser, A., & Abbott, C. M. (2021). Modelling epilepsy in the mouse: Challenges and solutions. *Disease Models & Mechanisms, 14*(3). https://doi.org/10.1242/DMM.047449

Matsuo, M., Maeda, T., Sasaki, K., Ishii, K., & Hamasaki, Y. (2010). Frequent association of autism spectrum disorder in patients with childhood onset epilepsy. *Brain and Development, 32*(9), 759–763. https://doi.org/10.1016/j.braindev.2010.05.005

McEvoy, J. D., Kossatz, U., Malek, N., & Singer, J. D. (2007). Constitutive turnover of cyclin E by Cul3 maintains quiescence. *Molecular and Cellular Biology, 27*(10), 3651–3666. https://doi.org/10.1128/MCB.00720-06/FORMAT/EPUB

Mervis, C. B., Dida, J., Lam, E., Crawford-Zelli, N. A., Young, E. J., Henderson, D. R., Onay, T., Morris, C. A., Woodruff-Borden, J., Yeomans, J., & Osborne, L. R. (2012). Duplication of GTF2I results in separation anxiety in mice and humans. *The American Journal of Human Genetics, 90*(6), 1064–1070. https://doi.org/10.1016/J.AJHG.2012.04.012

Mervis, C. B., Morris, C. A., Klein-Tasman, B. P., Velleman, S. L., & Osborne, L. R. (2021). 7q11.23 Duplication Syndrome. *GeneReviews®*. https://www.ncbi.nlm.nih.gov/books/NBK327268/

Mignot, C., von Stülpnage, C., Nava, C., Ville, D., Sanlaville, D., Lesca, G., Rastetter, A., Gachet, B., Marie, Y., Korenke, G. C., Borggraefe, I., Hoffmann-Zacharska, D., Szczepanik, E., Rudzka-Dybala, M., Yiş, U., Çağlayan, H., Isapof, A., Marey, I., Panagiotakaki, E., et al. (2016). Genetic and neurodevelopmental spectrum of SYNGAP1-associated intellectual disability and epilepsy. *Journal of Medical Genetics, 53*(8), 511–522. https://doi.org/10.1136/jmedgenet-2015-103451

Monteiro, P., & Feng, G. (2017). SHANK proteins: Roles at the synapse and in autism spectrum disorder. *Nature Reviews Neuroscience, 18*(3), 147–157. https://doi.org/10.1038/NRN.2016.183

Morris, C. A., Mervis, C. B., Paciorkowski, A. P., Abdul-Rahman, O., Dugan, S. L., Rope, A. F., Bader, P., Hendon, L. G., Velleman, S. L., Klein-Tasman, B. P., & Osborne, L. R. (2015). 7q11.23 duplication syndrome: Physical characteristics and natural history. *American Journal of Medical Genetics Part A, 167*(12), 2916–2935. https://doi.org/10.1002/AJMG.A.37340

Nakashima, M., Kato, M., Matsukura, M., Kira, R., Ngu, L. H., Lichtenbelt, K. D., van Gassen, K. L. I., Mitsuhashi, S., Saitsu, H., & Matsumoto, N. (2020). De novo variants in CUL3 are associated with global developmental delays with or without infantile spasms. *Journal of Human Genetics, 65*(9), 727–734. https://doi.org/10.1038/s10038-020-0758-2

Ohta, A., Schumacher, F. R., Mehellou, Y., Johnson, C., Knebel, A., Macartney, T. J., Wood, N. T., Alessi, D. R., & Kurz, T. (2013). The CUL3–KLHL3 E3 ligase complex mutated in Gordon's hypertension syndrome interacts with and ubiquitylates WNK isoforms: Disease-causing mutations in KLHL3 and WNK4 disrupt interaction. *Biochemical Journal, 451*(Pt 1), 111–122. https://doi.org/10.1042/BJ20121903

Ozkan, E. D., Creson, T. K., Kramár, E. A., Rojas, C., Seese, R. R., Babyan, A. H., Shi, Y., Lucero, R., Xu, X., Noebels, J. L., Miller, C. A., Lynch, G., & Rumbaugh, G. (2014). Reduced cognition in Syngap1 mutants is caused by isolated damage within developing forebrain excitatory neurons. *Neuron, 82*(6), 1317–1333. https://doi.org/10.1016/j.neuron.2014.05.015

Parker, M. J., Fryer, A. E., Shears, D. J., Lachlan, K. L., Mckee, S. A., Magee, A. C., Mohammed, S., Vasudevan, P. C., Park, S. M., Benoit, V., Lederer, D., Maystadt, I., Fitzpatrick, D. R., & study, D. (2015). De novo, heterozygous, loss-of-function mutations in SYNGAP1 cause a syndromic form of intellectual disability. *American Journal of Medical Genetics, Part A, 167*(10), 2231–2237. https://doi.org/10.1002/ajmg.a.37189

Peñagarikano, O., Abrahams, B. S., Herman, E. I., Winden, K. D., Gdalyahu, A., Dong, H., Sonnenblick, L. I., Gruver, R., Almajano, J., Bragin, A., Golshani, P., Trachtenberg, J. T., Peles, E., & Geschwind, D. H. (2011). Absence of CNTNAP2 leads to epilepsy, neuronal migration abnormalities, and core autism-related deficits. *Cell, 147*(1), 235–246. https://doi.org/10.1016/j.cell.2011.08.040

Pérez Jurado, L. A., Wang, Y. K., Peoples, R., Coloma, A., Cruces, J., & Francke, U. (1998). A duplicated gene in the breakpoint regions of the 7q11.23 Williams-Beuren syndrome deletion encodes the initiator binding protein TFII-I and BAP-135, a phosphorylation target of BTK. *Human Molecular Genetics, 7*(3), 325–334. https://doi.org/10.1093/HMG/7.3.325

Phelan, K., Boccuto, L., Powell, C. M., Boeckers, T. M., van Ravenswaaij-Arts, C., Rogers, R. C., Sala, C., Verpelli, C., Thurm, A., Bennett, W. E., Winrow, C. J., Garrison, S. R., Toro, R., & Bourgeron, T. (2022). Phelan-McDermid syndrome: A classification system after 30 years of experience. *Orphanet Journal of Rare Diseases, 17*(1), 1–4. https://doi.org/10.1186/S13023-022-02180-5/METRICS

Poliak, S., Gollan, L., Martinez, R., Custer, A., Einheber, S., Salzer, J. L., Trimmer, J. S., Shrager, P., & Peles, E. (1999). Caspr2, a new member of the neurexin superfamily, is localized at the juxtaparanodes of myelinated axons and associates with K+ channels. *Neuron, 24*(4), 1037–1047. https://doi.org/10.1016/S0896-6273(00)81049-1

Poliak, S., Salomon, D., Elhanany, H., Sabanay, H., Kiernan, B., Pevny, L., Stewart, C. L., Xu, X., Chiu, S.-Y., Shrager, P., Furley, A. J. W., & Peles, E. (2003). Juxtaparanodal clustering of Shaker-like K+ channels in myelinated axons depends on Caspr2 and TAG-1. *Journal of Cell Biology, 162*(6), 1149–1160. https://doi.org/10.1083/jcb.200305018

Portmann, T., Yang, M., Mao, R., Panagiotakos, G., Ellegood, J., Dolen, G., Bader, P. L., Grueter, B. A., Goold, C., Fisher, E., Clifford, K., Rengarajan, P., Kalikhman, D., Loureiro, D., Saw, N. L., Zhengqui, Z., Miller, M. A., Lerch, J. P., Henkelman, M., et al. (2014). Behavioral abnormalities and circuit defects in the basal ganglia of a mouse model of 16p11.2 deletion syndrome. *Cell Reports, 7*(4), 1077–1092. https://doi.org/10.1016/j.celrep.2014.03.036

Raveau, M., Shimohata, A., Amano, K., Miyamoto, H., & Yamakawa, K. (2018). DYRK1A-haploinsufficiency in mice causes autistic-like features and febrile seizures. *Neurobiology of Disease, 110*, 180–191. https://doi.org/10.1016/j.nbd.2017.12.003

Rodenas-Cuadrado, P., Pietrafusa, N., Francavilla, T., la Neve, A., Striano, P., & Vernes, S. C. (2016). Characterisation of CASPR2 deficiency disorder – A syndrome involving autism, epilepsy and language impairment. *BMC Medical Genetics, 17*(1), 8. https://doi.org/10.1186/s12881-016-0272-8

Rødgaard, E. M., Jensen, K., Miskowiak, K. W., & Mottron, L. (2021). Autism comorbidities show elevated female-to-male odds ratios and are associated with the age of first autism diagnosis. *Acta Psychiatrica Scandinavica, 144*(5), 475–486. https://doi.org/10.1111/ACPS.13345

Rosenfeld, J. A., Coppinger, J., Bejjani, B. A., Girirajan, S., Eichler, E. E., Shaffer, L. G., & Ballif, B. C. (2010). Speech delays and behavioral problems are the predominant features in individuals with developmental delays and 16p11.2 microdeletions and microduplications. *Journal of Neurodevelopmental Disorders, 2*(1), 26–38. https://doi.org/10.1007/s11689-009-9037-4

Rubenstein, J. L. R., & Merzenich, M. M. (2003). Model of autism: Increased ratio of excitation/inhibition in key neural systems. *Genes, Brain and Behavior, 2*(5), 255–267. https://doi.org/10.1034/j.1601-183X.2003.00037.x

Saemundsen, E., Ludvigsson, P., Hilmarsdottir, I., & Rafnsson, V. (2007a). Autism spectrum disorders in children with seizures in the first year of life-A population-based study. *Epilepsia, 48*(9), 1724–1730. https://doi.org/10.1111/j.1528-1167.2007.01150.x

Saemundsen, E., Ludvigsson, P., & Rafnsson, V. (2007b). Autism spectrum disorders in children with a history of infantile spasms: A population-based study. *Journal of Child Neurology, 22*(9), 1102–1107. https://doi.org/10.1177/0883073807306251

Saemundsen, E., Ludvigsson, P., & Rafnsson, V. (2008). Risk of autism spectrum disorders after infantile spasms: A population-based study nested in a cohort with seizures in the first year of life. *Epilepsia, 49*(11), 1865–1870. https://doi.org/10.1111/j.1528-1167.2008.01688.x

Santos, S., Ferreira, H., Martins, J., Gonçalves, J., & Castelo-Branco, M. (2022). Male sex bias in early and late onset neurodevelopmental disorders: Shared aspects and differences in Autism Spectrum Disorder, Attention Deficit/hyperactivity Disorder, and Schizophrenia. *Neuroscience & Biobehavioral Reviews, 135*, 104577. https://doi.org/10.1016/J.NEUBIOREV.2022.104577

Savic, I. (2014). Sex differences in human epilepsy. *Experimental Neurology, 259*, 38–43. https://doi.org/10.1016/J.EXPNEUROL.2014.04.009

Shinawi, M., Liu, P., Kang, S.-H. L., Shen, J., Belmont, J. W., Scott, D. A., Probst, F. J., Craigen, W. J., Graham, B. H., Pursley, A., Clark, G., Lee, J., Proud, M., Stocco, A., Rodriguez, D. L., Kozel, B. A., Sparagana, S., Roeder, E. R., McGrew, S. G., et al. (2010). Recurrent reciprocal 16p11.2 rearrangements associated with global developmental delay, behavioural problems, dysmorphism, epilepsy, and abnormal head size. *Journal of Medical Genetics, 47*(5), 332–341. https://doi.org/10.1136/jmg.2009.073015

Silva-Santos, S., van Woerden, G. M., Bruinsma, C. F., Mientjes, E., Jolfaei, M. A., Distel, B., Kushner, S. A., & Elgersma, Y. (2015). Ube3a reinstatement identifies distinct developmental windows in a murine Angelman syndrome model. *Journal of Clinical Investigation, 125*(5), 2069–2076. https://doi.org/10.1172/JCI80554

Spence, S. J., & Schneider, M. T. (2009). The role of epilepsy and epileptiform EEGs in autism spectrum disorders. *Pediatric Research, 65*(6), 599–606. https://doi.org/10.1203/PDR.0b013e31819e7168

Strauss, K. A., Puffenberger, E. G., Huentelman, M. J., Gottlieb, S., Dobrin, S. E., Parod, J. M., Stephan, D. A., & Morton, D. H. (2006). Recessive symptomatic focal epilepsy and mutant contactin-associated protein-like 2. *New England Journal of Medicine, 354*(13), 1370–1377. https://doi.org/10.1056/NEJMoa052773

Sullivan, B. J., Ammanuel, S., Kipnis, P. A., Araki, Y., Huganir, R. L., & Kadam, S. D. (2020). Low-dose perampanel rescues cortical gamma dysregulation associated with parvalbumin interneuron GluA2 upregulation in epileptic Syngap1+/− mice. *Biological Psychiatry, 87*(9), 829–842. https://doi.org/10.1016/j.biopsych.2019.12.025

Sullivan, B. J., Kipnis, P. A., Ammanuel, S. G., & Kadam, S. D. (n.d.). GABAergic signaling promotes early-life seizures in epileptic SYNGAP1 +/−mice. https://doi.org/10.1101/2020.11.14.381749

Supekar, K., Iyer, T., & Menon, V. (2017). The influence of sex and age on prevalence rates of comorbid conditions in autism. *Autism Research, 10*(5), 778–789. https://doi.org/10.1002/AUR.1741

Temme, S. J., Bell, R. Z., Fisher, G. L., & Murphy, G. G. (2016). Deletion of the mouse homolog of *CACNA1C* disrupts discrete forms of hippocampal-dependent memory and neurogenesis within the dentate gyrus. *Eneuro, 3*(6), ENEURO.0118-16.2016. https://doi.org/10.1523/ENEURO.0118-16.2016

Thomas, A. M., Schwartz, M. D., Saxe, M. D., & Kilduff, T. S. (2017). Cntnap2 knockout rats and mice exhibit epileptiform activity and abnormal sleep–wake physiology. *Sleep, 40*(1). https://doi.org/10.1093/sleep/zsw026

van der Sluijs, P. J., Jansen, S., Vergano, S. A., Adachi-Fukuda, M., Alanay, Y., AlKindy, A., Baban, A., Bayat, A., Beck-Wödl, S., Berry, K., Bijlsma, E. K., Bok, L. A., Brouwer, A. F. J., van der Burgt, I., Campeau, P. M., Canham, N., Chrzanowska, K., Chu, Y. W. Y., Chung, B. H. Y., et al. (2019). The ARID1B spectrum in 143 patients: From nonsyndromic intellectual disability to Coffin–Siris syndrome. *Genetics in Medicine, 21*(6), 1295–1307. https://doi.org/10.1038/S41436-018-0330-Z/ATTACHMENT/BB0A14B2-E49C-48D9-9421-C6BD5583C44D/MMC1.PDF

Vlaskamp, D. R. M., Shaw, B. J., Burgess, R., Mei, D., Montomoli, M., Xie, H., Myers, C. T., Bennett, M. F., Xiangwei, W., Williams, D., Maas, S. M., Brooks, A. S., Mancini, G. M. S., van de Laar, I. M. B. H., van Hagen, J. M., Ware, T. L., Webster, R. I., Malone, S., Berkovic, S. F., et al. (2019). SYNGAP1 encephalopathy: A distinctive generalized developmental and epileptic encephalopathy. *Neurology, 92*(2), E96–E107. https://doi.org/10.1212/WNL.0000000000006729

von Stülpnagel, C., Funke, C., Haberl, C., Hörtnagel, K., Jüngling, J., Weber, Y. G., Staudt, M., & Kluger, G. (2015). SYNGAP1 mutation in focal and generalized epilepsy: A literature overview and a case report with special aspects of the EEG. *Neuropediatrics, 46*(4), 287–291. https://doi.org/10.1055/s-0035-1554098. Hippokrates Verlag GmbH.

Yasuhara, A. (2010). Correlation between EEG abnormalities and symptoms of autism spectrum disorder (ASD). *Brain and Development, 32*(10), 791–798. https://doi.org/10.1016/j.braindev.2010.08.010

Chapter 8
Lipid-Related Pathophysiology of ASD

Kelly Noah and Elaine Tierney

8.1 Lipids Are Essential to Biomechanisms

Autism spectrum disorder (ASD) is the unitary diagnostic classification for a heterogeneous neurodevelopmental disorder for which the underlying etiologies have not yet been identified in most cases. The classification of ASD presently encompasses deficits in social-emotional reciprocity, nonverbal communicative behavior used for social interaction and relationships, and by the presence of restricted, repetitive patterns of behavior, interests, or activities. However, the diagnostic criteria of ASD and other disorders on the autism spectrum have varied over time, including the creation of a new diagnosis of social communication disorder with the Diagnostic and Statistical Manual, 5th edition (American Psychiatric Association, 2013). ASD has a heritability estimate of 0.83 (Sandin et al., 2017) and a sibling recurrence risk of 6.9–18.7% (Ozonoff et al., 2011; Grønborg et al., 2013), suggesting the significant impact of genetics on the development of ASD.

K. Noah
Smith-Lemli-Opitz Foundation, Fargo, ND, USA
e-mail: knoah@smithlemliopitz.org

E. Tierney (✉)
Kennedy Krieger Institute, Baltimore, MD, USA

Department of Psychiatry and Behavioral Sciences, Johns Hopkins University
School of Medicine, Baltimore, MD, USA
e-mail: tierney@kennedykrieger.org

A. El Idrissi, D. McCloskey (eds.), *Neurobiology of Autism Spectrum Disorders*,
https://doi.org/10.1007/978-3-031-42383-3_8

8.1.1 Lipids in the Periphery

8.1.1.1 Cholesterol Transport and Lipoproteins

Cholesterol is a sterol, a type of lipid, that is both synthesized in the body and obtained from dietary sources, playing essential roles during embryogenesis, myelination and brain development (Howard et al., 1969; Hamilton et al., 1992; Mutka et al., 2004; Saher et al., 2005). Cholesterol transport is carried out by lipoproteins, which are complex particles comprised of a hydrophilic surface and a hydrophobic core containing non-polar lipids that are transported in the bloodstream (Fig. 8.1a) (Amigo et al., 2016). In the periphery, high-density lipoprotein (HDL) and low-density lipoprotein (LDL) are the main lipoproteins responsible for cholesterol transport, with HDL transporting excess cholesterol from peripheral tissues to the liver for excretion from the body through a process called reverse cholesterol transport, and LDL transporting cholesterol from the liver to the body (Lund-Katz & Phillips, 2010). Apolipoproteins function as ligands for lipoprotein receptors within the periphery (Elliott et al., 2010). While it has many functions, in the periphery, apolipoprotein A-1 (ApoA-1), encoded by *APOA1* (OMIM 107680), serves as the primary structural protein of HDL (Alaupovic et al., 1972). Other structural apolipoproteins of HDL include the families of ApoA, ApoC, and ApoE (Lund-Katz & Phillips, 2010). Apolipoprotein B (ApoB) exists in two forms in the plasma, ApoB-100 and ApoB-48, both encoded by *APOB* (OMIM 107730). ApoB-100 is the primary structural protein of LDL (Seidel et al., 1970) (Fig. 8.1b). Cholesterol uptake into cells is mediated by LDL receptors (LDLR), which are located on the cell surface. Proprotein convertase subtilisin/kexin type 9 (PCSK9) is a serine protease predominantly expressed in the liver, intestines, and kidneys. In humans, the *PCSK9* gene (OMIM 607786) encodes an inactive glycoprotein that undergoes autocatalytic processing in the endoplasmic reticulum (Kosmas et al., 2018). PCSK9 binds to ApoB-100 on LDL particles and to LDLR, inducing receptor degradation (Abifadel et al., 2009). Thus, low PCSK9 levels increase LDLR recycling on the cell surface, which could lead to increased clearance of LDL and subsequently lower LDL cholesterol levels, resulting in hypocholesterolemia. The brain is dependent upon ApoA-1, where it also transports lipids and is involved in their metabolism (Elliott et al., 2010). LDLR is an important neuronal signaling mediator (Beffert et al., 2004).

8.1.1.2 Lipoproteins Transport Materials Including Steroids and miRNA

In the periphery, the adrenal glands and gonads obtain cholesterol from both HDL and LDL to synthesize neuroactive steroids (Borkowski et al., 1967; Bochem et al., 2013), which are then transported by HDL and LDL to the central nervous system (CNS) (Roy & Bélanger, 1990; Marx et al., 2006). In the brain, the steroids produce rapid, nongenomic effects, majorly through actions on ligand- or voltage-gated

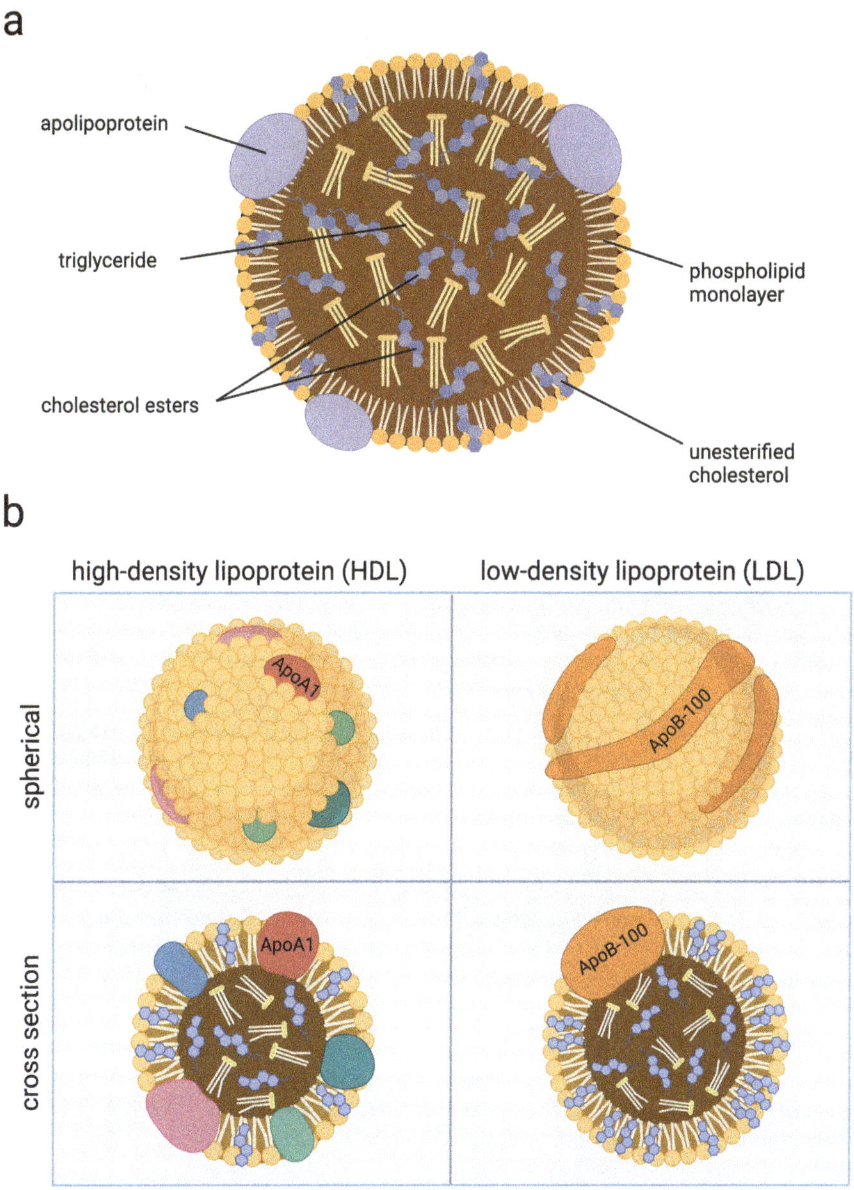

Fig. 8.1 Structure of lipoproteins. (**a**) General structure of a lipoprotein. Lipoproteins are comprised of a hydrophilic monolayer primarily consisting of phospholipids, unesterified cholesterol, and apolipoproteins, and a hydrophobic core primarily containing cholesterol esters and triglycerides. (**b**) Spherical and cross-sectional illustrations of a high-density lipoprotein (HDL) and low-density lipoprotein (LDL). The predominant apolipoprotein (Apo) of LDL is ApoB-100. ApoA-1 is one of several apolipoproteins in HDL. Other apolipoproteins of HDL include the families of ApoA, ApoC and ApoE (Lund-Katz & Phillips, 2010; Amigo et al., 2016)

channels (Stoffel-Wagner, 2001). Neuroactive steroids promote dendrite growth and synaptogenesis, exhibit neuroprotective and neurodevelopmental effects, and modulate neurotransmitter receptor activity (Charalampopoulos et al., 2008). Decreased neuroactive steroid levels have been reported in a subgroup with ASD (Strous et al., 2005). Other sterol-related abnormalities implicated in ASD involve cortisol, testosterone, estrogen, progesterone and vitamin D levels (Aneja & Tierney, 2008; Lee & Tierney, 2011; Gillberg et al., 2017).

HDL is able to come into contact with almost all cells of the body, giving it significant capabilities to affect a wide variety of systems (Vickers & Remaley, 2013). It transports proteins, hormones, carotenoids, vitamins, bioactive lipids, and small RNAs. Approximately 90% of extracellular microRNAs (miRNAs) are packaged with proteins, including HDL, Ago2, and other RNA-binding proteins (Zhu & Fan, 2011). MiRNAs are short (18–25 nucleotides in length), non-coding RNA molecules that influence gene expression at the post-transcriptional level and numerous cellular processes, including proliferation, differentiation, and apoptosis (Canfran-Duque et al., 2016), and have the ability to regulate broad molecular pathways in response to environmental stimuli without altering the genetic code (Qiu et al., 2012). The other 10% of extracellular miRNAs not transported by proteins are wrapped in small membranous particles like exosomes, micro-vesicles, and apoptotic bodies (Zhu & Fan, 2011). A reduction in HDL levels reduces the number of particles available for the transport of miRNA and changes gene expression (Vickers et al., 2011). Each miRNA has hundreds of potential messenger RNA (mRNA) targets and thus have the ability to modify entire gene networks (Cloonan, 2015). It is estimated that over 60% of human protein-coding genes are regulated by miRNAs (Friedman et al., 2009).

Dysregulated miRNA has been implicated in ASD (Abu-Elneel et al., 2008; Talebizadeh et al., 2008; Hicks & Middleton, 2016). Hicks and Middleton (2016) reviewed the findings of 12 studies comparing miRNA expression in cases of ASD to miRNA profiles in healthy controls. In the 12 blood, brain, and saliva studies, the authors identified 27 dysregulated miRNAs that were implicated in more than one ASD study. These 27 miRNAs are thought to target 1,245 mRNAs, impacting pathways involved in brain development, neurogenesis, neural and glial differentiation, synaptic development and function, and cognitive and motor function. In addition, 86 of the mRNAs overlapped with the 519 human ASD candidate genes listed by Simons Foundation Autism Research Initiative (https://gene.sfari.org/). An example is the SH3 and multiple ankyrin repeat domains 3 gene (*SHANK3*; OMIM 606230), which encodes for multidomain scaffold proteins of the postsynaptic density. Shank proteins connect neurotransmitter receptors, ion channels, and other membrane proteins to the actin cytoskeleton and G-protein coupled receptor (GPCR) signal pathways and function in synapse formation and dendritic spine maturation (Kreienkamp, 2008). A study of the cerebellar cortex of 13 adults with ASD identified 28 miRNAs with differential expression, 7 of which targeted the SHANK3 transcript (Abu-Elneel et al., 2008) and genome-wide miRNA expression profiling of brain tissue confirmed regulatory relationships between several mRNAs in the target mRNA in primary human neural progenitors (Wu et al., 2016). Dysregulated miRNA have also been implicated in glucose and lipid metabolism (Agbu et al., 2020).

8.1.2 Lipids in the CNS

8.1.2.1 Cholesterol Production in the Brain

Cholesterol is essential for neuronal physiology not just during the period of brain development, but also into adulthood. The brain is the most cholesterol-rich organ in the body, containing almost 25% of the body's cholesterol (Vitali et al., 2014). The majority of cholesterol in the brain and the rest the CNS is produced *in situ* rather than imported from the blood (Pfrieger, 2003). Brain cholesterol is synthesized by two types of glial cells: astrocytes and oligodendrocytes. The brain is able to sequester its cholesterol supply due to the blood-brain barrier (BBB), which separates the circulating blood from the brain and CNS (Vitali et al., 2014; Tietz & Engelhardt, 2015). Cholesterol homeostasis in brain regions is age- and sex-dependent (Segatto et al., 2013).

8.1.2.2 Cholesterol's Role in Development

Cholesterol fulfills an essential role during embryogenesis where it functions as a transporter-molecule that hedgehog signaling proteins require for normal morphogenesis (Porter et al., 1996). First discovered in *Drosophila*, hedgehog (Hh) proteins are composed of two distinct domains, the amino-terminal 'Hedge' domain (HhN) that has functions in signal activity, and the carboxy-terminal 'Hog' domain (HhC) that autocatalytically cleaves the Hh protein and adds a cholesterol moiety to HhN (Bürglin, 2008). Hh proteins have to be covalently linked to cholesterol to be active (Roux et al., 2000), and function to promote specification, proliferation, differentiation, and migration in various cells (Riobo, 2012). During the development of the vertebrate CNS, sonic hedgehog (SHH) proteins, encoded by *SHH* (OMIM 600725), function in a major signaling pathway that controls embryonic morphogenesis, affecting neurogenesis and neuronal patterning. The SHH pathway mediates ventral configuration, proliferation, and differentiation of precursor cells. It also regulates the proliferation of radial glial and intermediate precursor cells during neocortex development (Yang et al., 2021) and functions in BBB formation (Gozal et al., 2021). Dysregulation of the SHH pathway can lead to developmental disorders (Yang et al., 2021) and may play a pathological role in ASD (Al-Ayadhi, 2012).

8.1.2.3 Cholesterol's Role in Signaling

The production and function of the myelin sheath of neurons (Rawlins & Uzman, 1970), synapses, dendrites, axons, and cellular membranes all require the involvement of cholesterol. Cholesterol functions as a structural component of membrane lipid rafts, specialized microdomains involved in cell membrane organization and communication (Lee & Tierney, 2011). Cholesterol also modulates the activity of the serotonin 1A (5-HT_{1A}) receptor (Pucadyil & Chattopadhyay, 2004) a GPCR, encoded by *HTR1A* (OMIM 109760), that preferentially couples to Gi/o proteins to inhibit adenylyl cyclase and other second messenger cascades (Drago et al., 2008). The disturbance of serotonin signaling has previously been implicated as an etiological contributor to ASD (Chugani, 2004). Oxytocin receptors are GPCRs that couple with different G proteins, stimulating phospholipase C-β isoforms. In addition, oxytocin receptors can transduce growth-inhibitor or proliferator signals. The affinity of oxytocin receptors for ligands is strongly dependent on Mg^{2+} and cholesterol, which function as positive allosteric modulators (Gimpl et al., 2008). The amount of cholesterol available at GPCR transmembranes can affect GPCR structure and function, including serotonin receptors (Paila & Chattopadhyay, 2010) and oxytocin receptors (Gimpl et al., 2008). Blood oxytocin concentrations and polymorphisms in the oxytocin receptor gene (*OXTR*; OMIM 167055) have been shown to predict social impairment in children with ASD (Parker et al., 2014) and treatment with oxytocin boosted social behaviors in individuals with ASD in some but not in all trials (Parker et al., 2017; Bernaerts et al., 2020; Alaerts et al., 2021; Sikich et al., 2021).

8.1.2.4 Cholesterol's Role in Synaptic Function

Cholesterol reduction in neurons impairs synaptic vesicle exocytosis, neuronal activity, neurotransmission, and leads to dendritic spine and synapse degradation (Zhang & Liu, 2015). Variants in genes that alter synaptic function can lead to intellectual disability and ASD (Bourgeron, 2009). Mammalian target of rapamycin (mTOR), encoded by *MTOR* (OMIM 601231), is a serine/threonine kinase that is a major regulator of cell growth and metabolism. It functions as a signal integrator and shapes neuronal response from activation and inhibition signals that are generated by synaptic activity (Hoeffer & Klann, 2010). The mTOR protein is affected by cholesterol homeostatic changes (Xu et al., 2010), and the mTOR pathway is dysregulated in individuals with neurodevelopmental conditions including phosphatase and tensin homolog Hamartoma Tumor syndrome, tuberous sclerosis, neurofibromatosis, Fragile X syndrome (FXS), and Rett syndrome (RTT) (Magdalon et al., 2017).

8.1.2.5 Other Lipids in the Brain

HDL is not synthesized in the CNS but is able to traverse the BBB and the blood-cerebrospinal fluid barrier (BCSFB) via transcytosis, but cholesterol and apolipoprotein E (ApoE), encoded by *APOE* (OMIM 107741), must be produced *in situ* in the CNS. Low levels of HDL have been associated with cognitive impairment, impaired verbal memory, and cognitive decline in individuals with dementia (Vitali et al., 2014). Altered cholesterol, HDL and LDL metabolism have been implicated in neurological disorders including Parkinson disease (Huang et al., 2008), Huntington disease (Valenza et al., 2005; Karasinska & Hayden, 2011), Alzheimer disease (AD) (Benarroch, 2008), and amyotrophic lateral sclerosis (Zeng & Zhou, 2019).

Studies have shown that the ApoA-1 found in the brain are derived from small blood-HDL particles, which are able to enter the CNS via selective uptake. Since only HDL and ApoA-1 appear to be able to cross the BBB and BCSFB through transcytosis or selective uptake, all other lipoproteins found in the CNS must be produced *in situ* (Vitali et al., 2014). ApoE and apolipoprotein J (ApoJ), encoded by *CLU* (OMIM 185430), are the most abundant apolipoproteins in the brain and are mostly synthesized by astrocytes and found on HDL (Wang & Eckel, 2014). ApoB polymorphisms have been identified as risk factors for several neurological diseases (Elliott et al., 2010) and a psychiatric population presented with fourfold higher rate of abnormally low ApoB levels than the general population, although ApoA-1 levels were not assessed during the study (Cariou et al., 2018). The ApoE-4 allele is implicated in AD (Strittmatter et al., 1993) and increased levels of ApoE mRNA has been identified in the brains of subjects with autism (Purcell et al., 2001). ApoA-1 found in cerebrospinal fluid (CSF) is derived from blood HDL (Koch et al., 2001). ApoA-1 crosses the BBB through clathrin-independent and cholesterol-mediated endocytosis (Zhou et al., 2019).

The accumulation of amyloid beta (Aβ) peptides in the brain is a major factor in the etiology of AD. In an animal model of AD, the mice that lacked ApoA-1 fared poorly on cognitive tests and deposited more Aβ in cerebral blood vessels than the AD mice with ApoA-1 (Lefterov et al., 2010; Lewis et al., 2010). Previous research on the relationship between ApoA-1 and AD has been inconsistent (Song et al., 1997; Merched et al., 2000; Launer et al., 2001; Reitz et al., 2004; Saczynski et al., 2007).

8.2 Lipid-Related Abnormalities Have Been Found in ASD-Related Genetic Disorders

8.2.1 *Smith-Lemli-Opitz Syndrome*

Smith-Lemli-Opitz syndrome (SLOS) is an autosomal recessive disorder characterized by a wide and variable spectrum of phenotypic abnormalities, including intellectual disabilities and distinct physical malformations (Peng et al., 2018). These include microcephaly, cleft palate, cardiac and lung anomalies, soft tissue syndactyly between the second and third toes (toe webbing), and a variety of other structural defects (Nowaczyk & Irons, 2012). Reported SLOS prevalence estimates range from 1/20,000 to 1/60,000, varying significantly among populations. Current SLOS carrier frequency estimates rank individuals of Ashkenazi Jewish descent (1 in 43) and Northern European descent (1 in 54) highest, while individuals of African or Asian descent are rarely identified (Lazarin et al., 2017). SLOS is caused by variants in *DHCR7* (OMIM 602858) on chromosome 11 (Fitzky et al., 1998; Wassif et al., 1998; Waterham et al., 1998), which encodes for 7-dehydrocholesterol reductase (DHCR7), an enzyme that catalyzes the final step in cholesterol biosynthesis (Irons et al., 1993). Under normal conditions, DHCR7 reduces the C7-C8 double bond of 7DHC, the precursor molecule to cholesterol. A deficiency in DHCR7 activity experienced by individuals with SLOS often results in low cholesterol levels and abnormally elevated 7DHC or its isomer 8-dehydrocholesterol (8DHC) (Yu & Patel, 2005) (Fig. 8.2).

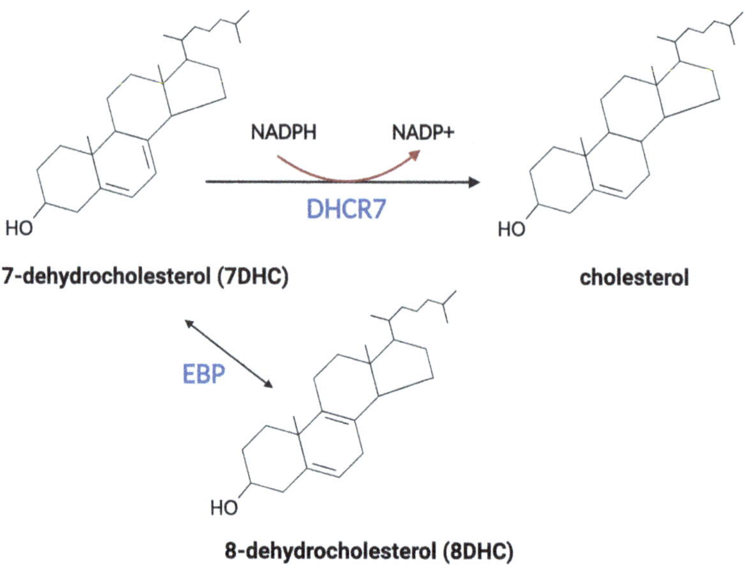

Fig. 8.2 Function of 7-dehydrocholesterol cholesterol reductase (DHCR7). DHCR7 catalyzes the final step in cholesterol biosynthesis by reducing 7-dehydrocholesterol (7DHC) to cholesterol. Isomerization of 7DHC by emopamil-binding protein (EBP) results in the formation of 8-dehydrocholesterol (8DHC) (Porter et al., 2011)

In SLOS, the degree of gross neurologic pathology is inversely correlated with cholesterol levels measured at birth (Kelley, 1995; Thurm et al., 2016) and the ratio of 7DHC+8DHC to cholesterol in both serum and CSF was significantly correlated with the SLOS anatomical severity score (Thurm et al., 2016). ASD is one of the most pervasive and severe behavioral traits observed in SLOS, associating with approximately 50% of cases (Tierney et al., 2001; Sikora et al., 2006; Thurm et al., 2016). In individuals with SLOS, both serum and CSF 7DHC levels negatively correlated with intelligence (IQ) and adaptive functioning while serum cholesterol levels positively correlated with IQ and adaptive functioning (Thurm et al., 2016). Individuals with severe SLOS had reduced levels of alpha-apolipoprotein (one study defined the protein as ApoA-1 and the other as ApoA without a subclassification) compared to individuals with mild SLOS (Behúlová et al., 2000; Oláh et al., 2013) and significantly less alpha-apolipoprotein than subjects with less severe SLOS (Oláh et al., 2013). Yet, levels of ApoB did not differ between the subjects with SLOS who were less affected and those subjects with SLOS who were more severely affected (Behúlová et al., 2000). Furthermore, it was found that individuals with SLOS present with an altered lipid subfraction profile and the decrease of HDL-associated paraoxonase-1 (PON1) enzyme, encoded by *PON1* (OMIM 168820), function was impaired, which might result in impaired antioxidant function (Lőrincz et al., 2015). HDL facilitates the secretion of PON1 and offers a hydrophobic harbor that is essential for optimal PON1 activity (Vekic et al., 2007). In addition, loss of ApoE function exacerbated neonatal lethality in a SLOS mouse model (Solcà et al., 2007).

There is no current cure that exists for individuals with SLOS. There are, however, several therapeutic options under development. The treatment of SLOS has focused on enhancing cholesterol production and decreasing the accumulation of 7DHC and 8DHC precursor molecules, with the goal of improving areas of feeding, growth, development and behaviors (Svoboda et al., 2012). Primary therapeutic treatments have concentrated on dietary or other forms of cholesterol supplementation in individuals with SLOS. Providing cholesterol through either egg yolk or cholesterol suspension raises cholesterol levels and lowers 7DHC and 8DHC levels in some individuals with SLOS (Chan et al., 2009; Tierney et al., 2010). A double-blind placebo-controlled study of simvastatin in individuals with SLOS who were being treated with cholesterol supplementation compared outcomes to 7DHC levels and Aberrant Behavior Checklist-Community scale (ABC-C) Irritability domain scores (Aman et al., 1985). The study showed that the individuals who received simvastatin had significantly lower plasma 7DHC levels and trended toward having a significant decrease in 7DHC CSF levels. The individuals treated with simvastatin also had a decrease in their irritability as measured by the ABC-C. However, the potential benefit of treatment of simvastatin is out-weighed by the potential adverse effects of the medication since one of the individuals in the trial developed cataracts after the trial ended, while the development of cataracts in SLOS is rare after infancy (Wassif et al., 2017). In an animal model of SLOS, treatment with antioxidant supplements led to dramatic protection from light-induced deleterious change (Vaughan et al., 2006), and in fibroblasts from individuals with SLOS, treatment with antioxidant supplements led to decreased levels of 7DHC-derived oxysterols (Korade et al., 2014). A clinical trial of cholesterol plus antioxidant treatment is currently ongoing (University of Colorado Denver, 2008). Other possible treatments such as supplementation with cholic acid, a major bile acid made in the body that improves fat absorption, is ongoing (University of Nebraska et al., 2021) and gene therapy has been investigated in a mouse model of SLOS (Pasta et al., 2015). Surgical intervention may also be necessary to correct some of the physical abnormalities associated with the disorder (Boland & Tatonetti, 2016).

8.2.2 *Fragile X Syndrome*

Fragile X syndrome (FXS), also known as Martin-Bell syndrome, is a genetic disorder characterized by intellectual disability, hyperactivity, anxiety, speech delay, and learning challenges, and various physical phenotypic abnormalities, including distinct facial features (elongated face, large ears), macroorchidism, hyperlaxity of joints, and hypotonia (Hagerman et al., 2017). It is classified as one of the most common causes of inherited intellectual disability (Berry-Kravis et al., 2015), with reported FXS worldwide prevalence estimates of the pathologic variant (full mutation) that results in FSX frequency of 1.4 per 10,000 males (1/7,000) and 0.9 per 10,000 females (1/11,000) (Hunter et al., 2014). FXS is known as the most common identifiable single gene disorder to cause ASD: 46% of males and 16% of females with FXS are diagnosed with ASD (Bailey et al., 2008). It is caused by an expansion of the CGG triplet repeat in the 5′ untranslated region of fragile X messenger ribonucleoprotein 1 gene (*FMR1*; OMIM 309550) on the X chromosome. A premutation is said to be present when the gene has between 55 and 200 repeats.

A premutation of FMR1 is associated with Fragile X-associated primary ovarian insufficiency (FXPOI), Fragile X-associated tremor and ataxia syndrome (FXTAS), and Fragile X-associated neuropsychiatric disorder (FXAND). Repeats greater than 200 result in the methylation of the promoter region and transcriptional silencing of *FMR1*, leading to decreased expression of fragile X messenger ribonucleoprotein (FMRP) (Pieretti et al., 1991; Fyke and Velinov, 2021). This results in dysregulation of synaptic signal transduction and channel function (Berry-Kravis et al., 2015).

FMRP controls lipid and glucose metabolism in both mice and humans (Leboucher et al., 2019). Berry-Kravis et al. (2015) found total cholesterol, LDL, and HDL levels to be significantly reduced in males with FXS relative to controls. Lisik et al. (2016) analyzed the serum of 23 males affected by FXS and 24 controls and determined that levels of HDL were lower and triglycerides were higher in the serum of individuals with FXS. An additional study analyzed blood samples from 25 individuals with FXS and 26 controls and found that the FXS patients had significantly lower levels of total cholesterol, HDL, LDL, and ApoA-1 when compared to the control group (Çaku et al., 2017). Although the PCSK9 plasma levels were not significantly different between the FXS and the typically developing control groups, PCSK9 levels did correlate with total cholesterol, LDL, and triglycerides in the control group but no significant association was detected in the FXS group. The investigators observed an inverse relationship between both total cholesterol and ApoB with aberrant behavior as measured by the ABC-C Total score. The investigators also determined that the lower the ApoA-1 level, the more impairment the individual with FSX had in adaptive functioning as measured by the Adaptive Behavior Assessment System-II (ABAS-II) and observed a similar result with the level of the lipid lipoprotein(a) (Lp(a)) (Çaku et al., 2017).

8.2.3 Rett Syndrome

Rett syndrome (RTT) is a progressive neurodevelopmental disorder characterized by loss of developed cognitive, motor, and social skills in a four-stage neurological regression (Smeets et al., 2012). It is one of the most common causes of intellectual disability in females, occurring almost exclusively in females, with reported RTT prevalence estimates ranging between 1/10,000 to 1/15,000. Individuals with RTT typically develop normally until 6–18 months of age, then gradually experience regression of acquired speech and motor skills and develop autistic features, intellectual disability, distinctive hand movements, problems with walking, seizures, and irregular breathing (Amir et al., 1999). It is caused by a variant of the methyl CpG binding protein 2 gene (*MECP2*; OMIM 300005) on the X chromosome. This results in changes to DNA binding, chromatin clustering ability, and the stability of MeCP2, a regulator protein of neural development (Good et al., 2021). MeCP2 alterations have also been identified in individuals with neural developmental disorders, including ASD, but who did not have distinct clinical features of RTT (Wen et al., 2017).

Individuals with RTT and the related cyclin-dependent kinase-like 5 (CDKL5) deficiency disorder, which results from gene variants of *CDKL5* (OMIM 300203), have elevated lipid levels (Nagy & Ackerman, 2013; Pecorelli et al., 2015). Analysis of the plasma lipoprotein profile of 151 individuals with RTT showed significant increases in total cholesterol, LDL, and HDL levels of RTT patients when compared to the control group (Sticozzi et al., 2013). Analysis of the lipid profile and expression of proteins involved in the cholesterol metabolism regulatory network in the fibroblasts and plasma of 15 female affected by RTT demonstrated significantly higher total cholesterol and LDL cholesterol levels in the plasma of RTT patients compared to healthy controls. The results from the protein analyses demonstrated that the activity and levels of the main proteins involved in the maintenance of cholesterol homeostasis are altered in RTT patients (Segatto et al., 2014). Furthermore, cholesterol metabolism was shown to be dysregulated in the brain and liver of MECP2-null male mice, suggesting that cholesterol metabolism could be altered in patients with RTT (Buchovecky et al., 2013).

8.3 Lipid Disorders with Known Genetic Variants Not Associated with ASD

8.3.1 Post-Squalene Disorders Involved with Cholesterol Biosynthesis

In 1993, SLOS was the first post-squalene disorder of cholesterol biosynthesis whose underlying defect was identified (Tint, 1993). Five other disorders involving enzyme defects in post-squalene cholesterol biosynthesis have since been identified: desmosterolosis, X-linked dominant chondrodysplasia punctata 2 (CDPX2), congenital hemidysplasia with ichthyosiform nevus and limb defects (CHILD) syndrome, lathosterolosis, and hydrops-ectopic calcification-moth-eaten (HEM) skeletal dysplasia (Herman, 2003; Porter, 2003; Platt et al., 2014). All 6 disorders include some degree of dysmorphia and/or physical deformity. Desmosterolosis is associated with intellectual disability (Dias et al., 2014), as is lathosterolosis (Yaplito-Lee et al., 2020), while those with CDPX2 and CHILD syndrome usually have normal IQ. There has not been enough research on HEM skeletal dysplasia to determine expected IQ levels or the presence of ASD (Porter & Herman, 2011).

8.3.2 Hypoapolipoprotein Disorders: Hypoalphalipoproteinemia, Hypobetalipoproteinemia, Abetalipoproteinemia, and Familial Combined Hypolipidemia

Primary hypocholesterolemia typically stems from genetic variants and includes diseases of cholesterol biosynthesis. Variants in the *APOA1* gene can result in primary hypoalphalipoproteinemia 2, combined ApoA-1 and ApoC-3 deficiency (via a *APOA1/APOC3* fusion gene) or amyloidosis (Rees et al., 1985; Genschel et al., 1998;

Yamakawa-Kobayashi et al., 1999). Familial hypobetalipoproteinemia 1, which is a dominant disorder whose frequency in a heterozygous form is estimated to be 1/1,000 to 1/3,000 (Tarugi et al., 2007), results from variants in *APOB* (Law et al., 1985) or *PCSK9* (Yue et al., 2006; Zhao et al., 2006), and can result in hepatic steatosis (Di Filippo et al., 2014); but may cause no known medical difficulties. Variants in the microsomal triglyceride transfer protein gene (*MTTP*; OMIM 157147) can result in abetalipoproteinemia, also known as MTTP deficiency or Bassen-Kornzweig syndrome (Wetterau et al., 1992; Sharp et al., 1993; Narcisi et al., 1995). MTTP deficiency can result in hypocholesterolemia with malabsorption of lipid soluble vitamins leading to retinal degeneration, neuropathy and coagulopathy; hepatic steatosis is also common. The root cause is the improper packaging and secretion of ApoB containing particles which results in the absence of ApoB. Variants in the angiopoietin-like protein 3 gene (*ANGPTL3*; OMIM 604774) cause familial hypobetalipoproteinemia 2 (FHBL2), also known as familial combined hypolipidemia (Minicocci et al., 2012), with low levels of LDL and HDL. Variants in the ATP-binding cassette subfamily A member (*ABCA1*) gene (OMIM 600046) can result in familial HDL deficiency 1, also known as Tangier disease (Pisciotta et al., 2004).

8.3.3 Hyperlipidemic Disorders Caused by Variants in Genes That Also Cause Hypolipidemic Disorders

Missense variants in the *APOB* gene can result in elevated cholesterol levels or familial hypercholesterolemia (Bertolini et al., 2013; Khera et al., 2016) Gain-of-function variants of *PCSK9* can result in familial hypercholesterolemia (Abifadel et al., 2003).

8.4 Lipid Abnormalities Associated with ASD that Are Not Observed with Known Genetic Disorders

Analysis of plasma samples from 100 individuals with ASD participating with the Autism Genetic Resource Exchange (AGRE), revealed no subject with SLOS but did show a higher-than-expected frequency of hypocholesterolemia (19%) with cholesterol levels below the fifth centile of cholesterol levels of typically developing children. Decreased lathosterol levels in the samples suggested that the hypocholesterolemia was due to a synthetic defect rather than decreased dietary intake or gastrointestinal loss (Tierney et al., 2006).

A case-control study conducted in South Korea with the plasma of 29 boys with autism and 29 healthy controls found no significant differences in the levels of total serum cholesterol and LDL as compared to controls, while the mean HDL level was significantly lower, and triglyceride levels were significantly higher in the boys with ASD (Kim et al., 2010). Another study conducted in Japan measured the lipid profile of subjects with high-functioning autism and found significantly low levels of serum total cholesterol and VLDL upon lipoprotein fractionation (Matsuzaki et al., 2010). In addition, increased plasma lipoprotein lipase activity but not increased lipoprotein lipase levels were detected in individuals with ASD (Hirai et al., 2020).

Like in SLOS, studies have also suggested an implication of the HDL associated enzyme PON1 in ASD when the bioavailability and catalytic activity of PON1 was found to be impaired in individuals with ASD (Paşca et al., 2010) .

ApoB-100 and ApoA-4 were found in a proteomic analysis to be reduced in the serum of children with ASD (n = 69) when compared to typically developing children (n = 35), although not statistically significant after correction for multiple comparison. However, when the most expressed identified proteins with the highest effect size (0.99 or greater), a p value <.05, a mascot score of greater than 30 were examined, the level of ApoB-100 precursor was significantly lower in the serum of the children with autism compared to the typically developing controls, as identified by 10 peptide components (Corbett et al., 2007).

Increased levels of cholesterol and LDL were identified in 22 individuals with Asperger syndrome in comparison to well matched controls after accounting for lifestyle variables and clinical symptomatology that could affect them (Dziobek et al., 2007) (the diagnosis of Asperger syndrome is no longer used. It is now considered to be the diagnosis of ASD with the Diagnostic and Statistical Manual, 5th edition (American Psychiatric Association, 2013)). In a proteomic analysis of serum in 30 individuals with Asperger syndrome and 29 controls, the ApoA-1 levels were significantly decreased in females with Asperger syndrome when controlled for false discovery rate. Additional proteins involved in lipid metabolism were found to be significantly elevated or decreased in females and males with Asperger syndrome (Steeb et al., 2014). A pathway-level analyses of ASD-implicated genes have reported an enrichment of lipid metabolism pathways (David et al., 2016), as has a precision medicine approach (Luo et al., 2020).

While the underlying etiologies of ASD have not been identified for most cases, recent research has sought to better understand how abnormalities of lipid metabolism may contribute to ASD pathogenesis. Lipids are essential to biomechanisms in both the periphery and CNS. Cholesterol, which plays critical roles during embryogenesis, myelination, and brain development, is transported in the bloodstream via lipoproteins. Lipid-related biomechanisms commonly associated with ASD include abnormalities in steroid and miRNA transport, dysregulation to the SHH and mTOR pathways, and disturbance to serotonin and oxytocin signaling. Many lipid- related abnormalities have been found in neurologic disorders including the following ASD-related genetic disorders: SLOS, FXS, and RTT. Lipid disorders with known genetic variants that are not associated with ASD have also been identified, including post-squalene disorders involving cholesterol biosynthesis, hypoapolipoprotein disorders, and hyperlipidemic disorders. More recent studies have identified lipid abnormalities associated with ASD that are not currently observed with known genetic disorders. Exploring abnormalities of lipid metabolism contributing to ASD pathogenesis, as well as ASD-related genetic disorders and lipid disorders, may aid in elucidating the underpinnings of this complex disorder in the future.

References

Abifadel, M., Varret, M., Rabes, J. P., Allard, D., Ouguerram, K., Devillers, M., Cruaud, C., Benjannet, S., Wickham, L., Erlich, D., Derre, A., Villeger, L., Farnier, M., Beucler, I., Bruckert, E., Chambaz, J., Chanu, B., Lecerf, J. M., Luc, G., Moulin, P., Weissenbach, J., Prat, A., Krempf, M., Junien, C., Seidah, N. G., & Boileau, C. (2003). Mutations in PCSK9 cause autosomal dominant hypercholesterolemia. *Nature Genetics, 34*, 154–156.

Abifadel, M., Rabes, J. P., Devillers, M., Munnich, A., Erlich, D., Junien, C., Varret, M., & Boileau, C. (2009). Mutations and polymorphisms in the proprotein convertase subtilisin kexin 9 (PCSK9) gene in cholesterol metabolism and disease. *Human Mutation, 30*, 520–529.

Abu-Elneel, K., Liu, T., Gazzaniga, F. S., Nishimura, Y., Wall, D. P., Geschwind, D. H., Lao, K., & Kosik, K. S. (2008). Heterogeneous dysregulation of microRNAs across the autism spectrum. *Neurogenetics, 9*, 153–161.

Agbu, P., Cassidy, J. J., Braverman, J., Jacobson, A., & Carthew, R. W. (2020). MicroRNA miR-7 regulates secretion of insulin-like peptides. *Endocrinology, 161*.

Alaerts, K., Steyaert, J., Vanaudenaerde, B., Wenderoth, N., & Bernaerts, S. (2021). Changes in endogenous oxytocin levels after intranasal oxytocin treatment in adult men with autism: An exploratory study with long-term follow-up. *European Neuropsychopharmacology, 43*, 147–152.

Alaupovic, P., Lee, D. M., & Mcconathy, W. J. (1972). Studies on the composition and structure of plasma lipoproteins. Distribution of lipoprotein families in major density classes of normal human plasma lipoproteins. *Biochimica et Biophysica Acta, 260*, 689–707.

Al-Ayadhi, L. Y. (2012). Relationship between Sonic hedgehog protein, brain-derived neurotrophic factor and oxidative stress in autism spectrum disorders. *Neurochemical Research, 37*, 394–400.

Aman, M. G., Singh, N. N., Stewart, A. W., & Field, C. J. (1985). The aberrant behavior checklist: A behavior rating scale for the assessment of treatment effects. *American Journal of Mental Deficiency, 89*, 485–491.

Amigo, N., Mallol, R., Heras, M., Martinez-Hervas, S., Blanco Vaca, F. F. F., Escola-Gil, J. C., Plana, N., Yanes, O., Masana, L., & Correig, X. (2016). Lipoprotein hydrophobic core lipids are partially extruded to surface in smaller HDL: "Herniated" HDL, a common feature in diabetes. *Sci Rep, 6*, 19249.

American Psychiatric Association. (2013). *Diagnostic and statistical manual of mental disorders*. American Psychiatric Publishing.

Amir, R. E., Van Den Veyver, I. B., Wan, M., Tran, C. Q., Francke, U., & Zoghbi, H. Y. (1999). Rett syndrome is caused by mutations in X-linked MECP2, encoding methyl-CpG-binding protein 2. *Nature Genetics, 23*, 185–188.

Aneja, A., & Tierney, E. (2008). Autism: The role of cholesterol in treatment. *International Review of Psychiatry, 20*, 165–170.

Bailey, D. B., Jr., Raspa, M., Olmsted, M., & Holiday, D. B. (2008). Co-occurring conditions associated with FMR1 gene variations: findings from a national parent survey. *Am J Med Genet A, 146A*, 2060–2069.

Beffert, U., Stolt, P. C., & Herz, J. (2004). Functions of lipoprotein receptors in neurons. *Journal of Lipid Research, 45*, 403–409.

Behúlová, D., Bzdúch, V., Skodová, J., Dello Russo, A., Corso, G., Ponec, J., & Kasanická, A. (2000). Serum lipids and apolipoproteins in children with the Smith-Lemli-Opitz syndrome. *Journal of Inherited Metabolic Disease, 23*, 413–415.

Benarroch, E. E. (2008). Brain cholesterol metabolism and neurologic disease. *Neurology, 71*, 1368–1373.

Bernaerts, S., Boets, B., Bosmans, G., Steyaert, J., & Alaerts, K. (2020). Behavioral effects of multiple-dose oxytocin treatment in autism: A randomized, placebo-controlled trial with long-term follow-up. *Molecular Autism, 11*, 6.

Berry-Kravis, E., Levin, R., Shah, H., Mathur, S., Darnell, J. C., & Ouyang, B. (2015). Cholesterol levels in fragile X syndrome. *American Journal of Medical Genetics. Part A, 167A*, 379–384.

Bertolini, S., Pisciotta, L., Rabacchi, C., Cefalù, A. B., Noto, D., Fasano, T., Signori, A., Fresa, R., Averna, M., & Calandra, S. (2013). Spectrum of mutations and phenotypic expression in patients with autosomal dominant hypercholesterolemia identified in Italy. *Atherosclerosis, 227*, 342–348.

Bochem, A. E., Holleboom, A. G., Romijn, J. A., Hoekstra, M., Dallinga-Thie, G. M., Motazacker, M. M., Hovingh, G. K., Kuivenhoven, J. A., & Stroes, E. S. G. (2013). High density lipoprotein as a source of cholesterol for adrenal steroidogenesis: A study in individuals with low plasma HDL-C. *Journal of Lipid Research, 54*, 1698–1704.

Boland, M. R., & Tatonetti, N. P. (2016). Investigation of 7-dehydrocholesterol reductase pathway to elucidate off-target prenatal effects of pharmaceuticals: A systematic review. *The Pharmacogenomics Journal, 16*, 411–429.

Borkowski, A. J., Levin, S., Delcroix, C., Mahler, A., & Verhas, V. (1967). Blood cholesterol and hydrocortisone production in man: Quantitative aspects of the utilization of circulating cholesterol by the adrenals at rest and under adrenocorticotropin stimulation. *The Journal of Clinical Investigation, 46*, 797–811.

Bourgeron, T. (2009). A synaptic trek to autism. *Current Opinion in Neurobiology, 19*, 231–234.

Buchovecky, C. M., Turley, S. D., Brown, H. M., Kyle, S. M., Mcdonald, J. G., Liu, B., Pieper, A. A., Huang, W., Katz, D. M., Russell, D. W., Shendure, J., & Justice, M. J. (2013). A suppressor screen in Mecp2 mutant mice implicates cholesterol metabolism in Rett syndrome. *Nature Genetics, 45*, 1013–1020.

Bürglin, T. R. (2008). The Hedgehog protein family. *Genome Biology, 9*, 241.

Çaku, A., Seidah, N. G., Lortie, A., Gagné, N., Perron, P., Dubé, J., & Corbin, F. (2017). New insights of altered lipid profile in Fragile X Syndrome. *PLoS One, 12*, e0174301.

Canfran-Duque, A., Lin, C. S., Goedeke, L., Suarez, Y., & Fernandez-Hernando, C. (2016). MicroRNAs and high-density lipoprotein metabolism. *Arteriosclerosis, Thrombosis, and Vascular Biology, 36*, 1076–1084.

Cariou, B., Challet-Bouju, G., Bernard, C., Marrec, M., Hardouin, J. B., Authier, C., Bach-Ngohou, K., Leux, C., Pichelin, M., & Grall-Bronnec, M. (2018). Prevalence of hypobetalipoproteinemia and related psychiatric characteristics in a psychiatric population: Results from the retrospective HYPOPSY study. *Lipids in Health and Disease, 17*, 249.

Chan, Y. M., Merkens, L. S., Connor, W. E., Roullet, J. B., Penfield, J. A., Jordan, J. M., Steiner, R. D., & Jones, P. J. (2009). Effects of dietary cholesterol and simvastatin on cholesterol synthesis in Smith-Lemli-Opitz syndrome. *Pediatric Research, 65*, 681–685.

Charalampopoulos, I., Remboutsika, E., Margioris, A. N., & Gravanis, A. (2008). Neurosteroids as modulators of neurogenesis and neuronal survival. *Trends in Endocrinology and Metabolism, 19*, 300–307.

Chugani, D. C. (2004). Serotonin in autism and pediatric epilepsies. *Mental Retardation and Developmental Disabilities Research Reviews, 10*, 112–116.

Cloonan, N. (2015). Re-thinking miRNA-mRNA interactions: Intertwining issues confound target discovery. *BioEssays, 37*, 379–388.

Corbett, B. A., Kantor, A. B., Schulman, H., Walker, W. L., Lit, L., Ashwood, P., Rocke, D. M., & Sharp, F. R. (2007). A proteomic study of serum from children with autism showing differential expression of apolipoproteins and complement proteins. *Molecular Psychiatry, 12*, 292–306.

David, M. M., Enard, D., Ozturk, A., Daniels, J., Jung, J. Y., Diaz-Beltran, L., & Wall, D. P. (2016). Comorbid analysis of genes associated with autism spectrum disorders reveals differential evolutionary constraints. *PLoS One, 11*, e0157937.

Di Filippo, M., Moulin, P., Roy, P., Samson-Bouma, M. E., Collardeau-Frachon, S., Chebel-Dumont, S., Peretti, N., Dumortier, J., Zoulim, F., Fontanges, T., Parini, R., Rigoldi, M., Furlan, F., Mancini, G., Bonnefont-Rousselot, D., Bruckert, E., Schmitz, J., Scoazec, J. Y., Charrière, S., Villar-Fimbel, S., Gottrand, F., Dubern, B., Doummar, D., Joly, F., Liard-Meillon, M. E., Lachaux, A., & Sassolas, A. (2014). Homozygous MTTP and APOB mutations may lead to hepatic steatosis and fibrosis despite metabolic differences in congenital hypocholesterolemia. *Journal of Hepatology, 61*, 891–902.

Dias, C., Rupps, R., Millar, B., Choi, K., Marra, M., Demos, M., Kratz, L. E., & Boerkoel, C. F. (2014). Desmosterolosis: An illustration of diagnostic ambiguity of cholesterol synthesis disorders. *Orphanet Journal of Rare Diseases, 9*, 94.

Drago, A., Ronchi, D. D., & Serretti, A. (2008). 5-HT1A gene variants and psychiatric disorders: A review of current literature and selection of SNPs for future studies. *The International Journal of Neuropsychopharmacology, 11*, 701–721.

Dziobek, I., Gold, S. M., Wolf, O. T., & Convit, A. (2007). Hypercholesterolemia in Asperger syndrome: Independence from lifestyle, obsessive-compulsive behavior, and social anxiety. *Psychiatry Research, 149*, 321–324.

Elliott, D. A., Weickert, C. S., & Garner, B. (2010). Apolipoproteins in the brain: Implications for neurological and psychiatric disorders. *Clinical Lipidology, 51*, 555–573.

Fitzky, B. U., Witsch-Baumgartner, M., Erdel, M., Lee, J. N., Paik, Y. K., Glossmann, H., Utermann, G., & Moebius, F. F. (1998). Mutations in the Delta7-sterol reductase gene in patients with the Smith-Lemli-Opitz syndrome. *Proceedings of the National Academy of Sciences of the United States of America, 95*, 8181–8186.

Friedman, R. C., Farh, K. K., Burge, C. B., & Bartel, D. P. (2009). Most mammalian mRNAs are conserved targets of microRNAs. *Genome Research, 19*, 92–105.

Fyke, W., & Velinov, M. (2021). FMR1 and autism, an intriguing connection revisited. *Genes (Basel)*, 12.

Genschel, J., Haas, R., Pröpsting, M. J., & Schmidt, H. H. (1998). Apolipoprotein A-I induced amyloidosis. *FEBS Letters, 430*, 145–149.

Gillberg, C., Fernell, E., Kocovska, E., Minnis, H., Bourgeron, T., Thompson, L., & Allely, C. S. (2017). The role of cholesterol metabolism and various steroid abnormalities in autism spectrum disorders: A hypothesis paper. *Autism Research, 10*, 1022–1044.

Gimpl, G., Reitz, J., Brauer, S., & Trossen, C. (2008). Oxytocin receptors: Ligand binding, signalling and cholesterol dependence. *Progress in Brain Research, 170*, 193–204.

Good, K. V., Vincent, J. B., & Ausió, J. (2021). MeCP2: The genetic driver of Rett syndrome epigenetics. *Frontiers in Genetics, 12*, 620859.

Gozal, E., Jagadapillai, R., Cai, J., & Barnes, G. N. (2021). Potential crosstalk between sonic hedgehog-WNT signaling and neurovascular molecules: Implications for blood-brain barrier integrity in autism spectrum disorder. *Journal of Neurochemistry, 159*, 15–28.

Grønborg, T. K., Schendel, D. E., & Parner, E. T. (2013). Recurrence of autism spectrum disorders in full- and half-siblings and trends over time: A population-based cohort study. *JAMA Pediatrics, 167*, 947–953.

Hagerman, R. J., Berry-Kravis, E., Hazlett, H. C., Bailey, D. B., JR., Moine, H., Kooy, R. F., Tassone, F., Gantois, I., Sonenberg, N., Mandel, J. L., & Hagerman, P. J. (2017). Fragile X syndrome. *Nat Rev Dis Primers, 3*, 17065.

Hamilton, J. J., Synnes, A., & Innis, S. M. (1992). Plasma cholesterol and lathosterol levels in term infants in the early neonatal period. *Pediatric Research, 31*, 396–400.

Herman, G. E. (2003). Disorders of cholesterol biosynthesis: Prototypic metabolic malformation syndromes. *Human Molecular Genetics, 12 Spec No 1*, R75–R88.

Hicks, S. D., & Middleton, F. A. (2016). A comparative review of microRNA expression patterns in autism spectrum disorder. *Frontiers in Psychiatry, 7*, 176.

Hirai, T., Usui, N., Iwata, K., Miyachi, T., Tsuchiya, K. J., Xie, M., Nakamura, K., Tsujii, M., Sugiyama, T., & Matsuzaki, H. (2020). Increased plasma lipoprotein lipase activity in males with autism spectrum disorder. *Research in Autism Spectrum Disorder, 77*, 101630.

Hoeffer, C. A., & Klann, E. (2010). mTOR signaling: At the crossroads of plasticity, memory and disease. *Trends in Neurosciences, 33*, 67–75.

Howard, E., Granoff, D. M., & Bujnovszky, P. (1969). DNA, RNA, and cholesterol increases in cerebrum and cerebellum during development of human fetus. *Brain Research, 14*, 697–706.

Huang, X., Abbott, R. D., Petrovitch, H., Mailman, R. B., & Ross, G. W. (2008). Low LDL cholesterol and increased risk of Parkinson's disease: Prospective results from Honolulu-Asia Aging Study. *Movement Disorders, 23*, 1013–1018.

Hunter, J., Rivero-Arias, O., Angelov, A., Kim, E., Fotheringham, I., & Leal, J. (2014). Epidemiology of fragile X syndrome: A systematic review and meta-analysis. *American Journal of Medical Genetics. Part A, 164A*, 1648–1658.

Irons, M., Elias, E. R., Salen, G., Tint, G. S., & Batta, A. K. (1993). Defective cholesterol biosynthesis in Smith-Lemli-Opitz syndrome. *Lancet, 341*, 1414.

Karasinska, J. M., & Hayden, M. R. (2011). Cholesterol metabolism in Huntington disease. *Nature Reviews. Neurology, 7*, 561–572.

Kelley, R. I. (1995). Diagnosis of Smith-Lemli-Opitz syndrome by gas chromatography/mass spectrometry of 7-dehydrocholesterol in plasma, amniotic fluid and cultured skin fibroblasts. *Clinica Chimica Acta, 236*, 45–58.

Khera, A. V., Won, H. H., Peloso, G. M., Lawson, K. S., Bartz, T. M., Deng, X., Van Leeuwen, E. M., Natarajan, P., Emdin, C. A., Bick, A. G., Morrison, A. C., Brody, J. A., Gupta, N., Nomura, A., Kessler, T., Duga, S., Bis, J. C., Van Duijn, C. M., Cupples, L. A., Psaty, B., Rader, D. J., Danesh, J., Schunkert, H., Mcpherson, R., Farrall, M., Watkins, H., Lander, E., Wilson, J. G., Correa, A., Boerwinkle, E., Merlini, P. A., Ardissino, D., Saleheen, D., Gabriel, S., & Kathiresan, S. (2016). Diagnostic yield and clinical utility of sequencing familial hypercholesterolemia genes in patients with severe hypercholesterolemia. *Journal of the American College of Cardiology, 67*, 2578–2589.

Kim, E. K., Neggers, Y. H., Shin, C. S., Kim, E., & Kim, E. M. (2010). Alterations in lipid profile of autistic boys: A case control study. *Nutrition Research, 30*, 255–260.

Koch, S., Donarski, N., Goetze, K., Kreckel, M., Stuerenburg, H. J., Buhmann, C., & Beisiegel, U. (2001). Characterization of four lipoprotein classes in human cerebrospinal fluid. *Journal of Lipid Research, 42*, 1143–1151.

Korade, Z., Xu, L., Harrison, F. E., Ahsen, R., Hart, S. E., Folkes, O. M., Mirnics, K., & Porter, N. A. (2014). Antioxidant supplementation ameliorates molecular deficits in Smith-Lemli-Opitz syndrome. *Biological Psychiatry, 75*, 215–222.

Kosmas, C. E., Sourlas, A., Bouza, K. V., Dejesus, E., Silverio, D., Montan, P. D., & Guzman, E. (2018). Proprotein convertase subtilisin/kexin type 9 (PCSK9) inhibition with evolocumab: Powerful low-density lipoprotein cholesterol (LDL-C) lowering and improved cardiovascular outcomes without an increase in the risk of diabetes mellitus. *The Annals of Translational Medicine, 6*, 130.

Kreienkamp, H. J. (2008). Scaffolding proteins at the postsynaptic density: Shank as the architectural framework. *Handbook of Experimental Pharmacology, 186*, 365–380.

Launer, L. J., White, L. R., Petrovitch, H., Ross, G. W., & Curb, J. D. (2001). Cholesterol and neuropathologic markers of AD: A population-based autopsy study. *Neurology, 57*, 1447–1452.

Law, S. W., Lackner, K. J., Hospattankar, A. V., Anchors, J. M., Sakaguchi, A. Y., Naylor, S. L., & Brewer, H. B. (1985). Human apolipoprotein B-100: cloning, analysis of liver mRNA, and assignment of the gene to chromosome 2. *Proceedings of the National Academy of Sciences, 82*, 8340–8344.

Lazarin, G. A., Haque, I. S., Evans, E. A., & Goldberg, J. D. (2017). Smith-Lemli-Opitz syndrome carrier frequency and estimates of in utero mortality rates. *Prenatal Diagnosis, 37*, 350–355.

Leboucher, A., Pisani, D. F., Martinez-Gili, L., Chilloux, J., Bermudez-Martin, P., Van Dijck, A., Ganief, T., Macek, B., Becker, J. A. J., Le Merrer, J., Kooy, R. F., Amri, E. Z., Khandjian, E. W., Dumas, M. E., & Davidovic, L. (2019). The translational regulator FMRP controls lipid and glucose metabolism in mice and humans. *Molecular Metabolism, 21*, 22–35.

Lee, R. W., & Tierney, E. (2011). Hypothesis: The role of sterols in autism spectrum disorder. *Autism Research and Treatment, 2011*, 653570.

Lefterov, I., Fitz, N. F., Cronican, A. A., Fogg, A., Lefterov, P., Kodali, R., Wetzel, R., & Koldamova, R. (2010). Apolipoprotein A-I deficiency increases cerebral amyloid angiopathy and cognitive deficits in APP/PS1DeltaE9 mice. *The Journal of Biological Chemistry, 285*, 36945–36957.

Lewis, T. L., Cao, D., Lu, H., Mans, R. A., Su, Y. R., Jungbauer, L., Linton, M. F., Fazio, S., Ladu, M. J., & Li, L. (2010). Overexpression of human apolipoprotein A-I preserves cognitive function and attenuates neuroinflammation and cerebral amyloid angiopathy in a mouse model of Alzheimer disease. *The Journal of Biological Chemistry, 285*, 36958–36968.

Lisik, M. Z., Gutmajster, E., & Sieroń, A. L. (2016). Low levels of HDL in fragile X syndrome patients. *Lipids, 51*, 189–192.

Lőrincz, H., Harangi, M., Oláh, A. V., Szabó, G. P., Fülöp, P., Somodi, S., Paragh, G., & Seres, I. (2015). Altered lipid subfraction profile and impaired antioxidant defense of high-density lipoprotein in Smith-Lemli-Opitz syndrome. *Pediatric Research, 77*, 703–709.

Lund-Katz, S., & Phillips, M. C. (2010). High density lipoprotein structure-function and role in reverse cholesterol transport. *Subcell Biochem, 51*, 183–227.

Luo, Y., Eran, A., Palmer, N., Avillach, P., Levy-Moonshine, A., Szolovits, P., & Kohane, I. S. (2020). A multidimensional precision medicine approach identifies an autism subtype characterized by dyslipidemia. *Nature Medicine, 26*, 1375–1379.

Magdalon, J., Sánchez-Sánchez, S. M., Griesi-Oliveira, K., & Sertié, A. L. (2017). Dysfunctional mTORC1 signaling: A convergent mechanism between syndromic and nonsyndromic forms of autism spectrum disorder? *International Journal of Molecular Sciences, 18*, 659.

Marx, C. E., Stevens, R. D., Shampine, L. J., Uzunova, V., Trost, W. T., Butterfield, M. I., Massing, M. W., Hamer, R. M., Morrow, A. L., & Lieberman, J. A. (2006). Neuroactive steroids are altered in schizophrenia and bipolar disorder: relevance to pathophysiology and therapeutics. *Neuropsychopharmacol, 31*, 1249–1263.

Matsuzaki, H., Iwata, K., Suda, S., Tsuchiya, K. J., Suzuki, K., Nakamura, K., Tsujii, M., Takei, N., & Mori, N. (2010). *Dyslipidemia in male patients with high-functioning autism*. International meeting for autism research 2010, Philadelphia, PA

Merched, A., Xia, Y., Visvikis, S., Serot, J. M., & Siest, G. (2000). Decreased high-density lipoprotein cholesterol and serum apolipoprotein AI concentrations are highly correlated with the severity of Alzheimer's disease. *Neurobiology of Aging, 21*, 27–30.

Minicocci, I., Montali, A., Robciuc, M. R., Quagliarini, F., Censi, V., Labbadia, G., Gabiati, C., Pigna, G., Sepe, M. L., Pannozzo, F., Lütjohann, D., Fazio, S., Jauhiainen, M., Ehnholm, C., & Arca, M. (2012). Mutations in the ANGPTL3 gene and familial combined hypolipidemia: A clinical and biochemical characterization. *The Journal of Clinical Endocrinology and Metabolism, 97*, E1266–E1275.

Mutka, A. L., Lusa, S., Linder, M. D., Jokitalo, E., Kopra, O., Jauhiainen, M., & Ikonen, E. (2004). Secretion of sterols and the NPC2 protein from primary astrocytes. *The Journal of Biological Chemistry, 279*, 48654–48662.

Nagy, G., & Ackerman, S. L. (2013). Cholesterol metabolism and Rett syndrome pathogenesis. *Nature Genetics, 45*, 965–967.

Narcisi, T. M., Shoulders, C. C., Chester, S. A., Read, J., Brett, D. J., Harrison, G. B., Grantham, T. T., Fox, M. F., Povey, S., De Bruin, T. W., et al. (1995). Mutations of the microsomal triglyceride-transfer-protein gene in abetalipoproteinemia. *American Journal of Human Genetics, 57*, 1298–1310.

Nowaczyk, M. J., & Irons, M. B. (2012). Smith-Lemli-Opitz syndrome: Phenotype, natural history, and epidemiology. *American Journal of Medical Genetics. Part C, Seminars in Medical Genetics, 160c*, 250–262.

Oláh, A. V., Szabó, G. P., Varga, J., Balogh, L., Csábi, G., Csákváry, V., Erwa, W., & Balogh, I. (2013). Relation between biomarkers and clinical severity in patients with Smith-Lemli-Opitz syndrome. *European Journal of Pediatrics, 172*, 623–630.

Ozonoff, S., Young, G. S., Carter, A., Messinger, D., Yirmiya, N., Zwaigenbaum, L., Bryson, S., Carver, L. J., Constantino, J. N., Dobkins, K., Hutman, T., Iverson, J. M., Landa, R., Rogers, S. J., Sigman, M., & Stone, W. L. (2011). Recurrence risk for autism spectrum disorders: A baby siblings research consortium study. *Pediatrics, 128*, e488–e495.

Paila, Y. D., & Chattopadhyay, A. (2010). Membrane cholesterol in the function and organization of G-protein coupled receptors. *Sub-Cellular Biochemistry, 51*, 439–466.

Parker, K. J., Garner, J. P., Libove, R. A., Hyde, S. A., Hornbeak, K. B., Carson, D. S., Liao, C. P., Phillips, J. M., Hallmayer, J. F., & Hardan, A. Y. (2014). Plasma oxytocin concentrations and OXTR polymorphisms predict social impairments in children with and without autism spectrum disorder. *Proceedings of the National Academy of Sciences of the United States of America, 111*, 12258–12263.

Parker, K. J., Oztan, O., Libove, R. A., Sumiyoshi, R. D., Jackson, L. P., Karhson, D. S., Summers, J. E., Hinman, K. E., Motonaga, K. S., Phillips, J. M., Carson, D. S., Garner, J. P., & Hardan, A. Y. (2017). Intranasal oxytocin treatment for social deficits and biomarkers of response in children with autism. *Proceedings of the National Academy of Sciences of the United States of America, 114*, 8119–8124.

Paşca, S. P., Dronca, E., Nemeş, B., Kaucsár, T., Endreffy, E., Iftene, F., Benga, I., Cornean, R., & Dronca, M. (2010). Paraoxonase 1 activities and polymorphisms in autism spectrum disorders. *Journal of Cellular and Molecular Medicine, 14*, 600–607.

Pasta, S., Akhile, O., Tabron, D., Ting, F., Shackleton, C., & Watson, G. (2015). Delivery of the 7-dehydrocholesterol reductase gene to the central nervous system using adeno-associated virus vector in a mouse model of Smith-Lemli-Opitz syndrome. *Molecular Genetics and Metabolism Reports, 4*, 92–98.

Pecorelli, A., Belmonte, G., Meloni, I., Cervellati, F., Gardi, C., Sticozzi, C., De Felice, C., Signorini, C., Cortelazzo, A., Leoncini, S., Ciccoli, L., Renieri, A., Jay Forman, H., Hayek, J., & Valacchi, G. (2015). Alteration of serum lipid profile, SRB1 loss, and impaired Nrf2 activation in CDKL5 disorder. *Free Radical Biology & Medicine, 86*, 156–165.

Peng, Y., Myers, R., Zhang, W., & Alexov, E. (2018). Computational investigation of the missense mutations in DHCR7 gene associated with Smith-Lemli-Opitz syndrome. *International Journal of Molecular Sciences, 19*, 141.

Pfrieger, F. W. (2003). Cholesterol homeostasis and function in neurons of the central nervous system. *Cellular and Molecular Life Sciences, 60*, 1158–1171.

Pieretti, M., Zhang, F. P., Fu, Y. H., Warren, S. T., Oostra, B. A., Caskey, C. T., & Nelson, D. L. (1991). Absence of expression of the FMR-1 gene in fragile X syndrome. *Cell, 66*, 817–822.

Pisciotta, L., Hamilton-Craig, I., Tarugi, P., Bellocchio, A., Fasano, T., Alessandrini, P., Bon, G. B., Siepi, D., Mannarino, E., Cattin, L., Averna, M., Cefalù, A. B., Cantafora, A., Calandra, S., & Bertolini, S. (2004). Familial HDL deficiency due to ABCA1 gene mutations with or without other genetic lipoprotein disorders. *Atherosclerosis, 172*, 309–320.

Platt, F. M., Wassif, C., Colaco, A., Dardis, A., Lloyd-Evans, E., Bembi, B., & Porter, F. D. (2014). Disorders of cholesterol metabolism and their unanticipated convergent mechanisms of disease. *Annual Review of Genomics and Human Genetics, 15*, 173–194.

Porter, F. D. (2003). Human malformation syndromes due to inborn errors of cholesterol synthesis. *Current Opinion in Pediatrics, 15*, 607–613.

Porter, F. D., & Herman, G. E. (2011). Malformation syndromes caused by disorders of cholesterol synthesis. *Journal of Lipid Research, 52*, 6–34.

Porter, J. A., Young, K. E., & Beachy, P. A. (1996). Cholesterol modification of hedgehog signaling proteins in animal development. *Science, 274*, 255–259.

Pucadyil, T. J., & Chattopadhyay, A. (2004). Cholesterol modulates ligand binding and G-protein coupling to serotonin(1A) receptors from bovine hippocampus. *Biochimica et Biophysica Acta, 1663*, 188–200.

Purcell, A. E., Jeon, O. H., & Pevsner, J. (2001). The abnormal regulation of gene expression in autistic brain tissue. *Journal of Autism and Developmental Disorders, 31*, 545–549.

Qiu, C., Chen, G., & Cui, Q. (2012). Towards the understanding of microRNA and environmental factor interactions and their relationships to human diseases. *Scientific Reports, 2*, 318.

Rawlins, F. A., & Uzman, B. G. (1970). Retardation of peripheral nerve myelination in mice treated with inhibitors of cholesterol biosynthesis. A quantitative electron microscopic study. *The Journal of Cell Biology, 46*, 505–517.

Rees, A., Stocks, J., Sharpe, C. R., Vella, M. A., Shoulders, C. C., Katz, J., Jowett, N. I., Baralle, F. E., & Galton, D. J. (1985). Deoxyribonucleic acid polymorphism in the apolipoprotein A-1-C-III gene cluster. Association with hypertriglyceridemia. *The Journal of Clinical Investigation, 76*, 1090–1095.

Reitz, C., Tang, M. X., Luchsinger, J., & Mayeux, R. (2004). Relation of plasma lipids to Alzheimer disease and vascular dementia. *Archives of Neurology, 61*, 705–714.

Riobo, N. A. (2012). Cholesterol and its derivatives in Sonic Hedgehog signaling and cancer. *Current Opinion in Pharmacology, 12*, 736–741.

Roux, C., Wolf, C., Mulliez, N., Gaoua, W., Cormier, V., Chevy, F., & Citadelle, D. (2000). Role of cholesterol in embryonic development. *The American Journal of Clinical Nutrition, 71*, 1270s–1279s.

Roy, R., & Bélanger, A. (1990). Nonpolar Pregnenolone metabolites are transported in lipoprotein fractions in human serum. *Annals of the New York Academy of Sciences, 595*, 432–434.
Saczynski, J. S., White, L., Peila, R. L., Rodriguez, B. L., & Launer, L. J. (2007). The relation between apolipoprotein A-I and dementia: The Honolulu-Asia aging study. *American Journal of Epidemiology, 165*, 985–992.
Saher, G., Brügger, B., Lappe-Siefke, C., Möbius, W., Tozawa, R. -I., Wehr, M. C., Wieland, F., Ishibashi, S., & Nave, K, -A. (2005). High cholesterol level is essential for myelin membrane growth. *Nature Neuroscience, 8*, 468–475.
Sandin, S., Lichtenstein, P., Kuja-Halkola, R., Hultman, C., Larsson, H., & Reichenberg, A. (2017). The heritability of autism spectrum disorder. *JAMA, 318*, 1182–1184.
Segatto, M., Di Giovanni, A., Marino, M., & Pallottini, V. (2013). Analysis of the protein network of cholesterol homeostasis in different brain regions: An age and sex dependent perspective. *Journal of Cellular Physiology, 228*, 1561–1567.
Segatto, M., Trapani, L., Di Tunno, I., Sticozzi, C., Valacchi, G., Hayek, J., & Pallottini, V. (2014). Cholesterol metabolism is altered in Rett syndrome: A study on plasma and primary cultured fibroblasts derived from patients. *PLoS One, 9*, e104834.
Seidel, D., Alaupovic, P., Furman, R. H., & Mcconathy, W. J. (1970). A lipoprotein characterizing obstructive jaundice. II. Isolation and partial characterization of the protein moieties of low density lipoproteins. *The Journal of Clinical Investigation, 49*, 2396–2407.
Sharifi, M., Futema, M., Nair, D., & Humphries, S. E. (2017). Genetic architecture of familial hypercholesterolaemia. *Curr Cardiol Rep, 19*, 44.
Sharp, D., Blinderman, L., Combs, K. A., Kienzle, B., Ricci, B., Wager-Smith, K., Gil, C. M., Turck, C. W., Boumas, M.-E., Rader, D. J., Aggerbeck, L. P., Gregg, R. E., Gordon, D. A., & Wetterau, J. R. (1993). Cloning and gene defects in microsomal triglyceride transfer protein associated with abetalipoproteinaemia. *Nature, 365*, 65–69.
Sikich, L., Kolevzon, A., King, B. H., Mcdougle, C. J., Sanders, K. B., Kim, S. J., Spanos, M., Chandrasekhar, T., Trelles, M. D. P., Rockhill, C. M., Palumbo, M. L., Witters Cundiff, A., Montgomery, A., Siper, P., Minjarez, M., Nowinski, L. A., Marler, S., Shuffrey, L. C., Alderman, C., Weissman, J., Zappone, B., Mullett, J. E., Crosson, H., Hong, N., Siecinski, S. K., Giamberardino, S. N., Luo, S., She, L., Bhapkar, M., Dean, R., Scheer, A., Johnson, J. L., Gregory, S. G., & Veenstra-Vanderweele, J. (2021). Intranasal oxytocin in children and adolescents with autism spectrum disorder. *The New England Journal of Medicine, 385*, 1462–1473.
Sikora, D. M., Pettit-Kekel, K., Penfield, J., Merkens, L. S., & Steiner, R. D. (2006). The near universal presence of autism spectrum disorders in children with Smith-Lemli-Opitz syndrome. *American Journal of Medical Genetics. Part A, 140*, 1511–1518.
Smeets, E. E., Pelc, K., & Dan, B. (2012). Rett Syndrome. *Molecular Syndromology, 2*, 113–127.
Solcà, C., Pandit, B., Yu, H., Tint, G. S., & Patel, S. B. (2007). Loss of apolipoprotein E exacerbates the neonatal lethality of the Smith-Lemli-Opitz syndrome mouse. *Molecular Genetics and Metabolism, 91*, 7–14.
Song, H., Saito, K., Seishima, M., Noma, A., Urakami, K., & Nakashima, K. (1997). Cerebrospinal fluid apo E and apo A-I concentrations in early- and late-onset Alzheimer's disease. *Neuroscience Letters, 231*, 175–178.
Steeb, H., Ramsey, J. M., Guest, P. C., Stocki, P., Cooper, J. D., Rahmoune, H., Ingudomnukul, E., Auyeung, B., Ruta, L., Baron-Cohen, S., & Bahn, S. (2014). Serum proteomic analysis identifies sex-specific differences in lipid metabolism and inflammation profiles in adults diagnosed with Asperger syndrome. *Molecular Autism, 5*, 4.
Sticozzi, C., Belmonte, G., Pecorelli, A., Cervellati, F., Leoncini, S., Signorini, C., Ciccoli, L., De Felice, C., Hayek, J., & Valacchi, G. (2013). Scavenger receptor B1 post-translational modifications in Rett syndrome. *FEBS Letters, 587*, 2199–2204.
Stoffel-Wagner, B. (2001). Neurosteroid metabolism in the human brain. *European Journal of Endocrinology, 145*, 669–679.
Strittmatter, W. J., Saunders, A. M., Schmechel, D., Pericak-Vance, M., Enghild, J., Salvesen, G. S., & Roses, A. D. (1993). Apolipoprotein E: High-avidity binding to beta-amyloid and

increased frequency of type 4 allele in late-onset familial Alzheimer disease. *Proceedings of the National Academy of Sciences of the United States of America, 90*, 1977–1981.

Strous, R. D., Golubchik, P., Maayan, R., Mozes, T., Tuati-Werner, D., Weizman, A., & Spivak, B. (2005). Lowered DHEA-S plasma levels in adult individuals with autistic disorder. *European Neuropsychopharmacology, 15*, 305–309.

Svoboda, M. D., Christie, J. M., Eroglu, Y., Freeman, K. A., & Steiner, R. D. (2012). Treatment of Smith-Lemli-Opitz syndrome and other sterol disorders. *American Journal of Medical Genetics. Part C, Seminars in Medical Genetics, 160c*, 285–294.

Talebizadeh, Z., Butler, M. G., & Theodoro, M. F. (2008). Feasibility and relevance of examining lymphoblastoid cell lines to study role of microRNAs in autism. *Autism Research, 1*, 240–250.

Tarugi, P., Averna, M., Di Leo, E., Cefalù, A. B., Noto, D., Magnolo, L., Cattin, L., Bertolini, S., & Calandra, S. (2007). Molecular diagnosis of hypobetalipoproteinemia: An ENID review. *Atherosclerosis, 195*, e19–e27.

Thurm, A., Tierney, E., Farmer, C., Albert, P., Joseph, L., Swedo, S., Bianconi, S., Bukelis, I., Wheeler, C., Sarphare, G., Lanham, D., Wassif, C. A., & Porter, F. D. (2016). Development, behavior, and biomarker characterization of Smith-Lemli-Opitz syndrome: An update. *Journal of Neurodevelopmental Disorders, 8*, 12.

Tierney, E., Nwokoro, N. A., Porter, F. D., Freund, L. S., Ghuman, J. K., & Kelley, R. I. (2001). Behavior phenotype in the RSH/Smith-Lemli-Opitz syndrome. *American Journal of Medical Genetics, 98*, 191–200.

Tierney, E., Bukelis, I., Thompson, R. E., Ahmed, K., Aneja, A., Kratz, L., & Kelley, R. I. (2006). Abnormalities of cholesterol metabolism in autism spectrum disorders. *American Journal of Medical Genetics. Part B, Neuropsychiatric Genetics, 141b*, 666–668.

Tierney, E., Conley, S. K., Goodwin, H., & Porter, F. D. (2010). Analysis of short-term behavioral effects of dietary cholesterol supplementation in Smith-Lemli-Opitz syndrome. *American Journal of Medical Genetics. Part A, 152a*, 91–95.

Tietz, S., & Engelhardt, B. (2015). Brain barriers: Crosstalk between complex tight junctions and adherens junctions. *The Journal of Cell Biology, 209*, 493–506.

Tint, G. S. (1993). Cholesterol defect in Smith-Lemli-Opitz syndrome. *American Journal of Medical Genetics, 47*, 573–574.

University of Colorado Denver. (2008). *Cholesterol and antioxidant treatment in patients with Smith-Lemli-Opitz Syndrome (SLOS)*. https://ClinicalTrials.gov/show/NCT01773278

University of Nebraska, University of Colorado, D., Children's Hospital Medical Center, C. & Pittsburgh, U. O. (2021). *Smith-Lemli-Opitz syndrome and cholic acid*. https://ClinicalTrials.gov/show/NCT03720990

Valenza, M., Rigamonti, D., Goffredo, D., Zuccato, C., Fenu, S., Jamot, L., Strand, A., Tarditi, A., Woodman, B., Racchi, M., Mariotti, C., Di Donato, S., Corsini, A., Bates, G., Pruss, R., Olson, J. M., Sipione, S., Tartari, M., & Cattaneo, E. (2005). Dysfunction of the cholesterol biosynthetic pathway in Huntington's disease. *The Journal of Neuroscience, 25*, 9932–9939.

Vaughan, D. K., Peachey, N. S., Richards, M. J., Buchan, B., & Fliesler, S. J. (2006). Light-induced exacerbation of retinal degeneration in a rat model of Smith-Lemli-Opitz syndrome. *Experimental Eye Research, 82*, 496–504.

Vekic, J., Kotur-Stevuljevic, J., Jelic-Ivanovic, Z., Spasic, S., Spasojevic-Kalimanovska, V., Topic, A., Zeljkovic, A., Stefanovic, A., & Zunic, G. (2007). Association of oxidative stress and PON1 with LDL and HDL particle size in middle-aged subjects. *European Journal of Clinical Investigation, 37*, 715–723.

Vickers, K. C., & Remaley, A. T. (2013). Functional diversity of HDL cargo. *Journal of Lipid Research, 55*, 4–12.

Vickers, K. C., Palmisano, B. T., Shoucri, B. M., Shamburek, R. D., & Remaley, A. T. (2011). MicroRNAs are transported in plasma and delivered to recipient cells by high-density lipoproteins. *Nature Cell Biology, 13*, 423–433.

Vitali, C., Wellington, C. L., & Calabresi, L. (2014). HDL and cholesterol handling in the brain. *Cardiovascular Research, 103*, 405–413.

Wang, H., & Eckel, R. H. (2014). What are lipoproteins doing in the brain? *Trends in Endocrinology and Metabolism, 25*, 8–14.

Wassif, C. A., Maslen, C., Kachilele-Linjewile, S., Lin, D., Linck, L. M., Connor, W. E., Steiner, R. D., & Porter, F. D. (1998). Mutations in the human sterol delta7-reductase gene at 11q12-13 cause Smith-Lemli-Opitz syndrome. *American Journal of Human Genetics, 63*, 55–62.

Wassif, C. A., Kratz, L., Sparks, S. E., Wheeler, C., Bianconi, S., Gropman, A., Calis, K. A., Kelley, R. I., Tierney, E., & Porter, F. D. (2017). A placebo-controlled trial of simvastatin therapy in Smith-Lemli-Opitz syndrome. *Genetics in Medicine, 19*, 297–305.

Waterham, H. R., Wijburg, F. A., Hennekam, R. C., Vreken, P., Poll-The, B. T., Dorland, L., Duran, M., Jira, P. E., Smeitink, J. A., Wevers, R. A., & Wanders, R. J. (1998). Smith-Lemli-Opitz syndrome is caused by mutations in the 7-dehydrocholesterol reductase gene. *American Journal of Human Genetics, 63*, 329–338.

Wen, Z., Cheng, T. L., Li, G. Z., Sun, S. B., Yu, S. Y., Zhang, Y., Du, Y. S., & Qiu, Z. (2017). Identification of autism-related MECP2 mutations by whole-exome sequencing and functional validation. *Molecular Autism, 8*, 43.

Wetterau, J. R., Aggerbeck, L. P., Bouma, M. E., Eisenberg, C., Munck, A., Hermier, M., Schmitz, J., Gay, G., Rader, D. J., & Gregg, R. E. (1992). Absence of microsomal triglyceride transfer protein in individuals with abetalipoproteinemia. *Science, 258*, 999–1001.

Wu, Y. E., Parikshak, N. N., Belgard, T. G., & Geschwind, D. H. (2016). Genome-wide integrative analysis implicates microRNA dysregulation in autism spectrum disorder. *Nature Neuroscience, 19*, 1463–1476.

Xu, J., Dang, Y., Ren, Y. R., & Liu, J. O. (2010). Cholesterol trafficking is required for mTOR activation in endothelial cells. *Proceedings of the National Academy of Sciences of the United States of America, 107*, 4764–4769.

Yamakawa-Kobayashi, K., Yanagi, H., Fukayama, H., Hirano, C., Shimakura, Y., Yamamoto, N., Arinami, T., Tsuchiya, S., & Hamaguchi, H. (1999). Frequent occurrence of Hypoalphalipoproteinemia due to mutant Apolipoprotein A-I gene in the population: A population-based survey. *Human Molecular Genetics, 8*, 331–336.

Yang, C., Qi, Y., & Sun, Z. (2021). The role of Sonic Hedgehog pathway in the development of the central nervous system and aging-related neurodegenerative diseases. *Frontiers in Molecular Biosciences, 8*, 711710.

Yaplito-Lee, J., Pai, G., Hardikar, W., Hong, K. M., Pitt, J., Marum, J., & Amor, D. J. (2020). Successful treatment of lathosterolosis: A rare defect in cholesterol biosynthesis-a case report and review of literature. *JIMD Reports, 56*, 14–19.

Yu, H., & Patel, S. B. (2005). Recent insights into the Smith-Lemli-Opitz syndrome. *Clinical Genetics, 68*, 383–391.

Yue, P., Averna, M., Lin, X., & Schonfeld, G. (2006). The c.43_44insCTG variation in PCSK9 is associated with low plasma LDL-cholesterol in a Caucasian population. *Human Mutation, 27*, 460–466.

Zeng, P., & Zhou, X. (2019). Causal effects of blood lipids on amyotrophic lateral sclerosis: A Mendelian randomization study. *Human Molecular Genetics, 28*, 688–697.

Zhang, J., & Liu, Q. (2015). Cholesterol metabolism and homeostasis in the brain. *Protein & Cell, 6*, 254–264.

Zhao, Z., Tuakli-Wosornu, Y., Lagace, T. A., Kinch, L., Grishin, N. V., Horton, J. D., Cohen, J. C., & Hobbs, H. H. (2006). Molecular characterization of loss-of-function mutations in PCSK9 and identification of a compound heterozygote. *American Journal of Human Genetics, 79*, 514–523.

Zhou, A. L., Swaminathan, S. K., Curran, G. L., Poduslo, J. F., Lowe, V. J., Li, L., & Kandimalla, K. K. (2019). Apolipoprotein A-I crosses the blood-brain barrier through Clathrin-independent and cholesterol-mediated endocytosis. *The Journal of Pharmacology and Experimental Therapeutics, 369*, 481–488.

Zhu, H., & Fan, G. C. (2011). Extracellular/circulating microRNAs and their potential role in cardiovascular disease. *The American Journal of Cardiovascular Disease, 1*, 138–149.

Chapter 9
Perinatal Insulin-Like Growth Factor as a Risk Factor for Autism

Gary Steinman and David Mankuta

9.1 Introduction

> *"My first-born child, Mikey, was fine during his first 18 months of infancy. He was growing normally and was cheerful within a loving attentive home environment. Mikey played with his toys and liked to share them with playmates. However, nearing his second birthday, we began to notice that he was not developing socially similar to other children his age. He now tended to avoid interaction with other children and seemed to prefer playing by himself. Mikey would spend hours arranging his blocks in a row, disassembling them, and then repeating the same action over and over again. We expected him to be speaking single words and two-word phrases by this time, but he only babbled incoherently without answering us when we called him by name. He acted as though we were not in the same room with him.*
>
> *"When our pediatrician was approached about this situation, he suggested that our son should be evaluated by a neuropsychologist. I told the specialist that I had no medical difficulties before or during my pregnancy with Mikey. None of the children in our extended families had problems comparable to what we were witnessing with our son. We lived in a comfortable, middle-class, pollution-free neighborhood. Mikey weighed 7 pounds at his 40th-week birth and cried soon after delivery, following an 8-hour, uncomplicated labor. My husband and I were both 28 years old at the time of Mikey's birth.*
>
> *"After a detailed evaluation, the specialist concluded that our son was autistic."*
>
> \- Signed, distressed mother (fictitious)

Autism is highly variable among those affected by it. The disorder in general is characterized in the patient bearing it as a neurodevelopmental malady marked by deficiencies in social interaction, communication, outside interests, and behavior. Usually, no signs or symptoms in the gravida during the pregnancy nor in the early months following birth give a warning in the child that autism will develop.

G. Steinman (✉) · D. Mankuta
Department of Obstetrics & Gynecology, Hadassah Hospital – Hebrew University, Jerusalem, Israel

A. El Idrissi, D. McCloskey (eds.), *Neurobiology of Autism Spectrum Disorders*,
https://doi.org/10.1007/978-3-031-42383-3_9

Prior to the release of the fifth edition of the Diagnostic and Statistical Manual of Mental Disorders (DSM-5), much confusion existed with differentiating Asperger's Syndrome, Pervasive Development Disorder – not otherwise specified (PDD-NOS), Childhood Disintegrative Disorder, and Autistic Disorder. The designation of PDD was often applied when all criteria for Asperger's and/or Autistic Disorder were not met.

Based on scientific advances, targeted research, and clinical experience, the classification scheme for these conditions was modified in 2013. The nomenclature in DSM-5 differentiated all cases deemed to be autism primarily based on severity. The general term of "autism spectrum disorder" (ASD) includes all of the earlier named maladies in this group and is classified by degree of behavioral involvement (but not intellectual abilities). In particular, ASD is seen as a psychological condition which had presentations ranging from mildly to severely affected (the "Spectrum").

As defined in the DSM-5 and ranked by degree, ASD is further diagnosed by at least three deficits in childhood and is caused by significant impairment in the customary social or occupational areas of:

1. Social and emotional reciprocity;
2. Non-verbal communication behaviors;
3. Developing and maintaining appropriate relationships;
4. Repetitive speech, movements, or employing of objects;
5. Excessive maintenance of routines, formalized patterns of verbal/nonverbal behavior, or excessive refusal to change;
6. Highly restricted, unusual, fixated interests (e.g., strong preoccupation with atypical objects); and
7. Hyper- or hypo-reactivity to sensory input or extraordinary interest in appreciated aspects of the environment (Steinman, 2013a, b, 2014; Steinman & Mankuta, 2013, 2014).

With this definition of ASD, it was found that impairment of normal motor control, behavior, and learning skills in autistic children characterize this neurological disorder. In test animals with ASD-like characteristics, synaptic plasticity, especially in the cerebellum, appears to be at the core of at least some of the cases of such behavior. For example, dysfunctional cerebellar Purkinje cells contribute to autism-like behavior in Shank2-deficient mice. It is proposed that the release of GABA (gamma-aminobutyric acid) by these cerebellar cells reduces nerve impulse transmission, as also observed in some children with autism or with Niemann-Pick disease type C. However, these changes are not seen in all autistic children, thereby challenging the premise that such genetic or anatomic changes can be primary causes of all cases of autism. Both possibilities now to be examined in more detail.

Therefore, in evaluating a child with behavioral problems, it is essential to carefully consider observations, analyze pertinence and credibility of actions, and distinguish between correlations and cause-and-effect associations. As noted in the opening maternal letter above, usually no signs or symptoms in the mother during the pregnancy or in the neonate in the early months following birth give a warning that autism may develop in the child (Steinman, 2014). Later in this discussion, the means for reducing the chance of autism development in the youngster, especially when the mother experienced a serious febrile course during the pregnancy, will be reviewed.

Confusion often results if a distinction is not clearly made between correlation and causation. Coincidence or association does not necessarily imply causation. It is important to question such terms as "appears to be", "implicated in", "associated with", "linked to", and "suggesting." The weakest of all such mental exercises is "unsubstantiated intuition." This is especially risky if a psychological overtone or a false sense of urgency is subjectively implanted into the analysis of research, diagnosis, or emotional findings. Uncorroborated claims advanced before a specific outcome are realized should be suspect. As a result, caution must be used in claiming the definitive etiology of a complex medical condition such as autism.

In 1883, Robert Koch proposed three essential requirements to define the basis of particular medical maladies:

1. Suspected causal facts must be constantly and recurrently associated with the disease;
2. Dysfunction must be identified within an affected individual;
3. When a susceptible host is modified appropriately, such as with genetic silencing, the same symptoms of the spontaneous disease must develop in the test case when occurring naturally.

Requirement #3 is especially difficult to realize in psychosocial conditions such as autism if a cause other than genetic alteration is the key factor. (Steinman, 2014).

9.2 Methods

(a) **Subject literature review**

Prior publications have emphasized the central role played by ***insulin-like growth factor-1*** (IGF1) in attempting to prevent autism (Steinman, 2013a, b, 2014, 2015, 2017, 2019a, b, c, 2020a, b; Steinman & Mankuta, 2013, 2014, 2018, 2019, 2020a, 2021a). IGF1, a major multipurpose growth factor, is produced primarily by the liver. In particular, this factor promotes the myelination of new nerves as the babies develop their fetal/neonatal central and peripheral nervous systems. Some published reports have proposed etiologies of autism emphasizing diminished development of active nerve circuits, especially in the newborn. The level of IGF1 in human milk is higher than cow's milk, thus enhancing neural development and maturation (Steinman & Mankuta, 2019, 2021a).

IGF1 is produced in many tissues, especially the liver. As early as 2001, Riikonen and then Patterson began drawing attention to juvenile cases with neurochemical deficits of cerebrospinal IGF1, which appeared to coexist with autism. For example, in obstetrical cases where the mother suffered from antepartum fever-inducing viral

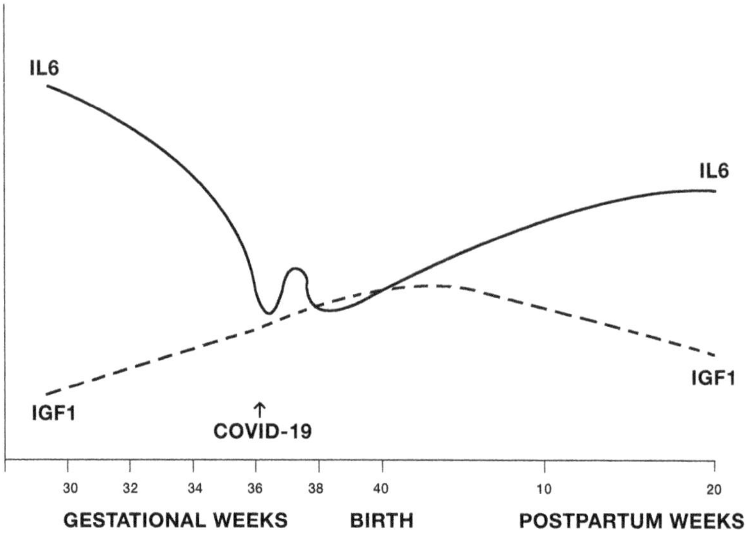

Fig. 9.1 Postulated relationship between serum IL6 and IGF1 levels to autism generation following a maternal covid-19 bout prior to parturition

Table 9.1 Key differences between Rett Syndrome and Autism

Trait	Rett Syndrome	Autism
Cranial abnormalities	Microcephaly	Macrocephaly
Primary effects	Cholinergic neurons	Serotonergic neurons
Brain abnormalities	Frontal cortex path.	Cerebellum-hippocampus
Age of onset	Postnatal	Early prenatal
Mutation site	MECP2[a]	Undetermined
Gender association	Almost all female	Male/female (4/1)

Based on Refs. Riikonen et al. (2006), Vanhala et al. (2001)
[a]*MECP2* methyl CpG binding protein-2

illnesses (e.g., coronavirus, influenza), increases in inflammatory interleukins and a decrease in IGF1 were frequently found (Fig. 9.1). This may be followed by birth of offspring destined to become autistic (Steinman, 2020a, b; Steinman & Mankuta, 2020a, 2021a, b; Vanhala et al., 2001; Riikonen et al., 2006; Riikonen, 2006, 2008, 2016, 2017; Makkonen et al., 2011; Patterson, 2011; Hsiao & Patterson, 2011; Hsiao et al., 2012; Leventhal et al., 1999).

This biochemical approach to explaining the etiology of autism is in marked contradistinction to often incorrectly cited genetic errors. For example, Rett Syndrome has been discussed as an example of an autism-like condition. The key differences are illustrated in Table 9.1.

However, over 100 various nuclear defects have been reported but have occurred in fewer than 10% of cases of autism (Leventhal et al., 1999; Geschwind, 2011). In contradistinction, a biochemical error such as deficient IGF1 is central to inadequate myelination of neo-neurons in the developing fetus and neonate. This results in reduced velocity (down to as little as 0.1% compared to similar myelinated neurons) of normally functioning nerve impulses in the brain and subnormal neurologic function in key brain centers (Herman, 2007).

Although the genetic approach to elucidating the etiology of autism has been less emphasized recently, several studies have reported epidemiological examinations linking ASD with exposure to various viruses such as rubella, measles and mumps, polyomarviruses, cytomegalovirus, and influenza (Shuid et al., 2021; Meltzer & Van de Water, 2017; Choi et al., 2016). This will be explained more fully later in the discussion about the interactions between IGF1 and IL-6 in particular.

As early as 1988, bovine colostrum and milk were found to contain insulin-like growth factor-1 (Francis et al., 1988). Supplying IGF1 via human breast milk exclusively for the first postpartum year has been reported in several large meta-analytical studies to reduce the incidence of autism significantly. After this, when autistic traits appear overtly in children (typically at ages 1–4 years old), it would be too late to correct the biochemical/behavioral problems, within the current limitations of available therapy. Low levels of IGF1 occur in the cerebrospinal fluid significantly more often in young children with autism than in normal individuals. Hence, it would be desirable to test all umbilical cord bloods at birth for insufficient IGF1 to identify the most at-risk neonates, in particular (Steinman, 2013a, b, 2019a, b, c; Schultz et al., 2006; Boucher et al., 2017; Al-Farsi et al., 2012; Soke et al., 2019; Ghozy et al., 2020; Shafai et al., 2014, 2017; Hong et al., 2014; Chen et al., 2021; Tseng et al., 2017; Huang et al., 2022; Lawrence, 2014; Shamsedine et al., 2020).

In babies born preterm, the serum IGF1 concentration often correlates with brain volume, unmyelinated white matter volume, cerebellar volume, and subnormal mental development (Hellstrom et al., 2016). Treating children already with idiopathic autism with IGF1 supplement induces transcriptional responses in their neurons, although complete cure is usually unattainable (Linker et al., 2020). Worldwide, almost 70% of all infants are not breastfed exclusively for at least 6 months postpartum.

Potentially more reliable would be to treat all newborn with milk rich in IGF1. Such is the case with human lactation. Many infants are not fed human milk, which has a higher natural provision of IGF1 than cow's milk. Children nursed on human milk exclusively for 6 months have 75% lower chance of developing autism than those fed by other means. To be discussed here in detail is the relatively equal dietary benefit for human milk to be derived from bovine milk or formula which has been augmented with synthetic IGF in this important regard. IGF1 is a linear polypeptide with 70 amino acid units joined together by peptide bonds. Although it is a polypeptide orally ingested, IGF1 apparently survives passage through the stomach (Shen & Xu, 2000; Philipps et al., 2000). In the current market, this polypeptide when isolated from synthetic or natural sources is expensive to purchase. This makes it unreachable by the broad market of women looking to feed their babies with this beneficial additive to food forms other than human milk. The historical trend to follow is the increasing ability to produce synthetic, effective IGF1 at a reasonable cost.

Beginning in 2012, nine meta-analytical studies have shown the connection between neonatal breast-feeding and the reduction of autism in young children (Boucher et al., 2017; Al-Farsi et al., 2012; Soke et al., 2019; Ghozy et al., 2020; Shafai et al., 2014, 2017; Chen et al., 2021; Tseng et al., 2017; Huang et al., 2022; Lawrence, 2014; Shamsedine et al., 2020). The key to this was apparently the higher level of IGF1 in human milk than in bovine milk or formula. The longer the mother continued exclusive breast-feeding, the lower was the typical occurrence of autism in the neonate. For example, a multi-center study in China of the benefit of breast-feeding for 6 months reported an autism percent prevalence of 2.4 in those with no breast-feeding and 0.7 with exclusive breast-feeding (Hellstrom et al., 2016; Linker et al., 2020; Huang et al., 2022). These results are comparable to those found in several other comparative studies. [Reported overall rates in a given population may depart from the exact value because of idiosyncratic characteristics of each community or country, and the reliability and completeness of methods used to collect the data. An additional essential factor for data analysis in a particular population is the medical definition of the condition under investigation.]

Gravidas who experience a severe febrile disease, such as coronavirus, during the pregnancy are often low in vitamin D3 and IGF1. As a result, babies fed formula much of the first year of life display a significantly higher occurrence of autism spectrum disorder later in childhood and beyond for the same reason. For those women unable to provide sufficient breast milk to their neonates, human milk supplies are commercially available. Similarly, milk augmented with commercially prepared IGF1 added to milk or formula should serve the same purpose (Steinman, 2020a, b; Steinman & Mankuta, 2020a, 2021b). If one child born to parents is autistic, the chance of a subsequent birth also yielding an affected child is 20%. If the maternal and neonatal diets were supplemented with extra vitamin D, the affected sibling rate dropped to as little as 5% (Stefenburg et al., 1989). These studies would support the advice of providing neonates with enough vitamin D3 and IGF1 from maternal milk for a year to reduce the chance of developing autism. This is especially important for babies with growth hormone deficiency (Linker et al., 2020). Alternatively, if women cannot breastfeed the full postpartum year, means should be developed to provide the equivalent IGF1 and D3 nourishment by an alternative mode.

Vitamin D enhances circulating IGF1 and helps reduce the level of interleukins in the presence of coexistent inflammatory processes. An example is treatment of otherwise normal individuals with 5000–7000 iu of vitamin D3/week; control volunteers received no D supplement. As a result, IGF1 was raised in the test group by 12.7–13.1 ng/ml, compared to no change in the control group. Thus, vitamin D was shown to increase IGF1 in these participants (Ameri et al., 2013).

(b) **Deficient biosynthesis**

The antepartum placenta has the capacity to generate a wide range of needed metabolic substances. By 1977, it had been determined that a large number of biochemicals necessary for fetal development were produced by the placenta (Ville, 1977). By 1980, the biosynthesis of additional placental proteins and peptide hormones were reported (Munro, 1980). Cultured placental cells are now employed commercially to prepare needed agents in specific medical situations.

Among these controlling biochemicals, IGF1 was studied for its many known functions and applications. Beginning in 2001, this agent was evaluated for its effect on the development of the nervous system in particular. Type I IGF1 receptors are found in much of the neonatal body, but especially in the vicinity of neural tissue. IGF1 is an important factor in axonal regeneration, astrocyte proliferation, and neurite sprouting, especially with neural myelination. The velocity of impulse transport in nerves may vary by whether they are myelinated or not (Herman, 2007). IGF1 directly affects the rate at which oligodendrocytes promote myelination and neural circuitry development in the infantile CNS. The level of IGF1 in the neonate may be reduced because of inadequate nutrition, advanced parental age, or immune activation. Insufficient myelination may be the result of genetic alterations, polymorphisms, or mutations in the baby's nuclear makeup, but this accounts for neurologic problems in only a small percentage of abnormal cases (Kanner, 1943; Diaz-Gomez et al., 1997).

Although juvenile autism was first observed in 11 children and reported in 1943 as a psychological disturbance by Leo Kanner (Kanner, 1943), the identification of IGF1 deficiency as the apparent cause was not expressed until the 1990s by D'Ercole (D'Ercole, 1993). Since the general definition and characterization of autism in the early 1940s was as a unique psychopathological entity with varying presentations, little attention was paid to it by the general public until the last 20 years or so.

In 2001, Riikonen and coworkers began reporting a series of studies on IGF1 deficiency in the cerebrospinal fluid of autistic youngsters compared to neurologically normal children (Vanhala et al., 2001; Riikonen et al., 2006; Riikonen, 2006, 2008, 2016, 2017; Makkonen et al., 2011). In 2011, Patterson studied maternal infection and immune involvement in autism generation (Patterson, 2011). He detailed the reactions in the placenta in the presence of maternal inflammatory reactions, which raised the level of IL-6 (interleukin 6) and activated JAK/STAT3, thereby causing a reduction of IGF1 production. (For the current relevance of this, see references Steinman & Mankuta, 2020b, 2021a, b).

The level of IGF1 at birth can also be proportional to the baby's size, apparently due to an underactive placenta or antepartum maternal infections. Therefore, more attention must be paid to underweight neonates specifically for possible inadequate IGF1 levels, which are often an early warning sign of potentially developing autism later.

As adults grow older, their serum IGF1 levels decrease. Thus, older birthing mothers would have reduced blood IGF1 in comparison to younger mothers. This may account in part for why older parents are more likely to give birth to babies who develop autism.

Another factor which has been found to be related to autism occurrence in young children is vitamin D supplementation. This is apparently because of the increased level of IGF1 and reduced levels of interleukins in the presence of coexistent inflammatory processes in children receiving this vitamin augmentation. Of interest is the metabolic increase of IGF1 in response to dietary vitamin D supplementation, especially during coexistent inflammatory processes in the body. This supports the contention that this vitamin in particular can be a stimulus to IGF1 enhancement (Wang et al., 2020; Stubbs et al., 2016). To contemplate ways to produce IGF1 artificially, peptide synthesis methods need review for their utility.

(c) **Extracorporeal peptide synthesis**

The first successful method for amino acid polymerization in the laboratory was the solid phase synthesis of a tetrapeptide by RB Merrifield, published in 1963. The approach utilized protected amino acids joined to an enlarging peptide chain, attached covalently to a resin particle. For the 10 years before this, the method at that time involved amino acid residues which created polymers that became increasing insoluble and unreactive with enlarging size as the reactions proceeded. The advantage of the Merrifield technique is that the polymer is attached to an insoluble resin component, thereby allowing for washing of unused reactants as the polypeptide grew (Merrifield, 1963). This would shorten the time to complete the peptide polymer production compared to earlier methods.

The next improvement in methodology was completed by Paul Berg in 1972. This process, named genetic engineering, involved removing DNA from a bacterium and placing it in a virus (a recombinant cell). A variation of this was to fuse a tumor cell with a cell able to produce proteins, now called a hydrioma (Jackson & Berg, 1972). By 1977, evidence for the production by the human placenta was reported for chorionic gonadotropin, lactogen, thyrotropin, follicle stimulating hormone, relaxin, oxytocin, vasopressin, alkaline phosphatase, and renin. By 1982, this recombinant method was approved for producing insulin in E. coli for use in human patients.

In 2014, Simon, Pentelute, and coworkers published a solid-based peptide synthesis technique which was rapid flow-based. Another amino acid could be added to the growing polymer every 2–3 minutes. Reagents and solvents were delivered to a solid support at high flow rate, thereby accelerating the production of polymer. The process was continuously monitored by UV detectors. Polypeptides as large as 130 residues could be produced by this method in relatively short time (Simon et al., 2014).

By 2016, polymerase chain reaction (PCR) had been applied in several biochemical studies of DNA. Using this technique small amounts of DNA could be rapidly reproduced to yield significant quantities of specific nucleic acids. Now linear DNA expression templates synthesized directly from PCR could be used to promote cell-free protein syntheses (Schinn et al., 2016).

Automated flow-based accelerated peptide synthesis was published by Mijalis, Pentelute, and associates in 2018 to execute rapid solid-phase polypeptide synthesis. The system was fully automated and created new peptide bonds every 40 seconds. In less than one hour, it can form a complete peptide polymer with up to 60 amino acids linked together with 99% efficiency. The mixture of reactants is briefly heated to 90 degrees Celsius to make them more reactive (Mijalis et al., 2018).

Polypeptide syntheses in the presence of competing factors in a water/dichloromethane biphasic system was reported in 2019. The aqueous phase was incorporated to extract impurities of the reaction. Polymerization of N-carboxyanhydrides in this reaction environment led to low-cost, large-scale synthesis of polypeptides (Song et al., 2019).

Currently, beneficial peptides can be produced with modified placental cells (See website of Pluristem Therapeutics for details).

IGF1 is biosynthesized primarily in the liver and is under the control of growth hormone for the quantity to be secreted into the general circulation. By IGF1 stimulating oligodendrocytes, myelination of neo-neurons is promoted in the baby. Axonal regeneration is induced by local deposition of the polypeptide. In addition, IGF1 is synthesized in peripheral tissues as an autocrine/paracrine agent for local needs. The hormone promotes its effects via IGF1-specific, cell membrane-bound receptors (Redford, 2014).

(d) **Increasing/decreasing IGF1 availability *in situ***

There are six different IGF-binding proteins (BP) in the human body, usually holding/storing 99% of the available but inactive IGF1 in the circulation. Binding protein #3 (BP_3) carries the largest amount of IGF1. The potential release of IGF1 attached to the bonding protein (BP_3) by cGP (cyclic glycyl-proline) may explain the mechanism ameliorating its deficiency. In the presence of added cGP, IGF1 is released. IGF1 attached to one of the six binding factors is more stable over time than free IGF1 (Steinman & Mankuta, 2013, 2014, 2020a).

$$\mathrm{IGFBP_3 - IGF_1 + cGP \rightarrow IGFBP_3 - cGP + IGF_1}$$

Therefore, it appears possible to utilize laboratory-based methods for mass-producing IGF1, IGF1-promoting agents, or IGF1-mimicking drugs economically. The product itself, IGF1, could be added by the caregiver to bovine milk or formula to reach the nutritional level afforded by human milk, once such a product is available at an acceptable price (Steinman, 2014).

The ability to proceed with exclusive breast-feeding for 6 months or more is often impeded by family size, work commitments outside the home, or maternal illness, among other reasons. Many parents circumvent these obstacles by feeding their infants bovine milk or commercial formula. As noted earlier, these sources typically have significantly less IGF1 and/or vitamin D than mother's milk. One approach to resolving this deficiency would be to secure human milk from commercial sources. Secondly, many lactating mothers use manual pumps to obtain and store their milk while away from the home. A third solution would be to add IGF1 powder to cow's milk or formula. However, this source of the biochemical is commonly used for laboratory research or treatment of childhood growth inadequacy, and is currently quite expensive.

Milk from humans is similar in most parameters to that from cows. Human milk is 4.1% fat totally and from cows 2.5%. Furthermore, bovine milk is 1.1%

unsaturated fat and human milk is 2.3% unsaturated fat, on average. As a result, lipids in human brain development play a greater role than in cattle. Furthermore, this is consistent with cow's whole milk containing, on average, 3.8 iu/L (international units/liter) of vitamin D, whereas human milk has 15 iu/L. The major source of this vitamin D is skin exposure to sunlight. It may be ingested from the diet and supplements as well.

If a method could be developed to make possible bulk commercial manufacturing of IGF1, the financial problem caused by an inadequate, expensive supply of this factor (now about $600/50 microgram) could be avoided. For example, earlier in this report a method for producing biochemicals using placental cells was described. If such a method could be developed to the point where IGF1 would be available in bulk for a reasonable price, it might serve to reduce the incidence of autism in young children. Autism in their child now can cause the parents a lifelong health-related excess expenditure of $3,000,000 or more for therapy, supervision, and medications (Steinman & Mankuta, 2013).

It is apparent that children who are breast-fed generally do not suffer from conditions resulting from excess IGF1 intake. However, if the missing IGF1 is available and added to cow's milk, care must be taken to measure the supplementary amounts so as to avoid abnormal growth of the skeleton. In children, gigantism causes bones to grow longer, resulting in very tall people with large feet and hands, or an error in lipid metabolism. A defect in lipid metabolism is also observed in some patients with acromegaly. The high level of IGF1 in acromegaly is related to an increased risk of some malignancies, particularly colon cancer and thyroid cancer. Excess administered IGF1 can lead to abnormal glucose regulation, causing diabetes mellitus. However, these side-effects are uncommon and may result from lack of proper care and caution using this dietary factor. In contradistinction, the added IGF1 may be enough to prevent autism spectrum disorder, a lifelong disability (Haniani & Vafa, 2019).

IGF1 availability is commonly reduced during and after an infectious cytokine storm, such as during pregnancy.

It would appear that the generation of some (but not necessarily all) cases of autism begin with the intrauterine fetus exposed before birth to elevated maternal temperatures (e,g., caused by influenza, SARS-CoV-1, H5N1, MERS-CoV), and groups of interleukins due to the disease. In the developing baby, pre- and postpartum, rising interleukin levels are matched with falling IGF1. In selected laboratory animals, a link has been demonstrated between maternal immune activation and autism-like outcomes. In a post-mortem study of human brains, elevated cytokines and infection states have been observed. Children with inflammatory diseases typically have reduced IGF1 and elevated IL-6. Overall, a balance between pro-inflammatory and anti-inflammatory function cytokines is need for good health (Steinman, 2019a, b, c; Pevsner, 2014; Turner et al., 2014; Zie et al., 2017).

Antepartum maternal infection can promote the release of specific cytokines such as IL-6 into the mother's bloodstream. In a meta-analysis of >40,000 autism cases, maternal infection during pregnancy was correlated with autism in the babies.

In contrast, IL-10 is anti-inflammatory. In a study of 69 severely affected coronavirus patients where IL-6 was used as a monitoring marker, elevated levels of LDH, C-reactive protein, ferritin and D-dimer were commonly found. Use of anti-pyretics reduced the incidence of autism in these cases (Vargas et al., 2005; Turner et al., 2014.)

9.3 Results/Discussion

9.3.1 Conclusions

It appears possible to utilize already available laboratory-developed methods for mass-producing IGF1 economically. Such a product could be added by the caregiver to bovine milk or formula to reach the nutritional level afforded by human milk, once such a product is available at an acceptable price. One might expect better long-term therapeutic results in patients who have not yet displayed symptoms than those who are overtly symptomatic with autistic characteristics.

Another approach to treating/preventing autism spectrum disorder is to employ the truncated terminal dipeptide and/or tripeptide found in bovine colostrum or in modified forms isolated from synthetic sources. The redesigned dipeptide has medicinal benefits similar to IGF1 in the case of autism-related maladies and is more stable than the tripeptide, once isolated. Studies are underway using recombinant IGF1 format on children with Rett Syndrome (Sahin, 2018). [This syndrome resembles autism in some aspects but differs from it in others – see Table 9.1].

Phelan-McDermid Syndrome, another autism-like neurologic disorder with some characteristics in common with Rett Syndrome, has been traced to a segmental deletion of chromosome 22q13.3. Promising clinical trials are now underway using a derivative of cGP previously uncovered and developed in the laboratory (Neuren, 2022; Glass et al., 2018; Guan et al., 2014, 2015).

Acknowledgements The completion of this review required no external financial support. The authors declare there is no conflict of interest, thank Roberta Zuckerman for her helpful discussions about this presentation, and acknowledge the expert help of Aviva Adler, librarian, for her cooperative assistance in locating relevant literature references.

References

Al-Farsi, Y. M., Al-Sharbatim, M. M., Waly, M. I., et al. (2012). Effect of suboptimal breastfeeding on occurrence of autism: A case-control study. *Nutrition, 28*(7–8), e27–e32.

Ameri, P., Giusti, A., Boschetti, M., et al. (2013). Vitamin D increases circulating IGF1 in adults: Potential implication for the treatment of GH deficiency. *European Journal of Endocrinology, 169*(6), 767–772.

Boucher, D., Julvez, J., Guxens, M., et al. (2017). Association between breastfeeding, in duration and cognitive development. Autistic traits and ADHD symptoms: A multicenter study in Spain. *Pediatric Research, 81*(3), 434–442.

Chen, J., Strodl, E., Huang, L., et al. (2021). The mediating role of breastfeeding in associations between prenatal education and autistic-like behaviors in preschoolers. *Children, 8*. https://doi.org/10.3390/xxxxx

Choi, G. B., Yim, Y. S., Wong, H., et al. (2016). The maternal interleukin-17a pathway in mice promotes autism-like phenotypes in offspring. *Science, 351*, 933–939.

D'Ercole, A. J. (1993). Expression of insulin-like growth factor-1 in transgenic mice. *Annals of the New York Academy of Sciences, 692*, 149.

Diaz-Gomez, N. M., Domenech, E., & Barroso, F. (1997). Breastfeeding and growth factors in preterm newborn infants. *Journal of Pediatric Gastroenterology and Nutrition, 24*(3), 322–327.

Francis, G. L., Upton, F. M., Ballard, F. J., et al. (1988). Insulin-like growth factors 1 and 2 in bovine colostrum. Sequences and biological activities compared with those of a potent truncated form. *Biochemical Journal, 251*(1), 95–103.

Geschwind, D. H. (2011). Genetics of autism spectrum disorder. *Trends in Cognitive Sciences, 15*(9), 409–416.

Ghozy, S., Tran, L., Navee, S., et al. (2020). Association of breastfeeding status with risk of autism spectrum disorder: A systematic review, dose-response analysis and meta-analysis. *Asian Journal of Psychiatry, 48*, 101916.

Glass, L. I., Bickerdike, M. J., Snape, M. F., et al. (2018). *Neuroprotective bicyclic compounds and methods for their use in treating autism spectrum disorders and neurodevelopmental disorders*. US Patent #9,867,823. Jan16, 2018

Guan, J., Gluckman, P., Yang, P., et al. (2014). Cyclic glycine-proline regulates IGF-1 homeostasis by altering the binding of IGFBP-3 to IGF-1. *Scientific Reports, 4*, 4388.

Guan, J., Harris, P., Brimble, M., et al. (2015). The role for IGF-1 derived small neuropeptides as a therapeutic target for neurological disorders. *Expert Opinion on Therapeutic Targets, 29*(6), 785–793.

Haniani, N. A., & Vafa, M. (2019). The role of IGF-1 in obesity, cardiovascular and cancer. *Medical Journal of the Islamic Republic of Iran, 33*, 56.

Hellstrom, A., Ley, D., Hansen-Pupp, I., et al. (2016). Role of insulin-like growth factor 1 in fetal development and in the early postnatal life of premature infants. *American Journal of Perinatology, 33*(11), 1067–1071.

Herman, I. P. (2007). *Physics of the human body* (p. 721). Springer.

Hong, L., Ziegler, J., & Broady, R. (2014). Breastfeeding and autism spectrum disorders. *Topics in Clinical Nutrition, 29*(3), 278–285.

Hsiao, E. Y., & Patterson, P. H. (2011). Activation of the maternal immune system induces endocrine changes in the placenta via IL-6. *Brain, Behavior, and Immunity, 25*(4), 604–615.

Hsiao, E. Y., McBride, S. W., Chow, J., et al. (2012). Modeling an autism risk factor in mice leads to permanent immune dysregulation. *PNAS, 109*(31), 12776–12781.

Huang, S., Wang, X., Sun, T., et al. (2022). Association of breastfeeding in the first 6 months of life and autism spectrum disorders: A national multi-center study in China. *Nutrients, 14*(1), 45. https://doi.org/10.3390/nu14010045

Jackson, D. A., & Berg, P. (1972). Biochemical method for inserting new genetic information into DNA containing lambda phage genes and the galactose operon of Escherichia coli. *Proceedings of the National Academy of Sciences USA, 69*, 2904–2911.

Kanner, L. (1943). Autistic disturbances of affective contact. *The Nervous Child, 2*, 217–250.

Lawrence, R. A. (2014). The risks of not breastfeeding: New associations. *Breastfeeding Medicine, 9*(5), 237–238.

Leventhal, P. S., Russell, J. W., & Feldman, E. L. (1999). IGFs and the nervous system, chap.19. In R. G. Rosenfeld & C. T. Roberts (Eds.), *The IGF system*. Humana Press.

Linker, S. B., Mendes, A. P. D., & Marchetto, M. C. (2020). IGF-1 treatment causes unique transcriptional response in neurons from individuals with idiopathic autism. *Molecular Autism.* https://doi.org/10.1186/s13229-020-00359-w

Makkonen, I., Kokki, H., Kuikka, J., et al. (2011). Effects of fluoxetine treatment on striatal dopamine transporter binding and cerebrospinal fluid insulin-like growth factor-1 in children with autism. *Neuropediatrics, 42*(5), 207–209.

Meltzer, A., & Van de Water, J. (2017). The role of immune system in autism spectrum disorder. *Neuropsychopharmacology, 42*, 284–298.

Merrifield, R. B. (1963). Solid phase peptide synthesis. I. The synthesis of a tetrapeptide. *Journal of the American Chemical Society, 85*, 2149–2162.

Mijalis, A. J., Steinauer, A., Schepartz., & Pentelute, B. L. (2018). Pushing the limits of solid phase peptide synthesis with continuous flow. Chap.1.

Munro, H. N. (1980). Placental protein and peptide hormone synthesis: Impact of maternal nutrition. *Federation Proceedings, 39*(2), 255–260.

Neuren Pharmaceuticals, Australia. (2022). ASX Announcement – "Neuren receives FDA approval for Phelan-McDermid IND and Phase 2 trial", 3/23/2022

Patterson, P. H. (2011). Maternal infection and immune involvement. *Trends in Molecular Medicine, 17*(7), 389–394.

Pevsner, J. (2014). Gene expression in autism. In M. L. Bauman & T. L. Kemper (Eds.), *Neurobiology of autism* (2nd ed., pp. 207–216). Johns Hopkins Press.

Philipps, A. F., Dvorak, B., Gling, P. J., et al. (2000). Absorption of milk-borne insulin-like growth factor-I into portal blood of suckling rats. *Journal of Pediatric Gastroenterology and Nutrition, 31*, 128–135.

Redford, K. (2014). Myelination, chapter 6A. In G. Steinman (Ed.), *The cause of autism – concepts and misconceptions*. Baffin Books Publishing.

Riikonen, R. (2006). Insulin-like growth factor delivery across the blood-brain barrier. *Chemotherapy, 52*, 279–281.

Riikonen, R. (2008). Insulin-like growth factors-neurobiological regulators of brain growth in autism? In A. W. Zimmerman (Ed.), *Autism-current theories and evidence* (p. 233–244). Humana Press.

Riikonen, R. (2016). Treatment of autistic spectrum disorder with insulin-like growth factor. *European Journal of Paediatric Neurology, 20*, 816–823.

Riikonen, R. (2017). Insulin-like growth factor I the pathogenesis of neurological disease of children. *International Journal of Molecular Sciences, 18*(10), 2056–2073.

Riikonen, R., Maddonen, I., Vanhala, R., et al. (2006). Cerebrospinal fluid insulin-like growth factors IGF1 and IGF2 in infantile autism. *Developmental Medicine and Child Neurology, 48*, 751–755.

Sahin, M. (2018). *Treatment of Rett Syndrome with recombinant human IGF-1. Clinical Trial NCT01777542*. Boston Children's Hospital, 3/26/2018.

Schinn, S. M., Broadbent, A., Bradley, W. T., et al. (2016). Protein synthesis directly from PCR: Progress and applications of cell-free protein synthesis with linear DNA. *New Biotechnology, 33*(4), 480–487.

Schultz, S. T., Glonoff-Cohen, H. S., Wingard, D. L., et al. (2006). Breastfeeding, infant formula supplementation, and autistic disorder. *International Breastfeeding Journal*. https://doi.org/10.1186/1746-4358-216zx

Shafai, T., Mustafa, M., Hild, T., et al. (2014). The association of early weaning and formula feeding with autism spectrum disorders. *Breastfeeding Medicine, 9*(5), 275–276.

Shafai, T., Mustafa, M., Mulay, J., et al. (2017). Impact of infant feeding methods on the development of autism spectrum disease. In *Autism - paradigms, recent research and clinical applications*. https://doi.org/10.5772/67624

Shamsedine, L., Mailhac, A., Badaous, B., et al. (2020). Breastfeeding association with autism spectrum disorder: A case-control study from Lebanon. *Research in Autism Spectrum Disorder, 78*, 101651.

Shen, W. H., & Xu, R. J. (2000). Stability of insulin-like growth factor I in the gastrointestinal lumen in neonatal pigs. *Journal of Pediatric Gastroenterology and Nutrition, 30*, 299–304.
Shuid, A. N., Javusman, P. A., Shuid, N., et al. (2021). Association between viral infections and risk of autistic disorder: An overiew. *International Journal of Environmental Research and Public Health, 18*(6), 2817.
Simon, M. D., Heider, P. L., & Adamo, A. (2014). Rapid-flow-based peptide synthesis. *Chembiochem, 15*(5), 713–720.
Soke, G. N., Maenner, M., Winham, G., et al. (2019). Association between breastfeeding initiation and duration, and autism spectrum disorder in preschool children enrolled in the study to explore early development. *Autism Research, 12*(5), 816–829.
Song, Z., Tan, Z., & Cheng, J. (2019). Recent advances and future perspectives of synthetic polypeptides from N-carboxyanhydrides. *Macromolecules, 52*, 8521–8539.
Stefenburg, S., Gillberg, C., Helligrew, L., et al. (1989). A twin study in Denmark, Finland, Iceland, Norway, & Sweden. *Journal of Child Psychology and Psychiatry, 30*(13), 405–416.
Steinman, G. (2013a). Predicting autism at birth. *Medical Hypotheses, 81*(1), 21–25.
Steinman, G. (2013b). Breastfeeding as a possible deterrent to autism. *Medical Hypotheses, 81*(6), 999–1001.
Steinman, G. (Ed.). (2014). *The cause of autism – concepts and misconceptions*. Baffin Publishing Co..
Steinman, G. (2015). Plausible etiology of brain dysconnectivity in autism – review and prospectus. *Medical Hypotheses, 85*(4), 405–407.
Steinman, G. (2017). Reduction of neonatal neuropathology. *SM Autism and Epilepsy, 1*(1), 1001.
Steinman, G. (2019a). IGF – autism prevention/amelioration. *Medical Hypotheses, 122*, 45–47.
Steinman, G. (2019b). Prenatal identification of autism propensity. *Medical Hypotheses, 122*, 210–211.
Steinman, G. (2019c). Autism neospectrum – at birth. *International Journal of Neurobiology, 1*(1), 101–104.
Steinman, G. (2020a). Covid-19 and autism. *Medical Hypotheses, 142*, 109797.
Steinman, G. (2020b). The putative etiology and prevention of autism, Chap.4. In *Progress in molecular biology and translation science* (Vol. 173). Academic.
Steinman, G., & Mankuta, D. (2013). Insulin-like growth factor and the etiology of autism. *Medical Hypotheses, 80*(4), 475–480.
Steinman, G., & Mankuta, D. (2014). Umbilical cord biomarkers in autism determination. *Biomarkers in Medicine, 8*(3), 317–319.
Steinman, G., & Mankuta, D. (2018). Gene polymorphism in the genesis of autism. *BAOJ Neurology, 4*(2), 58–60.
Steinman, G., & Mankuta, D. (2019). Molecular biology of autism's etiology – an alternative mechanism. *Medical Hypotheses, 130*, 109272.
Steinman, G., & Mankuta, D. (2020a). The role of oligopeptides in preventing autism. *Medical Hypotheses, 138*, 109604.
Steinman, G., & Mankuta, D. (2020b). Covid-19 and autism – part 2. *Integrated Gynecology and Obstetrics Journal, 3*(3), 1–3.
Steinman, G., & Mankuta, D. (2021a). Febrile corona virus and autism. *Journal of Pediatric Neurology and Neuroscience, 5*(2), 190–198.
Steinman, G., & Mankuta, D. (2021b). Antepartum Covid-19 and postpartum autism. *Australian Journal of Clinical Neuroscience, 8*(2), 1149–1151.
Stubbs, G., Henley, K., & Green, J. (2016). Autism: Will vitamin D supplementation during pregnancy and early childhood reduce the recurrence rate of autism in newborn siblings? *Medical Hypotheses, 88*, 74–78.
Tseng, P. T., Chen, Y.-W., Stubbs, B., et al. (2017). Maternal breastfeeding and autism disorder: A systematic review and meta-analysis. *Nutritional Neuroscience*, 13885978.
Turner, M. D., Nedjai, B., Hurst, T., et al. (2014). Cytokines and chemokines: At the crossroads of cell signaling and inflkammatory disease. *Biochimica et Biophysica Acta, 1843*, 2563–2582.

Vanhala, R., Turpeinen, U., & Riidonene, R. (2001). Neurotrophic factors in the pathogenesis of Rett syndrome. *Developmental Medicine and Child Neurology, 43*(9), 614–616.
Vargas, D. L., Nascimbene, C., Krishnan, C., Zimmerman, A. W., & Pardo, C. A. (2005). Neuroglial activation and neuroinflammation in the brain of patients with autism. *Annals of Neurology, 57*, 67–81.
Ville, C. D. (1977). Synthesis of proteins in the placenta. *Gynecologic Investigation, 8*, 145–161.
Wang, Z., Ding, R., & Wang, J. (2020). The association between vitamin D status and autism spectrum disorder (ASD): a systematic review and meta-analysis. *Nutrients, 13*(1), 86.
Zie, J., Huang, L., Xiaohong, L., et al. (2017). Immunological cytokine profiling identified TNF-a as a key molecule dysregulated in autistic children. *Oncotarget, 8*, 82d390–82d398.

Chapter 10
Prophylactic Treatment of ASD Based on Sleep-Wake Circadian Rhythm Formation in Infancy to Early Childhood

Teruhisa Miike, Makiko Toyoura, Kentaro Oniki, Shiro Tonooka, and Seiki Tajima

Abbreviations

ADHD	Attention-deficit/hyperactivity disorder
ASD	Autism spectrum disorder
OR	Odds ratio
SCN	Suprachiasmatic nuclei

T. Miike (✉)
Professor Emeritus Kumamoto University, Kumamoto, Japan

Japanese Society Min-Iku (Sleep Education) Research and Promotion, Tokyo, Japan

M. Toyoura
Hyogo Rehabilitation Central Hospital, Children's Sleep and Development Medical Research Center, Kobe, Japan

Department of Pediatrics, Hyogo Children's Sleep and Development Medical Research Center, Kobe, Japan

K. Oniki
Division of Pharmacology and Therapeutics, Graduate School of Pharmaceutical Sciences, Kumamoto University, Kumamoto, Japan
e-mail: oniken@kumamoto-u.ac.jp

S. Tonooka
Kagoshima Comprehensive Clinic for Disabled Children, Kagoshima, Japan
e-mail: shirotono@po5.synapse.ne.jp

S. Tajima
Department of Pediatrics, Hyogo Children's Sleep and Development Medical Research Center, Kobe, Japan

A. El Idrissi, D. McCloskey (eds.), *Neurobiology of Autism Spectrum Disorders*,
https://doi.org/10.1007/978-3-031-42383-3_10

10.1 Introduction

The prevalence of autism spectrum disorder (ASD) has been increasing, and has become a "serious problem" in modern society (Boyle et al., 2011; Christensen et al., 2016; Sheldrick & Carter, 2018). However, the pathogenesis underlying this condition and the cause(s) of the rapid increase remains unclear. Given the increasing incidence of ASD in modern society, not only genetic background but also environmental factors will be a major issue, because many pathological conditions are known to develop with the addition of living environmental factors to a genetic background. For example, some type of ornithine transcarbamylase (OTC) deficiencies, which have a well-defined genetic predisposition, can be prevented by dietary protein restriction.

From this point of view, we have been thinking that it makes the most sense to focus on the formation of the biological clock that controls the life-sustaining function as an environmental factor having a great influence on the children's mental and physical development. It is well known that children and adolescents with ASD suffer from sleep disorders (Allik et al., 2006; Cortesi et al., 2010). We have summarized the contents of reported ASD sleep disorders as follows: (1) sleep-onset insomnia, (2) frequent nighttime awakening (sleep fragmentation), (3) low sleep efficiency, (4) difficulty in sustaining sleep, and (5) long night awakening and shortening of total sleep time (Limoges et al., 2005; Allik et al., 2006; Miano et al., 2007). These are suggested as mainly chronobiological abnormalities. Since around 2000, a recent theory suggests that biological rhythm disorders, especially the sleep-wake rhythm of ASD patients, may not only be a related symptom but may also be closely associated with the etiology of ASD (Wimpory et al., 2002; Nicholas et al., 2007; Bourgeron, 2007; Tordjman et al., 2015; Pinato et al., 2019; Hoang et al., 2021). Against this background, we have focused on chronobiological rhythm formation, especially the sleep-wake rhythm, of ASD patients for over 25 years.

Although it has been discussed for a long time whether the onset of ASD is primary matter and the sleep disorder is only a secondary concomitant symptom (Verhoeff et al., 2018), some recent papers reported that sleep disorders of ASD patients are the early symptom that precedes the onset of ASD and are more likely to be a major causative background (MacDuffie et al., 2020; Miike et al., 2020).

In drawing conclusions, we would like to introduce some important reports that emphasized that—under any circumstances—children's sleep disorders require quick treatment, and there is recognition of the fact that sleep disorders have a negative effect on children's growth and development (Jan et al., 2010; Ophoff et al., 2018; Licis, 2020). It is important to note that the sleep disorders in early childhood are significantly associated with future adolescents' anxiety, depression, and attention-deficit/hyperactivity disorder (ADHD) (Lam & Lam, 2021). It is also important to understand that this sleep disorder should be recognized as the abnormality of the circadian rhythm which is one of life-sustaining functions including autonomic function (Noda et al., 1998), hormone secretion/core body temperature rhythm (Tomoda et al., 1997), energy production such as glucose metabolism

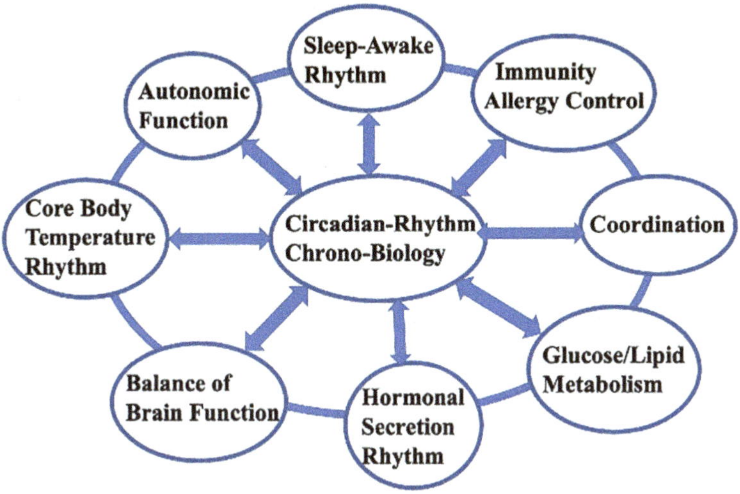

Fig. 10.1 Circadian rhythm and life-sustaining function

(Iwatani et al., 1997; Poggiogalle et al., 2018; Toyoura et al., 2020), coordinated movement and immune function of 37 trillion cells throughout the human body (Qian & Scheer, 2016) (refer to Fig. 10.1). In any case, it must be recognized that the people involved in pediatric medicine are not permitted to leave children's sleep disorder unattended.

Sleep is associated with the development, growth, damage repair, protection of function and further evolution of a brain. In modern society, however, the sleeping environment has changed drastically, and the sleeping environment of children continues to be exacerbated. For example, the majority of sleep-onset time in infancy was around 20:00 in the 1960s in Japan, but recently more than 30% of children fall asleep after 22:00 indicating more than 2 h shift of daily life in early childhood (Konishi et al., 2019).

We would like to express the concepts for ASD from chronobiological viewpoint (Wimpory et al., 2002; Nicholas et al., 2007; Bourgeron, 2007; Tordjman et al., 2015; Geoffray et al., 2016) in the present report. The disfunction of chronobiological theory (chronodisruption theory) (Qian & Scheer, 2016; Erren & Reiter, 2009; Oldham et al., 2016; Touitou et al., 2017) is logical to understand and/or explain various clinical symptoms in ASD, such as sleep disorders, epilepsy, enuresis, autonomic disorders, language problems, intellectual disabilities, gastrointestinal symptoms, diabetes, depression, etc. In particular, the ASD symptoms associated with the disfunction of mitochondria, which exist in a large number in the cell group of the entire body, are diverse neurological symptoms including easy fatigue, gastrointestinal disorders, and motor dysfunction. Accordingly, the mitochondrial dysfunction has been directly linked to a variety of medical disorders, particularly neuromuscular disorders, and increasing number of evidences have linked mitochondrial dysfunction to neurodegenerative and neurodevelopmental disorders, such as

Alzheimer's disease, Parkinson's disease, Rett syndrome and ASD.(Rossignol & Frye, 2012; Cardinali et al., 2013).

In addition, it is well known that the mitochondrial dysfunction deeply relates to circadian rhythm chronobiology. In other word, chronodisruption theory is the most appropriate theory to explain these diverse symptoms over the entire body.

At present, there is no suitable medical treatment to directly correct the disturbed biological clock of children. Therefore, one possible way is to modify the sleep-wake rhythm, which is the only visible circadian rhythm, and form a suitable biological clock for children. In this chapter, the authors would like to emphasize the importance of starting treatment for circadian rhythm sleep disorders during infancy and early childhood, which we believe to be the prevention for future ASD, before the central biological clock is fixed in their suprachiasmatic nuclei (SCN).

Circadian rhythms protect life-sustaining functions by maintaining the harmony and coordination of the 37 trillion cells which form human tissues and organs.

10.2 Circadian Rhythm Formation

Before discussing the actual treatment, it may be important to know when and how the children's circadian rhythms are formed. The formation of chronobiological rhythms deeply relates to the relationship between mother and fetus during pregnancy (Rivkees, 1997; Mirmiran et al., 2003; Serón-Ferré et al., 2007, 2012). It has been reported that formation of the center regulating ultradian rhythms, which is the base of future circadian-rhythm formation, begins at 29–33 weeks (Wakayama et al., 1993) and/or 28–30 weeks of gestation (Fukushima et al., 2008). The ultradian rhythm of fetuses at this time can be classified into four groups by period: (Group I, 5–30 min; Group II, 30–60 min; Group III, 60–100 min; Group IV, >100 min) (Wakayama et al., 1993). The period of Group II, which seems to be the clinically basic rhythm in neonates, (Fukushima et al., 2008) shows that there is a critical period in the development of central nervous activity at 29–33 weeks of gestation.

We have reported abnormal sleep-wake rhythms in neonates, and the so-called irritable types (Ornitz, 1973; Ornitz et al., 1983), such as (1) frequent awakening, (2) difficulty in falling asleep, (3) short sleep time, and (4) continuous crying and grumpiness, may be considered as the precursors to future development of ASD (Miike et al., 2020). In the study, sleep-wake rhythm abnormalities in neonatal period suggest immature or disrupted formation of ultradian rhythm during the fetal period in the pons and medulla oblongata (Fukushima et al., 2008) and also suggest immature or disrupted circadian rhythm formation in the SCN mainly after birth (Wimpory et al., 2002; Nicholas et al., 2007; Bourgeron, 2007; Tordjman et al., 2015). We have also reported that the risk of ASD development is significantly greater at <35 weeks (odds ratio [OR], 24.67) compared to the risks at 35–36 weeks (0.32), 37–39 weeks (1.11) and ≥40 weeks (1.00) of gestation (Miike et al., 2020). Although the estimates of the prevalence vary widely across studies, it has been reported that extreme-preterm (<26 weeks) (Johnson et al., 2010) and early-preterm infants (Darcy-Mahoney et al., 2016; Harel-Gadassi et al., 2018) (<33 6/7 weeks) are at an increased risk of developing ASD. These results suggest that immaturity of the ultradian rhythm center is considered as one of the main causes of ultradian rhythm abnormalities in newborn period.

A newborn's daily life is controlled mainly by the ultradian cycle with a sleep-wake rhythm of approximately 3 (Bueno & Menna-Barreto, 2016) and/or 4 h (Meier-Koll et al., 1978; Stephenson et al., 2012), which is considered to be a combination of REM and non-REM cycles of 3–4 times (Fukushima et al., 2008).

In addition, formation of the chronobiological circadian clock starts in the fetal period (Rivkees, 1997; Mirmiran et al., 2003; Serón-Ferré et al., 2007) is thought to be completed in about 1 year and 6 months after birth (Serón-Ferré et al., 2012; Touchette et al., 2013). Interestingly, Bruni et al. (2014) reported that sleep patterns change during the first year of life but most sleep variables (i.e., sleep latency and duration) show little variation from 6 to 12 months. The data provide a context for clinicians to discuss sleep issues with parents and suggest that the efforts to prevent sleep disorder of infants should be devoted to the first 3–6 months after birth, since sleep patterns of infants show stability from that time point to 12 months. This data is considered to indicate the importance of dealing with the abnormal sleep/wake rhythm of children ideally within the 3 months to 1 year of age. There is also a report of a chronobiologist describing that a period determining the synchrony of the biological (circadian) clock (critical period) exists within the first year of life. In this report the author warns that if artificial and irregular living environment does not create biological clock synchronism during the critical period, people will be forced to live in a state where the daily life rhythm is out of synchrony for the rest of social life (Inoue, 2006).

It is important to understand that the circadian rhythm is created based on the appropriate ultradian rhythm (Bueno & Menna-Barreto, 2016). In other words, the immaturity of the ultradian rhythm during the fetal and neonatal periods makes subsequent formation of the circadian chronobiological rhythm difficult.

10.3 Age-related Changes in Sleep Development

We will examine the data on the transition of sleep-wake rhythm formation of Japanese children and the similar data in the reports from other countries and compare those data with the data on sleep characteristics of children with ASD.

10.3.1 Sleep Duration

It has been reported that nocturnal sleep duration of a typically growing infant is expected to be approximately 5–6 h at the age of 2 months (Pinilla & Birch, 1993), 8–9 h at 4 months (Henderson et al., 2011) and 8–12 h at 6–7 months (Paul et al., 2016). Another report describes that the children can maintain a good night's sleep have high self-soothing ability (Dönmez & Temel, 2019; Mohr et al., 2019), and thus they cry less at night. For making the self-soothing habit, it is necessary to improve the posture of the guardian who reaches out as soon as the baby cries (Sadeh et al., 2016).

One of the authors of the present report conducted a study on 5000 Japanese children attending daycare centers and aged from 2 months to 6 years and 7 months by utilizing 14 days sleep record tables during the period from 2013 to 2016 (Study 1). The data from the study showed that the average nocturnal sleep duration required by these children mainly ranged from 9 to 11 h (about 10 h on average; unpublished data). Then, it was defined as "Night-time Basic Sleep Duration = NBSD", because decrease in their total sleep time with age was dependent on decreased daytime sleep duration, while their nocturnal sleep duration stays almost constant from infant to 6 years of age without change (refer to Fig. 10.2). Almost similar results have been reported by Iglowstein et al. (2003) and Blair et al. (2012). Although neither report pays particular attention to this phenomenon, the data in these reports showed that nocturnal sleep duration does not change significantly from early childhood to 6 years of age and daytime sleep duration gradually decrease to disappear at the age of 5–6. The first report from Switzerland shows that nocturnal sleep duration was fairly constant within the range from 11 h and 6 min to 11 h and 42 min, and gradually decreased slightly with the age between 1 and 5 years (Iglowstein et al., 2003). The later report from UK shows a nearly constant nocturnal sleep duration within the range from 11 h and 8 min to 11 h and 20 min on average at the age between 18 and 81 months (Blair et al., 2012). Both reports show that the average nocturnal sleep duration is in the 11-h range in infancy and early childhood. In addition, a study on children aged 6 months to 4 years in Canada reported that the percentage of children taking 10-h nocturnal sleep duration (48%)

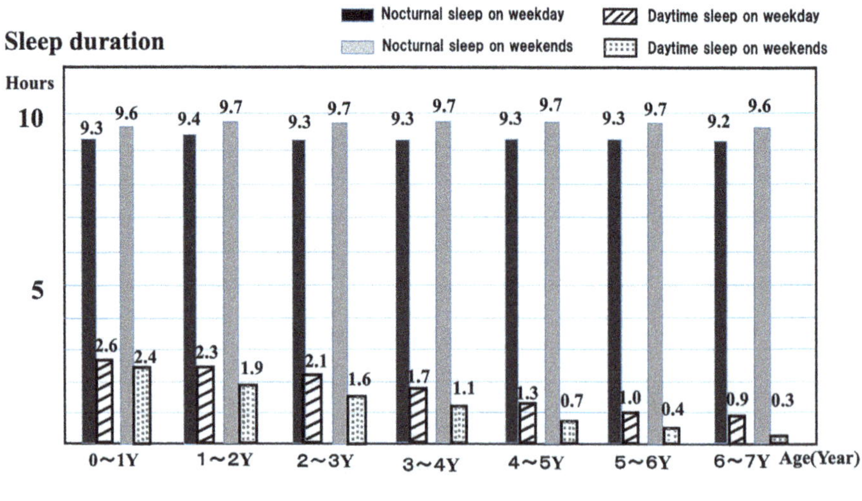

Fig. 10.2 Nocturnal sleep duration and daytime sleep duration of the age groups on weekdays and weekends

was about the same as those taking 11-h nocturnal sleep duration (42%) (Touchette et al., 2013). Compared to the data from these reports in Western countries, the average nocturnal sleep duration of Japanese children in our data is a little shorter than that of children (Fig. 10.2). This phenomenon is also found in the comparison of the total sleep time of children in each country (Mindell et al., 2010).

Although their nocturnal sleep duration did not change with age ($p = 0.079$) from 0 year to 6 years old, their nocturnal sleep duration on weekends was significantly longer than that on weekdays ($p < 0.001$).

10.3.2 *Sleep Characteristics of Children with ASD*

Recently, a paper has reported that highly irritable children have shorter continuous sleep periods than less irritable children and they behaved in a considerably stereotyped manner (Yavuz-Kodat et al., 2020). We have reported that sleep-wake rhythm abnormalities of neonates may be significant precursors to future development of ASD (Miike et al., 2020). In particular, the irritable types (Hiscock & Jordan, 2004; Keefe et al., 1996) of sleep-wake rhythm abnormalities, such as (1) frequent waking up, (2) difficulty in falling asleep, (3) short sleep hours, and (4) continuous crying and grumpiness, which cause difficulty in raising infants, were observed more often in the group of infants with ASD than in the control group (Miike et al., 2020). Following this research, we conducted an additional study on sleep-wake rhythms of children from infancy to early childhood (Study 2) (refer to Table 10.1). Participants in this study were 190 children from four nurseries (in Tokushima, Japan) as the control and 166 children with ASD from two hospitals (113 from a hospital in Kagoshima, Japan represented as K-ASD, and 53 from a hospital in Hyogo, Japan represented as H-ASD). There is no time difference between these cities. The study was certified by informed consent of all collaborators and inspection by the ethics committee of the hospitals.

The results of Study-2 showed that the sleep characteristics found in infancy and early childhood were the same as that of the future ASD patients. The sleep characteristics are (1) short sleep (<8 h), (2) sleep onset time after 22:00 on

Table 10.1 Characteristics of the subjects at the time of the study

	Control		K-ASD		H-ASD	
Age (years old)	Male	Female	Male	Female	Male	Female
1	21	24		0	3	3
2	20	18	3	0	3	0
3	20	22	14	3	4	2
4	17	17	24	8	16	3
5	13	9	30	11	5	5
6	5	4	17	3	7	2
Total	96	94	88	25	38	15
Overall total	190		113		53	

weekdays and 23:00 on holidays, (3) waking time after 07:00 on weekdays and 08:00 on holidays, and (4) frequent awakening (>3 times) and/or long-time awakening (>60 min) (refer to Figs. 10.3 and 10.4). The figures show that the shifting to late sleep and late awakening with age from infancy to early childhood significantly increases the risk of developing ASD in the future. In other words, the biological clock shifts (daily rhythm shifts) are closely associated with ASD. Interestingly, several researchers studied the relation between maternal depression in the perinatal period and poor infant sleep by using data from a population-based birth cohort and reported that the main outcome variable was the poor infant sleep at 12 months of age, defined as >3 night waking, >1h nocturnal wakefulness or <9h total sleep duration (Halal et al., 2021). Another study reported that children with ASD began to sleep shorter than children without ASD after the age of 30 months and the difference reached 40 min at the age of 81 months, though no significant difference in total sleep duration was found between the children with and without ASD during

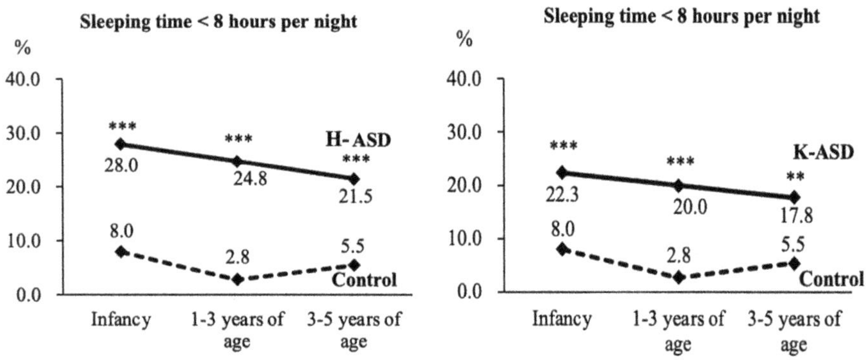

Fig. 10.3 Sleep status of children with ASD and of control group from infancy to 5 years of age. Chi-squared test. *P < 0.05, **P<0.01, ***P<0.001

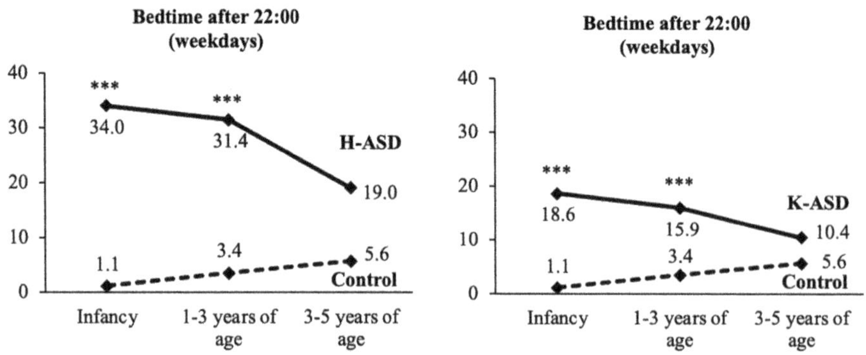

Fig. 10.4 Bedtime of children with ASD and of control group from infancy to 5 years of age. Chi-squared test. *P < 0.05, **P<0.01, ***P<0.001

their infancy. The authors suggest that the night-time sleep duration was shortened by later bedtime and earlier waking time (Humphreys et al., 2014).

10.3.3 Night Awakening (Sleep Fragmentation)

In addition to the time zone and duration of sleep, the importance of continuous sleep throughout the night without waking up should be recognized. Thus, attention must be paid to the state of frequent awakening (Limoges, 2005), especially three or more times of awakening at night or nocturnal wakefulness longer than 1 h (Humphreys et al., 2014; Halal et al., 2021). According to our data, three or more times of awakening and nocturnal wakefulness longer than 60 min need to be defined as sleep fragmentation and/or sleep persistence disorder, and especially attention must be paid to such awakening and wakefulness found in the period from the latter half of infancy to early childhood. Although International Classification of Sleep Disorders, Third Edition (ICSD-3) states that awakening time of 20 min or more is judged to be a sleep persistence disorder, our data show no significant difference between the children with ASD and the control group in sustained arousal within 60 min of infants and toddlers.

10.3.4 Difference in Sleep Duration Between Weekdays and Weekends

In our study-1 (unpublished data), the morning wake-up time on weekdays was almost the same between the age groups of the children, and their morning wake-up time on weekends (7:08 ± 46 min) was clearly later than that on weekdays (6:46 ± 34 min). The difference increased with age, and the children of all the age groups exhibited longer sleep time on weekends (9 h and 39 min ± 44 min) than that on weekdays (9 h and 19 min ± 35 min) ($p < 0.001$). Analyzing the data, we conceived that children slightly lack sleep on weekdays and supplement the shortage on holidays. Although their nocturnal sleep duration did not change with age ($p = 0.079$) from 0 year to 6 years old, their nocturnal sleep duration on weekends was significantly longer than that on weekdays ($p < 0.001$). We also found that children diagnosed with ASD tended to sleep longer on holidays than on weekdays.

This data suggested that children are obliged to wake up around 6:00 a.m. on weekdays for attending nursery school and/or depending on their parents' daily life to create some sleep deprivation. We estimate that the difference in the average morning wake-up time of the children between weekdays and weekends reflects the fact that they compensate their sleep deprivation on weekdays by their late wake up time on weekends. This phenomenon is now known worldwide to be commonly found in childhood (Hansen et al., 2005; Wing et al., 2009; Maeda et al., 2019).

10.3.5 Social Jet Lag Tendency of Infants and Young Children

Although the difference in wake-up time between weekdays and weekends in our study-1 is only about 20 min, this is an average value and we must suppose that the social jet lag (SJL) tendency is widespread among children. Adults with 120 min or more difference in wake-up time between weekdays and weekends (Wittmann et al., 2006; Malone et al., 2016; Hena & Garmy, 2020) and children with 90 min or more difference (Maeda, 2019) in that are judged to have high risks of developing "social jet lag". Reports from many countries put stress on the relation between weight gain or obesity and compensation or catch-up sleep on weekends (Wing et al., 2009; Maeda et al., 2019; Kim et al., 2012; Chaput et al., 2006; Danielsen et al., 2010; Marshall et al., 2008). In addition, the authors would like to claim that the compensation sleep relates to the shift of biological clock of entire body organs that may cause future "social jetlag" as a systemic serious problem (Qian & Scheer, 2016; Wittmann et al., 2006; Hart & Jelalian, 2008). Repeated weekend compensation sleep leads to social jet lag which results in impaired mental and physical health of children and may become a background factor of their withdrawal from school and social life (Tomoda et al., 1994; Miike et al., 2004; Sivertsen et al., 2013; Hochadel et al., 2014; Hysing et al., 2015).

It draws our strong interest that children with ASD and neurodevelopmental impairments in social communication often refuse to go to school because of difficulties in waking up in the morning and learning in class (Higuchi, 2017; Bond & Oliphant, 2018; Bitsika et al., 2021; Adams et al., 2022). According to our data, 60–80% of children who refuse to attend school are diagnosed with ASD.

10.4 Appropriate Period of Nocturnal Sleep

In the analysis of our study-2 data, we found that children greatly deviating from the good sleep habits described below frequently develop ASD-like symptoms (Couturier et al., 2005; Limoges et al., 2005; Allik et al., 2006; Krakowiak et al., 2008). From various information including the above, the authors have come to consider that the good sleep habits are indispensable for the balanced mental and physical development of children.

The good sleep habits shown below promise a healthy life for children.

1. Children should sleep between 7:00 p.m. and 7:00 a.m.
2. The sleep onset time should not be later than 10:00 p.m.
3. Less frequent or no awakening during the night sleep (<2 times, <3 days/week: and/or <20 min awakening duration)
4. The nocturnal sleep duration is at least 9 h and may be shorter than 12 h.
5. The difference in the sleep onset time and awakening time between weekdays and weekends must not be longer than 60 min.

The background of the requirement for the good sleep habits includes (1) formation of the daily life rhythm necessary for future school social life, (2) necessity to secure 10h sleeping in average as NBSD, (3) securing of continuous sleep complying with the information from past data and international classification for chronic insomnia, (4) the lower and upper limits of NBSD, and (5) protection of the regularity of the chronobiological rhythm and prevention of social jet lag (SJL) (Roenneberg et al., 2019; Wittmann et al., 2006, Remi, 2015; Maeda et al., 2019).

10.5 Direction of ASD Treatment

The goal of the treatment is to correct the (1) shift of sleep-wake rhythm (backward shift) and (2) sleep fragmentation (persistent disorder) during nighttime sleep.

As mentioned above, sleep disorders of children with ASD is one of circadian rhythm abnormalities. Since the main symptom of ASD has been considered as a social communication disorder, the diagnosis is generally confirmed after the age of two to three. However, sleep disorders, in other words, chronodisruption (Erren & Reiter, 2009; Oldham et al., 2016; Touitou, 2017) has already been found before the onset of ASD, (MacDuffie et al., 2020) even in neonates (Miike et al., 2020). Some paper also reports that ASD patients show some prodromal symptoms by 6–12 months after birth (Elsabbagh et al., 2013; Wolff et al., 2014; Estes et al., 2015; Iverson et al., 2019). We would like to recommend pediatricians to recognize the circadian rhythm of the children with the prodromal symptoms. In any case, the treatment should be based on the chronodisruption hypothesis, which can be called a theory now. At present, however, no definitive methods are available for correcting the biological clock and the treatment is concentrated on the correction of sleep/wake rhythm which represents a clinically visible circadian rhythm (Fig. 10.1). The question is when to start treatment. In addition, it is necessary to determine whether the treatment of their sleep-wake rhythm is needed or not.

10.6 Appropriate Age for Treatment

With the accumulation of such data, we considered that inconsistent sleep-wake circadian rhythm formation in infancy and early childhood may be a significant risk factor for the future development of ASD. As a prophylactic therapy for future ASD, sleep-awake rhythms should be corrected, if possible, around 3–6 months after birth, at least by the age of 2, before the circadian rhythm is completely formed and fixed.

(i) Future onset of ASD can be estimated from sleep-wake rhythm abnormality during the neonatal period.
(ii) However, the sleep-wake rhythm abnormality in the newborn period does not adversely affect all children, and much attention has not been paid to the sleep

and waking state of newborn babies yet. In other words, no proper treatment for sleep-wake rhythm abnormality is given to neonates so far.

(iii) Although circadian rhythm formation starts during the fetal period, the formation of the clinically visible circadian rhythms begins 2 months of age and is almost completed by about one to one and a half years of age (2 years of age clinically). Thus, the daily living conditions of neonates from 3 to 18 months after their birth is important for forming the base of circadian rhythm.

(iv) In conclusion, the treatment for sleep-wake rhythm abnormalities should be started ideally between the ages of 3 and 18 months from theoretical point of view.

Considering the formation and fixation of the circadian biological clock, response to and treatment of sleep-wake rhythm abnormalities in early childhood, before entering primary school at the latest, is important and unavoidable. On the other hand, it cannot be denied that the treatment started after the age of 2 is less effective than that started before the age. For starting the treatment, the health condition of the parents must be taken into consideration. If the parents are exhausted due to the problems with their children's sleep-wake rhythm, treatment of the parents should be considered simultaneously with the treatment of the children (Goldberg et al., 2013; Cockshaw et al., 2014; Tikotzky, 2016; Malish et al., 2016).

10.7 Therapy for Sleep Disorders in Infancy to Early Childhood

The correspondence and treatments for sleep disorders in infancy for the purpose of preventing future onset of ASD may be limited to the adjustment of the sleep-awake circadian rhythm and correction of nocturnal awakening,

10.7.1 Sleep Hygiene

10.7.1.1 Sleep-Wake Rhythm Adjustment

Based on the finding that the circadian rhythm biological clock is almost completed after 1 year to 1 year and a half of age, a report recommended that the sleep onset time should be set within the period from 7:00 to 8:00 p.m. by 1–4 months after birth (Paul et al., 2016). It is an important suggestion for the formation of an appropriate circadian rhythm, and the formation of an appropriate biological clock is suitable for future school life. Our study-2 also suggested that the habit of a natural awakening from 6:00 to 7:00 a.m. in the morning is extremely important. For securing a proper length of NBSD, the appropriate sleep onset time should be 7:00 p.m. for children requiring 11 h of NBSD and 8 p.m. for children requiring 10 h of

NBSD. The proper sleep-wake habit with sufficient NBSD protects comfortable daily life, balances the development of childhood and prevents the future onset of ASD, and this is estimated to be one of the reasons why the night sleep is recommended to be taken from 7:00 p.m. to 7:00 a.m. From the newborn period, some children show difficulty in falling asleep, and it can be attributed to their proneness to shift in the biological clock. For such children, the whole family should cooperate to develop their habit of falling asleep from 7:00 pm to before 9:00 pm during infancy. In order to maintain NBSD (10 h on average) until waking at a time between 6:00 and 7:00 a.m., appropriate sleep onset time is naturally within the range from 7:00 p.m. to before 9:00 p.m. It is recommended that the whole family finish their meals, take a bath and do other activities until 7:00 or 8:00 p.m., then turn off the lights at an appropriate time and make the infants sleep for 10–14 days with the whole family. In many cases, this method works.

10.7.1.2 Awakening During Night

Discontinuing (Breast) Feeding at Night (Fig. 10.5)

The sleep state at night during the period from infancy to early childhood is changing because their biological rhythm is in a transition period from ultradian rhythm to circadian rhythm. Thus, their awakening at night for a certain length of time (1–4 h) is likely to occur physiologically. At the awakening of infants at night, many Japanese mothers tend to make them fall asleep again by breast-feeding. In other words, breast-feeding is often used to deal with midnight awakening of infants and young children.

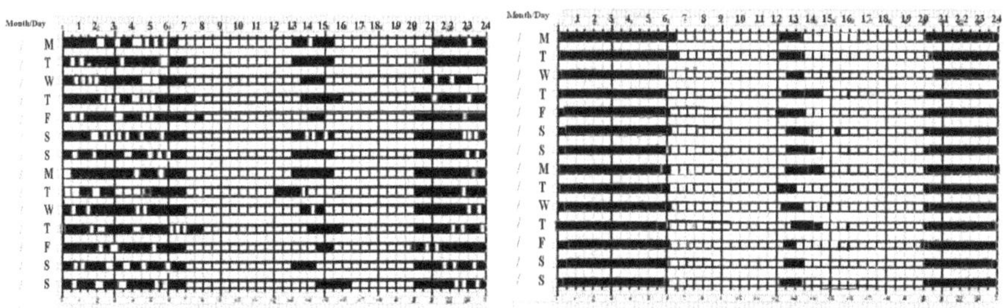

(A)1y2m F (B) 1y8m F

Fig. 10.5 Example showing the relation between night feeding and sleep fragmentation. (**a**) Although the sleep onset time is appropriate, frequent awakening of the infant followed by breast-feeding was observed during night-time sleep. Breast-feeding during the night seemed to cause the infant to have a regular arousal response during sleep. Thus, the discontinuation of breastfeeding was proposed to the mother. (**b**): A few days later, the sleep fragmentation tended to improve significantly, and a few months later, an appropriate sleep-wake rhythm was formed. Her development is well-balanced and has no problems

It is known that children with night-time (breast) feeding habits tend to be more awakening. Some reports have drawn attention with the finding that frequent caring for children who awaken during the night, especially caring by food intake (breast milk, milk, water, etc.), stimulates the peripheral (visceral) clock and causes an arousal reaction and sleep disorders (Sadeh et al., 2010; Hiscock et al., 2014; Kahn et al., 2018). In our study-2, breast-feeding to infants every time of their awakening showed considerable tendency of developing future ASD ($P < 0.01 \sim 0.001$).

For frequent (>3) night awakenings of infants aged 2 months or more, caretakers better not respond to them immediately and should watch over them to help them fall asleep naturally (self- soothing) (Paul et al., 2016; Öztürk et al., 2019; Adams et al., 2020). Therefore, it is recommendable to discontinue night feeding after neonate period or start to reduce the number of times of feeding from the age of 2 months and discontinue feeding by 3–4 months of age. It is known that nighttime sleep duration extends to 8–9 h at 4 months of age, and feeding is not always necessary physically and nutritionally (Henderson, 2011; Tordjman et al., 2015; Paul et al., 2016).

10.7.2 Pharmacotherapy (Figs. 10.6, 10.7 and 10.8)

10.7.2.1 Delayed Sleep Onset (Sleep Onset Insomnia)

Melatonin

Sleep inducers, such as zolpidem and eszopiclone, for children with ADHD showed no improvement in sleep problems or ADHD ratings (Ballester et al., 2020) and in these of ASD (in our experience). This is due to the background of ASD's sleep disorders, mainly circadian rhythm abnormality, we estimate. In other words, sleep inducers cannot influence circadian rhythm itself. On the other hand, melatonin is the first choice for setting the sleep onset time (Miike et al., 2004; Reiter, 2014; Claustrat & Leston, 2015), especially for setting the sleep onset time of ASD patients (Tordjman et al., 2015; McDonagh et al. 2019). It has been reported that the role of melatonin in the ontogenetic establishment of circadian rhythms and the synchronization of the circadian clock network suggests that the hormone might be also involved in the synchrony of motor, emotional, and interpersonal rhythms. In the author's clinical experience, the hormone has brought a favorable result on language development. (data not shown). Recently, it has increasingly become known that melatonin is an important substance for the prevention and treatment of ASD (Wu et al., 2020).

Unfortunately, however, it is difficult to achieve the expected effects by melatonin in infancy. This may be attributed to the immature development of the melatonin receptor of infants in addition to their underdeveloped melatonin secretion. A study reported that the mean night-time melatonin concentration in serum was lower during the first 6 months of life than that after the period (Waldhauser et al., 1988).

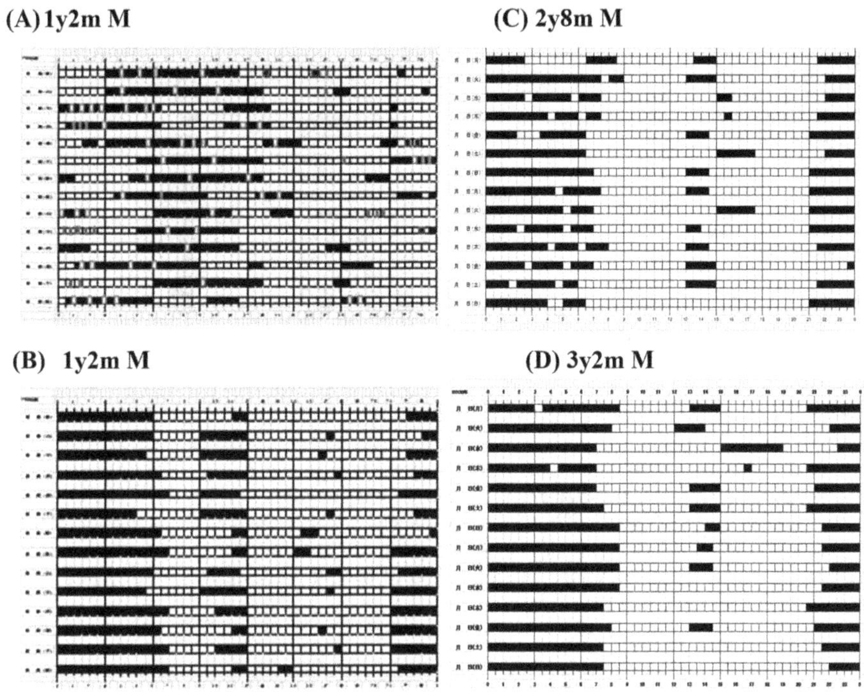

Fig. 10.6 Sleep disorder and ASD. (**a**) From the newborn period, his sleep-wake rhythm is restless to suggest irritable type. The pediatrician who examined him ordered the follow-up, saying that his development was normal range. When asked for an opinion, the author recommended a melatonin (1.5 mg) therapy because his sleep chart clearly suggested circadian rhythm sleep disorders (chrono-disruption). (**b**) One month later, the therapy was stopped when an almost appropriate daily life rhythm was obtained. The grumpiness before going to bed disappeared. (**c**) After that, he did not visit to our hospital due to his problems in family life until 2 years and 8 months old. He returned to our clinic complaining the following symptoms and was diagnosed with ASD at another clinic. He focuses only on the items of his interest and tends to be hyperactive. He woke up frequently and kept awake for a long time. When he woke up, he became wild and panicked, and yelled. 1.5 mg of melatonin and 0.4 mg of risperidone 1.0 mg were administered. (**d**) The nursery school life has become stable, panic has decreased, and the mood has improved. The hyperactive tendency has also improved. The progress since then has been smooth, and he is now living his school life (3rd grade of primary school) without any problems

Therefore, if melatonin does not give sleep-inducing effect, other drugs are often administered additionally. In other words, melatonin is quite useful for children after 1 year of age.

Currently, melatonin has been identified not only in the pineal gland, but also in extra-pineal tissues (Claustrat & Leston, 2015), that is, retina, gut mucosa, cerebellum, airway epithelium, liver, kidney, adrenals, thymus, thyroid, pancreas, ovary, carotid body, placenta and endometrium as well as in non-neuroendocrine cells like mast cells, natural killer cells, eosinophilic leukocytes, platelets and endothelial cells. Although functions of this hormone in human bodies are known mainly based

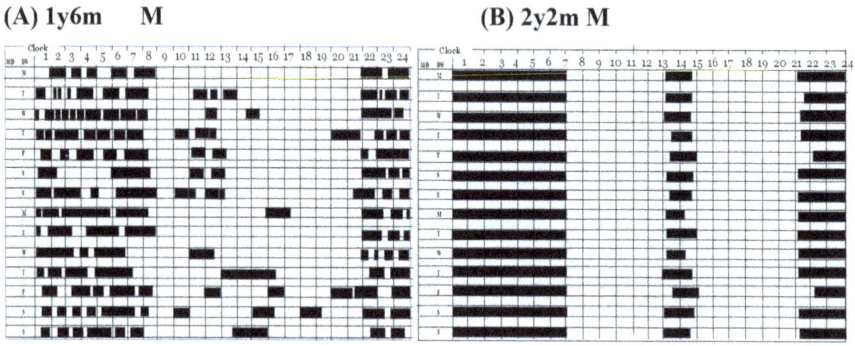

Fig. 10.7 Melatonin and language development. (**a**) The child could not take continuous sleep at night and showed grumpiness and frequent crying during the day. As a result, the mother complained chronic sleep deprivation and was exhausted. The language development of the child was also delayed. (**b**) Administration of 1.5 mg of melatonin was not effective and 0.5 tablets (75 mg) of clonidine, 12.5 mg of hydroxyzine, 2 mg of nitrazepam and 0.4 mg of risperidone were administered at 8:00 p.m. (**c**) Eight months after the treatment, a proper daily life rhythm that was almost regular and could be adapted to future school social life was obtained. The child became emotionally stable and made rapid development of language. The mother also became emotionally stable and has recovered from fatigue

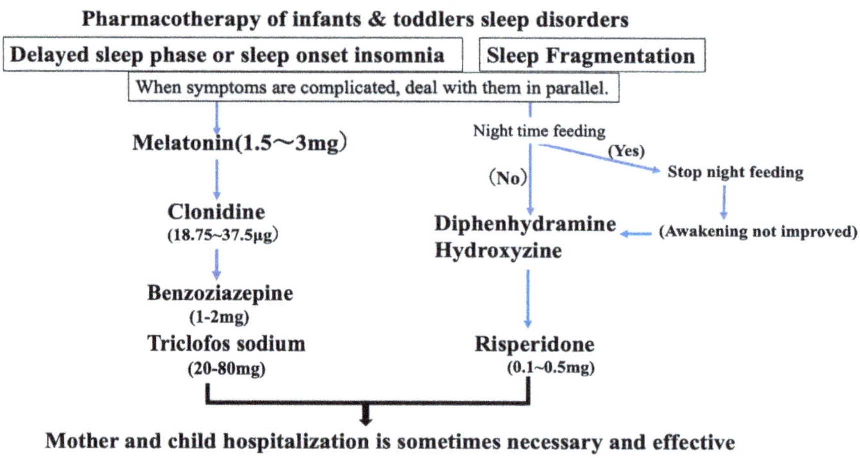

Fig. 10.8 Pharmacotherapy for sleep disorders of infants and toddlers

on the correlation between clinical observations and melatonin secretion, there is some evidence that melatonin stabilizes and strengthens the coupling of circadian rhythms, especially of core temperature and sleep-wake rhythms. The circadian organization of other physiological functions also depend on the melatonin signal, for instance, immunity, antioxidant defenses, hemostasis and glucose regulation (refer to Fig. 10.1). As mentioned above, melatonin is present in all tissues throughout the human body and is understood as a substance promoting the harmonization of the functions of systemic cells. It is considered that this adjustment of the rhythm of the whole body is related to the development of the language system as a rhythm.

Depending on the situation of the patients, 3–20 mg of melatonin may be used safely without problem, (Ninomiya et al., 2001; Foley & Steel, 2019) but in general, 1–1.5 mg is sufficient regardless of age. It is necessary to instruct them to take the medicine 30–60 min before the scheduled time of falling asleep and to go to bed soon and/or within 30 min after taking the medicine.

Clonidine

Clonidine is a well-known drug for pediatricians because it was used as a growth hormone-secreting reagent for children in the past. Clonidine, along with antihistamines, has become a commonly prescribed drug for sleep disorders of children, (Schnoes, 2006) especially for neurodevelopmental disorders (Blackmer, 2016; Bruni et al., 2019), such as ADHD (Anand et al., 2017). In particular, the author has preferred to use it in combination with melatonin for the treatment of sleep onset disorders since the 1980s. For infants, 0.25–0.5 tablets (75 µg) of clonidine in combination with melatonin can help them fall asleep at scheduled time. These two prescriptions give almost acceptable results for sleep onset but includes the problem of waking up several times after falling asleep at midnight. In this case, an additional prescription is required. It should be used, of course, with caution because clinical data is insufficient and there are still uncertainties regarding the dosing guideline and tolerability (Schnoes, 2006). However, the author's 35 years of clinical experience with infants indicates that both safety and efficacy are satisfactory and there are no cases to worry about, so far.

Benzodiazepine

If the above prescriptions are unsuccessful or unusable, the use of benzodiazepines may be unavoidable. For example, 1 to 2-mg dose of Nitrazepam or Diazepam, = 30 to 60 min before scheduled bedtime is favorable to patients in infancy to early childhood.

Triclofos Sodium(TFS)

TFS has been used for many years in children as a sedative for painless medical procedures including EEG examination and ophthalmic evaluation. It is physiologically and pharmacologically similar to chloral hydrate, the use of which in children with neurocognitive disorders has been criticized. TFS administered in a controlled and monitored environment may be safe for use in children including those with underlying neurocognitive disorders (Kaplan et al., 2019). Since the drug is familiar to pediatricians, many doctors use it for sleep onset and sustaining sleep. It is stated that the standard dose is 20–80 mg (0.2–0.8 mL)/kg depending on the age and condition of patients, purpose of treatment, etc., and the total amount up to 2 g (20 mL) is safe and effective for children younger than 5 years of age who undergo visual examination (Sethi et al., 2021).

10.7.2.2 Awakening (Sleep Fragmentation)

Antihistamine

As a prescription for promoting continuous sleep, we recommend the use of the first-generation antihistamines (e.g., Diphenhydramine, Hydroxyzine) that cause drowsiness as a side effect. Although many pediatricians avoid the use of the medicine for the generally known unfavorable side effect, it is recommended for the treatment of sleep disorders, especially the disorders associated with itching due to dermatitis (Kelsay, 2006). Appropriate doses in such use poses almost no side effects. The clinical dose is 7–20 mg once before intended bedtime but most of the doses do not exceed 15 mg. Although other generally known side effects, such as exacerbation of epilepsy, make some pediatricians avoid the use of the medicine, such worry can be solved by using an appropriate amount to properly choose patients with caution.

Risperidone

Risperidone is not so familiar medicine for pediatricians, and only risperidone and aripiprazole are currently approved by the US Food and Drug Administration (FDA) for the treatment of irritability associated with ASD of children and adolescents (Shafiq & Pringsheim, 2018; Goel et al., 2018; Lamy & Erickson, 2018; Turner, 2020). There is evidence to support the short-term efficacy of risperidone and aripiprazole for the irritability associated with ASD and the short-term efficacy of risperidone for the aggressive and disruptive behavior found in disruptive behavior disorders. Some authors, however, concerned that antipsychotics should be considered as a short-term strategy while psychosocial and behavioral therapies are required to be continuously employed.

Only a few structured review studies of pharmacological interventions for sleep disorders of patients with ASD have been published (Ballester et al., 2020). In this report, the authors recommend a small amount of risperidone (0.1–0.5 mg of dose at a time 30–60 min before scheduled bedtime) for frequent interruptions of sleep at night (>3) and awakening for 20 min or more. Risperidone in this dose is a safe and highly useful medicine and pediatricians should not hesitate to use it in the treatment of sleep disorders.

10.7.3 *Hospitalization*

If the treatments mentioned above cannot improve frequent arousal of children at night, the hospitalization of the children and their mothers is necessary and effective. When a mother feel strong fatigability due to her child's sleep problems, it is recommendable to hospitalize the mother and child as soon as possible. Persistent crying and sleep disorders of children increase the frustration of parents and may lead to unexpected abusive behavior. Therefore, they should not be left unattended and overlooked (Li et al., 2018; Halal, 2021; Richey, 2022).

10.8 Summary

1. At first, it is important to understand the relationship between the onset of ASD and the formation of the biological clock, especially the circadian rhythm. The concept is based on the theory that disturbed formation of chronological circadian rhythm causes the circadian rhythm disruption of systemic tissues and cells which is the cause of the ASD with imbalanced systemic function.
2. Next, it is important to know that this circadian rhythm is almost completed by the age between 1 year and 1 and a half years, and then gradually fixed in the SCN as the central circadian clock.
3. Given the formation and fixation of the circadian biological clock, the treatment for sleep-wake rhythm abnormalities should be started ideally in the period between the age of 3 months and early childhood and at the latest before entering elementary school.
4. The goal of the treatment is to correct two disorders, i.e., the shift of sleep-wake rhythm (backward shift) and the fragmentation of nighttime sleep (persistent disorder).
5. Four medicines are enough for the pharmacotherapy of sleep disorders and they should be carefully used in a proper combination. For example, melatonin and clonidine are effective for sleep onset settings, and antihistamines or risperidone are useful for curing sleep fragmentation (persistent disorders).
6. The treatment period may range from several months to several years. When the sleep-wake rhythm is corrected and the night-time arousal disappears, the time to stop the treatment should be considered. It is recommendable to use a sleep table to assess the sleep-wake rhythms of patients.
7. If a mother complains of sleep deprivation followed by fatigue, her child often has a sleep-wake rhythm disorder that needs to be corrected.
8. The hospitalization of children and their mothers is necessary and effective when the frequent arousal of the children at night is not improved the treatments.

Acknowledgments Part of this study has been supported by the Grant-in Aid for Scientific Research on Innovative Areas "Constructive Developmental Science" (No. 24119004) from The Ministry of Education, Culture, Sports, Science and Technology, Japan.

References

Adams, E. L., Master, L., Buxton, O. M., et al. (2020). Patterns of infant-only wake bouts and night feeds during early infancy: An exploratory study using actigraphy in mother-father-infant triads. *Pediatric Obesity, 15*(10), e12.

Adams, D., McLucas, R., Mitchelson, H., et al. (2022). Form, function and feedback on the refusal assessment scale-revised in children on the autism spectrum. *Journal of Autism and Developmental Disorders, 52*(5), 2156–2167.

Allik, H., Larson, J. O., & Smedje, H. (2006). Insomnia in school-age children with Asperger syndrome or high-function autism. *BMC Psychiatry, 6*, 18.

Anand, S., Tong, H., Besag, F. M. C., et al. (2017). Safety, tolerability. and efficacy of drugsfor treating behavioural insomnia in children with attention-deficit/hyperactivity disorder: A systematic review with methodological quality assessment. *Paediatric Drugs, 19*(3), 235–250.

Ballester, P., Richdale, A. L., Baker, E. K., et al. (2020). Sleep in autism: A biomolecular approach to aetiology and treatment. *Sleep Medicine Reviews, 54*, 101357.

Bitsika, V., Heyne, D. A., & Sharpley, C. F. (2021). Is bullying associated with emerging school refusal in autistic boys? *Journal of Autism and Developmental Disorders, 51*(4), 1081–1092.

Blackmer, A. B., & Feinstein, J. A. (2016). Management of sleep disorders in children with neurodevelopmental disorders: A review. *Pharmacotherapy, 36*(1), 84–98.

Blair, P. S., Humphreys, J. S., Gringras, P., et al. (2012). Childhood sleep duration and associated demographic characteristics in an English cohort. *Sleep, 35*(3), 353–360.

Bond, E. C., & Oliphant, R. Y. O. (2018). Pervasive refusal syndrome in autistic spectrum disorder. *Case Reports in Psychiatry, 2018*, 5049818.

Bourgeron, T. (2007). The possible interplay of synaptic and clock genes in autism spectrum disorders. *Cold Spring Harbor Symposia on Quantitative Biology, 72*, 645–654.

Boyle, C. A., Boulet, S., Schieve, L. A., et al. (2011). Trends in the prevalence of developmental disabilities in US children, 1997–2008. *Pediatrics, 127*(6), 1034–1042.

Bruni, O., Baumgartner, E., Sette, S., et al. (2014). Longitudinal study of sleep behavior in normal infants during the first year of life. *Journal of Clinical Sleep Medicine, 10*(10), 1119–1127.

Bruni, O., Angriman, M., Melegari, M. G., et al. (2019). Pharmacotherapeutic management of sleep disorders in children with neurodevelopmental disorders. *Expert Opinion on Pharmacotherapy, 20*(18), 2257–2271.

Bueno, C., & Menna-Barreto, L. (2016). Development of sleep/wake, activity and temperature rhythm in newborns maintained in a neonatal intensive care unit and the impact of feeding schedules. *Infant Behavior & Development, 44*, 212–218.

Cardinali, D. P., Pagano, E. S., Scacchi Bernasconi, P. A., et al. (2013). Melatonin and mitochondrial dysfunction in the central nervous system. *Hormones and Behavior, 63*(2), 322–330.

Chaput, J. P., Brunet, M., & Tremblay, A. (2006). Relationship between short sleeping hours and childhood overweight/obesity: Results from the 'Québec en Forme' Project. *International Journal of Obesity, 30*(7), 1080–1085.

Christensen, D. L., Baio, J., Van Naarden Braun, K., et al. (2016). Centers for Disease Control and Prevention (CDC). Prevalence and characteristics of autism spectrum disorder among children aged 8 years--Autism and developmental disabilities monitoring network, 11 sites, United States, 2012. *MMWR Surveillance Summaries, 65*(3), 1–23.

Claustrat, B., & Leston, J. (2015). Melatoni physiological effects in humans. *Neuro-Chirurgie, 61*(2–3), 77–84.

Cockshaw, W. D., Muscat, T., Obst, P. L., et al. (2014). Paternal postnatal depressive symptoms, infant sleeping and feeding behaviors, and rigid parental regulation: A correlational study. *Journal of Psychosomatic Obstetrics and Gynaecology, 35*(4), 124–131.

Cortesi, F., Giannotti, F., Ivanenko, A., et al. (2010). Sleep in children with autistic spectrum disorder. *Sleep Medicine, 11*, 659–664.

Couturier, J. L., Speechley, K. N., Steele, M., et al. (2005). Parental perception of sleep problems in children of normal intelligence with pervasive developmental disorders: Prevalence, severity, and pattern. *Journal of the American Academy of Child and Adolescent Psychiatry, 44*, 815–822.

Danielsen, Y. S., Pallesen, S., Stormark, K. M., et al. (2010). The relationship between school day sleep duration and body mass index in Norwegian children (aged 10–12). *International Journal of Pediatric Obesity, 5*(3), 214–220.

Darcy-Mahoney, A., Minter, B., Higgins, M., et al. (2016). Probability of an autism diagnosis by gestational age. *Newborn and Infant Nursing Reviews, 16*(4), 322–332.

Dönmez, R. Ö., & Temel, A. B. (2019). Effect of soothing techniques on infants' self-regulation behaviors (sleeping, crying, feeding): A randomized controlled study. *Japan Journal of Nursing Science, 50*(4), 407–419.

Elsabbagh, M., Fernandes, J., Webb, S. J., et al. (2013). Disengagement of visual attention in infancy is associated with emerging autism in toddlerhood. *Biological Psychiatry, 74*(3), 189–194.

Erren, T. C., & Reiter, R. J. (2009). Defining chronodisruption. *Journal of Pineal Research, 46*(3), 245–247.

Estes, A., Zwaigenbaum, L., Gu, H., et al. (2015). Behavioral, cognitive, and adaptive development in infants with autism spectrum disorder in the first 2 years of life. *Journal of Neurodevelopmental Disorders, 7*(1), 24. https://doi.org/10.1186/s11689-015-9117-6. Epub 2015 Jul 16.

Foley, H. M., & Steel, A. E. (2019). Adverse events associated with oral administration of melatonin: A critical systematic review of clinical evidence. *Complementary Therapies in Medicine, 42*, 65–81.

Fukushima, K., Morokuma, S., & Nakano, H. (2008). Fetal behavior: Ontogenesis and transition to neonate. In A. Kurjak et al. (Eds.), *Donald school textbook* (2nd ed., pp. 641–655). Jaypee Brothers Medical Publishers.

Geoffray, M. M., Nicolas, A., Speranza, M., et al. (2016). Are circadian rhythms new pathways to understand Autism Spectrum Disorder? *Journal of Physiology, Paris, 110*(4 Pt B), 434–438.

Goel, R., Hong, J. S., & Findling, R. L. (2018). An update on pharmacotherapy of autism spectrum. disorder in children and adolescents. *International Review of Psychiatry, 30*(1), 78–95.

Goldberg, W. A., Lucas-Thompson, R. G., Germo, G. R., et al. (2013). Eye of the beholder? Maternal mental health and the quality of infant sleep. *Social Science & Medicine, 79*, 101–108.

Halal, C. S., Bassani, D. G., Santos, I. S., et al. (2021). Maternal perinatal depression and infant sleep problems at 1 year of age: Subjective and actigraphy data from a population-based birth cohort study. *Journal of Sleep Research, 30*(2), e13047.

Hansen, M., Janssen, I., Schiff, A., et al. (2005). The impact of school daily schedule on adolescent sleep. *Pediatrics, 115*(6), 1555–1561.

Harel-Gadassi, A., Friedlander, E., Yaari, M., et al. (2018). Risk for ASD in preterm infants: A three-year follow-up study. *Autism Research and Treatment, 2018*, 8316212. https://doi.org/10.1155/2018/8316212. eCollection 2018. PMID: 30534432.

Hart, C. N., & Jelalian, E. (2008). Shortened sleep duration is associated with pediatric overweight. *Behavioral Sleep Medicine, 6*(4), 251–267.

Hena, M., & Garmy, P. (2020). Social jetlag and its association with screen time and. nighttime texting among adolescents in Sweden: A cross-sectional study. *Frontiers in Neuroscience, 14*, 122. https://doi.org/10.3389/fnins.2020.00122. eCollection 2020.

Henderson, J. M., France, K. G., & Blampied, N. M. (2011). The consolidation of infants' nocturnal sleep across the first year of life. *Sleep Medicine Reviews, 15*(4), 211–220.

Higuchi, T., Ishizaki, Y., Noritake, A., et al. (2017). Spatiotemporal characteristics of gaze of children with autism spectrum disorders while looking at classroom scenes. *PLoS One, 12*(5), e0175912.

Hiscock, H., & Jordan, B. (2004). 1. Problem crying in infancy. *The Medical Journal of Australia, 181*(9), 507–512.

Hiscock, H., Cook, F., Bayer, J., et al. (2014). Preventing early infant sleep and crying problems and postnatal depression: A randomized trial. *Pediatrics, 133*(2), e346–e354.

Hoang, N., Yuen, R. K. C., Howe, J., et al. (2021). Sleep phenotype of individuals with autism spectrum disorder bearing mutations in the PER2 circadian rhythm gene. *American Journal of Medical Genetics. Part A, 185*(4), 1120–1130.

Hochadel, J., Frölich, J., Wiater, A., et al. (2014). Prevalence of sleep problems and relationship between sleep problems and school refusal behavior in school-aged children's and parent's ratings. *Psychopathology, 47*(2), 119–126.

Humphreys, J. S., Gringras, P., Blair, P. S., et al. (2014). Sleep patterns in children with autistic spectrum disorders: A prospective cohort study. *Archives of Disease in Childhood, 99*, 114–118.

Hysing, M., Haugland, S., Stormark, K. M., et al. (2015). Sleep and school attendance in adolescence: Results from a large population-based study. *Scandinavian Journal of Public Health, 43*(1), 2–9.

Iglowstein, I., Jenni, O. G., Molinari, L., et al. (2003). Sleep duration from infancy to adolescence: Reference values and generational trends. *Pediatrics, 111*(2), 302–307.

Inoue, S. (2006). *Science of sleep (in Japanese)* (p. 38). Asakura-shoten.

Iverson, J. M., Shic, F., Carla, A., et al. (2019). Early motor abilities in infants at heightened versus low risk for ASD: A Baby Siblings Research Consortium (BSRC) study. *Journal of Abnormal Psychology, 128*(1), 69–80.

Iwatani, N., Miike, T., Kai, Y., et al. (1997). Glucoregulatory disorders in school-refusal students. *Clinical Endocrinology, 47*(3), 273–278.

Jan, J. E., Reiter, R. J., Bax, M. C., et al. (2010). Long-term sleep disturbances in children: A cause of neuronal loss. *European Journal of Paediatric Neurology, 14*, 380–390.

Johnson, S., Hollis, C., Kochhar, P., et al. (2010). Autism spectrum disorders in extremely preterm children. *The Journal of Pediatrics, 156*(4), 525–531.

Kahn, M., Budinger, Y., Volkovich, E., et al. (2018). Links between infant sleep and parental tolerance for infant crying: Longitudinal assessment from pregnancy through six months postpartum. *Sleep Medicine, 50*, 72–78.

Kaplan, E., Daka, A., Weissbach, A., et al. (2019). Triclofos sodium for pediatric sedation in non-painful neurodiagnostic studies. *Paediatric Drugs, 21*(5), 371–378.

Keefe, M. R., Kotzer, A. M., Froese-Fretz, A., et al. (1996). A longitudinal comparison of irritable and nonirritable infants. *Nursing Research, 45*(1), 4–9.

Kelsay, K. (2006). Management of sleep disturbance associated with atopic dermatitis. *The Journal of Allergy and Clinical Immunology, 118*(1), 198–201.

Kim, C. W., Choi, M. K., Im, H. J., et al. (2012). Weekend catch-up sleep is associated with decreased risk of being overweight among fifth-grade students with short sleep duration. *Journal of Sleep Research, 21*(5), 546–551.

Konishi, Y., Toyoura, M., Konishi, Y., et al. (2019). New developments in the autism spectrum. *Journal of Daycare Research Sick Children* (in Japanese), *10*(separate volume), 14–21.

Krakowiak, P., Goodlin-Jones, B., Hertz-Picciotto, I., et al. (2008). Sleep problems in children with autism spectrum disorders, developmental delays, and typical development: A population-based study. *Journal of Sleep Research, 17*, 197–206.

Lam, L. T., & Lam, M. K. (2021). Sleep disorders in early childhood and the development of mental health problems in adolescents: A systematic review of longitudinal and prospective studies. *International Journal of Environmental Research and Public Health, 18*(22), 11782.

Lamy, M., & Erickson, C. A. (2018). Pharmacological management of behavioral. disturbances in children and adolescents with autism spectrum disorders. *Current Problems in Pediatric and Adolescent Health Care, 48*(10), 250–264.

Li, T., Horta, M., Mascaro, J. S., et al. (2018). Explaining individual variation in paternal brain responses to infant cries. *Physiology & Behavior, 193*(Pt A), 43–54.
Licis, A. (2020). Sleep-wake disorders in childhood. *Continuum (Minneap Minn), 26*(4), 1034–1069.
Limoges, É., Mottron, L., Bolduc, C., et al. (2005). Atypical sleep architecture and the autism phenotype. *Brain, 128*, 1049–1061.
MacDuffie, K. E., Shen, M. D., Dager, S. R., et al. (2020). Sleep onset problems and subcortical development in infants later diagnosed with autism spectrum disorder. *The American Journal of Psychiatry, 177*(6), 518–525.
Maeda, T., Oniki, K., & Miike, T. (2019). Sleep education in primary school prevents future school refusal behavior. *Pediatrics International, 61*, 1036–1042.
Malish, S., Arastu, F., & O'Brien, L. M. (2016). A preliminary study of new parents, sleep disruption, and driving: A population at risk? *Maternal and Child Health Journal, 20*(2), 290–297.
Malone, S. K., Zemel, B., Compher, C., et al. (2016). Characteristics associated with sleep duration, chronotype, and social jet lag in adolescents. *The Journal of School Nursing, 32*(2), 120–131.
Marshall, N. S., Glozier, N., & Grunstein, R. R. (2008). Is sleep duration related to obesity? A critical review of the epidemiological evidence. *Sleep Medicine Reviews, 12*(4), 289–298.
McDonagh, M. S., Holmes, R., & Hsu, F. (2019). Pharmacologic treatments for sleep disorders in children: A systematic review. *Journal of Child Neurology, 34*(5), 237–247.
Meier-Koll, A., Hall, U., Hellwig, U., et al. (1978). A biological oscillator system and the development of sleep-waking behavior during early infancy. *Chronobiologia, 5*(4), 425–440.
Miano, S., Bruni, O., Elia, M., et al. (2007). Sleep in children with autistic spectrum disorder: A questionnaire and polysomnographic study. *Sleep Medicine, 9*(1), 64–70.
Miike, T., Tomoda, A., Jhodoi, T., et al. (2004). Learning and memorization impairment in childhood chronic fatigue syndrome manifesting as school phobia in Japan. *Brain Dev, 26*, 442–447.
Miike, T., Toyoura, M., Tonooka, S., et al. (2020). Neonatal irritable sleep-wake rhythm as a predictor of autism spectrum disorders. *Neurobiology of Sleep and Circadian Rhythms, 9*, 100053.
Mindell, J. A., Sadeh, A., Wiegand, B., et al. (2010). Cross-cultural differences in infant and toddler sleep. *Sleep Medicine, 11*(3), 274–280.
Mirmiran, M., Maas, Y. G., & Ariagno, R. L. (2003). Development of fetal and neonatal sleep and circadian rhythms. *Sleep Medicine Reviews, 7*(4), 321–334.
Mohr, C., Gross-Hemmi, M. H., & Meyer, A. H. (2019). Temporal patterns of infant regulatory behaviors in relation to maternal mood and soothing strategies. *Child Psychiatry and Human Development, 50*(4), 566–579.
Nicholas, B., Rudrasingham, V., Nash, S., et al. (2007). Association of Per1 and Per2 with autistic disorder: Support for the clock Genes/social timing hypothesis. *Molecular Psychiatry, 12*, 581–589.
Ninomiya, T., Iwatani, N., Tomoda, A., et al. (2001). Effects of exogenous melatonin on pituitary hormones in humans. *Clinical Physiology, 21*(3), 292–299.
Noda, A., Yasuma, F., Okada, T., et al. (1998). Circadian rhythm of autonomic activity in patients with obstructive sleep apnea syndrome. *Clinical Cardiology, 21*, 271–276.
Oldham, M. A., Lee, H. B., & Desan, P. H. (2016). Circadian rhythm disruption in the critically ill: An opportunity for improving outcomes. *Critical Care Medicine, 44*(1), 207–217.
Ophoff, D., Slaats, M. A., Boudewyns, A., et al. (2018). Sleep disorders during childhood: A practical review. *European Journal of Pediatrics, 177*(5), 641–648.
Ornitz, E. M. (1973). Childhood autism. A review of the clinical and experimental literature. *California Medicine, 118*(4), 21–47.
Ornitz, E. M., Yanagisita, M., & Koshino, Y. (1983). Infantile autism. *Shonika (In Japanese), 24*(13), 1581–1591.
Öztürk Dönmez, R., & Temel, A. B. (2019). Effect of soothing techniques on infants' self-regulation behaviors (sleeping, crying, feeding): A randomized controlled study. *Japan Journal of Nursing Science, 16*(4), 407–419.

Paul, I. M., Savage, J. S., Anzman-Frasca, S., et al. (2016). INSIGHT responsive parenting intervention and infant sleep. *Pediatrics, 138*(1), e20160762.

Pinato, L., Galina Spilla, C. S., Markus, R. P., et al. (2019). Dysregulation of circadian rhythms in autism spectrum disorders. *Current Pharmaceutical Design, 25*(41), 4379–4393.

Pinilla, T., Birch, L. L. (1993). Help me make it through the night: behavioral entrainment of breast-fed infants' sleep patterns. *Pediatrics, 91*(2), 436–444.

Poggiogalle, E., Jamshed, H., & Peterson, C. M. (2018). Circadian regulation of glucose, lipid, and energy metabolism in humans. *Metabolism, 84*, 11–27.

Qian, J., & Scheer, F. A. J. L. (2016). Circadian system and glucose metabolism: Implications for physiology and disease. *Trends in Endocrinology and Metabolism, 27*(5), 282–293.

Reiter, R. J., Tan, D. X., Korkmaz, A., et al. (2014). Melatonin and stable circadian rhythms optimize maternal, placental and fetal physiology. *Human Reproduction Update, 20*(2), 293–307.

Remi, J. (2015). Humans entrain to sunlight – Impact of social jet lag on disease and implications for critical illness. *Current Pharmaceutical Design, 21*(24), 3431–3437.

Richey, L., Li, T., & Rilling, J. K. (2022). Perception of male and female infant cry aversiveness by adult men. *Journal of Reproductive and Infant Psychology, 40*(1), 76–90.

Rivkees, S. A. (1997). Developing circadian rhythmicity. Basic and clinical aspect. *Pediatric Clinics of North America, 44*(2), 467–487.

Roenneberg, T., Pilz, L. K., Zerbini, G., et al. (2019). Chronotype and social jetlag: A (self-) critical review. *Biology (Basel), 8*(3), 54.

Rossignol, D. A., & Frye, R. E. (2012). Mitochondrial dysfunction in autism spectrum disorders: A systematic review and meta-analysis. *Molecular Psychiatry, 17*(3), 290–314.

Sadeh, A., Tikotzky, L., & Scher, A. (2010). Parenting and infant sleep. *Sleep Medicine Reviews, 14*(2), 89–96.

Sadeh, A., Juda-Hanael, M., Livne-Karp, E., et al. (2016). Low parental tolerance for infant crying: An underlying factor in infant sleep problems? *Journal of Sleep Research, 25*(5), 501–507.

Schnoes, C. J., Kuhn, B. R., Workman, E. F., et al. (2006). Pediatric prescribing practices for clonidine and other pharmacologic agents for children with sleep disturbance. *Clinical Pediatrics (Phila), 45*(3), 229–238.

Serón-Ferré, M., Valenzuela, G. J., & Torres-Farfan, C. (2007). Circadian clocks during embryonic and fetal development. *Birth Defects Research. Part C, Embryo Today: Reviews, 81*(3), 204–214.

Serón-Ferré, M., Mendez, N., Abarzua-Catalan, L., et al. (2012). Circadian rhythms in the fetus. *Molecular and Cellular Endocrinology, 349*(1), 68–75.

Sethi, S., Joshi, G., Bloria, S. D., et al. (2021). Safety and efficacy of oral Triclofos in the ophthalmic evaluation of children with pediatric glaucoma: An observational study. *Indian Journal of Ophthalmology, 69*(1), 83–86.

Shafiq, S., & Pringsheim, T. (2018). Using antipsychotics for behavioral problems in children. *Expert Opinion on Pharmacotherapy, 19*(13), 1475–1488.

Sheldrick, R. C., & Carter, A. S. (2018). State-level trends in the prevalence of autism spectrum disorder (ASD) from 2000 to 2012: A reanalysis of findings from the autism and developmental disabilities network. *Journal of Autism and Developmental Disorders, 48*(9), 3086–3092.

Sivertsen, B., Pallesen, S., Stormark, K. M., et al. (2013). Delayed sleep phase syndrome in adolescents: Prevalence and correlates in a large population based study. *BMC Public Health, 13*, 1163.

Stephenson, R., Lim, J., Famina, S., et al. (2012). Sleep-wake behavior in the rat: Ultradian rhythms in a light-dark cycle and continuous bright light. *Journal of Biological Rhythms, 27*(6), 490–501.

Tikotzky, L. (2016). Postpartum maternal sleep, maternal depressive symptoms and self-perceived mother-infant emotional relationship. *Behavioral Sleep Medicine, 14*(1), 5–22.

Tomoda, A., Miike, T., Uezono, K., et al. (1994). A school refusal with biological rhythm disturbance and melatonin therapy. *Brain and Development, 16*(1), 71–76.

Tomoda, A., Miike, T., Yonamine, K., et al. (1997). Disturbed circadian core body temperature rhythm and sleep disturbance in school refusal children and adolescents. *Biological Psychiatry, 41*, 810–813.

Tordjman, S., Davlantis, K. S., Georgieff, N., et al. (2015). Autism as a disorder of biological and behavioral rhythms: Toward new therapeutic perspectives. *Frontiers in Pediatrics, 3*, 1–15.

Touchette, E., Dionne, G., Forget-Dubois, N., et al. (2013). Genetic and environmental influences on daytime and nighttime sleep duration in early childhood. *Pediatrics, 131*(6), e1874–e1880.

Touitou, Y., Reinberg, A., & Touitou, D. (2017). Association between light at night, melatonin secretion, sleep deprivation, and the internal clock: Health impacts and mechanisms of circadian disruption. *Life Sciences, 173*, 94–106.

Toyoura, M., Miike, T., Tajima, S., et al. (2020). Inadequate sleep as acontributor to impaired glucose tolerance: A cross-sectional study in children, adolescents, and young adults with circadian rhythm sleep-wake disorder. *Pediatric Diabetes, 21*(4), 557–564.

Turner, M. (2020). The role of drugs in the treatment of autism. *Australian Prescriber, 43*(6), 185–190.

Verhoeff, M. E., Blanken, M. L. E., Kocevska, D., et al. (2018). The bidirectional association between sleep problems and autism spectrum disorder: A population-based cohort study. *Molecular Autism, 9*, 8.

Wakayama, K., Ogawa, T., Goto, K., et al. (1993). Development of ultradian rhythm of EEG activities in premature babies. *Early Human Development, 32*(1), 11–30.

Waldhauser, F., Weiszenbacher, G., Tatzer, E., et al. (1988). Alterations in nocturnal serum melatonin levels in humans with growth and aging. *The Journal of Clinical Endocrinology and Metabolism, 66*(3), 648–652.

Wimpory, D., Nicholas, B., & Nash, J. (2002). Social timing, clock genes and autism: A new hypothesis. *Journal of Intellectual Disability Research, 46*, 352–358.

Wing, Y. K., Li, S. X., Li, A. M., et al. (2009). The effect of weekend and holiday sleep compensation on childhood overweight and obesity. *Pediatrics, 124*(5), e994–e1000.

Wittmann, M., Dinich, J., Merrow, M., et al. (2006). Social jetlag: Misalignment of biological and social time. *Chronobiology International, 23*(1–2), 497–509.

Wolff, J. J., Botteron, K. N., Dager, S. R., et al. (2014). Longitudinal patterns of repetitive behavior in toddlers with autism. *Journal of Child Psychology and Psychiatry, 55*(8), 945–953.

Wu, Z. Y., Huang, S. D., Zou, J. J., et al. (2020). Autism spectrum disorder (ASD): Disturbance of the melatonin system and its implications. *Biomedicine & Pharmacotherapy, 130*, 110496.

Yavuz-Kodat, E., Reynaud, E., Geoffray, M. M., et al. (2020). Disturbances of continuous sleep and circadian rhythms account for behavioral difficulties in children with autism spectrum disorder. *Journal of Clinical Medicine, 9*(6), 1978.

Chapter 11
Imbalances of Inhibitory and Excitatory Systems in Autism Spectrum Disorders

Reed C. Carroll

11.1 The Excitation/Inhibition Imbalance Model and Autism

Autism is a neurodevelopmental disorder characterized by challenges with social interactions and communications, restricted interests and repetitive behaviors (American Psychiatric Association, 2022). Autism Spectrum Disorder (ASD) encompasses a wide range of phenotypically related conditions, with individuals on the spectrum exhibiting varied behavioral characteristics and severity of symptoms. The etiology of ASD is also complex with numerous identified genetic as well as environmental risk factors. The existence of a spectrum of phenotypically related disorders with such heterogeneity in cause and expression, suggests that diverse factors may impact a common developmentally regulated cortical property resulting in these similar conditions. An influential model proposed by Rubenstein and Mereznich suggested an imbalance in the relative levels of excitation and inhibition in the CNS may be a common factor underlying ASD. With an elevation of relative excitation, they proposed development of a cortex that is poorly differentiated and hyperexcitable, which could contribute to deficits in perception, and memory such as can be seen in ASD (Rubenstein & Merzenich, 2003).

The balance of excitation and inhibition has been recognized as important to normal brain functions, and imbalances have been linked to cognitive disorders and seizures. Information processing in the brain is the consequence of integrated information not only propagated by excitatory signals, but also by integration and regulation by inhibitory signals which shape the output of neural circuits. Inhibitory neurons increase the computational complexity of a neural circuit, regulate the initiation and timing of spike activity, and serve as a brake to prevent runaway excitation (Isaacson & Scanziani, 2011). The proper balance of excitation and inhibition

R. C. Carroll (✉)
Biology Department, New Jersey City University, Jersey City, NJ, USA
e-mail: RCarroll@njcu.edu

A. El Idrissi, D. McCloskey (eds.), *Neurobiology of Autism Spectrum Disorders*,
https://doi.org/10.1007/978-3-031-42383-3_11

(E/I) is highly regulated at the cellular and circuit level. Consequently, disruption of multiple factors which impact the levels of either excitation or inhibition relative to each other could lead to an imbalance of excitability, possibly consistent with the complex genetic and environmental factors associated with ASD.

At the cellular level, the relative strength and number of excitatory and inhibitory inputs is subject to modification developmentally and through activity driven plasticity. Regulation of synaptic connections is a key factor in the modification of neural function that occurs during development and with learning. It has been established that homeostatic mechanisms maintain the balance of inhibition and excitation even with ongoing changes in the relative strength and numbers of individual synaptic inputs (Turrigiano, 2012). Inhibition and excitation are also modulated at the circuit level, with the relative excitatory and inhibitory drives of converging circuits contributing to the overall balance of the system (Sohal & Rubenstein, 2019). As such, even with a unifying model of E/I imbalance as a contributing factor to ASD, a complex combination of factors may be involved, influencing both excitatory and inhibitory drive. Disruptions or alterations in synapse development, synaptic transmission, plasticity, regulatory signaling pathways, as well as developmental factors which impact the formations and interaction of circuits all can impact overall E/I balance (Lee et al., 2017).

Since the initial proposal of E/I imbalance as an underlying factor in ASD, numerous studies have investigated evidence for links. Correlative support for altered E/I in ASD has been found in clinical studies and postmortem analysis. With the identification of possible genetic factors, animal models have provided significant insight into how multiple alterations in excitation and inhibition may contribute to similar specific behavioral characteristics associated with ASD. Importantly these findings and new studies have provided evidence that there are promising interventions which may prove to be beneficial for therapeutic approaches in the future.

11.2 Clinical Evidence for E/I Imbalance in ASD

Multiple lines of evidence from studies of ASD individuals support a possible role of altered E/I balance. Existence of common comorbidities, behavioral and clinical analysis, as well as genetic studies suggest, directly and indirectly, that E/I may be altered in autism.

The high rate of comorbidity of seizures and epilepsy in ASD and associated disorders such as Fragile X syndrome (FXS) is one factor suggestive of an underlying E/I imbalance. From 12% to 30% of autistic individuals have been reported to exhibit seizures (Viscidi et al., 2013; Lukmanji et al., 2019). Evidence suggests that the excessive neural activity associated with seizures may have links to altered inhibition in cortical circuits. Alterations in the synthesis and release of gamma amino butyric acid (GABA), the primary inhibitory neurotransmitter in the cortex, as well as GABA receptors (GABARs) themselves can all impact cellular inhibition.

Reduced GABAR expression has been observed in epilepsy patients (Princivalle et al., 2003) as has reduced GABA binding, and GABA levels (Treiman, 1989). Glutamic acid decarboxylase (GAD), an enzyme essential for GABA synthesis and whose deficiency results in epilepsy in mice (Kash et al., 1997), has also been reported to be lower in epilepsy patients. (Neuray et al., 2020; Gospe Jr. et al., 1994). The high level of co-occurrence between ASD and epilepsy is suggestive of a common underlying basis, and in support of this similar evidence for altered inhibitory signaling have been observed in ASD individuals.

Studies of inhibitory neurotransmitters, enzymes, and receptors as well as investigations of neural activity in inhibitory circuits have provided further evidence for possible alterations in E/I with ASD. Results from these studies have been varied, with evidence supporting either reduced inhibition or increased excitation, depending on the brain areas and ages of individuals studied. Examinations in tissues have reported alterations in GABA type A receptors ($GABA_ARs$) levels in multiple ASD brain regions with $GABA_AR$ density reported to be reduced in cerebellar, hippocampal, parietal, and frontal regions (Blatt & Fatemi, 2011; Fatemi et al., 2009, 2014; Oblak et al., 2009, 2011). Levels of GAD 65 and 67 were also reduced in the cerebellum and parietal cortex of individuals with ASD (Fatemi et al., 2002; Yip et al., 2007). Magnetic Resonance Spectroscopy (MRS) has allowed for a means to indirectly measure neurotransmitter levels from patients. In several studies, GABA levels have been found to be reduced in the pre/post central gyrus (Gaetz et al., 2014; Puts et al., 2017; Harada et al., 2011). While others have reported no effects, a possible age-dependence has been suggested as young children and adults show little alteration in GABA levels while children around 10 years old show decreased levels in ASD (Cochran et al., 2015; Kubas et al., 2012). Examinations of glutamatergic signaling also have reported differences in ASD, however the effects have been variable. While these studies do not provide definitive evidence for a uniform deficiency as a common underlying factor in ASD, they do provide overall support for the concept that multiple factors that link to altered E/I might contribute broadly to the disorder.

Additional evidence for possible alterations in E/I balance within ASD patients come from investigations of cortical minicolumns. Minicolumns are vertically organized groups of pyramidal cells rising in cortical layers VI and II (Dickinson et al., 2016). Minicolumns are considered to be the smallest essential circuit module capable of information processing. Minicolumns of pyramidal cells are surrounded by GABAergic interneurons in the surrounding neuropil which effectively prevent excitatory information spreading from neighboring columns. The minicolumn neuropil has been found to be reduced in ASD individuals along with overall narrower minicolumns (Casanova, 2004; Casanova et al., 2002). This is observed particularly in the prefrontal cortex (PFC), a brain region involved in social behavior and communication, behaviors commonly disrupted in autism. One apparent implication of a reduced neuropil could be a reduction in inhibition resulting in a hyperexcitability in the minicolumns. However, the precise impact of such changes remains uncertain.

Direct analysis of inhibitory neuron numbers and types have supported changes consistent with reduced inhibition in autism. A broad classification of inhibitory neurons groups cells by their expression of parvalbumin (PV), somatostatin (SST) and serotonin receptor 5HT3a (5HT3Ra) (Rudy et al., 2011). Of these, the most common are the fast-spiking PV interneurons which include basket cells and chandelier cells. Basket cells form synapses on the soma and proximal dendrites, while chandelier cells form synapses on the axon initial segment. A key role of the basket cells is in the modulation of spike timing in nearby excitatory neurons (Pouille & Scanziani, 2001; Wehr & Zador, 2003). The high level of connectivity of basket cells with surrounding pyramidal neurons provides them with a potent capability of modulating local circuit activity through feedforward and feedback inhibition. Studies of ASD individuals showed that *PVALB* mRNA, encoding parvalbumin, is the most strongly downregulated of measured transcripts in human ASD samples (Schwede et al., 2018). Additionally, reduction on PV-stained neurons compared to matched controls, with no changes in other types of interneurons. (Ariza et al., 2018; Hashemi et al., 2018). It is not completely clear whether these changes represent a loss of PV neurons themselves, a loss of PV expression, or dysfunction or death of PV neurons in response to reduced *PVALB*. In any case, a likely outcome of reduced PV-stained interneurons is an altered balance in the E/I ratio in affected brains.

Further evidence for altered E/I in ASD has been found in EEG recordings of gamma oscillations. Gamma activity patterns occur as oscillations in the 30–100 Hz frequency. PV interneurons, via their ability to provide strong feed forward inhibition have been implicated in the generation of gamma oscillations (Buzsáki & Draguhn, 2004). Due to an essential role of inhibition in generating gamma oscillations, it is thought to be a good indicator of E/I balance in local cortical circuits (Whittington et al., 2000; Buzsaki & Wang, 2012). Multiple studies of gamma activity by EEG analysis have suggested a decrease in gamma activity in autism patients in response to simple sensory stimuli (Rojas & Wilson, 2014; Brown et al., 2005).

A correlation between altered perceptual behavior with autism and altered GABAergic signaling has also investigated (Robertson et al., 2016). Binocular rivalry occurs when the two eyes are presented with different images and the brain alternates for perceptual dominance from each eye. This alteration has been tied to a balance of excitation and inhibition allowing for the alternating suppression of perceptual processing from each eye. In control patients the dynamics of the alternation of perceptual processing was found to be tightly linked to GABA and glutamate levels detected by MRS in the visual cortex. Autistic patients showed a defect in binocular rivalry as well as an absence of linkage to neurotransmitter levels. Results support a possible connection between disrupted GABAergic signaling and altered visual perception in ASD patients.

Phenotypic evidence for E/I imbalance is further supported by the complex genetic linkages which are reported with ASD. The genetic nature of ASD is supported by twin studies exhibiting 70–90% concordance in monozygotic twins versus 0–10% in dizygotic twins, and a high rate of heritability (above 90%) (Abrahams & Geschwind, 2008; Toro et al., 2010). Observations have highlighted involvement of genes that have direct and indirect impact on the development and modification of both excitatory and inhibitory neurons, synapses, and circuits.

Genetic linkage analysis has identified multiple autism risk genes as well as multiple copy number variation (CNV) loci. A gender bias (4:1 male to female) of the disorders suggests involvement of the X chromosome (Abrahams & Geschwind, 2008). Identified chromosomal rearrangements with ASD include 16p11.2 deletion and 15q11-q13 duplications. 15q11 contains a cluster of three GABAR subunit genes including *GABRB3*, *GABRA5*, and *GABRG3* which encode the $GABA_A$ receptor β3, α5, and γ3 subunits (Cook Jr. et al., 1998; Boyar et al., 2001; Menold et al., 2001). While ASD is generally considered to be of mulitigenetic origin, a proportion of syndromic ASD with known genetic causes are of monogenetic nature and impact genes which can influence neural excitability. Fragile X syndrome (FXS), the most common form of syndromic autism (2% of ASD individuals), results from deficiencies in *FMR1*, an mRNA binding protein linked to synapse development and modulation (Richter & Zhao, 2021; Tran et al., 2019). Other disorders associated with autism result from alterations in genes which directly or indirectly also impact the formation or function of excitatory and inhibitory synapses and circuits including Tuberous sclerosis (*TSC1* or *2* mutation), and Rett Syndrome (*MECP2* mutation) (Rylaarsdam & Guemez-Gamboa, 2019). Furthermore, analysis ASD has also identified links to mutations and copy number variations (CNVs) in genes critically linked to synaptic structure at excitatory and/or inhibitory synapses including SHANK3, Neuroligin3, Neuroligin4, Neurexin1 and CNTNAP (Ecker et al., 2013). These diverse genetic linkages highlight the complexity of the underlying factors of ASD, but they also have provided for promising directions for studies in model systems which have supported possible connections between behavioral abnormalities associated with ASD and defects altering the balance of neuronal excitation and inhibition.

11.3 Links Between E/I Imbalance and Autism-Like Behavior in Model Systems

Mouse models of autism and autism-related disorders have been valuable in the characterization of genetic effects on both behavior and neural functions. These model systems include mutations or deletions in single genes linked to syndromic autism or other identified risk genes in ASD as well as environmental or genetic models identified by behavioral characteristics associated with autism.

Rett syndrome is one of the monogenetic disorders associated with autism. Linked to mutations in the X linked *MECP2* gene encoding the transcriptional regulator Methyl-CpG-binding protein2 (MeCP2). Rett syndrome is characterized by delayed developmental regression, motor dysfunctions and seizures as well as key features of autism. Mouse models lacking MECP2 display reduced GABAergic signaling, reduced GAD expression, and lowered GABA levels (Chao et al., 2010). This altered GABAergic signaling is accompanied by phenotypic behaviors common to autism including stereotyped behaviors, anxiety phenotypes, altered social behavior, and sensory deficits (Chao et al., 2010; Orefice et al., 2016).

FXS, resulting from mutation in the *FMR1* gene, is the most common monogenetic disorder known to cause autism (Santoro et al., 2012). Studies of mouse models of FXS indicate significant alterations in excitatory, rather than inhibitory drive. FMRP is an RNA binding protein linked to protein synthesis in dendrites. *FMR1* mutant mice exhibit altered excitatory dendritic structures and deficits in mTOR signaling that is involved in activity driven neuronal protein synthesis (Banerjee et al., 2018). Additionally, *FMR1* deficient mice exhibit a reduced excitatory drive on fast spiking neurons resulting in a decrease in inhibitory drive and hyperexcitability (Gibson et al., 2008).

Tuberous sclerosis complex (TSC) is a severe disorder affecting multiple body systems and is caused by non-cancerous growths (Haji et al., 2020). It is sometimes associated with cognitive disabilities and seizures and autism. Tuberous sclerosis is caused by mutations in the *TSC1* or *TSC2* genes which encode TSC 1 and 2, repressors of the mTOR signaling pathway. Deletion of *TSC1* in a mouse model results in reduced inhibitory signaling, hyperexcitability (Bateup et al., 2013) and enhanced excitatory transmission (Bateup et al., 2011). Mice with haploinsufficiency of *TSC2* were found to exhibit characteristics of autism including reductions in social behaviors (Sato et al., 2012). In animals in which *TSC1* was deleted in the cerebellum, animals also exhibited altered social interactions as well as repetitive behaviors. Interestingly the cerebellar physiology appeared to demonstrate a reduction in excitability (Tsai et al., 2012).

The vast majority of ASD disorders do not have a monogenic cause, and many genetic studies have been conducted to find other associated risk genes, with as many as 800 identified so far (Masini et al., 2020). Among these are several which may have links, directly or indirectly to altered synaptic, neuronal, and circuit development which could impact the overall E/I balance. Notable among these are the $GABA_AR$ genes whose alterations can be directly linked to changes in inhibitory signaling (Cook Jr. et al., 1998; Boyar et al., 2001; Menold et al., 2001), while other genes have been shown to have significant effects on neural circuit excitability and behaviors in model animals. A group of functionally related proteins which have received significant attention are the ASD risk genes which fall into a group of cell adhesion and associated molecules. Among their roles, these molecules are involved in the development and stabilization of synapses.

Neurexins found in the presynaptic terminals interact with neuroligins in the postsynaptic membrane. Neurexins are found in both excitatory and inhibitory synapses (Südhof, 2008). Mutations and CNVs of NRXN1α have been found to be associated with ASD (Kirov, 2015). Mutations in this gene have been linked to deficits in excitatory transmitter release at synapses (Etherton et al., 2009; Pak et al., 2015; Davatolhagh & Fuccillo, 2021). While deficiencies in NRXN1 have been reported to be associated with reduced social affiliative behavior in mice (Armstrong et al., 2020; Twining et al., 2017). Contactin associated protein-like 2 (CNTNAP2) is another member of the neurexin family which has been found to be mutated or exhibit CNVs in ASD (Alarcon et al., 2008; Arking et al., 2008; Bakkaloglu et al., 2008). CNTNAP2 deficient mice exhibit reductions in inhibitory neurons, altered neural synchronization, and increased seizure-like activity (Thomas et al., 2017)

(Peñagarikano et al., 2011). Behaviorally, the knockout mice showed less behavioral flexibility and increased levels of repetitive motor activities consistent with behaviors associated with ASD (Peñagarikano et al., 2011; Sacai et al., 2020).

Neurexins interact with neuroligin proteins located in the postsynaptic membrane, which in turn bind to a number of intracellular postsynaptic proteins (Südhof, 2008). Neuroligins include four family members, NLGN1–4, with NLGN1 found in excitatory synapses, NLGN2 in inhibitory synapses, and NLGN 3 in both. NLGN4 is generally found in glycinergic synapses (Taylor et al., 2020; Hoon et al., 2011). Despite their differences in expression and function, mutations or deletions of each of these have been found in ASD (Jamain et al., 2003). Neuroligins have been reported to influence synapse maturation, morphology, and number, although the results are varied depending on the model studied (Taylor et al., 2020). Deletion of neuroligins 1–3 alters E/I balance by reducing inhibitory signaling (Varoqueaux et al., 2006). Knockout of *NRLGN1* appears to affect the ratio of striatal NMDA and AMPA type glutamate receptors, along with a reduction in excitatory synaptic plasticity in hippocampus (Blundell et al., 2010). Knockout of *NRLGN1* was also found to result in decreased social affiliation in mice, as did knock in of a deficient mutant form (Blundell et al., 2010; Nakanishi et al., 2017). Conditional knockout of *NLGN2* in the medial PFC (mPFC) resulted in a reduction of inhibitory signaling as well as decreases in social affiliation (Liang et al., 2015). Overexpression of a *NRLGN3* mutation found in ASD (Arg451Cys) has in some studies been reported to result in decreased social affiliative behavior (Kumar et al., 2014; Etherton et al., 2011; Jaramillo et al., 2014) and was found to increase excitatory signaling in the hippocampus (Etherton et al., 2011). There have also been studies of *NRLGN4* deficiencies in the behavior and properties in mouse models consistent with alterations in synaptic signaling and autism-like behaviors (Jamain et al., 2008). However, there is divergence in the evolution of this gene, which is X-linked in humans and somatic in rodents, that may make evaluation of these results in a mouse model more complicated (Nguyen et al., 2020).

Another family of synaptic proteins found in ASD includes the SH3 and multiple ankyrin repeat domains proteins (SHANKs), SHANK1, SHANK2, and SHANK3 (Monteiro & Feng, 2017). *SHANK* deletions, mutations and/or duplications have been found in ASD patients for genes encoding all SHANK proteins (Leblond et al., 2014). *SHANK3* haploinsufficiency has been found in patients with the autism disorder Phelan–McDermid syndrome associated with chromosome 22q13 deletion (Phelan & Mcdermid, 2012). SHANK proteins are synaptic scaffold proteins which serve to link receptors and their activity to various elements critical to synaptic signaling. Among proteins SHANKs can interact with are the neuroligins. Alterations in SHANK protein expression result in disrupted synaptic properties, particularly dysregulation of excitatory synapses and signaling.

SHANK 1 and 2 proteins promote the maturation of synapses. *SHANK1* knockouts exhibit smaller and fewer dendritic spines and reduced excitatory synaptic transmission (Hung et al., 2008). *Shank2* mutants had variable effects resulting in either no effect or reduced spine density (Schmeisser et al., 2012; Ha et al., 2016; Won et al., 2012). SHANK 3 is associated with the enlargement of dendritic spines.

A *SHANK3* mutant with deficiency in exon 9 exhibited reduced hippocampal excitatory transmission with increased spontaneous inhibitory synaptic events. In the mPFC there a decrease in frequency of spontaneous inhibitory synaptic events suggest alterations in E/I balance (Lee et al., 2015). Loss of SHANK 3 reduced excitatory transmission, spine density and levels of synaptic proteins that interact with SHANK3 (Monteiro & Feng, 2017). Together with alterations in excitability, mice deficient in the SHANK proteins exhibit characteristics similar to ASD (Jiang & Ehlers, 2013). With knockout of *SHANK1,* mice exhibit elevated anxiety, decreased vocal communication, reduced locomotion and as well as deficits in memory (Hung et al., 2008; Wöhr et al., 2011). *SHANK2* knock-out mice were reported to be hyper-excitable, exhibiting increased anxiety, as well as repetitive grooming behaviors, and abnormal vocalizations and social behaviors (Schmeisser et al., 2012; Won et al., 2012). *SHANK3* knock-out mice also showed repetitive grooming, and deficits in social interaction and vocalizations (Peça et al., 2011; Wang et al., 2011; Jaramillo et al., 2017).

Multiple other mutations have been found in genes which can directly or indirectly impact the development, formation, and function of synapses and neural circuits (Lee et al., 2017). The common ability of each of these to impact the excitatory, or inhibitory signaling or circuitry provides further support for a common link of E/I imbalance as a contributing factor to autism.

Two additional mouse models characterized by their phenotypic characteristics shared with autism, have provided further insights into possible disruptions of E/I balance in ASD; VPA and BTBR mice. Prenatal exposure to valproic acid (VPA) is associated with an increased incidence of autism in children (Christensen et al., 2013). Injection of VPA in pregnant mice results in offspring with increased social behavior abnormalities (Lucchina & Depino, 2014; Schneider et al., 2008). This environmental model of autism in rats has been reported to exhibit evidence of altered E/I balance paralleling what has been observed with ASD. Along with a reduction in inhibitory transmission (Banerjee et al., 2013), VPA mice exhibited reductions in PV neurons in the parietal cortex (Lauber et al., 2016) as has been observed in some cases of ASD. Several other mouse models exhibit reductions in PV neurons in multiple brain areas including FMRP (Lee et al., 2019), MeCP2 (Chen et al., 2017) and NLGN-3 (Gogolla et al., 2009).

The functional significance of PV neurons and their activation on social behaviors impacted in ASD is highlighted by a striking study used optogenetics to directly manipulate E/I balance in prefrontal cortex (Yizhar et al., 2011). Increased excitability through an optically stimulated depolarization of neurons resulted in social behavior defects in treated mice. Together with these behavior effects there was an increase in the extent of gamma rate activity. Significantly, compensatory restoration of inhibition in these animals with elevated optical activation of PV neurons results in a partial rescue of the behavioral deficits (Yizhar et al., 2011).

The VPA mouse model also exhibits an additional interesting characteristic relevant to the proper EI balance. An essential stage in the development of the inhibitory signaling in the CNS involves a developmental switch in the signaling of synaptic GABAergic receptors. Early in development activation of GABAergic receptors is excitatory as the chloride currents through the GABAergic channels are depolarizing. This effect is the consequence of the relatively high levels of Cl^- inside the cell compared to what is observed in older animals. At around 1 week postnatal in rodents there is a switch in the balance of Cl^- ion concentrations as the result of decreased expression of NKCC1, a chloride importer, and an increased expression of the KCC2 transporters which export Cl^- ions. With this developmentally regulated switch, GABAergic signaling shifts to become inhibitory (Represa & Ben-Ari, 2005). Several factors, including activation of depolarizing GABARs themselves, contribute to this developmental switch. It is believed that factors which impact the timing of this developmental switch to GABAergic transmission could have profound effects on the subsequent levels of inhibition in the brain. VPA mice (Tyzio et al., 2014; Roux et al., 2018), exhibit a delay in the excitatory-inhibitory shift. Such a delayed shift has also been observed in *FMR1* and *MECP2* deficient mice (Lozovaya et al., 2019). A developmentally delayed switch from excitatory to inhibitory GABAergic signaling has been proposed to contribute to some forms of ASD. For example, TSC patient tissues exhibited a change in the expression of ratio of NKCC2 to KCC2 which could result in a reduced inhibitory drive of GABAergic transmission, and recordings from neurons TSC tubers exhibited depolarizing GABAergic currents (Talos et al., 2012).

An additional mouse model that has provided interesting insight into links between excitatory balance and autisms is the BTBR mouse, originally bred for studies on insulin-resistance and diabetes (Meyza & Blanchard, 2017; Nadler et al., 2006). Behavioral studies revealed that these animals exhibited characteristics consistent with ASD including altered behavioral interactions, reduced vocalizations, and repetitive grooming (Meyza et al., 2013). Transcriptome analysis has shown that the expression of hundreds of genes is altered in this model. Protein analysis has, identified some interesting candidates associated with synaptic structure and function including spinophilin, Synapsin 1, PSD95 and NeuN (Jasien et al., 2014). Physiological analysis has supported apparent alterations in excitatory and inhibitory signaling. Increased GABAergic and reduced glutamatergic synaptic currents were recorded in the hippocampus of early postnatal mice (Cellot et al., 2016). In contrast, increased spontaneous excitatory and reduced inhibitory synaptic currents were reported in adult hippocampus (Han et al., 2014) suggesting a more complicated developmental or compensatory nature of the imbalance.

11.4 Implications for Therapeutic Treatments Targeting E/I Imbalance

Due to the complex etiology of autism, treatments have been limited and focus greatly on behavioral therapies. Currently, there are no approved medical treatments for broadly improving or preventing symptoms of autism. However, identification of links between E/I imbalances and ASD has led to the investigation of possible therapeutic interventions in both animal models and humans, several of which have shown promise for positive outcomes.

Pharmacological therapeutic approaches to treating specific symptoms of autism have included treatments which directly impact inhibitor signaling. Regulators of GABAergic signaling, a treatment for epilepsy, has been used in part due to the high incidence of seizures in ASD. Limited studies of anticonvulsants have not clearly shown an advantage of these treatments for behavioral symptoms in patients (Limbu et al., 2022). However, in some instances reports suggest possible improvements such as in the case of a small study of tuberous sclerosis patients who showed dramatic improvements in autistic behaviors with vigabatrin treatment (Jambaqué et al., 2000).

More promising results have been reported for bumetanide, an inhibitor of the NKCC1 transporter, which enhances the inhibitory drive of GABA. In two mouse models for autism, VPA and FRX1 mice, deficits in social behaviors, vocalization induced by maternal separation, were largely restored in offspring of mothers treated with bumetanide (Lemonnier et al., 2012; Tyzio et al., 2014). Effects appeared to last into adult animals (Eftekhari et al., 2014). Clinical studies have suggested possible ameliorative effects of the drug as well. In neonates and children with autism, improvements have been reported for a range of autistic symptoms in multiple studies (Lemonnier et al., 2012; Lemonnier et al., 2017; Du et al., 2015; Hadjikhani et al., 2015; Zhang et al., 2020). In support of the importance of this treatment impacting excitatory balance, GABA levels were found to be enhanced in treated children (Lemonnier et al., 2012).

A possible imbalance in E/I underlying behavioral deficits in autism suggests the possibility that restoration of excitatory balance through direct stimulation of cortical circuits might alleviate symptoms of the disorder. In fact, in studies of CNTNAP2 deficient mice, optogenetic stimulation of mPFC neurons was able to restore a more normal E/I ratio. Stimulations of PV interneurons in the mPFC, to restore otherwise reduced inhibition, resulted in an acute rescue of reduced social interactions in these animals. Similarly, inhibition of excitatory drive from pyramidal neurons also resulted in a behavioral rescue (Selimbeyoglu et al., 2017). While such manipulations are not currently possible in humans, transcranial direct stimulation(tDCs) and TMS may serve as an intervention that can modulate inhibitory balance in patients. Each of these methods have been found to alter neural excitability levels. Several limited studies have in fact found that these stimulation methods resulted in improvements of measures in behavioral aspects of autism including sociability, irritability, and repetitive behaviors (Khaleghi et al., 2020). One study has suggested that TMS could in fact alter inhibition in a limited trial of patients on the spectrum (Enticott et al., 2010).

With extensive studies and growing understanding of the diverse genetic targets linked to autism and autism-related behaviors, there is potential for future interventions as gene therapy technology develops. Syndromic autism with monogenic origins, holds promise for successful interventions. Rescue of behavioral deficits have been reported in mouse models of FXS with viral reintroduction of *FMR1* gene (Hooper et al., 2021). Reintroduction of *MECP2* in somatosensory neurons also rescued some behavioral alterations in Rett syndrome model mice (Orefice et al., 2016). A viral based gene therapy approach is being developed for treatment of Rett syndrome. Due to the diverse nature of the underlying genetics of idiopathic ASD, effective treatments would require both improved understanding of the complexities of multiple gene effects over development as well as suitable testing to enable personalized treatments. Interventions which might provide compensatory changes to alleviate effects altered E/I balance may also have potential. While such treatments are not immediately possible, the rapid development of approaches to gene therapies provides promise for increasing opportunities in the coming decades.

References

Abrahams, B. S., & Geschwind, D. H. (2008). Advances in autism genetics: On the threshold of a new neurobiology. *Nature Reviews. Genetics, 9*, 341–355.

Alarcon, M., Abrahams, B. S., Stone, J. L., Duvall, J. A., Perederiy, J. V., Bomar, J. M., Sebat, J., Wigler, M., Martin, C. L., Ledbetter, D. H., Nelson, S. F., Cantor, R. M., & Geschwind, D. H. (2008). Linkage, association, and gene-expression analyses identify CNTNAP2 as an autism-susceptibility gene. *American Journal of Human Genetics, 82*, 150–159.

American Psychiatric Association. (2022). *Diagnostic and statistical manual of mental disorders: DSM-5-TR*. American Psychiatric Association Publishing.

Ariza, J., Rogers, H., Hashemi, E., Noctor, S. C., & Martínez-Cerdeño, V. (2018). The number of chandelier and basket cells are differentially decreased in prefrontal cortex in autism. *Cerebral Cortex, 28*, 411–420.

Arking, D. E., Cutler, D. J., Brune, C. W., Teslovich, T. M., West, K., Ikeda, M., Rea, A., Guy, M., Lin, S., Cook, E. H., & Chakravarti, A. (2008). A common genetic variant in the neurexin superfamily member CNTNAP2 increases familial risk of autism. *American Journal of Human Genetics, 82*, 160–164.

Armstrong, E. C., Caruso, A., Servadio, M., Andreae, L. C., Trezza, V., Scattoni, M. L., & Fernandes, C. (2020). Assessing the developmental trajectory of mouse models of neurodevelopmental disorders: Social and communication deficits in mice with Neurexin 1alpha deletion. *Genes, Brain, and Behavior, 19*, e12630.

Bakkaloglu, B., O'Roak, B. J., Louvi, A., Gupta, A. R., Abelson, J. F., Morgan, T. M., Chawarska, K., Klin, A., Ercan-Sencicek, A. G., Stillman, A. A., Tanriover, G., Abrahams, B. S., Duvall, J. A., Robbins, E. M., Geschwind, D. H., Biederer, T., Gunel, M., Lifton, R. P., & State, M. W. (2008). Molecular cytogenetic analysis and resequencing of contactin associated protein-like 2 in autism spectrum disorders. *American Journal of Human Genetics, 82*, 165–173.

Banerjee, A., García-Oscos, F., Roychowdhury, S., Galindo, L. C., Hall, S., Kilgard, M. P., & Atzori, M. (2013). Impairment of cortical GABAergic synaptic transmission in an environmental rat model of autism. *The International Journal of Neuropsychopharmacology, 16*, 1309–1318.

Banerjee, A., Ifrim, M. F., Valdez, A. N., Raj, N., & Bassell, G. J. (2018). Aberrant RNA translation in fragile X syndrome: From FMRP mechanisms to emerging therapeutic strategies. *Brain Research, 1693*, 24–36.

Bateup, H. S., Takasaki, K. T., Saulnier, J. L., Denefrio, C. L., & Sabatini, B. L. (2011). Loss of Tsc1 in vivo impairs hippocampal mGluR-LTD and increases excitatory synaptic function. *The Journal of Neuroscience, 31*, 8862–8869.

Bateup, H. S., Johnson, C. A., Denefrio, C. L., Saulnier, J. L., Kornacker, K., & Sabatini, B. L. (2013). Excitatory/inhibitory synaptic imbalance leads to hippocampal hyperexcitability in mouse models of tuberous sclerosis. *Neuron, 78*, 510–522.

Blatt, G. J., & Fatemi, S. H. (2011). Alterations in GABAergic biomarkers in the autism brain: Research findings and clinical implications. *The Anatomical Record (Hoboken), 294*, 1646–1652.

Blundell, J., Blaiss, C. A., Etherton, M. R., Espinosa, F., Tabuchi, K., Walz, C., Bolliger, M. F., Sudhof, T. C., & Powell, C. M. (2010). Neuroligin-1 deletion results in impaired spatial memory and increased repetitive behavior. *The Journal of Neuroscience, 30*, 2115–2129.

Boyar, F. Z., Whitney, M. M., Lossie, A. C., Gray, B. A., Keller, K. L., Stalker, H. J., Zori, R. T., Geffken, G., Mutch, J., Edge, P. J., Voeller, K. S., Williams, C. A., & Driscoll, D. J. (2001). A family with a grand-maternally derived interstitial duplication of proximal 15q. *Clinical Genetics, 60*, 421–430.

Brown, C., Gruber, T., Boucher, J., Rippon, G., & Brock, J. (2005). Gamma abnormalities during perception of illusory figures in autism. *Cortex, 41*, 364–376.

Buzsáki, G., & Draguhn, A. (2004). Neuronal oscillations in cortical networks. *Science, 304*, 1926–1929.

Buzsaki, G., & Wang, X. J. (2012). Mechanisms of gamma oscillations. *Annual Review of Neuroscience, 35*, 203–225.

Casanova, M. F. (2004). White matter volume increase and minicolumns in autism. *Annals of Neurology, 56*, 453; author reply 454.

Casanova, M. F., Buxhoeveden, D. P., Switala, A. E., & Roy, E. (2002). Minicolumnar pathology in autism. *Neurology, 58*, 428–432.

Cellot, G., Maggi, L., Di Castro, M. A., Catalano, M., Migliore, R., Migliore, M., Scattoni, M. L., Calamandrei, G., & Cherubini, E. (2016). Premature changes in neuronal excitability account for hippocampal network impairment and autistic-like behavior in neonatal BTBR T+tf/J mice. *Scientific Reports, 6*, 31696.

Chao, H. T., Chen, H., Samaco, R. C., Xue, M., Chahrour, M., Yoo, J., Neul, J. L., Gong, S., Lu, H. C., Heintz, N., Ekker, M., Rubenstein, J. L., Noebels, J. L., Rosenmund, C., & Zoghbi, H. Y. (2010). Dysfunction in GABA signalling mediates autism-like stereotypies and Rett syndrome phenotypes. *Nature, 468*, 263–269.

Chen, Z., Li, X., Zhou, J., Yuan, B., Yu, B., Tong, D., Cheng, C., Shao, Y., Xia, S., Zhang, R., Lyu, J., Yu, X., Dong, C., Zhou, W. H., & Qiu, Z. (2017). Accumulated quiescent neural stem cells in adult hippocampus of the mouse model for the MECP2 duplication syndrome. *Scientific Reports, 7*, 41701.

Christensen, J., Gronborg, T. K., Sorensen, M. J., Schendel, D., Parner, E. T., Pedersen, L. H., & Vestergaard, M. (2013). Prenatal valproate exposure and risk of autism spectrum disorders and childhood autism. *JAMA, 309*, 1696–1703.

Cochran, D. M., Sikoglu, E. M., Hodge, S. M., Edden, R. A., Foley, A., Kennedy, D. N., Moore, C. M., & Frazier, J. A. (2015). Relationship among glutamine, γ-aminobutyric acid, and social cognition in autism spectrum disorders. *Journal of Child and Adolescent Psychopharmacology, 25*, 314–322.

Cook, E. H., Jr., Courchesne, R. Y., Cox, N. J., Lord, C., Gonen, D., Guter, S. J., Lincoln, A., Nix, K., Haas, R., Leventhal, B. L., & Courchesne, E. (1998). Linkage-disequilibrium mapping of autistic disorder, with 15q11-13 markers. *American Journal of Human Genetics, 62*, 1077–1083.

Davatolhagh, M. F., & Fuccillo, M. V. (2021). Neurexin1 differentially regulates synaptic efficacy within striatal circuits. *Cell Reports, 34*, 108773.

Dickinson, A., Jones, M., & Milne, E. (2016). Measuring neural excitation and inhibition in autism: Different approaches, different findings and different interpretations. *Brain Research, 1648*, 277–289.

Du, L., Shan, L., Wang, B., Li, H., Xu, Z., Staal, W. G., & Jia, F. (2015). A pilot study on the combination of applied behavior analysis and bumetanide treatment for children with autism. *Journal of Child and Adolescent Psychopharmacology, 25*, 585–588.

Ecker, C., Spooren, W., & Murphy, D. G. (2013). Translational approaches to the biology of autism: False dawn or a new era? *Molecular Psychiatry, 18*, 435–442.

Eftekhari, S., Shahrokhi, A., Tsintsadze, V., Nardou, R., Brouchoud, C., Conesa, M., Burnashev, N., Ferrari, D. C., & Ben-Ari, Y. (2014). Response to comment on "Oxytocin-mediated GABA inhibition during delivery attenuates autism pathogenesis in rodent offspring". *Science, 346*, 176.

Enticott, P. G., Rinehart, N. J., Tonge, B. J., Bradshaw, J. L., & Fitzgerald, P. B. (2010). A preliminary transcranial magnetic stimulation study of cortical inhibition and excitability in high-functioning autism and Asperger disorder. *Developmental Medicine and Child Neurology, 52*, e179–e183.

Etherton, M. R., Blaiss, C. A., Powell, C. M., & Sudhof, T. C. (2009). Mouse neurexin-1alpha deletion causes correlated electrophysiological and behavioral changes consistent with cognitive impairments. *Proceedings of the National Academy of Sciences of the United States of America, 106*, 17998–18003.

Etherton, M., Foldy, C., Sharma, M., Tabuchi, K., Liu, X., Shamloo, M., Malenka, R. C., & Sudhof, T. C. (2011). Autism-linked neuroligin-3 R451C mutation differentially alters hippocampal and cortical synaptic function. *Proceedings of the National Academy of Sciences of the United States of America, 108*, 13764–13769.

Fatemi, S. H., Halt, A. R., Stary, J. M., Kanodia, R., Schulz, S. C., & Realmuto, G. R. (2002). Glutamic acid decarboxylase 65 and 67 kDa proteins are reduced in autistic parietal and cerebellar cortices. *Biological Psychiatry, 52*, 805–810.

Fatemi, S. H., Reutiman, T. J., Folsom, T. D., & Thuras, P. D. (2009). GABA(A) receptor downregulation in brains of subjects with autism. *Journal of Autism and Developmental Disorders, 39*, 223–230.

Fatemi, S. H., Reutiman, T. J., Folsom, T. D., Rustan, O. G., Rooney, R. J., & Thuras, P. D. (2014). Downregulation of GABAA receptor protein subunits α6, β2, δ, ε, γ2, θ, and ρ2 in superior frontal cortex of subjects with autism. *Journal of Autism and Developmental Disorders, 44*, 1833–1845.

Gaetz, W., Bloy, L., Wang, D. J., Port, R. G., Blaskey, L., Levy, S. E., & Roberts, T. P. (2014). GABA estimation in the brains of children on the autism spectrum: Measurement precision and regional cortical variation. *NeuroImage, 86*, 1–9.

Gibson, J. R., Bartley, A. F., Hays, S. A., & Huber, K. M. (2008). Imbalance of neocortical excitation and inhibition and altered UP states reflect network hyperexcitability in the mouse model of fragile X syndrome. *Journal of Neurophysiology, 100*, 2615–2626.

Gogolla, N., Leblanc, J. J., Quast, K. B., Südhof, T. C., Fagiolini, M., & Hensch, T. K. (2009). Common circuit defect of excitatory-inhibitory balance in mouse models of autism. *Journal of Neurodevelopmental Disorders, 1*, 172–181.

Gospe, S. M., Jr., Olin, K. L., & Keen, C. L. (1994). Reduced GABA synthesis in pyridoxine-dependent seizures. *Lancet, 343*, 1133–1134.

Ha, S., Lee, D., Cho, Y. S., Chung, C., Yoo, Y. E., Kim, J., Lee, J., Kim, W., Kim, H., Bae, Y. C., Tanaka-Yamamoto, K., & Kim, E. (2016). Cerebellar Shank2 regulates excitatory synapse density, motor coordination, and specific repetitive and anxiety-like behaviors. *The Journal of Neuroscience, 36*, 12129–12143.

Hadjikhani, N., Zürcher, N. R., Rogier, O., Ruest, T., Hippolyte, L., Ben-Ari, Y., & Lemonnier, E. (2015). Improving emotional face perception in autism with diuretic bumetanide: A proof-of-concept behavioral and functional brain imaging pilot study. *Autism, 19*, 149–157.

Haji, N., Riebe, I., Aguilar-Valles, A., Artinian, J., Laplante, I., & Lacaille, J. C. (2020). Tsc1 haploinsufficiency in Nkx2.1 cells upregulates hippocampal interneuron mTORC1 activity, impairs pyramidal cell synaptic inhibition, and alters contextual fear discrimination and spatial working memory in mice. *Molecular Autism, 11*, 29.

Han, S., Tai, C., Jones, C. J., Scheuer, T., & Catterall, W. A. (2014). Enhancement of inhibitory neurotransmission by GABAA receptors having α2,3-subunits ameliorates behavioral deficits in a mouse model of autism. *Neuron, 81*, 1282–1289.

Harada, M., Taki, M. M., Nose, A., Kubo, H., Mori, K., Nishitani, H., & Matsuda, T. (2011). Non-invasive evaluation of the GABAergic/glutamatergic system in autistic patients observed by MEGA-editing proton MR spectroscopy using a clinical 3 tesla instrument. *Journal of Autism and Developmental Disorders, 41*, 447–454.

Hashemi, E., Ariza, J., Rogers, H., Noctor, S. C., & Martínez-Cerdeño, V. (2018). The number of parvalbumin-expressing interneurons is decreased in the prefrontal cortex in autism. *Cerebral Cortex, 28*, 690.

Hoon, M., Soykan, T., Falkenburger, B., Hammer, M., Patrizi, A., Schmidt, K. F., Sassoe-Pognetto, M., Lowel, S., Moser, T., Taschenberger, H., Brose, N., & Varoqueaux, F. (2011). Neuroligin-4 is localized to glycinergic postsynapses and regulates inhibition in the retina. *Proceedings of the National Academy of Sciences of the United States of America, 108*, 3053–3058.

Hooper, A. W. M., Wong, H., Niibori, Y., Abdoli, R., Karumuthil-Melethil, S., Qiao, C., Danos, O., Bruder, J. T., & Hampson, D. R. (2021). Gene therapy using an ortholog of human fragile X mental retardation protein partially rescues behavioral abnormalities and EEG activity. *Molecular Therapy - Methods & Clinical Development, 22*, 196–209.

Hung, A. Y., Futai, K., Sala, C., Valtschanoff, J. G., Ryu, J., Woodworth, M. A., Kidd, F. L., Sung, C. C., Miyakawa, T., Bear, M. F., Weinberg, R. J., & Sheng, M. (2008). Smaller dendritic spines, weaker synaptic transmission, but enhanced spatial learning in mice lacking Shank1. *The Journal of Neuroscience, 28*, 1697–1708.

Isaacson, J. S., & Scanziani, M. (2011). How inhibition shapes cortical activity. *Neuron, 72*, 231–243.

Jamain, S., Quach, H., Betancur, C., Rastam, M., Colineaux, C., Gillberg, I. C., Soderstrom, H., Giros, B., Leboyer, M., Gillberg, C., Bourgeron, T., & Paris Autism Research International Sibpair, S. (2003). Mutations of the X-linked genes encoding neuroligins NLGN3 and NLGN4 are associated with autism. *Nature Genetics, 34*, 27–29.

Jamain, S., Radyushkin, K., Hammerschmidt, K., Granon, S., Boretius, S., Varoqueaux, F., Ramanantsoa, N., Gallego, J., Ronnenberg, A., Winter, D., Frahm, J., Fischer, J., Bourgeron, T., Ehrenreich, H., & Brose, N. (2008). Reduced social interaction and ultrasonic communication in a mouse model of monogenic heritable autism. *Proceedings of the National Academy of Sciences of the United States of America, 105*, 1710–1715.

Jambaqué, I., Chiron, C., Dumas, C., Mumford, J., & Dulac, O. (2000). Mental and behavioural outcome of infantile epilepsy treated by vigabatrin in tuberous sclerosis patients. *Epilepsy Research, 38*, 151–160.

Jaramillo, T. C., Liu, S., Pettersen, A., Birnbaum, S. G., & Powell, C. M. (2014). Autism-related neuroligin-3 mutation alters social behavior and spatial learning. *Autism Research, 7*, 264–272.

Jaramillo, T. C., Speed, H. E., Xuan, Z., Reimers, J. M., Escamilla, C. O., Weaver, T. P., Liu, S., Filonova, I., & Powell, C. M. (2017). Novel Shank3 mutant exhibits behaviors with face validity for autism and altered striatal and hippocampal function. *Autism Research, 10*, 42–65.

Jasien, J. M., Daimon, C. M., Wang, R., Shapiro, B. K., Martin, B., & Maudsley, S. (2014). The effects of aging on the BTBR mouse model of autism spectrum disorder. *Frontiers in Aging Neuroscience, 6*, 225.

Jiang, Y. H., & Ehlers, M. D. (2013). Modeling autism by SHANK gene mutations in mice. *Neuron, 78*, 8–27.

Kash, S. F., Johnson, R. S., Tecott, L. H., Noebels, J. L., Mayfield, R. D., Hanahan, D., & Baekkeskov, S. (1997). Epilepsy in mice deficient in the 65-kDa isoform of glutamic acid decarboxylase. *Proceedings of the National Academy of Sciences of the United States of America, 94*, 14060–14065.

Khaleghi, A., Zarafshan, H., Vand, S. R., & Mohammadi, M. R. (2020). Effects of non-invasive neurostimulation on autism spectrum disorder: A systematic review. *Clinical Psychopharmacology and Neuroscience, 18*, 527–552.

Kirov, G. (2015). CNVs in neuropsychiatric disorders. *Human Molecular Genetics, 24*, R45–R49.

Kubas, B., Kułak, W., Sobaniec, W., Tarasow, E., Lebkowska, U., & Walecki, J. (2012). Metabolite alterations in autistic children: A 1H MR spectroscopy study. *Advances in Medical Sciences, 57*, 152–156.

Kumar, M., Duda, J. T., Hwang, W. T., Kenworthy, C., Ittyerah, R., Pickup, S., Brodkin, E. S., Gee, J. C., Abel, T., & Poptani, H. (2014). High resolution magnetic resonance imaging for characterization of the neuroligin-3 knock-in mouse model associated with autism spectrum disorder. *PLoS One, 9*, e109872.

Lauber, E., Filice, F., & Schwaller, B. (2016). Prenatal valproate exposure differentially affects parvalbumin-expressing neurons and related circuits in the cortex and striatum of mice. *Frontiers in Molecular Neuroscience, 9*, 150.

Leblond, C. S., Nava, C., Polge, A., Gauthier, J., Huguet, G., Lumbroso, S., Giuliano, F., Stordeur, C., Depienne, C., Mouzat, K., Pinto, D., Howe, J., Lemière, N., Durand, C. M., Guibert, J., Ey, E., Toro, R., Peyre, H., Mathieu, A., Amsellem, F., Rastam, M., Gillberg, I. C., Rappold, G. A., Holt, R., Monaco, A. P., Maestrini, E., Galan, P., Heron, D., Jacquette, A., Afenjar, A., Rastetter, A., Brice, A., Devillard, F., Assouline, B., Laffargue, F., Lespinasse, J., Chiesa, J., Rivier, F., Bonneau, D., Regnault, B., Zelenika, D., Delepine, M., Lathrop, M., Sanlaville, D., Schluth-Bolard, C., Edery, P., Perrin, L., Tabet, A. C., Schmeisser, M. J., Boeckers, T. M., Coleman, M., Sato, D., Szatmari, P., Scherer, S. W., Rouleau, G. A., Betancur, C., Leboyer, M., Gillberg, C., Delorme, R., & Bourgeron, T. (2014). Meta-analysis of SHANK mutations in autism spectrum disorders: A gradient of severity in cognitive impairments. *PLoS Genetics, 10*, e1004580.

Lee, J., Chung, C., Ha, S., Lee, D., Kim, D. Y., Kim, H., & Kim, E. (2015). Shank3-mutant mice lacking exon 9 show altered excitation/inhibition balance, enhanced rearing, and spatial memory deficit. *Frontiers in Cellular Neuroscience, 9*, 94.

Lee, E., Lee, J., & Kim, E. (2017). Excitation/inhibition imbalance in animal models of autism spectrum disorders. *Biological Psychiatry, 81*, 838–847.

Lee, F. H. F., Lai, T. K. Y., Su, P., & Liu, F. (2019). Altered cortical Cytoarchitecture in the Fmr1 knockout mouse. *Molecular Brain, 12*, 56.

Lemonnier, E., Degrez, C., Phelep, M., Tyzio, R., Josse, F., Grandgeorge, M., Hadjikhani, N., & Ben-Ari, Y. (2012). A randomised controlled trial of bumetanide in the treatment of autism in children. *Translational Psychiatry, 2*, e202.

Lemonnier, E., Villeneuve, N., Sonie, S., Serret, S., Rosier, A., Roue, M., Brosset, P., Viellard, M., Bernoux, D., Rondeau, S., Thummler, S., Ravel, D., & Ben-Ari, Y. (2017). Effects of bumetanide on neurobehavioral function in children and adolescents with autism spectrum disorders. *Translational Psychiatry, 7*, e1056.

Liang, J., Xu, W., Hsu, Y. T., Yee, A. X., Chen, L., & Sudhof, T. C. (2015). Conditional neuroligin-2 knockout in adult medial prefrontal cortex links chronic changes in synaptic inhibition to cognitive impairments. *Molecular Psychiatry, 20*, 850–859.

Limbu, B., Deb, S., Roy, M., Lee, R., Roy, A., & Taiwo, O. (2022). Randomised controlled trials of mood stabilisers for people with autism spectrum disorder: Systematic review and meta-analysis. *BJPsych Open, 8*, e52.

Lozovaya, N., Nardou, R., Tyzio, R., Chiesa, M., Pons-Bennaceur, A., Eftekhari, S., Bui, T. T., Billon-Grand, M., Rasero, J., Bonifazi, P., Guimond, D., Gaiarsa, J. L., Ferrari, D. C., & Ben-Ari, Y. (2019). Early alterations in a mouse model of Rett syndrome: The GABA developmental shift is abolished at birth. *Scientific Reports, 9*, 9276.

Lucchina, L., & Depino, A. M. (2014). Altered peripheral and central inflammatory responses in a mouse model of autism. *Autism Research, 7*, 273–289.

Lukmanji, S., Manji, S. A., Kadhim, S., Sauro, K. M., Wirrell, E. C., Kwon, C. S., & Jetté, N. (2019). The co-occurrence of epilepsy and autism: A systematic review. *Epilepsy & Behavior, 98*, 238–248.

Masini, E., Loi, E., Vega-Benedetti, A. F., Carta, M., Doneddu, G., Fadda, R., & Zavattari, P. (2020). An overview of the main genetic, epigenetic and environmental factors involved in autism spectrum disorder focusing on synaptic activity. *International Journal of Molecular Sciences, 21*, 8290.

Menold, M. M., Shao, Y., Wolpert, C. M., Donnelly, S. L., Raiford, K. L., Martin, E. R., Ravan, S. A., Abramson, R. K., Wright, H. H., Delong, G. R., Cuccaro, M. L., Pericak-Vance, M. A., & Gilbert, J. R. (2001). Association analysis of chromosome 15 gabaa receptor subunit genes in autistic disorder. *Journal of Neurogenetics, 15*, 245–259.

Meyza, K. Z., & Blanchard, D. C. (2017). The BTBR mouse model of idiopathic autism – Current view on mechanisms. *Neuroscience and Biobehavioral Reviews, 76*, 99–110.

Meyza, K. Z., Defensor, E. B., Jensen, A. L., Corley, M. J., Pearson, B. L., Pobbe, R. L., Bolivar, V. J., Blanchard, D. C., & Blanchard, R. J. (2013). The BTBR T+ tf/J mouse model for autism spectrum disorders-in search of biomarkers. *Behavioural Brain Research, 251*, 25–34.

Monteiro, P., & Feng, G. (2017). SHANK proteins: Roles at the synapse and in autism spectrum disorder. *Nature Reviews. Neuroscience, 18*, 147–157.

Nadler, J. J., Zou, F., Huang, H., Moy, S. S., Lauder, J., Crawley, J. N., Threadgill, D. W., Wright, F. A., & Magnuson, T. R. (2006). Large-scale gene expression differences across brain regions and inbred strains correlate with a behavioral phenotype. *Genetics, 174*, 1229–1236.

Nakanishi, M., Nomura, J., Ji, X., Tamada, K., Arai, T., Takahashi, E., Bućan, M., & Takumi, T. (2017). Functional significance of rare neuroligin 1 variants found in autism. *PLoS Genetics, 13*, e1006940.

Neuray, C., Maroofian, R., Scala, M., Sultan, T., Pai, G. S., Mojarrad, M., Khashab, H. E., Deholl, L., Yue, W., Alsaif, H. S., Zanetti, M. N., Bello, O., Person, R., Eslahi, A., Khazaei, Z., Feizabadi, M. H., Efthymiou, S., El-Bassyouni, H. T., Soliman, D. R., Tekes, S., Ozer, L., Baltaci, V., Khan, S., Beetz, C., Amr, K. S., Salpietro, V., Jamshidi, Y., Alkuraya, F. S., Houlden, H., & Group, S. Y. S. (2020). Early-infantile onset epilepsy and developmental delay caused by bi-allelic GAD1 variants. *Brain, 143*, 2388–2397.

Nguyen, T. A., Lehr, A. W., & Roche, K. W. (2020). Neuroligins and neurodevelopmental disorders: X-linked genetics. *Frontiers in Synaptic Neuroscience, 12*, 33.

Oblak, A., Gibbs, T. T., & Blatt, G. J. (2009). Decreased GABAA receptors and benzodiazepine binding sites in the anterior cingulate cortex in autism. *Autism Research, 2*, 205–219.

Oblak, A. L., Gibbs, T. T., & Blatt, G. J. (2011). Reduced GABAA receptors and benzodiazepine binding sites in the posterior cingulate cortex and fusiform gyrus in autism. *Brain Research, 1380*, 218–228.

Orefice, L. L., Zimmerman, A. L., Chirila, A. M., Sleboda, S. J., Head, J. P., & Ginty, D. D. (2016). Peripheral mechanosensory neuron dysfunction underlies tactile and behavioral deficits in mouse models of ASDs. *Cell, 166*, 299–313.

Pak, C., Danko, T., Zhang, Y., Aoto, J., Anderson, G., Maxeiner, S., Yi, F., Wernig, M., & Sudhof, T. C. (2015). Human neuropsychiatric disease modeling using conditional deletion reveals synaptic transmission defects caused by heterozygous mutations in NRXN1. *Cell Stem Cell, 17*, 316–328.

Peça, J., Feliciano, C., Ting, J. T., Wang, W., Wells, M. F., Venkatraman, T. N., Lascola, C. D., Fu, Z., & Feng, G. (2011). Shank3 mutant mice display autistic-like behaviours and striatal dysfunction. *Nature, 472*, 437–442.

Peñagarikano, O., Abrahams, B. S., Herman, E. I., Winden, K. D., Gdalyahu, A., Dong, H., Sonnenblick, L. I., Gruver, R., Almajano, J., Bragin, A., Golshani, P., Trachtenberg, J. T., Peles, E., & Geschwind, D. H. (2011). Absence of CNTNAP2 leads to epilepsy, neuronal migration abnormalities, and core autism-related deficits. *Cell, 147*, 235–246.

Phelan, K., & Mcdermid, H. E. (2012). The 22q13.3 deletion syndrome (Phelan-McDermid syndrome). *Molecular Syndromology, 2*, 186–201.

Pouille, F., & Scanziani, M. (2001). Enforcement of temporal fidelity in pyramidal cells by somatic feed-forward inhibition. *Science, 293*, 1159–1163.

Princivalle, A. P., Duncan, J. S., Thom, M., & Bowery, N. G. (2003). GABA(B1a), GABA(B1b) AND GABA(B2) mRNA variants expression in hippocampus resected from patients with temporal lobe epilepsy. *Neuroscience, 122*, 975–984.

Puts, N. A. J., Wodka, E. L., Harris, A. D., Crocetti, D., Tommerdahl, M., Mostofsky, S. H., & Edden, R. A. E. (2017). Reduced GABA and altered somatosensory function in children with autism spectrum disorder. *Autism Research, 10*, 608–619.

Represa, A., & Ben-Ari, Y. (2005). Trophic actions of GABA on neuronal development. *Trends in Neurosciences, 28*, 278–283.

Richter, J. D., & Zhao, X. (2021). The molecular biology of FMRP: New insights into fragile X syndrome. *Nature Reviews. Neuroscience, 22*, 209–222.

Robertson, C. E., Ratai, E. M., & Kanwisher, N. (2016). Reduced GABAergic action in the autistic brain. *Current Biology, 26*, 80–85.

Rojas, D. C., & Wilson, L. B. (2014). γ-band abnormalities as markers of autism spectrum disorders. *Biomarkers in Medicine, 8*, 353–368.

Roux, S., Lohof, A., Ben-Ari, Y., Poulain, B., & Bossu, J. L. (2018). Maturation of GABAergic transmission in cerebellar Purkinje cells is sex dependent and altered in the valproate model of autism. *Frontiers in Cellular Neuroscience, 12*, 232.

Rubenstein, J. L., & Merzenich, M. M. (2003). Model of autism: Increased ratio of excitation/inhibition in key neural systems. *Genes, Brain, and Behavior, 2*, 255–267.

Rudy, B., Fishell, G., Lee, S., & Hjerling-Leffler, J. (2011). Three groups of interneurons account for nearly 100% of neocortical GABAergic neurons. *Developmental Neurobiology, 71*, 45–61.

Rylaarsdam, L., & Guemez-Gamboa, A. (2019). Genetic causes and modifiers of autism spectrum disorder. *Frontiers in Cellular Neuroscience, 13*, 385.

Sacai, H., Sakoori, K., Konno, K., Nagahama, K., Suzuki, H., Watanabe, T., Watanabe, M., Uesaka, N., & Kano, M. (2020). Autism spectrum disorder-like behavior caused by reduced excitatory synaptic transmission in pyramidal neurons of mouse prefrontal cortex. *Nature Communications, 11*, 5140.

Santoro, M. R., Bray, S. M., & Warren, S. T. (2012). Molecular mechanisms of fragile X syndrome: A twenty-year perspective. *Annual Review of Pathology, 7*, 219–245.

Sato, A., Kasai, S., Kobayashi, T., Takamatsu, Y., Hino, O., Ikeda, K., & Mizuguchi, M. (2012). Rapamycin reverses impaired social interaction in mouse models of tuberous sclerosis complex. *Nature Communications, 3*, 1292.

Schmeisser, M. J., Ey, E., Wegener, S., Bockmann, J., Stempel, A. V., Kuebler, A., Janssen, A. L., Udvardi, P. T., Shiban, E., Spilker, C., Balschun, D., Skryabin, B. V., Dieck, S., Smalla, K. H., Montag, D., Leblond, C. S., Faure, P., Torquet, N., Le Sourd, A. M., Toro, R., Grabrucker, A. M., Shoichet, S. A., Schmitz, D., Kreutz, M. R., Bourgeron, T., Gundelfinger, E. D., & Boeckers, T. M. (2012). Autistic-like behaviours and hyperactivity in mice lacking ProSAP1/Shank2. *Nature, 486*, 256–260.

Schneider, T., Roman, A., Basta-Kaim, A., Kubera, M., Budziszewska, B., Schneider, K., & Przewłocki, R. (2008). Gender-specific behavioral and immunological alterations in an animal model of autism induced by prenatal exposure to valproic acid. *Psychoneuroendocrinology, 33*, 728–740.

Schwede, M., Nagpal, S., Gandal, M. J., Parikshak, N. N., Mirnics, K., Geschwind, D. H., & Morrow, E. M. (2018). Strong correlation of downregulated genes related to synaptic transmission and mitochondria in post-mortem autism cerebral cortex. *Journal of Neurodevelopmental Disorders, 10*, 18.

Selimbeyoglu, A., Kim, C. K., Inoue, M., Lee, S. Y., Hong, A. S. O., Kauvar, I., Ramakrishnan, C., Fenno, L. E., Davidson, T. J., Wright, M., & Deisseroth, K. (2017). Modulation of prefrontal cortex excitation/inhibition balance rescues social behavior in CNTNAP2-deficient mice. *Science Translational Medicine, 9*, eaah6733.

Sohal, V. S., & Rubenstein, J. L. R. (2019). Excitation-inhibition balance as a framework for investigating mechanisms in neuropsychiatric disorders. *Molecular Psychiatry, 24*, 1248–1257.

Südhof, T. C. (2008). Neuroligins and neurexins link synaptic function to cognitive disease. *Nature, 455*, 903–911.

Talos, D. M., Sun, H., Kosaras, B., Joseph, A., Folkerth, R. D., Poduri, A., Madsen, J. R., Black, P. M., & Jensen, F. E. (2012). Altered inhibition in tuberous sclerosis and type IIb cortical dysplasia. *Annals of Neurology, 71*, 539–551.

Taylor, S. C., Ferri, S. L., Grewal, M., Smernoff, Z., Bucan, M., Weiner, J. A., Abel, T., & Brodkin, E. S. (2020). The role of synaptic cell adhesion molecules and associated scaffolding proteins in social affiliative behaviors. *Biological Psychiatry, 88*, 442–451.

Thomas, A. M., Schwartz, M. D., Saxe, M. D., & Kilduff, T. S. (2017). Cntnap2 knockout rats and mice exhibit epileptiform activity and abnormal sleep-wake physiology. *Sleep, 40*. https://doi.org/10.1093/sleep/zsw026

Toro, R., Konyukh, M., Delorme, R., Leblond, C., Chaste, P., Fauchereau, F., Coleman, M., Leboyer, M., Gillberg, C., & Bourgeron, T. (2010). Key role for gene dosage and synaptic homeostasis in autism spectrum disorders. *Trends in Genetics, 26*, 363–372.

Tran, S. S., Jun, H. I., Bahn, J. H., Azghadi, A., Ramaswami, G., Van Nostrand, E. L., Nguyen, T. B., Hsiao, Y. E., Lee, C., Pratt, G. A., Martínez-Cerdeño, V., Hagerman, R. J., Yeo, G. W., Geschwind, D. H., & Xiao, X. (2019). Widespread RNA editing dysregulation in brains from autistic individuals. *Nature Neuroscience, 22*, 25–36.

Treiman, D. M. (1989). gamma Vinyl GABA: Current role in the management of drug-resistant epilepsy. *Epilepsia, 30*(Suppl 3), S31–S35.

Tsai, P. T., Hull, C., Chu, Y., Greene-Colozzi, E., Sadowski, A. R., Leech, J. M., Steinberg, J., Crawley, J. N., Regehr, W. G., & Sahin, M. (2012). Autistic-like behaviour and cerebellar dysfunction in Purkinje cell Tsc1 mutant mice. *Nature, 488*, 647–651.

Turrigiano, G. (2012). Homeostatic synaptic plasticity: Local and global mechanisms for stabilizing neuronal function. *Cold Spring Harbor Perspectives in Biology, 4*, a005736.

Twining, R. C., Vantrease, J. E., Love, S., Padival, M., & Rosenkranz, J. A. (2017). An intra-amygdala circuit specifically regulates social fear learning. *Nature Neuroscience, 20*, 459–469.

Tyzio, R., Nardou, R., Ferrari, D. C., Tsintsadze, T., Shahrokhi, A., Eftekhari, S., Khalilov, I., Tsintsadze, V., Brouchoud, C., Chazal, G., Lemonnier, E., Lozovaya, N., Burnashev, N., & Ben-Ari, Y. (2014). Oxytocin-mediated GABA inhibition during delivery attenuates autism pathogenesis in rodent offspring. *Science, 343*, 675–679.

Varoqueaux, F., Aramuni, G., Rawson, R. L., Mohrmann, R., Missler, M., Gottmann, K., Zhang, W., Sudhof, T. C., & Brose, N. (2006). Neuroligins determine synapse maturation and function. *Neuron, 51*, 741–754.

Viscidi, E. W., Triche, E. W., Pescosolido, M. F., Mclean, R. L., Joseph, R. M., Spence, S. J., & Morrow, E. M. (2013). Clinical characteristics of children with autism spectrum disorder and co-occurring epilepsy. *PLoS One, 8*, e67797.

Wang, X., Mccoy, P. A., Rodriguiz, R. M., Pan, Y., Je, H. S., Roberts, A. C., Kim, C. J., Berrios, J., Colvin, J. S., Bousquet-Moore, D., Lorenzo, I., Wu, G., Weinberg, R. J., Ehlers, M. D., Philpot, B. D., Beaudet, A. L., Wetsel, W. C., & Jiang, Y. H. (2011). Synaptic dysfunction and abnormal behaviors in mice lacking major isoforms of Shank3. *Human Molecular Genetics, 20*, 3093–3108.

Wehr, M., & Zador, A. M. (2003). Balanced inhibition underlies tuning and sharpens spike timing in auditory cortex. *Nature, 426*, 442–446.

Whittington, M. A., Traub, R. D., Kopell, N., Ermentrout, B., & Buhl, E. H. (2000). Inhibition-based rhythms: Experimental and mathematical observations on network dynamics. *International Journal of Psychophysiology, 38*, 315–336.

Wöhr, M., Roullet, F. I., Hung, A. Y., Sheng, M., & Crawley, J. N. (2011). Communication impairments in mice lacking Shank1: Reduced levels of ultrasonic vocalizations and scent marking behavior. *PLoS One, 6*, e20631.

Won, H., Lee, H. R., Gee, H. Y., Mah, W., Kim, J. I., Lee, J., Ha, S., Chung, C., Jung, E. S., Cho, Y. S., Park, S. G., Lee, J. S., Lee, K., Kim, D., Bae, Y. C., Kaang, B. K., Lee, M. G., & Kim, E. (2012). Autistic-like social behaviour in Shank2-mutant mice improved by restoring NMDA receptor function. *Nature, 486*, 261–265.

Yip, J., Soghomonian, J. J., & Blatt, G. J. (2007). Decreased GAD67 mRNA levels in cerebellar Purkinje cells in autism: Pathophysiological implications. *Acta Neuropathologica, 113*, 559–568.

Yizhar, O., Fenno, L. E., Prigge, M., Schneider, F., Davidson, T. J., O'Shea, D. J., Sohal, V. S., Goshen, I., Finkelstein, J., Paz, J. T., Stehfest, K., Fudim, R., Ramakrishnan, C., Huguenard, J. R., Hegemann, P., & Deisseroth, K. (2011). Neocortical excitation/inhibition balance in information processing and social dysfunction. *Nature, 477*, 171–178.

Zhang, L., Huang, C. C., Dai, Y., Luo, Q., Ji, Y., Wang, K., Deng, S., Yu, J., Xu, M., Du, X., Tang, Y., Shen, C., Feng, J., Sahakian, B. J., Lin, C. P., & Li, F. (2020). Symptom improvement in children with autism spectrum disorder following bumetanide administration is associated with decreased GABA/glutamate ratios. *Translational Psychiatry, 10*, 9.

Chapter 12
Shared Developmental Neuropathological Traits Between Autism and Environmental Lead Exposures: Insights into Convergent Sulfur-Dependent Neurobiological Mechanisms

Lorenz S. Neuwirth, Michelle A. Vasquez, Mohammad Mian, Angelina M. Gagliardi, Bright U. Emenike, and Morri E. Markowitz

In the last two decades, there has been an increase in reports aimed at elucidating the relationship between the neurodevelopment of autism spectrum disorder (ASD) and environmental lead exposures (Amadi et al., 2022; Nakhaee et al., 2022; Goel & Aschner, 2021; Ijomone et al., 2020; Hauptman et al., 2019; Saghazadeh & Rezaei, 2017; Ye et al., 2017; Mostafa et al., 2016; Gorini et al., 2014; Yassa, 2014; Adams et al., 2007, 2013; Grabrucker, 2013; Seneff et al., 2013; Ko et al., 2012; George et al., 2010; Landrigan, 2010; Tian et al., 2009; Kern et al., 2007; Lidsky & Schneider, 2005; Millichap, 1996). These reports show evidence of a consistently emerging dataset spanning multiple countries that collectively suggest that children from both low- and middle-income countries are at elevated risks of lead-exposures

L. S. Neuwirth (✉) · A. M. Gagliardi
Department of Psychology, SUNY Old Westbury, Old Westbury, NY, USA

SUNY Neuroscience Research Institute, Old Westbury, NY, USA
e-mail: neuwirthl@oldwestbury.edu

M. A. Vasquez
SUNY Neuroscience Research Institute, Old Westbury, NY, USA

Department of Chemistry & Physics, SUNY Old Westbury, Old Westbury, NY, USA

M. Mian
SUNY Neuroscience Research Institute, Old Westbury, NY, USA

Department of Biology, SUNY Old Westbury, Old Westbury, NY, USA

B. U. Emenike
Department of Chemistry & Physics, SUNY Old Westbury, Old Westbury, NY, USA

M. E. Markowitz
Lead Poisoning Treatment and Prevention Program, Montefiore Medical Center, Bronx, NY, USA

A. El Idrissi, D. McCloskey (eds.), *Neurobiology of Autism Spectrum Disorders*,
https://doi.org/10.1007/978-3-031-42383-3_12

(Attina & Trasande, 2013). Notably, the removal of lead from gasoline and paint prior to 1980 failed to eliminate childhood lead poisoning (Finkelman, 1996), and it has been more concerning that other sources of lead exposure (i.e., e-waste recycling) continue to harm children. Moreover, there appears to be an elevated risk for children to develop or be identified as having ASD in developing low- and middle-income countries (Marlow et al., 2019; Stewart & Lee, 2017; Samms-Vaughan, 2014). However, screening for ASD may be more varied in developing countries given the lack of standardized diagnostic tools in these countries (Stewart & Lee, 2017) or perhaps more universal when Western diagnostic tools are employed across countries (Bauer et al., 2022). The latter raises some important ethical and clinical concerns when conducting research in low- and middle-income countries for reliability and generalizability purposes on both diagnostic and outcome measures of ASD (Daley et al., 2013).

As investigations continue, environmental influences on ASD should be guided by toxicological principles as best practices for clinical patient care and follow-up, as well as epidemiological outcomes (Lawler et al., 2005). Notably, the epidemiological profiles for children with ASD in developing countries has shown an increased prevalence rate of 1.5% along with increased comorbidity for intellectual disability that arise from epigenetic (i.e., gene x environment) interactions (Lyall et al., 2017; Kim & Leventhal, 2015). Further, not only has there been reports linking environmental lead exposures with ASD as a risk factor for children in developing countries (Saghazadeh & Rezaei, 2017; Hassanien & Shahawy, 2011), but in developed countries as well (Kim et al., 2016; Gorini et al., 2014). Furthermore, reports of meta-analyses and systematic reviews of the literature that utilized random-effects modeling techniques (in evaluating differences in blood, hair, and urine samples in children with ASD compared to children with lead exposure) showed significant increases in lead-levels in the blood and hair, but not urine (Nakhaee et al., 2022). These findings are consistent with reports from Kuwait on boys with ASD having elevated levels of lead, mercury, and uranium in their hair samples, when compared to age- and gender-matched control children (Fido & Al-Saad, 2005) as well as other studies solely screening for lead through hair samples (Betrand et al., 2003; Rubin et al., 2002; Esteban et al., 1999; Grandjean, 1984; Reeves et al., 1975).

Interestingly, a recent pre-clinical report used Swiss Webster mice to evaluate the relationship between perinatal lead exposures (i.e., 30 ppm and 230 ppm through the drinking water) with the offspring developing ASD-like behavioral phenotypes and found that the higher lead exposure reduced pup weight and litter size, but found no significant differences in the marble burying test, grooming assessment, nor the three-chambered social interaction test (Islam et al., 2021). However, this study's results should be evaluated with caution because the sample size used in evaluating the behavioral testing was too small (i.e., $n = 5$), and there was no control to measure litter effects, maternal care, and reliability. Additionally, the grooming test used a beaker instead of a cage that artificially constrains the rodent and biases their posture to groom, and their claims of lead exposure causing a reduced litter size relative to their control counterparts was not statistically validated. For example, there are reports that utilize maternal weight gain measures in Swiss Webster mice to predict the average litter size, which was between 6.4–6.9 pups (Murray et al., 2010) to 7.8–8.2 pups (Finlay et al., 2015; i.e., in the latter the differences were reported to be due to genetic drift). This is a critical point as the litter sizes reported by Islam et al. (2021) were not statistically different from the reports by Murray et al. (2010) and Finlay et al. (2015). Furthermore, Islam et al. (2021) did not assess pup or maternal weight in their studies, which would have been helpful in fully understanding the extent to which lead exposure might influence offspring weight. In this regard, the lead exposure of 230 ppm was found to have the largest impact on pup size, but the 30 ppm exposure group was too little to have much of an effect. Lastly, the amount of time that the mice were given lead in the drinking water *ad libitium* was insufficient to cause an alteration in the offspring behaviors consistent with other reports in the literature (Cruz et al., 2022; Neuwirth et al., 2019a, b, 2022; Vasquez et al., 2022). Therefore, more pre-clinical work in this area is warranted to assess lead-induced deficits in social interaction behaviors.

The distribution of children with lead poisoning can significantly vary by country with more modernized high-income countries having consistent public health approaches that result in effective clinical management (e.g., Canada [Clark et al., 2010] and Hong Kong [Ko et al., 2012]) to a more variable approach in middle-income countries (e.g., China [Ko et al., 2012], Egypt [El Baz Mohamed et al., 2015], Iraq [Alawad et al., 2019], and Malaysia [Wahil et al., 2022]), with even worse situations remaining amongst the lower-income and seasonal income-dependent countries, such as those in the Caribbean and Latin America: Barbados (Schroeder et al., 2013; Handler et al., 1986), Columbia and Latin America (Alvarez-Ortega et al., 2019, 2017; Olympio et al., 2017; Romieu et al., 1997), Cuba (Trepka et al., 2005), the Dominican Republic (Kaul et al., 1999), Jamaica (Rahbar et al., 2015a, b; Brownie et al., 2009; Lalor et al., 2007; Lowe et al., 2002; Matte et al., 1989), and Trinidad (Mohammed et al., 1996; Chang-Yen et al., 1995). It seems counterintuitive that countries in the Caribbean (that rely heavily on tourism for their gross domestic product by maintaining a beautiful and uncontaminated environment) are identified as having different patterns of contamination from industrialization, e-waste recycling plants, and other forms of metallurgy causing toxicity to their native peoples (e.g., Kaul et al., 1999; Matte et al., 1989).

In some cases, select islands in the Caribbean use dogs and goats as biological detectors of environmental lead exposures to evaluate risks for children (Brownie et al., 2009). Using animals in the Caribbean as biological detectors of lead poisoning is strikingly similar to Bruce Lanphear's analogy of the Center for Disease Control and Prevention (CDC) in the United States uses children as canaries in the coal mine to assess neighborhood levels of lead toxicity in the US (Lanphear, 2000). Moreover, Jamaica and Barbados have experienced increasing cases of ASD and in their local culture they seek out research programs to create more positive family support structures to improve their quality-of-life outcomes in the Caribbean (Christian et al., 2018; Sealy, 2017; Samms-Vaughan et al., 2017; Samms-Vaughan, 2014; Holder, 2013; Mann, 2013; Rahbar et al., 2012). These e-waste problems are also not unique to the Caribbean, but also occur in more modernized high-income countries (Chen et al., 2019).

12.1 Brief Literature Review of the Last Decade

Searching for "autism" AND "lead poison*" as keywords in *Psych Articles, Psych Info, Science Direct*, and *Pub Med* databases over the last decade (i.e., 2012–2022) yielded ($N = 1504$) the returned articles and the results are summarized in Fig. 12.1. *Science Direct* had the most relevant information and largest number of articles returned (n = 1080; 71.80%) followed by *Pub Med* (n = 385; 25.53%), then *Psych Articles* (n = 34; 2.26%), and finally *Psych Info* (n = 5; 0.33%), respectively (see Fig. 12.1).

The scant number of returned articles in *Psych Info* and *Psych Articles* might be due to the *Diagnostic and Statistical Manual of Mental Disorders 5th Ed.* (*DSM-5*) failing to recognize lead poisoning as a neurodevelopmental disorder (for review, see American Psychiatric Association [APA], 2013; Neuwirth et al., 2020). Thus, greater numbers of publications on lead and ASD can be found in the *Science Direct* and *Pub Med* databases, respectively. Further, the frequency of publications from 2012 to 2022 are shown in Fig. 12.2a, b for comparisons of trends (i.e., stippled lines with R^2 equations) between *Psych Articles* and *Psych Info* as well as *Science Direct* and *Pub Med*, showing a steady growth (R^2 = 0.53 as a moderately strong relationship) solely in the *Science Direct* database over the last decade (i.e., note that 2022 is incomplete as it is the present year and data are still being populated).

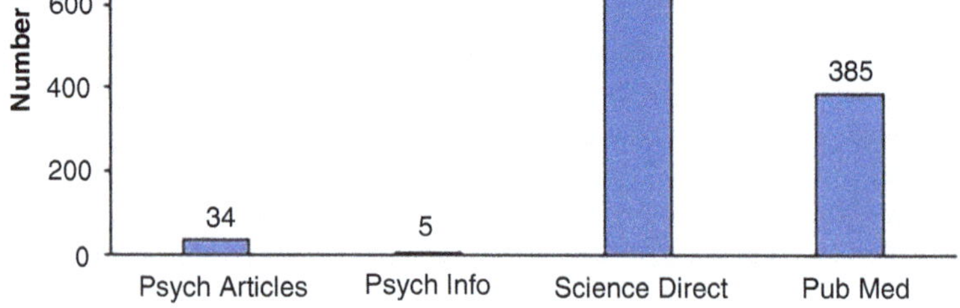

Fig. 12.1 Total number of articles returned when searching the key words "autism" AND "lead poison*" from 2012 to 2022. *Science Direct* was noted to be the best database to use for peer-reviewed articles on this topic

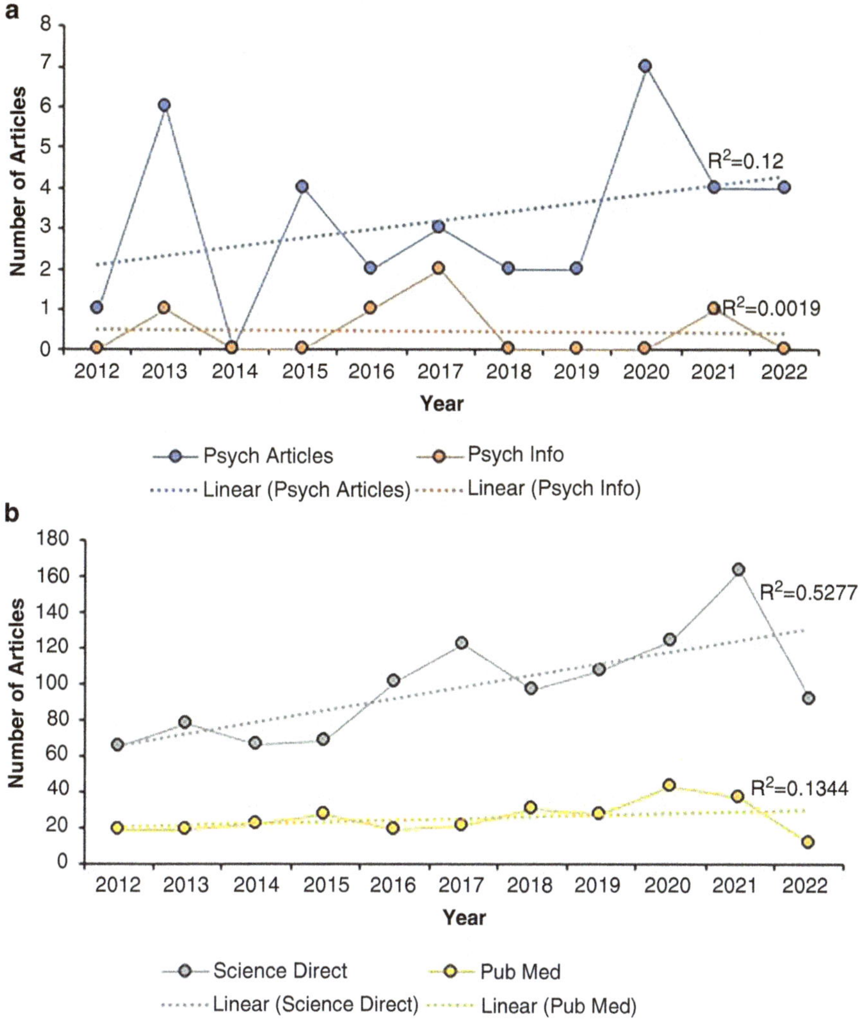

Fig. 12.2 Total number of articles returned when searching the key words "autism" AND "lead poison*" from 2012 to 2022 in *Psych Articles* and *Psych Info* (**a**) and *Science Direct* and *Pub Med* (**b**). The stippled lines show the trends with the R^2 for each database, in which *Science Direct* was noted to be the database with the greatest growth over the last decade with peer-reviewed articles on this topic

Taken together, there is an emerging literature evaluating the associations between autism and lead poisoning that has been steadily growing over the last decade. Since these are associations from the search terms one must be cautious of over interpreting these relationships until clear neurobiological mechanisms for elucidating these associations beyond correlations are identified and fully described.

12.2 Glutathione as a Shared Neurobiological Target for Understanding Late-to-Early Neurodevelopmental Disorders and Its Implications on the GABA-Shift

Niklowitz (1975) identified Alzheimer's-type neurofibrillary tangles in animals following acute experimental lead poisoning. More recently, Adeyiga (2021) reported that globally, early pediatric lead poisoning/exposures could contribute to and develop into neurodegenerative diseases (e.g., Alzheimer's and Parkinson's diseases). Consistent with those reports, Aljelehawy (2022) noted that lead and other trace metals can disrupt neuronal function as a contributing factor to the onset of later neurodevelopmental disorders such as Alzheimer's and Parkinson's diseases. As such, much attention has been directed towards understanding how trace metals, including lead, can cause neurodegenerative diseases (i.e., of both late and early types; Bagheri & Saboury, 2021). The relationship between early lead exposures and the development of neurodegenerative disorders brings to light the ethical considerations regarding children's quality of life across the lifespan (Whitehouse, 2022) and the 'avoidable' economic burdens (Grosse et al., 2002).

Research has identified the antioxidant properties of glutathione (GSH; $C_{10}H_{17}N_3O_6S$) as a well-established neuronal mechanism for cellular detoxification of reactive oxygen species (ROS; Fig. 12.3) for free radicals, lipid peroxidases, and trace metals (for review see, Agledal et al., 2010; Dringen & Hirrlinger, 2003) with particular vulnerabilities to the hippocampus and prefrontal cortex during aging when GSH are at low-levels (Venkateshappa et al., 2012). Further, studies show that neuronal GSH depletion is frequently observed in clinical patients with neurodegenerative disorders (e.g., Alzheimer's and Parkinson's diseases; Shukla et al., 2018; Chiang et al., 2017; Rae & Williams, 2017; Mandal et al., 2015; Aoyama & Nakaki, 2013; Venkateshappa et al., 2012; Jenner, 1994; Sian et al., 1994) as well as senescence (Venkateshappa et al., 2012), with larger GSH reductions occurring in the gray matter over white matter (~ difference of 0.35 mM; Slivka et al., 1987).

On the early phase of the developmental spectrum, reports on neurotypical pediatric patients (i.e., aged 6–14 years old) showed no changes in the brain regarding GABA and GSH levels (Saleh et al., 2020), which suggests that GSH levels decline much later in development. However, it is possible that when the brain undergoes synaptic pruning in establishing early brain neural networks that perhaps GSH levels may be elevated as GABA is shifting from excitation-to-inhibition (for review, see Gifford, 2021; Gifford et al., 2020; Ben-Ari, 2002) to mitigate any improper brain excitability that may trigger more oxidative stress reactions in the brain. The GABA-shift has been identified as a critical period in neurodevelopment that causes neurodevelopmental disorders when disrupted (e.g., seizure, ASD, intellectual impairment; Ben-Ari, 2015; Ben-Ari et al., 2012). Further, in preclinical models, lead exposure has been shown to disrupt the GABA-shift with long-term effects on behavior that resemble autistic-like phenotypes (i.e., learning and memory impairments, hyperactivity, decreased sensory thresholds, anxiety, and reduced GABA; Neuwirth, 2018; Neuwirth et al., 2017; Neuwirth, 2014; Vasquez et al., 2022). The

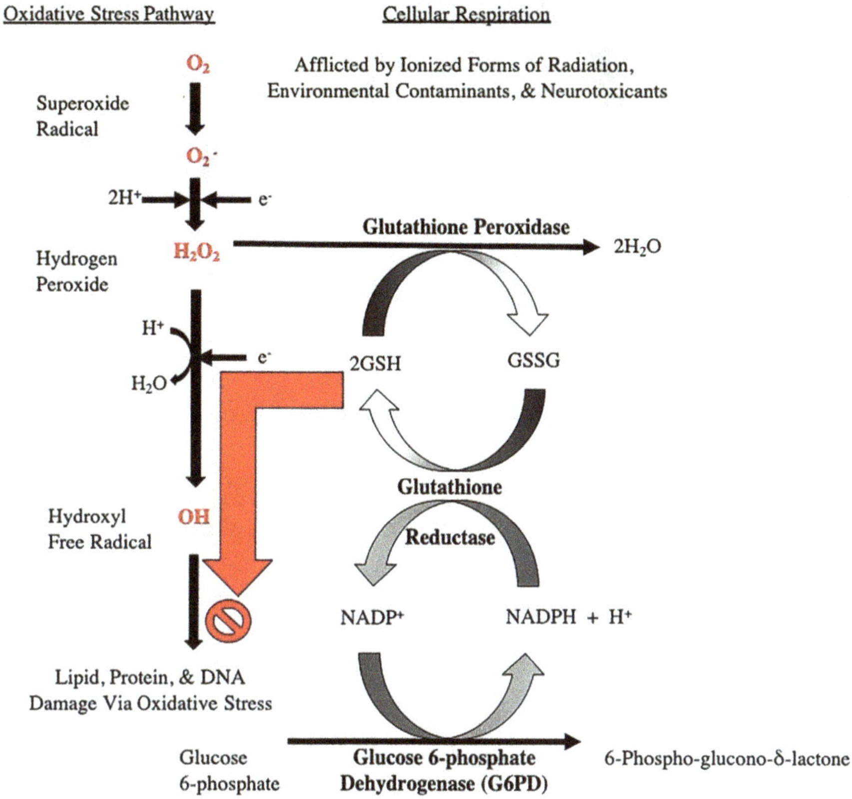

Fig. 12.3 Neuroprotective pathways mitigating free radical damage due to *reactive oxygen species* (ROS). Cellular respiration is impaired ROS induced by ionized forms of radiation, environmental contaminants, and neurotoxicants (i.e., heavy metals/metalloids) that trigger the oxidative stress pathways via superoxide radical formations, hydrogen peroxide, lipid peroxides, and hydroxyl free radicals. These pathways can be mitigated via glutathione peroxidase through *glutathione disulfide* (GSSH) and *glutathione* (2GSH) as well as downstream *glutathione reductase* through *nicotinamide adenine dinucleotide phosphate* ($NADP^+$) and *glucose 6-phosphate dehydrogenase* (G6PD) to protect against lipid, protein, and DNA damage caused by oxidative stress in neurons

effects of lead exposure altering the GABAergic system were reported by Silbergeld (1979) resulting in lowered threshold for and sensitivity to evoke seizures. A recent report by Pangrazzi et al. (2020) thoroughly evaluated the literature and showed that patients with ASD have elevated levels of oxidative stress, which may also be related to neuroimmune dysfunction contributing to their developmental neuropathologies (i.e., as a form of early neurodegeneration in a child's brain). Taken together, a common oxidative stress pathological trait resulting in ASD may arise from several neurotoxicants such as lead, phthalates, polychlorinated biphenyl (PCBs), and estrogen disruptors (Fig. 12.4).

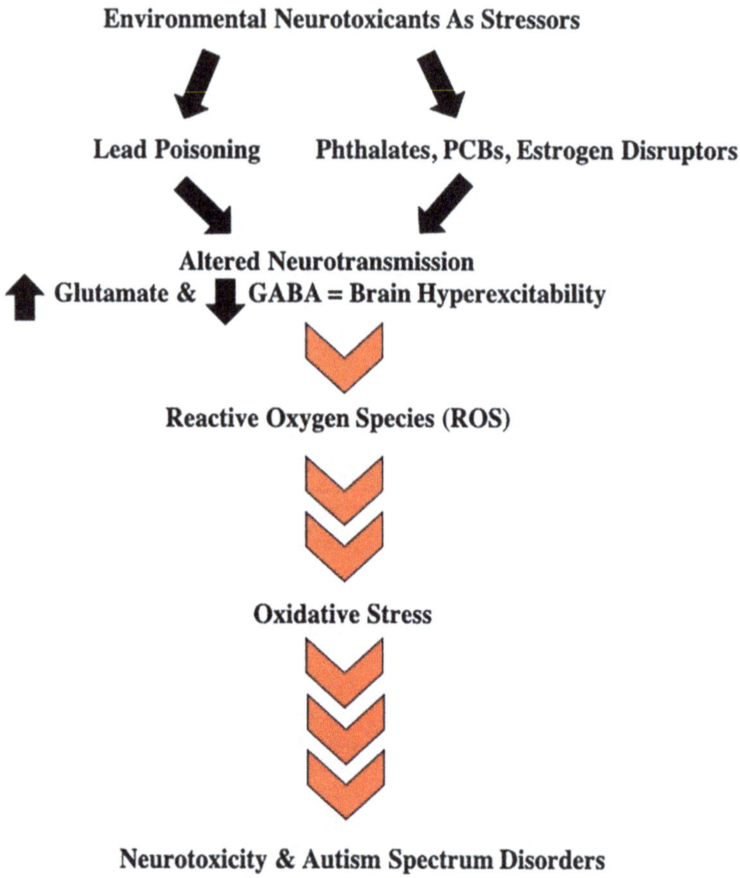

Fig. 12.4 Common cascade of environmental stressors such as lead, phthalates, polychlorinated biphenyl (PCBs), and estrogen disruptors can serve as inducing events for the neurodevelopment of autism spectrum disorders (ASD) in children. The exposure of environmental neurotoxicants can altered the brain's excitation-to-inhibition balancing resulting in hyperexcitability that causes an increase in *reactive oxygen species* (ROS) followed by oxidative stress, and subsequently identifies one aspect of the neurotoxicity profiles shared in ASD between these neurotoxicant exposures. The severity and/or intensity of the neurotoxic effects at each stage are depicted as an increasing number of red chevron arrows

12.3 Lead Poisoning and Autism Alter Brain Glutathione Levels

The tripeptide GSH—i.e., *L- γ-glytamyl-L-cysteinyl-glycine*, which is synthesized from the precursors amino acids glutamate, cysteine, and glycine in the cytosol with concentrations up to 10 μM serves as an important anti-oxidant agent, scavenging ROS to protect neurons from glutamate excitotoxicity (Schubert & Piasecki, 2001). There appears to be a consistent observation when organisms undergo stress in response to environmental challenges in which a negative linear relationship of increased oxidative stress and reductions in GSH contribute to early neurodevelopmental disorders caused by neurotoxicant exposures (Nishimura et al., 2021; Fig. 12.4). Likewise, adult workers that were exposed to lead (i.e., as an environmental stressor and neurotoxicant) had decreased blood GSH levels when compared to age-matched controls indicating an elevated physiological stress response (Gürer-Orhan et al., 2004). Other reports have shown that the increased production of superoxide dismutase in lead-exposed workers offered a neuroprotective response against oxygen free-radical neurotoxicity and iron-dependent lipid peroxidation (Hermes-Lima et al., 1990). Paradoxically, this neuroprotective response was coupled with *5-aminolaevulinic acid* (ALA)-generated oxygen radicals, that in turn, may contribute to the neuropsychiatric symptoms of lead poisoning (Hermes-Lima et al., 1990). However, it is also possible that ALA can generate free radicals that can induce superoxide dismutase. Further, ALA dehydratase activity is reduced by lead in children and adults which would increase its substrate, ALA (Fowler, 2016). Conversely, in rodent dams exposed to 0.5% lead acetate in their drinking water during pregnancy and lactation had pups with reduced body weight, increased liver, kidney, spleen, and brain weights with increased GSH in erythrocytes, liver, and kidney post-4-weeks of exposure without observable changes in *glutathione reductase* (GR) and *glutathione peroxidase* (GPx; Hsu, 1981). This suggests that different compensatory mechanisms can produce excess GSH in the liver and kidney that may scavenge lead-induced free radicals (Hsu, 1981). However, more recent studies have shown conversely that lead-induced oxidative stress causes an increase in ROS, a reduction in antioxidant defense via GSH reduction, inhibition of sulfhydryl (SH)-dependent enzymes — i.e., *sulfhydryl oxidases* (SOXs) that can oxidize free SH groups in proteins and thiol-containing small molecules using molecular oxygen as an electron acceptor —interference with essential metal metabolism for prooxidant functions, and altered membrane integrity (Gürer & Ercal, 2000). Moreover, when children with lead poisoning undergo chelation therapy to remove lead from their bodies, the duration of lead exposure showed a positive linear relationship ($r = 0.249$) with liver stiffness/fibrosis and elevated GSH (+9.2 ± 3.7), when compared to an age-matched control group (Teerasarntipan et al., 2020). Consistent with these reports, storage battery workers that were compared to age-matched healthy controls exhibited elevated blood lead levels (+36.4 μg/dL ± 2.07) and a 28% lower erythrocytes GSH (−3.3 μmole/gm Hb ± 0.40) and 56% lower plasma GSH (−0.49 μmole/L ± 0.15); that all three measures were improved by combining 1000 mg/day vitamin C (ascorbic acid) and 200 mg/day vitamin E (tocopherol), but did not reverse damage done to the brain or other bodily tissues (Al-Ubaidy et al., 2006). It has also been reported that in-vitro cell culture models of human neuroblastoma cells undergo a lead-induced (i.e., 0.01 μM- 10 μM exposures for 48 hrs) 50% reduction in proliferation at 5 μM and decreases in GSH across all lead concentrations that resulted in cell death (Chetty et al., 2005). Interestingly, exposure to 600 ppm lead acetate in drinking water during pregnancy and lactation caused oxidative stress in BALB mice (Wang et al., 2006). Moreover, the dam's offspring at 21 days exhibited decreased superoxide dismutase, GPx, and GR in the corpora quadrigemina (i.e., critical brain region responsible for connecting the forebrain and hindbrain reflexive centers to coordinate vision and hearing and when perturbed may contribute to sensory processing disorders), the striatum (i.e., regulates sensory motor coordination), and hypothalamus (i.e., regulates neuroendocrine functions), respectively (Wang et al., 2006).

In contrast, studies on children with ASD have shown that early oxidative stress that induces ROS as an etiological factor caused: (1) reduced capacity for methylation, decreases in sulfate levels and total GSH levels; (2) that was accompanied by increased oxidized GSSG, and an increased ratio of GSSG to reduced GSH in ASD; and (3) there is a reduction in GPx, superoxide dismutase, and catalase lowering the antioxidant defense system in ASD (Ghanizadeh et al., 2012). Further, due to the GSH redox imbalance in ASD, the GSH-mediated regulation of the *n-methyl-D-aspartate* (NMDA) receptors (i.e., glutamatergic-dependent learning and memory receptors), require glutamate as a GSH substrate which when disrupted or challenged, may contribute in part to glutamatergic excitotoxicity as a neurotoxicity susceptibility pathway (Bjørklund et al., 2020). It has been reported that GSH redox imbalance (e.g., decreased GSH) and increased GSSG with altered expression of GSH-related enzymes in the blood plasma and brain (i.e., neurons and astrocytes; Bjørklund et al., 2020). Further, Bjørklund et al. (2021) reported that the increased GSSG reacts with SH groups inducing proteotoxic stress in both the blood and brain as well as antagonistic interactions between glutamate and GSH causing neuronal dysfunction in glutamatergic signaling. Interestingly, in the brain of patients with ASD, the cerebellum and temporal cortices showed decreased GSH levels (i.e., 34.2% and 44.6%) and increased GSSG levels (38.2% and 45.5%; a nearly matched reversal of the other) when compared to an age-matched control group with a decreased redox ratio of 52.8% in the cerebellum and 60.8% in the temporal cortex, respectively (Chauhan et al., 2012). These findings in the cerebellum and temporal cortex are thought to be indicators of oxidative stress in the brains of people with autism, suggesting that the decreased GSH and increased GSSG produce altered redox and antioxidant capacities (Rose et al., 2012). Consistently, altered redox can contribute to autism developmental neuropathology as a chronic inflammatory response, elevated mitochondrial superoxide mutase production, and protein and DNA damage via free ROS (Rose et al., 2012). Thus, there is a shared relationship between the altered redox states experienced in the $NMDA_{R}$ learning and memory profiles in both autism and lead poisoning that warrant further investigation.

12.4 Understanding the Physical Chemistry Between Lead and Sulfur Compounds

To better understand the relationship between lead and sulfur compounds, the physical interactions between these two chemicals is required. There are only a few well described reports in the literature on lead interactions with sulfur compounds that attempt to address and resolve the ways in which lead may be interacting with sulfur to form noncovalent complexes with dynamic properties. As such, lead seems to form dynamic complexes with GSH through a lead-sulfide (Pb-S) bond formation (Bottari et al., 2020; Mah & Jalilehvand, 2012). Interestingly, through the use of nuclear magnetic resonance (NMR) to investigate these Pb-S bonds, these NMR studies are less definitive (i.e., despite observing a complex formation that was evident by the upfield shift in the proton NMR spectroscopy; Bottari et al., 2020; Mah & Jalilehvand, 2012), the electrospray ionization method detected two stable complexes through the Pb-S bond formation consistent with 1:1 Pb^{2+}/GSH complexes (Mah & Jalilehvand, 2012). As a polydentate (i.e., attaching to the central atom through such complexes via two or more bonds that is characteristic of ligands and chelating groups), GSH can also form complexes with other cations (Baloun et al., 2010). However, the Pb-S bond formation with Pb^{2+} can produce one of two complexes: (1) Pb–S and (2) short Pb–O bond (i.e., whereby they can further form into Pb(SO), $Pb(SO_2)$, $Pb(SO_3$, $Pb(SO_4)$, etc.) which can occur at distances at 2.64 (± 0.04 Å) and 2.28 (± 0.04 Å), as confirmed by X-ray absorption fine structural analyses (Mah & Jalilehvand, 2012). Notably, studies have investigated the relationships that disulfide bonds have in stabilizing protein structures and when disrupted are associated with dysregulated protein function and/or activity (Karimi et al., 2016). Additionally, several oxidants positive charges react with sulfur via electrostatic stabilization by remote groups or neighboring conformations aligned by their orbits to promote efficient free radical scavenging and highly stable proteins (Karimi et al., 2016). Given the shared chemical relationships that produced altered oxidative stress states between lead poisoning and autism, more work is required to further investigate how these individual lead complexes create neurodevelopmental disorders in cases both independent of and comorbid with autism. It would be predictive that more chronic and/or high blood lead level exposures would produce increased risk for comorbidity for autism. However, it may also be possible that given the neurodevelopmental time-period of exposure, that perhaps an acute and/or low lead level exposure could cause autism during a very sensitive and vulnerable stage in early brain development. Thus, both the concentration of exposure and the time-period during development in which a child's brain is exposed must be known for proper etiological assessment.

As a function of the concentration of these chemicals (i.e., which would be extrapolated from an experimental solute to biological tissues and/or blood), two or three thiolate ligands can form stable complexes with Pb^{2+} through PbS_2N_2 and PbS_3 bonding mechanisms (For Review, see Mah & Jalilehvand, 2012). Furthermore, balancing the prooxidants and antioxidants for cellular protection is a key factor in developing new treatments for developmental neuropathologies in which GSH has become a major target of investigation (Curpan et al., 2021) as evidenced by the case of schizophrenia (For review, see Kumar et al., 2020). Similarly, the regulation of disulfide bond during the production of normal and aberrant proteins (e.g., mutant, misfolded, etc.) can lead to excess ROS that GSH can regulate and further balance redox reactions to mitigate against cellular oxidative stress (Chakravarthi et al., 2006). Thus, there appears to be much dynamic and versatile complexes that can occur between lead and GSH, which suggests that sulfur-containing compounds may serve as atypical chelating agents as a potential complimentary alternative medicine (CAM) for treating lead poisoning and ASD. This is consistent with reports of SH compounds functioning as neuroprotective and cytoprotective agents through the dynamic range of modifications that can occur with SH groups, as well as the oxidation of or binding of SH groups that mitigate stress in animal models (Martin et al., 2000, pp. 419–456); which is the basis for how succimer functions as the gold standard for lead chelation. One such sulfur-containing compound that has shown intriguing neuroprotection in both neurodevelopmental disorders is taurine (2-aminoethanesulfonic acid; $C_2H_7NO_3S$), which merits further study and consideration for ameliorating both developmental neuropathological syndromes.

12.5 Sulfur-Dependent Approaches for Ameliorating Lead Poisoning and Autism: Taurine Neuroprotection Through Glutathione

Taurine has been reported to provide many neuroprotective properties, of which, regulating GSH and upregulating cellular and neuronal anti-oxidant responses has been effective in treating various disorders (For Review, see Baliou et al., 2021). In rats, 5% taurine in normal saline administered through intragastric intubations for 60 days at 50 mg, 250 mg, and 500 mg/kg of body weight resulted in increased GSH levels in the liver and stomach with reductions in GSH levels in the kidney (Anand et al., 2011). Further, GR activity increased in the kidney and decreased in the liver and plasma, while GPx activity was increased in the liver, heart, stomach, and plasma, respectively (Anand et al., 2011). These organ specific differences in GSH, GR, and GPx were modulated by taurine differentially by organ, attributed to differing levels of enzymatic activity due to the long-term oral administration of taurine. Taurine has also been shown to reverse the reductions in GSH, GPx, and glutathione-5-transferase (GST) in the plasma and tissues of ethanol-fed rats as a modulator of redox homeostasis (Pushpakiran et al., 2004). Moreover, in rat hepatocytes the production of cysteine, sulfate, taurine, and GSH shared a positive linear relationship with increases in cysteine or methionine. However, when cysteine levels were lowered GSH levels increased, and when cysteine levels were elevated it resulted in increased catabolic activity to sulfate and taurine (Stipanuk et al., 1992). Also, under low cysteine availability 90% of GSH was produced when compared to high cysteine availability only 19% GSH was produced suggesting a biphasic relationship (Stipanuk et al., 1992). In astrocyte cultures, the formation of hypotaurine and taurine is an important pathway regulating cysteine metabolism (Brand et al., 2002).

Interestingly, Chinese Hamster Ovary (CHO) cells exposed to lead, and then treated with taurine showed that taurine increased GSH levels that were lowered by lead, decreased G6PD activity that was elevated by lead exposure, and increased cell viability which was lowered by lead exposure (Gürer et al., 2001). For a review of the molecular mechanisms of lead neurotoxicity, see Bressler et al. (1999), Lidsky and Schneider (2003), Virgolini and Aschner (2021). Interestingly, taurine has been shown to improve hippocampal CA1 synaptic plasticity in both prenatal and perinatal lead exposures in rats where lead exposures reduced the fEPSP slopes and paired-pulse facilitation (Yu et al., 2007) and similar synaptic plasticity improvements in Fragile X Syndrome mice (i.e., single most inherited intellectual disorder that comprises approximately 10% of ASD) were observed in these autistic-like mice when treated with taurine (El Idrissi et al., 2010). Later in life, the WT mice paired-pulse facilitation switched to depression and in the Fmr1-KO mice this shift was not observed due to an altered GABAergic system (El Idrissi et al., 2010). Further, Zhu et al. (2005) showed that taurine, supplemented in the diet of lead exposed rats, could reverse the fEPSP and population spike depotentiation induced by lead in the dentate gyrus *in vivo* and suggested it as a potential treatment for reversing lead neurotoxicity. However, caution is needed in understanding the differences of chronic versus acute lead exposures on long-term potentiation of CA1 in the hippocampus fEPSPs and fiber volleys, as taurine produces opposite effects, perhaps due to a unique exposure length-dependent neurotoxicity profile and site of action (Yu et al., 2005).

Moreover, taurine has been shown to reduce anxiety-like behaviors in rats neurodevelopmentally exposed to lead acetate in their drinking water (150 ppm and 1000 ppm) that reduced the hyperexcitability and emotionally disrupted behavioral phenotypes (Cruz et al., 2022; Neuwirth et al., 2017, 2019a, 2022; Neuwirth, 2014), mitigating seizure-related brain excitability (Vasquez et al., 2022), as well as recovering the fronto-executive deficits induced by early neurodevelopmental lead exposures when treated with taurine concurrently (Neuwirth et al., 2019b, c; For Review, see Tait et al., 2018). These preclinical studies have led to the development of new taurine-derived drugs to be used as sulfur-dependent treatments for ameliorating childhood lead poisoning (Neuwirth & Emenike, 2020), as a way to address the fronto-executive dysfunctions that persist across the lifespan (Neuwirth et al., 2018, 2020).

Similarly, taurine treatment of Fragile X Disorder in mice has shown to reduce autistic-like behavioral traits (El Idrissi et al., 2012, 2009; Santora et al., 2013), reduce seizure susceptibility (Shukurova et al., 2022; El Idrissi & L'Amoreaux, 2008; Dobkin et al., 2001), and learning and memory deficits (Neuwirth et al., 2015, 2013; El Idrissi et al., 2010). Given the preclinical therapeutic effects of taurine as a sulfur-dependent treatment, clinical studies on children with ASD have attempted to evaluate the prospects of other sulfur drug treatments in mitigating the neurobehavioral symptoms of ASD in children. One such drug, sulforaphane (SFN; $C_6H_{11}NOS_2$) a nutraceutical from the cruciferous vegetables (Houghon, 2019), is the product of myrosinase catalysis of glucoraphanin (i.e., a glucosinolate) into SFN, that serves as both an antioxidant and an anti-inflammatory agent through the Nrf2 and NF-kB pathways with notable reductions in aging and neurodegenerative processes (Santín-Márquez et al., 2019). Clinical reports seeking to evaluate biomarkers of ASD in children noted reduced GSH, total GSH + GSSG, the ratio of GSH/GSSG, GR, glutathione-s-transferase (GST), thioredoxin (Trx), thioredoxin reductase (TrxR), and peroxidants (Prxs I and III) were observed in 20 male children with autism aged 3–15 when compared to 20 age and gender matched healthy children (Al-Yafee et al., 2011); suggesting oxidative stress biomarkers as key pathways in the etiology of ASD. Further, meta-analyses that evaluated five clinical trials that used SFN in the treatment of autism showed in their systematic review of the literature (i.e., in sample sizes ranging from 6 to 30 participants over a period of 12–30 weeks), despite SFN not acting as a direct anti- or pro-oxidant, it can increase antioxidant capacities and SFN can directly raise GSH levels by stimulating the Nrf2 pathways (McGuinness & Kim, 2020). Taken together, both lead poisoning and ASD share similar oxidative stress pathways and biomarkers that may direct and usher in the next decade of pre-clinical and clinical research to resolve how well sulfur-dependent derivatives, traditional pharmaceuticals, nutraceuticals, and/or CAMs work in ameliorating these neurodevelopmental disorders that negatively impact children's lives.

12.6 Discussion and Conclusions

There is more known about redox imbalances in neurodegenerative and pathological states, but less is known about the typical redox balancing properties and its functions in early development, and these latter points are critical for developing novel therapeutics for ameliorating disorders caused by oxidative stress in children (Sbodio et al., 2019). This is important, as some sulfur-dependent compounds like taurine also serve as unique yet essential neuromodulators in perinatal cortical development (Klib & Fukuda, 2017). Additionally, new insights into early infant memories show that GSH metabolism is necessary for initiating a rapid and consistent increase in neuronal GR (Bessières et al., 2021). The increase in neuronal GR serves to regenerate reduced GSH from GSSG and inhibition of GR can impair hippocampal GSH-mediated long-term memory formation in infant rat, but not in juvenile and adult rats (Bessières et al., 2021). This suggests yet another area of concern and caution in how GSH typically regulates neuronal activity in non-pathological states. Further, to what extent is this particular pathway perturbed that it becomes pathological in nature. This is a challenging predicament for researchers to parse as these pathways typically undergo rapid changes in redox signaling states. The redox signaling states are required for the dynamic metabolic demands of neurons to meditate its cellular and physiological adaptive requirements for survival (i.e., nutrient metabolism, antioxidant defense, regulating gene expression, DNA and protein synthesis to signal transduction, proliferation, and apoptosis, etc.; Aquilano et al., 2014). Consistent with these ideas are the utilization of redox pathways to forestall the progression of neurodegenerative diseases and developmental neuropathological disorders through cystamine, the oxidized form of cysteamine, in a proactive effort to mitigate against oxidative stress as other possible approaches (Paul & Snyder, 2019).

Although the neurobiology of autism is still being unraveled, there appears to be a shared convergent GSH pathway in which environmental neurotoxicants can induce oxidative stress, and in turn, cause neurotoxicity. The effects of lead poisoning on this shared convergent GSH pathway offers intriguing insights into both lead and autism novel therapeutics and neuropsychopharmacological approaches to providing both neuroprotection and reducing the level of brain damage that would otherwise ensue. Novel drugs that contain SH groups, target the GSH pathways directly and indirectly, and ultimately serve prooxidant and antioxidant properties by increasing GSH and decreasing GSSG in the brain and other organs are warranted. As a result, it may reduce the degree to which one may experience neurotoxicity in early neurodevelopment consistent with childhood lead poisoning and ASD. Further investigations will be needed to understand how lead interacts with GSH and SH groups, and, a similar assessment of the suggested environmental contaminants associated with the onset of ASD would also be required in order to understand more about the brains of children exposed to neurotoxicants that cause ASD and lead toxicity. Thus, more research at the pre-clinical and clinical levels are required to fully explore the potentials of such treatment options as CAMs and nutraceuticals become more favorable treatment options in the twenty-first century.

References

Adams, J. B., Romdalvik, J., Sadagopa-Ramanujam, V. M., & Legator, M. S. (2007). Mercury, lead, and zinc in baby teeth of children with autism versus controls. *Journal of Toxicology and Environmental Health, Part A, 70*(2), 1046–1051.

Adams, J. B., Audhya, T., McDonough-Means, S., Rubin, R. A., Quig, D., Geis, E., Gehn, E., Loresto, M., Mitchell, J., Atwood, S., Barnhouse, S., & Lee, W. (2013). Toxicological status of children with autism vs. neurotypical children and the association with autism severity. *Biological Trace Element Research, 151*, 171–180.

Adeyiga, A. B. (2021). Lead poisoning in pediatric groups: A global burden of disease. *International Journal of Pediatric Research, 7*, 083.

Agledal, L., Niere, M., & Ziegler, M. (2010). The phosphate makes a difference: Cellular functions of NADP. *Redox Report, 15*(1), 2–10. https://doi.org/10.1179/174329210X12650506623122

Alawad, Z. M., Al-Jobouri, S. M., & Majid, A. Y. (2019). Lead among children with autism in Iraq. Is it a potential factor? *Journal of Clinical and Analytical Medicine, 10*(2), 215–219.

Aljelehawy, Q. H. A. (2022). Effects of the lead, cadmium, manganese heavy metals, and magnesium oxide nanoparticles on never cell function in Alzheimer's and Parkinson's disease. *Central Asian Journal of Medical and Pharmaceutical Sciences Innovation, 2*(1), 25–36.

Al-Ubaidy, B., Al-Khashali, D. K., & Numan, N. A. (2006). The role of oxidative stress in lead poisoning. *Iraqi Journal of Pharmaceutical Sciences, 15*(1), 68–73. https://doi.org/10.31351/vol15iss1pp68-73

Alvarez-Ortega, N., Caballero-Gallardo, K., & Olivero-Verbel, J. (2017). Low blood lead levels impair intellectual and hematological function in children from Cartagena, Caribbean coast of Columbia. *Journal of Trace Elements in Medicine and Biology, 44*, 233–240. https://doi.org/10.1016/j.jtemb.2017.08.006

Alvarez-Ortega, N., Caballero-Gallardo, K., & Olivero-Verbel, J. (2019). Toxicological effects in children exposed to lead: A cross-sectional study at the Columbian Caribbean coast. *Environment International, 130*(1–11), 104809. https://doi.org/10.1016/j.envint.2019.05.003

Al-Yafee, Y. A., Al-Ayadhi, L. Y., Haq, S. H., & El-Ansary, A. K. (2011). Novel metabolic biomarkers related to sulfur-dependent detoxification pathways in autistic patients of Saudi Arabia. *BMC Neurology, 11*, 139. https://doi.org/10.1186/1471-2377-11-139

Amadi, C. N., Orish, C. N., Frazzoli, C., & Orisakwe, O. E. (2022). Association of autism with toxic metals: A systematic review of case-control studies. *Pharmacology Biochemistry and Behavior, 212*, 173313.

Anand, P., Rajakumar, D., Jeraud, M., Felix, A. J., & Balasubramanian, T. (2011). Effects of taurine on glutathione peroxidase, glutathione reductase and reduced glutathione levels in rats. *Pakistan Journal of Biological Sciences, 14*(3), 219–225.

Aoyama, K., & Nakaki, T. (2013). Impaired glutathione synthesis in neurodegeneration. *International Journal of Molecular Sciences, 14*, 21021–21044. https://doi.org/10.3390/ijms141021021

Aquilano, K., Baldelli, S., & Ciriolo, M. R. (2014). Glutathione: Ne roles in redox signaling for an old antioxidant. *Frontiers in Pharmacology, 5*. https://doi.org/10.3389/fphar.2014.00196

American Psychiatric Association [APA]. (2013). *Diagnostic and statistical manual of mental disorders* (5th ed.). Washington, DC: Author.

Attina, T. M., & Trasande, L. (2013). Economic costs of childhood lead exposure in low-and middle-income countries. *Environmental Health Perspectives, 121*(9), 1097–1102.

Bagheri, S., & Saboury, A. A. (2021). What role do metals play in Alzheimer's disease? *Journal of the Iranian Chemical Society, 18*(9), 2199–2213.

Baliou, S., Adamaki, M., Ioannou, P., Pappa, A., Panayiotidis, M. I., Spandidos, D. A., Christodoulou, I., Kyriakopoulos, A. M., & Zoumpourlis, V. (2021). Protective role of taurine against oxidative stress (Review). *Molecular Medicine Reports, 24*(2), 1–19. https://doi.org/10.3892/mmr.2021.12242

Baloun, J., Adam, V., Trnkova, L., Beklova, M., Svobodova, Z., Zeman, L., & Kizek, R. (2010). Complexes of glutathione with heavy metal ions as a new biochemical marker of aquatic environment pollution. *Environmental Toxicology and Chemistry, 29*(3), 497–500.

Bauer, K., Morin, K. L., Renz, T. E., III, & Zungu, S. (2022). Autism assessment in low- and middle-income countries: Feasibility and usability of western tools. *Focus on Autism and Other Developmental Disabilities*. https://doi.org/10.1177/10883576211073691

Ben-Ari, Y. (2002). Excitatory actions of GABA during development: The nature or the nurture. *Nature Reviews Neuroscience, 9*, 728–739.

Ben-Ari, Y. (2015). Is birth a critical period in the pathogenesis of autism spectrum disorders? *Nature Reviews Neuroscience, 16*(8), 498–505.

Ben-Ari, Y., Khalilov, I., Kahle, K. T., & Cherubini, E. (2012). The GABA excitatory/inhibitory shift in brain maturation and neurological disorders. *The Neuroscientist, 18*(5), 467–486.

Bessières, B., Cruz, E., & Alberini, C. M. (2021). Metabolomic profiling reveals a differential role for hippocampal glutathione reductase in infantile memory formation. *eLife, 10*, e68590. https://doi.org/10.7554/eLife.68590

Betrand, L., Doucet, J., Siminovici, A., Tsoucaris, G., & Wlater, P. (2003). Lead-revealed lipid organization in human hair. *Biochimica et Biophysica Acta (BBA)-General Subjects, 1620*(1–3), 218–224.

Bjørklund, G., Tinkov, A. A., Hosnedlova, B., Kizek, R., Ajsuvakova, O. P., Chirumbolo, S., Skalnaya, M. G., Peana, M., Dadar, M., El-Ansary, A., Qasem, H., Adams, J. B., Aaset, J., & Skalny, A. V. (2020). The role of glutathione redox imbalance in autism spectrum disorders: A review. *Free Radical Biology and Medicine, 160*, 149–162.

Bjørklund, G., Dosa, M. D., Maes, M., Dadar, M., Frye, R. E., Peana, M., & Chirumbolo, S. (2021). The impact of glutathione metabolism in autism spectrum disorder. *Pharmacological Research, 166*, 105437. https://doi.org/10.1016/j.phrs.2021.105437

Bottari, E., De Tommaso, G., Festa, M. R., Iuliano, M., & Zennaro, G. (2020). Behavior of glutathione as a ligand of lead (II). *Chemosphere, 246*, 125718.

Brand, A., Leibfritz, D., Hamprecht, B., & Dringen, R. (2002). Metabolism of cysteine in astroglial cells: Synthesis of hypotaurine and taurine. *Journal of Neurochemistry, 71*(2), 827–832.

Bressler, J., Kim, K. A., Chakraborti, T., & Goldstein, G. (1999). Molecular mechanisms of lead neurotoxicity. *Neurochemical Research, 24*(4), 595–600.

Brownie, C. F., Brownie, C., Lopez, V., & Cadogan, P. (2009). Dogs and goats as sentinels for environmental lead burden in Caribbean basin islands: Jamaica West Indies. *West Indian Veterinary Journal, 9*(2), 17–20.

Chakravarthi, S., Jessop, C. E., & Bulleid, N. J. (2006). The role of glutathione in disulphide bond formation and endoplasmic-reticulum-generated oxidative stress. *EMBO Reports, 7*, 271–275.

Chang-Yen, I., Emrit, C., & Hosein-Rahaman, A. (1995). Incidence of severe lead poisoning in children in Trinidad resulting from battery recycling operations. In J. J. Breen & C. R. Stroup (Eds.), *Lead poisoning* (pp. 63–67). CRC Press.

Chauhan, A., Audhya, T., & Chauhan, V. (2012). Brain region-specific glutathione redox imbalance in autism. *Neurochemical Research, 37*, 1681–1689.

Chen, Y., Xu, X., Zeng, Z., Lin, X., Qin, Q., & Huo, X. (2019). Blood lead and cadmium levels associated with hematological and hepatic functions in patients from an e-waste-polluted area. *Chemosphere, 220*, 531–538. https://doi.org/10.1016/j.chemosphere.2018.12.129

Chetty, C. S., Vemuri, M. C., Campbell, K., & Suresh, C. (2005). Lead-induced cell death of human neuroblastoma cells involves GSH deprivation. *Cellular & Molecular Biology Letters, 10*, 413–423.

Chiang, G. C., Mao, X., Kang, G., Chang, E., Pandya, S., Vallabhajosula, S., Isaacson, R., Ravdin, L. D., & Shungu, D. C. (2017). Relationships among Cortical Glutathione Levels, Brain Amyloidosis, and Memory in Healthy Older Adults Investigated In Vivo with 1H-MRS and Pittsburgh Compound-B PET. *AJNR. American Journal of Neuroradiology, 38*, 1130–1137.

Christian, M. A., Samms-Vaughan, M., Lee, M., Bressler, J., Hessabi, M., Grove, M. L., et al. (2018). Maternal exposures associated with autism spectrum disorder in Jamaican children. *Journal of Autism and Developmental Disorders, 48*(8), 2766–2778.

Clark, B., Vandermeer, B., Simonetti, A., & Buka, I. (2010). Is lead a concern in Canadian autistic children? *Paediatrics & Child Health, 15*(1), 17–22.

Cruz, G. B., Vasquez, M. A., Cabañas, E., Joseph, J. N., Skeen, J. C., Lynch, K. P., Ahmed, I., Khairi, E. B., Bonitto, J. R., Clarke, E. G., Rubi, S., Hameed, N., Kaur, S., Mathew, N., Dacius, T. F., Jr., Jose, T. J., Handford, G., Wolfe, S., Feher, A., Tidwell, K., Tobin, J., Ugalde, E., Fee, S., Choe, A., Gillenwater, K., Hindi, B., Pilout, S., Natale, N. R., Domahoski, N. P., Kent, M. H., Jacob, J. C., Lambert, K. G., & Neuwirth, L. S. (2022). Developmental lead exposure in rats causes sex-dependent changes in neurobiological and anxiety-like behaviors that are improved by taurine co-treatment. In S. W. Shaffer, S. Murakami, & A. E. El Idrissi (Eds.), *Taurine 12* (Advances in experimental medicine & biology) (Vol. 1370, pp. 461–479). Springer Cham. https://doi.org/10.1007/978-3-030-93337-1_43

Curpan, A. S., Luca, A. C., & Ciobica, A. (2021). Potential novel therapies for neurodevelopmental diseases targeting oxidative stress. *Oxidative Medicine and Cellular Longevity*, 6640206. https://doi.org/10.1155/2021/6640206

Daley, T. C., Singhal, N., & Krishnamurthy, V. (2013). Ethical considerations in conducting research on autism spectrum disorders in low- and middle-income countries. *Journal of Autism and Developmental Disorders, 43*(9), 2002–2014.

Dobkin, C. S., Ding, X., Brown, W. T., & El Idrissi, A. (2001). Reduced hippocampal GABAa receptors and increased kainic acid seizure susceptibility in the Fragile X mouse. *American Journal of Human Genetics, 69*(4), 635–635. 1427 E 60th St, Chicago, IL 60637-2954 USA: Univ Chicago Press.

Dringen, R., & Hirrlinger, J. (2003). Glutathione pathways in the brain. *Biological Chemistry, 384*(4), 505–516. https://doi.org/10.1515/BC.2003.059

El Baz Mohamed, F., Zaky, E. A., El-Sayed, A. B., Elhossieny, R. M., Zahra, S. S., Eldin, W. S., Youssef, W. Y., Khaled, R. A., & Youssef, A. M. (2015). Assessment of hair aluminum, lead, and mercury in a sample of autistic Egyptian children: Environmental risk factors of heavy metals in autism. *Behavioural Neurology*, 545674, 1–9. https://doi.org/10.1155/2015/545674

El Idrissi, A., & L'Amoreaux, W. J. (2008). Selective resistance of taurine-fed mice to isoniazide-potentiated seizures: In vivo functional test for the activity of glutamic acid decarboxylase. *Neuroscience, 156*(3), 693–699.

El Idrissi, A., Boukarrou, L., Heany, W., Malliaros, G., Sangdee, C., & Neuwirth, L. S. (2009). Effects of taurine on anxiety-like and locomotor behavior of mice. In J. Azuma et al. (Eds.), *Taurine 7: Taurine for the future healthcare* (Advances in experimental medicine & biology) (Vol. 643, pp. 207–215). Springer Press. https://doi.org/10.1007/978-0-387-75681-3_21

El Idrissi, A., Neuwirth, L. S., & L'Amoreaux, W. L. (2010). Taurine regulation of short term synaptic plasticity in Fragile X mice. *Journal of Biomedical Science, 17*(Suppl 1), S15: 1–8. https://doi.org/10.1186/1423-0127-17-S1-S15

El Idrissi, A., Iskra, B., & Neuwirth, L. S. (2012). Neurobehavioral effects of taurine in fragile X syndrome. In A. El Idrissi & W. L'Amoreaux (Eds.), *Taurine in health and disease* (pp. 293–328). Transworld Research Network.

Esteban, E., Rubin, C. H., Jones, R. L., & Noonan, G. (1999). Hair and blood substrates for screening children for lead poisoning. *Archives of Environmental Health: An International Journal, 54*(6), 436–440.

Fido, A., & Al-Saad, S. (2005). Toxic trace elements in the hair of children with autism. *Autism, 9*(3), 290–298.

Finkelman, J. (1996). Phasing out leaded gasoline will not end lead poisoning in developing countries. *Editorial: Environmental Health Perspectives, 104*(1), 10–11.

Finlay, J. B., Liu, X., Ermel, R. W., & Adamson, T. W. (2015). Maternal weight gain as a predictor of litter size in Swiss Webster, C57BL/6J and BALB/cJ mice. *Journal of the American Association for Laboratory Animal Science, 54*(6), 694–699.

Fowler, B. A. (2016). Molecular biological markers for toxicology and risk assessment. In B. A. Fowler (Ed.), *Molecular biological markers for toxicology and risk assessment* (pp. 1–16). Academic.

George, M., Heeney, M. M., & Woolf, A. D. (2010). Encephalopathy from lead poisoning masquerading as a flu-like syndrome in an autistic child. *Pediatric Emergency Care, 26*(5), 370–373.

Ghanizadeh, A., Akhondzadeh, S., Hormozi, M., Makarem, A., Abotorabi-Zarchi, M., & Firoozabadi, A. (2012). Glutathione-related factors and oxidative stress in autism, a review. *Current Medicinal Chemistry, 19*(23), 4000–4005.

Gifford, J. J. (2021). Assessing the GABA shift hypothesis through markers for oxidative stress. Neuronal activity and microglia number following early VPA exposures (Doctoral dissertation, Rutgers The State University of New Jersey, School of Graduate Studies.

Gifford, J. J., Norton, S. A., Kusnecov, A. W., & Wagner, G. C. (2020). Valproic acid induces nuclear factor erythroid 2-related factor 2 expression in fetal and neonatal brains but not in adult brain: Evidence of the gamma-aminobutyric acid-shift hypothesis. *Neuroreport, 31*(6), 433–436.

Goel, A., & Aschner, M. (2021). The effect of lead exposure on autism development. *International Journal of Molecular Sciences, 22*(4), 1637.

Gorini, F., Muratori, F., & Morales, M. A. (2014). The role of heavy metal pollution in neurobehavioral disorders: A focus on autism. *Review Journal of Autism and Developmental Disorders, 1*(4), 354–372.

Grabrucker, A. M. (2013). Environmental factors in autism. *Frontiers in Psychiatry, 3*, 118.

Grandjean, P. (1984). Lead poisoning: Hair analysis shows the calendar of events. *Human Toxicology, 3*(3), 223–228.

Grosse, S. D., Matte, T. D., Schwartz, J., & Jackson, R. J. (2002). Economic gains resulting from the reduction of children's exposure to lead in the United States. *Environmental Health Perspectives, 110*(6), 563–569.

Gürer, H., & Ercal, N. (2000). Can antioxidants be beneficial in the treatment of lead poisoning? *Free Radical Biology and Medicine, 29*(10), 927–945.

Gürer, H., Özgünes, H., Saygin, E., & Ercal, N. (2001). Antioxidant effect of taurine against lead-induced oxidative stress. *Archives of Environmental Contamination and Toxicology, 41*, 397–402.

Gürer-Orhan, H., Sabir, H. U., & Özgünes, H. (2004). Correlation between clinical indicators of lead poisoning and oxidative stress parameters in controls and lead-exposed workers. *Toxicology, 195*(2–3), 147–154.

Handler, J., Aufderheide, A., Corruccini, R., Brandon, E., & Wittmers, L. (1986). Lead contact and poisoning in Barbados slaves: Historical, chemical, and biological evidence. *Social Science History, 10*(4), 399–425. https://doi.org/10.1017/S014555320001556X

Hassanien, M. A., & Shahawy, A. M. E. (2011). Environmental heavy metals and mental disorders of children in developing countries. In L. Simeonov, M. Kochubovski, & B. Simeonova (Eds.), *Environmental heavy metal pollution and effects on child mental development* (NATO science for peace and security series C: Environmental security (1)) (pp. 1–25). Springer.

Hauptman, M., Stierman, B., & Woolf, A. D. (2019). Children with autism spectrum disorder and lead poisoning: Diagnostic challenges and management complexities. *Clinical Pediatrics, 58*(6), 605–612.

Hermes-Lima, M., Pereira, B., & Bechara, E. J. H. (1990). Are free radicals involved in lead poisoning? *Xenobiotica, 21*(8), 1085–1090.

Holder, D. E. (2013). Voices from the spectrum: The positive impact of autism on the family in three societies. *Caribbean Journal of Psychology, 5*(1), 40–51.

Houghon, C. A. (2019). Sulforaphane: Its "Coming of Age" as a clinically relevant nutraceutical in the prevention and treatment of chronic disease. *Oxidative Medicine and Cellular Longevity*, 2716870. https://doi.org/10.1155/2019/2716870

Hsu, J. M. (1981). Lead toxicity as related to glutathione metabolism. *The Journal of Nutrition, 11*(1), 26–33.

Ijomone, O. M., Olung, N. F., Akingbade, G. T., Okoh, C. O. A., & Aschner, M. (2020). Environmental influence on neurodevelopmental disorders: Potential association of heavy metal exposure and autism. *Journal of Trace Elements in Medicine and Biology, 62*, 126638.

Islam, M. S., Sachi, S., & Hasan, M. R. (2021). An investigation of chronic exposure of lead to the pregnant laboratory animal and its effects on their offspring: Autism investigation to the laboratory animal model in Bangladesh. *World Journal of Biology Pharmacy and Health Sciences, 5*(3), 085–094.

Jenner, P. (1994). Oxidative damage in neurodegenerative disease. *Lancet, 344*, 796–798.

Karimi, M., Ignasiak, M. T., Chan, B., Croft, A. K., Radom, L., Schiesser, C. H., Pattison, D. I., & Davies, M. J. (2016). Reactivity of disulfide bonds is markedly affected by structure and environment: Implications for protein modification and stability. *Scientific Reports, 6*, 38572. https://doi.org/10.1038/srep38572

Kaul, B., Sandhu, R. S., Depratt, C., & Reyes, F. (1999). Follow-up screening of lead-poisoned children near an auto battery recycling plant, Haina, Dominican Republic. *Environmental Health Perspectives, 107*(11), 917–920.

Kern, J. K., Grannemann, B. D., Trivedi, M. H., & Adams, J. B. (2007). Sulfhydryl-reactive metals in autism. *Journal of Toxicology and Environmental Health, Part A, 70*(8), 715–721.

Kim, Y. S., & Leventhal, B. (2015). Genetic epidemiology and insights into interactive genetic and environmental effects in autism spectrum disorders. *Biological Psychiatry, 77*(1), 66–74.

Kim, K. N., Kwon, H. J., & Hong, Y. C. (2016). Low-level lead exposure and autistic behaviors in school-aged children. *Neurotoxicology, 53*, 193–200.

Klib, W., & Fukuda, A. (2017). Taurine as an essential neuromodulator during perinatal cortical development. *Frontiers in Cellular Neuroscience*. https://doi.org/10.3389/fncel.2017.00328

Ko, L. Y. Y., Qin, Y. Y., & Wong, M. H. (2012). Heavy metal overloads and autism in children from mainland China and Hong Kong: A preliminary study. In *Environmental contamination-health risks and ecological restoration* (pp. 35–46). CRC Press.

Kumar, J., Liddle, E. B., Fernandes, C. C., Palaniyappan, L., Hall, E. L., Robson, S. E., Simmonite, M., Fiesal, J., Katshu, M. Z., Qureshi, A., Skelton, M., Christodoulou, N. G., Brookes, M. J., Morris, P. G., & Liddle, P. F. (2020). Glutathione and glutamate in schizophrenia: A 7T MRS study. *Molecular Psychiatry, 25*, 873–882.

Lalor, G., Vutchkov, M., & Bryan, S. (2007). Blood lead levels of Jamaican children Island-wide. *Science of the Total Environment, 374*(2–3), 235–241.

Landrigan, P. J. (2010). What causes autism? Exploring the environmental contribution. *Current Opinion in Pediatrics, 22*(2), 219–225.

Lanphear, B. (2000). Letters to the editor. *Editorial. Public Health Reports, 115*, 498. file:///Users/user/Downloads/cdc_64935_DS1.pdf.

Lawler, C. P., Croen, L. A., Grether, J. K., & Van de Water, J. (2005). Indentifying environmental contributions to autism: Provocative clues and false leads. *Mental Retardation and Developmental Disabilities Research Reviews: Special Issue: An Update on Autism Research, 10*(4), 292–302.

Lidsky, T. I., & Schneider, J. S. (2003). Lead neurotoxicity in children: Basic mechanisms and clinical correlates. *Brain, 126*(1), 5–19.

Lidsky, T. I., & Schneider, J. S. (2005). Autism and autistic symptoms associated with childhood lead poisoning. *The Journal of Applied Research, 5*(1), 80–87.

Lowe, H., Smith, R., Campbell, N., & Morrison, E. Y. (2002). Lead pollution and amelioration measures in the community of Frazers content, St Catherine, Jamaica. *The West Indian Medical Journal, 51*(3), 160–163.

Lyall, K., Croen, L., Daniels, J., Fallin, M. D., Ladd-Acosta, C., Lee, B. K., Park, B. Y., Snyder, N. W., Schendel, D., Volk, H., Windham, G. C., & Newschaffer, C. (2017). The changing epidemiology of autism spectrum disorders. *Annual Review of Public Health, 38*, 81–102.

Mah, V., & Jalilehvand, F. (2012). Lead(II) complex formation with glutathione. *Inorganic Chemistry, 51*(11), 6285–6298.

Mandal, P. K., Saharan, S., Tripathi, M., & Murari, G. (2015). Brain glutathione levels-a novel biomarker for mild cognitive impairment and Alzheimer's disease. *Biological Psychiatry, 78*, 702–710.

Mann, A. (2013). *The experiences of mothers of children with autism in Jamaica: An exploratory study of their journey*. University of South Florida.

Marlow, M., Servili, C., & Tomlinson, M. (2019). A review of screening tools for the identification of autism spectrum disorders and developmental delay in infants and young children: Recommendations for use in low- and middle-income countries. *Autism Research, 12*(2), 176–199.

Martin, M. J., Alarcón De La Lastra, C., Motilva, V., & La Casa, C. (2000). Antiulcer and gastroprotective activity of flavonic compounds: Mechanisms involved. In Atta-ur-Rahman (Ed.), *Studies in natural products chemistry* (Vol. 22 Part C, pp. 419–456). Elsevier. https://doi.org/10.1016/S1572-5995(00)80032-0

Matte, T. D., Figueroa, J. P., Ostrowski, S., Burr, G., Jackson-Hunt, L., Keenlyside, R. A., & Baker, E. L. (1989). Lead poisoning among household members exposed to lead-acid battery repair shops in Kingston, Jamaica. *International Journal of Epidemiology, 18*(4), 874–881.

McGuinness, G., & Kim, Y. (2020). Sulforaphane treatment for autism spectrum disorder: A systematic review. *EXCLI Journal, 19*, 892–903. https://doi.org/10.17179/excli2020-2487

Millichap, J. (1996). Lead intoxication in children with autism. *Pediatric Neurology Briefs, 10*(5), 39–40.

Mohammed, T. I., Chang-Yen, I., & Bekele, I. (1996). Lead pollution in East Trinidad resulting from lead recycling and smelting activities. *Environmental Geochemistry and Health, 18*(3), 123–128.

Mostafa, G. A., Bjørklund, G., Urbina, M. A., & Al-Ayadhi, L. Y. (2016). The positive association between elevated blood lead levels and brain-specific autoantibodies in autistic children from low lead-polluted areas. *Metabolic Brain Disease, 31*(5), 1047–1054.

Murray, S. A., Morgan, J. L., Kane, C., Sharma, Y., Heffner, C. S., Lake, J., & Donahue, L. R. (2010). Mouse gestation length is genetically determined. *PLoS One, 5*, e12418.

Nakhaee, S., Amirabaizadeh, A., Farnia, V., Azadi, N. A., Mansouri, B., & Radmehr, F. (2022). Association between biological lead concentrations and autism spectrum disorder (ASD) in children: A systematic review and meta-analysis. *Biological Trace Element Research*. https://doi.org/10.1007/s12011-022-03265-9

Neuwirth, L.S. (2014). The characterization of Pb2+ toxicity in rat neural development: An assessment of Pb2+ effects on the GABA shift in neural networks and implications for learning and memory disruption. UMI Proquest Dissertations & Theses 3612469. DAI/B 75-06(E), April 2014.

Neuwirth, L. S. (2018). Resurgent lead poisoning and renewed public attention towards environmental social justice issues: A review of current efforts and call to revitalize primary and secondary lead poisoning prevention for pregnant women, lactating mothers, and children within the U.S. *International Journal of Occupational and Environmental Health, 24*(3–4), 86–100.

Neuwirth, L. S., & Emenike, B. U. (2020). Compositions and methods of treating a subject with taurine and derivatives thereof. Inv. Ref. 200-2067US01. Serial No. 16/880,796 U.S. Patent Application.

Neuwirth, L. S., Volpe, N. P., & El Idrissi, A. (2013). Taurine effects on emotional learning and memory in aged mice: Neurochemical alterations and differentiation in auditory cued fear and context conditioning. In A. El Idrissi & W. L'Amoreaux (Eds.), *Taurine 8: Physiological roles and mechanisms of action* (Advances in experimental medicine & biology) (Vol. 775, pp. 195–214). Springer Press. https://doi.org/10.1007/978-1-4614-6130-2_17

Neuwirth, L. S., Volpe, N. P., Ng, S., Marsillo, A., Corwin, C., Madan, N., Ferraro, A. M., & El Idrissi, A. (2015). Taurine recovers mice emotional learning and memory disruptions associated with Fragile X Syndrome in context fear and auditory cued-conditioning. In J. Marcinkiewicz & S. W. Schaffer (Eds.), *Taurine 9* (Advances in experimental medicine & biology) (Vol. 803, pp. 425–438). Springer Press. https://doi.org/10.1007/978-3-319-15126-7_33

Neuwirth, L. S., Volpe, N. P., Corwin, C., Ng, S., Madan, N., Ferraro, A. M., Furman, Y., & El Idrissi, A. (2017). Taurine recovery of learning deficits induced by developmental Pb2+ exposure. In D. H. Lee, S. Shaffer, E. Park, & H. W. Kim (Eds.), *Taurine 10* (Advances in experimental medicine and biology) (Vol. 975, pp. 39–55). https://doi.org/10.1007/978-94-024-1079-2_4

Neuwirth, L. S., Phillips, G. R., & El Idrissi, A. (2018). Perinatal Pb2+ exposure alters the expression of genes related to the neurodevelopmental GABA-shift in postnatal rats. *Journal of Biomedical Science, 25*(45), 1–11. https://doi.org/10.1186/s12929-018-0450-4

Neuwirth, L. S., Kim, Y. J., Barrera, E. D., Jo, C., Chrisphonte, J. M., Hameed, N., Rubi, S., Dacius, T. F., Jr., Skeen, J. C., Bonitto, J. R., Khairi, E., Iqbal, A., Ahmed, I., Masood, S., Tranquille, B., & Thiruverkadu, V. (2019a). Early neurodevelopmental exposure to low lead levels induces fronto-executive dysfunctions that are recovered by taurine co-treatment in the rat attention set-shift test: Implications for taurine as a psychopharmacotherapy against neurotoxicants. In J. Hu, F. Piao, S. W. Schaffer, A. El Idrissi, & J. Y. Wu (Eds.), *Taurine 11* (Advances in experimental medicine and biology) (Vol. 1155, pp. 821–846). https://doi.org/10.1007/978-981-13-8023-5_70

Neuwirth, L. S., Emenike, B. U., Barrera, E. D., Hameed, N., Rubi, S., Dacius, T. F., Jr., Skeen, J. C., Bonitto, J. R., Khairi, E., Iqbal, A., Ahmed, I., Jose, T. J., Lynch, K. P., Khan, M., Alvira, A. L., Mathew, N., Kaur, S., Masood, S., Tranquilee, B., & Thiruverkadu, V. (2019b). Assessing the anxiolytic properties of taurine-derived compounds in rats following developmental lead exposure: A neurodevelopmental and behavioral pharmacological pilot study. In J. Hu, F. Piao, S. W. Schaffer, A. El Idrissi, & J. Y. Wu (Eds.), *Taurine 11* (Advances in experimental medicine and biology) (Vol. 1155, pp. 801–819). https://doi.org/10.1007/978-981-13-8023-5_69

Neuwirth, L. S., Masood, S., Anderson, D. W., & Schneider, J. S. (2019c). The attention set-shifting test is sensitive for revealing sex-based impairments in executive functions following developmental lead exposure in rats. *Behavioural Brain Research, 366*, 126–134. https://doi.org/10.1016/j.bbr.2019.03.022

Neuwirth, L. S., Lopez, O. E., Schneider, J. S., & Markowitz, M. E. (2020). Low-level lead exposure impairs fronto-executive functions: A call to update the DSM-5 with lead poisoning as a neurodevelopmental disorder. *Psychology & Neuroscience, 13*(3), 299.

Neuwirth, L. S., Emenike, B. U., Cruz, G. B., Cabañas, E., Vasquez, M. A., Joseph, J. N., Ayaz, Z., Mian, M., Ali, M. M., Clarke, E. G., Barrera, E. D., Hameed, N., Rubi, S., Dacius, T. F., Jr., Skeen, J. C., Bonitto, J. R., Khairi, E. B., Iqbal, A., Ahmed, I., Jose, T. J., Lynch, K. P., Alivira, A., Mathew, N., Kaur, S., Masood, S., Tranquile, B., & Thiruverkadu, V. (2022). Taurine-derived compounds produce anxiolytic effects in rats following developmental lead exposure. In S. W. Shaffer, S. Murakami, & A. E. El Idrissi (Eds.), *Taurine 12* (Advances in experimental medicine & biology) (Vol. 1370, pp. 445–460). Springer Cham. https://doi.org/10.1007/978-3-030-93337-1_42

Niklowitz, W. J. (1975). Neurofibrillary changes after acute experimental lead poisoning. *Neurology, 25*(1), 927.

Nishimura, Y., Kanda, Y., Sone, H., & Aoyama, H. (2021). Oxidatie stress as a common key event in developmental neurotoxicity. *Oxidative Medicine and Cellular Longevity*, 6685204. https://doi.org/10.1155/2021/6685204

Olympio, M. P. K., Goncalves, C. G., Salles, F. J., da Silva Ferreira, P. S., Soares, A. S., Buzzalaf, M. A. R., Cardoso, M. R. A., & Bechara, E. J. H. (2017). What are the blood lead levels of children living in Latin America and the Caribbean? *Environment International, 101*, 46–58. https://doi.org/10.1016/j.envint.2016.12.022

Pangrazzi, L., Balasco, L., & Bozzi, Y. (2020). Oxidative stress and immune system dysfunction in autism spectrum disorders. *International Journal of Molecular Sciences, 21*(9), 3293. https://doi.org/10.3390/ijms21093293

Paul, B. D., & Snyder, S. H. (2019). Therapeutic applications of cysteamine and cystamine in neurodegenerative and neuropsychiatric diseases. *Frontiers in Neurology, 12*. https://doi.org/10.3389/fneur.2019.01315

Pushpakiran, G., Mahalakshmi, K., & Venkataraman, A. C. (2004). Protective effects of taurine on glutathione and glutathione-dependent enzymes in ethanol-fed rats. *Pharmazie, 59*(11), 869–872.

Rae, C. D., & Williams, S. R. (2017). Glutathione in the human brain: Review of its roles and measurement by magnetic resonance spectroscopy. *Analytical Biochemistry, 529*, 127–143.

Rahbar, M. H., Samms-Vaughan, M., Loveland, K. A., Pearson, D. A., Bressler, J., Chen, Z., et al. (2012). Maternal and paternal age are jointly associated with childhood autism in Jamaica. *Journal of Autism and Developmental Disorders, 42*(9), 1928–1938.

Rahbar, M. H., Samms-Vaughan, M., Dickerson, A. S., Loveland, K. A., Arjomand-Hessabi, M., Bressler, J., Shakespeare-Pellington, S., Grove, M. L., Pearson, D. A., & Boerwinkle, E. (2015a). Blood lead concentrations in Jamaican Children with and without autism spectrum disorder. *International Journal of Environmental Research and Public Health, 12*(1), 83–105.

Rahbar, M. H., Samms-Vaughan, M., Dickerson, A. S., Loveland, K. A., Ardjomand-Hessabi, M., Bressler, J., et al. (2015b). Factors associated with blood lead concentrations of children in Jamaica. *Journal of Environmental Science and Health, Part A, 50*(6), 529–539.

Reeves, R. D., Jolley, K. W., & Buckley, P. D. (1975). Lead in human hair: Relation to age, sex, and environmental factors. *Bulletin of Environmental Contamination and Toxicology, 14*(5), 579–587.

Romieu, I., Lacasana, M., & McConnell, R. (1997). Lead exposure in Latin America and the Caribbean. Lead research group of the Pan-American health organization. *Environmental Health Perspectives, 105*(4), 398–405. https://doi.org/10.1289/ehp.97105398

Rose, S., Melynk, S., Pavliv, O., Bai, S., Nick, T. G., Frye, R. E., & James, S. J. (2012). Evidence of oxidative damage and inflammation associated with low glutathione redox status in the autism brain. *Translational Psychiatry, 2*, e134. https://doi.org/10.1038/tp.2012.61

Rubin, C. H., Esteban, E., Reissman, D. B., Daley, W. R., Noonan, G. P., Karpati, A., et al. (2002). Lead poisoning among young children in Russia: Concurrent evaluation of childhood lead exposure in Ekaterinburg, Krasnouralsk, and Volgograd. *Environmental Health Perspectives, 110*(6), 559–562.

Saghazadeh, A., & Rezaei, N. (2017). Systemic review and meta-analysis links autism and toxic metals and highlights the impact of country development status: Higher blood and erythrocyte levels for mercury and lead, and higher hair antimony, cadmium, lead, and mercury. *Progress in Neuro-Psychopharmacology and Biological Psychiatry, 79*, 340–368.

Saleh, M. G., Papantoni, A., Mikkelsen, M., Hui, S. C. N., Oeltzschner, G., Puts, N. A., Edden, R. A. E., & Carnell, S. (2020). Effect on age on GABA+ and glutathione in a pediatric sample. *American Journal of Neuroradiology, 41*, 1099–1104.

Samms-Vaughan, M. E. (2014). The status of early identification and early intervention in autism spectrum disorders in lower-and middle-income countries. *International Journal of Speech-Language Pathology, 16*(1), 30–35.

Samms-Vaughan, M., Rahbar, M. H., Dickerson, A. S., Loveland, K. A., Hessabi, M., Pearson, D. A., et al. (2017). The diagnosis of autism and autism spectrum disorder in low-and middle-income countries: Experience from Jamaica. *Autism, 21*(5), 564–572.

Santín-Márquez, R., Alarcón-Aguilar, A., López-Diazguerrero, N. E., Chondrogianni, N., & Königsberg, M. (2019). Sulforaphane – Role in aging and neurodegeneration. *GeroScience, 41*(5), 655–670.

Santora, A., Neuwirth, L. S., L'Amoreaux, W. J., & El Idrissi, A. (2013). The effects of chronic taurine supplementation on motor learning. In A. El Idrissi & W. L'Amoreaux (Eds.), *Taurine 8: Physiological roles and mechanisms of action* (Advances in experimental medicine & biology) (Vol. 775, pp. 177–185). Springer Press. https://doi.org/10.1007/978-1-4614-6130-2_15

Sbodio, J. I., Snyder, S. H., & Paul, B. D. (2019). Disease outcomes to therapeutic opportunities. *Antioxidants & Redox Signaling, 30*(11), 1450–1499. https://doi.org/10.1089/ars.2017.7321

Schroeder, H., Shuler, K. A., & Chenery, S. R. (2013). Childhood lead exposure in an enslaved African community in Barbados: Implications for birthplace and health status. *American Journal of Physical Anthropology, 150*(2), 203–209.

Schubert, D., & Piasecki, D. (2001). Oxidative glutamate toxicity can be a component of the excitotoxicity cascade. *Journal of Neuroscience, 21*(19), 7455–7462.

Sealy, J. (2017). Parent-child psychotherapeutic interventions with vulnerable clients: Are they within the scope of social work practitioners in Barbados? *Caribbean Journal of Social Work, 12*, 165–190.

Seneff, S., Lauritzen, A., Davidson, R. M., & Lentz-Marion, L. (2013). Is encephalopathy a mechanism to renew sulfate in autism? *Entropy, 15*(1), 372–406.

Shukla, D., Mandal, P. K., Ersland, L., Gruner, E. R., Tripathi, M., Raghunathan, P., Sharma, A., Chaithya, G. R., Punjabi, K., & Splaine, C. A. (2018). Multi-Center study on human brain glutathione conformation using magnetic resonance spectroscopy. *Journal of Alzheimer's Disease, 66*, 517–532.

Shukurova, S., Sadek, R., Mekawy, N., Bendaoud, M., Yachou, Y., Mamchyn, A., & El Idrissi, A. (2022). Electrophysiological evidence for anti-epileptic property of taurine. In S. W. Shaffer, S. Murakami, & A. E. El Idrissi (Eds.), *Taurine 12* (Advances in experimental medicine & biology) (Vol. 1370, pp. 333–340). Springer Cham. https://doi.org/10.1007/978-3-030-93337-1_43

Sian, J., Dexter, D. T., Lees, A. J., Daniel, S., Agid, Y., Javoy-Agid, F., Jenner, P., & Marsden, C. D. (1994). Alterations in glutathione levels in Parkinson's disease and other neurodegenerative disorders affecting basal ganglia. *Annals of Neurology, 36*, 348–355.

Silbergeld, E. K. (1979). Lead, GABA, and seizures: Effects of subencephalopathic lead exposure on seizure sensitivity and GABAergic function. *Environmental Research, 19*(2), 371–382.

Slivka, A., Spina, M. B., & Cohen, G. (1987). Reduced and oxidized glutathione in human and monkey brain. *Amino Acids, 42*, 207–219. https://doi.org/10.1016/0304(87)900061-9

Stewart, L. A., & Lee, L. C. (2017). Screening for autism spectrum disorder in low-and middle-income countries: A systematic review. *Autism, 21*(5), 527–539.

Stipanuk, M. H., Coloso, R. M., Garcia, R. A. G., & Banks, M. F. (1992). Cysteine concentration regulates cysteine metabolism to glutathione, sulfate and taurine in rat hepatocytes. *The Journal of Nutrition, 122*(3), 420–427.

Tait, D. S., Bowman, E. M., Neuwirth, L. S., & Brown, V. J. (2018). Assessment of attentional set-shifting testing in rats using manual and automated data analysis procedures. *Neuroscience and Biobehavioral Reviews, 89*, 72–84. https://doi.org/10.1016/j.neubiorev.2018.02.013

Teerasarntipan, T., Chaiteerakij, R., Prueksapanich, P., & Werawatganon, D. (2020). Changes in inflammatory cytokines, antioxidants and liver stiffness after chelation therapy in indivisuals with chronic lead poisoning. *BMC Gasteroenterology, 20*, 263. https://doi.org/10.1186/s12876/s12876-020-01386-w

Tian, Y., Green, P. G., Stamova, B., Hertz-Picciotto, I., Pessah, I. N., Hansen, R., Yang, X., Gregg, J. P., Ashwood, P., Jickling, G., Van de Water, J., & Sharp, F. R. (2009). Correlations of gene expression with blood lead levels in children with autism compared to typically developing controls. *Neurotoxicity Research, 19*, 1–13.

Trepka, M. J., Pekovic, V., Santana, J. C., & Zhang, G. (2005). Risk factors for lead poisoning among Cuban refugee children. *Public Health Reports, 120*(2), 179–185.

Vasquez, M. A., Cruz, G. B., Cabañas, E., Joseph, J. N., Mian, M., Madhira, S. K. V., Akintunde, C. A., Clarke, E. G., Skeen, J. C., Bonitto, J. R., Khairi, E. B., Lynch, K. P., Mekawy, N. H., El Idrissi, A., Kim, Y. J., Emenike, B. U., & Neuwirth, L. S. (2022). In-vivo sex-dependent effects of perinatal Pb^{2+}-exposure on pilocarpine & kainic acid induced seizure susceptibility & taurine neuropharmacology. In S. W. Shaffer, S. Murakami, & A. E. El Idrissi (Eds.), *Taurine 12* (Advances in experimental medicine & biology) (Vol. 1370, pp. 481–496). Springer Cham. https://doi.org/10.1007/978-3-030-93337-1_44

Venkateshappa, C., Harish, G., Mahadevan, A., Srinivas Bharath, M. M., & Shankar, S. K. (2012). Elevated oxidative stress and decreased antioxidant function in the human hippocampus and frontal cortex with increasing age: Implications for neurodegeneration in Alzheimer's' disease. *Neurochemical Research, 37*, 1601–1614. https://doi.org/10.1007/s11064-102-0755-8

Virgolini, M. B., & Aschner, M. (2021). Molecular mechanisms of lead neurotoxicity. *Advances in Neurotoxicology, 5*, 159–213.

Wahil, M. S. A., Ja'afar, M. H., & Isa, Z. M. (2022). Assessment of urinary lead (Pb) and essential trace elements in autism spectrum disorder: A case-control study among preschool children in Malaysia. *Biological Trace Element Research, 200*(1), 97–121.

Wang, J., Wu, J., & Zhang, Z. (2006). Oxidative stress in mouse brain exposed to lead. *The Annals of Occupational Hygiene, 50*(4), 405–409.

Whitehouse, P. J. (2022). Ethical issues in early diagnosis and prevention of Alzheimer's disease. *Dialogues in Clinical Neuroscience, 21*(1), 101–108.

Yassa, H. A. (2014). Autism: A form of lead and mercury toxicity. *Environmental Toxicology and Pharmacology, 38*(3), 1016–1024.

Ye, B. S., Leung, A. O. W., & Wong, M. H. (2017). The association of environmental toxicants and autism spectrum disorders in children. *Environmental Pollution, 227*, 234–242.

Yu, K., Yu, S. S., & Ruan, D. Y. (2005). Opposite effects of lead exposure on taurine- and HFS-induced LTP in rat hippocampus. *Brain Research Bulletin, 64*(6), 525–531.

Yu, S. S., Wang, M., Li, X. M., Chen, W. H., Chen, J. T., Wang, H. L., & Ruan, D. Y. (2007). Influences of different developmental periods of taurine supplements on synaptic plasticity in hippocampal CA1 area of rats following prenatal and perinatal lead exposure. *BMC Developmental Biology, 7*, 51. https://doi.org/10.1186/1471-213X-7-51

Zhu, D. M., Wang, M., She, J. Q., Yu, K., & Ruan, D. Y. (2005). Protection by a taurine supplemented diet from lead-induced deficits of long-term potentiation/depotentiation in dentate gyrus of rats in vivo. *Neuroscience, 134*(1), 215–224.

Chapter 13
Epidemiological Surveys of ASD: Current Findings and New Directions

Eric Fombonne

13.1 Introduction

Since the first autism survey conducted in England (Lotter, 1966), epidemiological surveys have increased in number and complexity. Contrasting with the first studies that were simple head counts of children already diagnosed with a severe autism phenotype and residing in small, circumscribed geographical areas, current surveys now include large populations, multiple sites, stratified samples and rely on intricate sets of screening procedures followed by some form of diagnostic confirmation. However, no agreed-upon formula exists for planning and conducting a survey, and there is no standardization of autism survey methodology. As a result, differences in methodologies account for substantial heterogeneity in survey findings. Each survey has unique design features that reflect the local educational and health services infrastructure and that are influenced by current social policies for children with disabilities in the local region or country. Moreover, survey protocols vary in whether they include parents, teachers and subjects with Autism Spectrum Disorder (ASD) as participants, and rely on variable screening instruments and diagnostic confirmation procedures. As such, prevalence differences between studies are hazardous to evaluate and whether observed discrepancies are due to method factors or true differences in population parameters cannot usually be determined.

We first provide a review of 141 surveys published from 1966 to 2020. Then, we review specific areas of survey methodology (case definition and case identification,

E. Fombonne (✉)
Department of Psychiatry, Oregon Health & Science University, Portland, OR, USA

Department of Behavioral Neuroscience, Oregon Health & Science University, Portland, OR, USA

Departments of Psychiatry, Pediatrics & Behavioral Neuroscience, Oregon Health & Science University, Portland, OR, USA
e-mail: fombonne@ohsu.edu

A. El Idrissi, D. McCloskey (eds.), *Neurobiology of Autism Spectrum Disorders*,
https://doi.org/10.1007/978-3-031-42383-3_13

surveys of parent reports, school screening, cross-cultural issues) where advances have been made or progress is still needed, and consider recent trends for developing surveillance programs. This Chapter is an abbreviated and slightly modified version of our most recent review (Fombonne et al., 2022 (in press)) and many issues have already been discussed in other reviews (Fombonne et al., 2021a, b; MacFarlane et al., 2022; Chiarotti et al., 2020; Zeidan et al., 2022) and commentaries (Fombonne, 2018, 2019). In this Chapter, unless indicated otherwise, we use the terms 'autism' or 'ASD' interchangeably as reflecting a single broad diagnostic category as defined recently in DSM-5 (American Psychiatric Association, 2013). It also encompasses the class of Pervasive Developmental Disorders (PDD) as defined in DSM-IV (American Psychiatric Association, 1994), ICD-10 (World Health Organization, 1992) and DSM-IV-TR (American Psychiatric Association, 2000).

13.2 Review of Prevalence Surveys

We identified 141 prevalence surveys of autism conducted between 1966 and 2020. Table 13.1 lists the 109 surveys conducted since 2000. The remaining 32 surveys conducted between 1966 and 2001 are described in a previous publication (Fombonne, 2003) to which the reader is referred for specific study details. The two sets of studies are mutually exclusive. We included all published cross-sectional surveys of population-based samples aiming at estimating ASD prevalence. Studies designed to test the efficacy of screening tools or screening programs were excluded. We included cohort studies that generated cumulative incidence proportions that can often approximate prevalence. We also excluded studies with a target population size of fewer than 5,000 individuals (they lack precision and are more prone to sampling biases), studies where no professional diagnostic confirmation was available (e.g. surveys relying exclusively on parent reports), studies published in a different language than English or as abstracts only, and duplicate studies conducted on identical/overlapping samples.

Summary statistics that follow must be interpreted with caution in light of the substantial heterogeneity of methods across surveys and over time. Two major features must be borne in mind: case definition and case ascertainment. First, the definition of autism used in surveys has changed over time. As illustrated in Table 13.1 (and in Table 1 in Fombonne, 2003), numerous diagnostic criteria were employed in epidemiological studies, starting with relatively narrow definitions in Kanner-derived criteria in the first survey (Lotter 1966), followed by ICD-9 (ICD-9; World Health Organization [WHO], 1977) or DSM-III (American Psychiatric Association [APA], 1980) diagnostic nomenclatures, and in the early 1990s, by the broader concepts of autism embraced in both ICD-10 (WHO, 1992) and DSM-IV (APA, 1994). The earliest diagnostic criteria reflected the more qualitatively deviant forms of the autism behavioral phenotype, usually associated with severe delays in language and cognitive skills. With ICD-10 and DSM-IV, less severe forms of autism were recognized and separate diagnostic categories (Pervasive Developmental Disorders Not

Table 13.1 Prevalence surveys of ASD since 2000

Pub Year	Study Year	Authors	Country	Income level[a]	Area	Population	Age	Number Affected	Diagnostic Criteria	% with Normal IQ	Gender Ratio (M:F)	Prevalence Rate/10,000	95% CI
2000	–	Baird et al.	UK	HI	South East Thames	16,235	7	94	*ICD-10*	60	15.7 (83:11)	57.9	46.8; 70.9
2000	1991–1996	Powell et al.	UK	HI	West Midlands	58,654[e]	1–5	122	*DSM-IV, ICD-10*	–	–	20.8	17.3; 24.9
2001	–	Fombonne et al.	UK	HI	England and Wales	10,438	5–15	27	*DSM-IV, ICD-10*	55.5	8 (24:3)	26.1	16.2; 36.0
2001	1998	Bertrand et al.	US	HI	New Jersey	8896	3–10	60	*DSM-IV*	51	2.7 (44:16)	67.4	51.5; 86.7
2001	1999–2000	Chakrabarti and Fombonne	UK	HI	Stafford	15,500	2.5–6.5	96	*ICD-10*	74.2	3.8 (77:20)	61.9	50.2; 75.6
2002	–	Scott et al.	UK	HI	Cambridge	33,598	5–11	196	*ICD-10*	–	4	58.3[e]	50.7; 67.1[e]
2002	1987–1994	Croen et al.	USA	HI	California	4,590,333	5–12	5037	*DSM-III/ DSM-IV 'full syndrome autism' in CDERs from DDS*	62.8	4.47 (4116:921)	11.0	10.7;11.3
2003	1996	[CDC] [d] Yeargin-Allsopp et al.	US	HI	Atlanta, GA	289,456	3–10	987	*DSM-IV*	31.8	4 (787:197)	34.0	32; 36
2003	2000	Lingam et al.	UK	HI	North East London	186,206	5–14	567	*ICD-10*	–	4.8 (469:98)	30.5[e]	27.9; 32.9[e]
2003	2001–2002	Gurney et al.	US	HI	Minnesota	787,308[e]	6–11	4094	*DSM-IV-R*	–	–	52.0	50.4; 53.6[e]
2004	2002	Icasiano et al.	Australia	HI	Barwon	45,153[e]	2–17	177	*DSM-IV*	53.4	8.3 (158:19)	39.2	33.8; 45.4[e]

(continued)

Table 13.1 (continued)

Pub Year	Study Year	Authors	Country	Income level[a]	Area	Population	Age	Number Affected	Diagnostic Criteria	% with Normal IQ	Gender Ratio (M:F)	Prevalence Rate/10,000	95% CI
2005	1999–2003	Chakrabarti and Fombonne	UK	HI	Stafford	10,903	4–6	64	*ICD-10*	70.2	6.1 (55:9)	58.7	45.2; 74.9
2005	2000	Zhang and Ji	China	UMI	Tianjin	7345	2–6	8	*CARS-CV, DSM-IV*	0	7 (7:1)	11	3.4; 25.4
2006	–	Baird et al.	UK	HI	South Thames	56,946	9–10	158	*ICD-10*	45	3.3 (121:37)	116.1	90.4; 141.8
2006	2001	Gillberg et al.	Sweden	HI	Göteborg	32,568	7–12	262	*DSM-III, DSM-IV, Gillberg's criteria for AS*	–	3.6 (205:57)	80.4	71.3; 90.3
2006	2001	Harrison et al.	UK	HI	Scotland	134,661	0–15	443	*DSM-IV, ICD-10*	–	7.0 (369:53)	44.2	39.5; 48.9
2006	2002	Ouellette-Kuntz et al.	Canada	HI	Manitoba and Prince Edward Island	227,526	1–14	657	*DSM-IV*	–	4.1 (527:130)	28.9[e]	26.8; 31.2[e]
2006	2003–2004	Fombonne et al.	Canada	HI	Montreal	27,749	5–17	180	*DSM-IV*	–	4.8 (149:31)	64.9	55.8; 75.0
2007	1988–2004	Latif and Williams	UK	HI	South Wales	39,220	0–17	240	*DSM-IV, ICD-10, Kanner's and Gillberg's criteria*	–	6.8	61.2	53.9; 69.4[e]
2007	2000	Oliveira et al.	Portugal	HI	Mainland Azores Islands	59,478 8317	6–9	107 13	*DSM-IV*	17	2.9	9.2 15.6	8.1; 10.0 8.0; 23.0
2007	2002	Ellefsen et al.	Denmark	HI	Faroe islands	7689	8–17	41	*ICD-10*	70.7[e]	5.83[e] (35:6)	53	36;70

2007a	2000	[CDC] ADDM 2000	US	HI	6 states	187,761	8	1252	*DSM-IV-TR*	38–60	2.8–5.5	67.0	63.1; 70.5[e]
2007b	2002	[CDC] ADDM 2002	US	HI	14 states	407,578	8	2685	*DSM-IV-TR*	55.4	3.4–6.5	66.0	63; 68
2007	2005	Croen et al.	US	HI	Northern California	132,844	5–10	593	*ICD-9-CM*	–	5.5 (501:92)	45	41.2; 48.4[e]
2008	1986–2005	Wong and Hui	China	UMI	Hong Kong Registry	4,247,206	0–14	682	*DSM-IV*	30	6.6 (592:90)	16.1	14.9; 17.3[e]
2008	2002	Kawamura et al.	Japan	HI	Toyota	12,589	5–8	228	*DSM-IV*	66.4	2.8 (168:60)	181.1	159.2; 205.9[e]
2008	2003	Williams et al.	UK	HI	Avon	14,062	11	86	*ICD-10*	85.3	6.8 (75:11)	61.9	48.8; 74.9
2008	2006	Montiel-Nava and Pena	Venezuela	LMI	Maracaibo	254,905	3–9	430	*DSM-IV-TR*	–	3.3 (329:101)	17	13; 20
2008	2006	Parner et al.	Denmark	HI	Nationwide	135,413	11	889	*ICD-10*	–	–	74.0	68.3; 80.2
2009	1997–2003	van Balkom et al.	Netherlands	HI	Aruba	13,109	0–13	69	*DSM-IV*	58.8	6.7 (60:9)	52.6	41.0; 66.6
2009	2003–2004	Baron-Cohen et al.	UK	HI	Cambridgeshire	8824	5–9	83	*ICD-10*	–	–	94	75; 116
2009	2006	[CDC] ADDM 2006	US	HI	11 states	308,038	8	2757	*DSM-IV*	59	4.5	90	86; 93
2009	2006	Nicholas et al.	US	HI	South Carolina	8156	4	65	*DSM-IV-TR*	44.2	4.7	80	61; 99
2010	–	Posserud et al.	Norway	HI	Bergen	9430	7–9	16	*DSM-IV, ICD-10*	–	7 (14:2)	87	–

(continued)

Table 13.1 (continued)

Pub Year	Study Year	Authors	Country	Income level[a]	Area	Population	Age	Number Affected	Diagnostic Criteria	% with Normal IQ	Gender Ratio (M:F)	Prevalence Rate/10,000	95% CI
2010	2002–2008	Maenner and Durkin	US	HI	Wisconsin	428,030	Elementary school–aged	3831	*DSM-IV like criteria for WI special education services (by school district)*	–	–	90	86.7; 92.4[e]
2010	2005–2008	Fernell and Gillberg	Sweden	HI	Stockholm	24,084	6	147	*DSM-IV, DSM-IV-TR, ICD-10*	33	5.1(123:24)	62	52; 72
2010	2008	Lazoff et al.	Canada	HI	Montreal	23,635	5–17	187	*DSM-IV*	–	5.4 (158:29)	79.1	67.8; 90.4
2010	2009	Barnevik-Olsson et al.	Sweden	HI	Stockholm	113,391	6–10	250	*DSM-IV*	0	–	22	19.4; 25.0[e]
2011	–	Windham et al.	US	HI	San Francisco Bay Area	80,249	9	374	*DSM-IV*	–	6.2 (324:50)	47	42; 52
2011	1994–2004	Parner et al.	Australia	HI	Western Australia	152,060	0–10	678	*DSM-IV, DSM-IV-TR*	–	4.1	51	47; 55.3
2011	2000–2002	Mattila et al.	Finland	HI	Northern Ostrobothnia County	5484	8	37	*DSM-IV-TR*	65	1.8	84	61; 115
2011	2005–2009	Kim et al.	S. Korea	HI	Goyang City	55,266	7–12	201	*DSM-IV*	31.5	3.8	264	191; 337
2011	2005	Chien et al.	Taiwan	HI	National Health Research Institute	229,457[e]	0–18	659	*ICD-9*	–	3.7	28.7	26.6; 31[e]
2011	2007	Brugha et al.	UK	HI	England	7333	16–98	72	*ICD-10*	100	3.8	98.2	30; 165
2011	2009	Al-Farsi et al.	Oman	HI	National Register	528,335	0–14	113	*DSM-IV-TR*	–	2.9 (84:29)	1.4	1.2; 1.7

2012	–	Isaksen et al.	Norway	HI	Oppland and Hedmark	31,015	6–12	158	*ICD-10*	–	4.27 (128:30)	51	43; 59
2012	–	Parner et al.	Denmark	HI	National Register	1,311,736	6–29	9556	*ICD-8, ICD-9, ICD-10*	–	4.1	72.9[e]	71.4; 74.3[e]
2012	2005–2009	Zeglam & Maouna	Libya	UMI	Tripoli	38,508	0–16	128	*DSM-IV*	31	4.4[e]	33.2	28; 39[e]
2012	2006–2009	Samadi et al.	Iran		National Register	1,320,334	5	826	*ICD*	–	4.3	6.4	5.84; 6.70
2012	2007	Idring et al.	Sweden	HI	Stockholm County Register	444,154	0–17	5100	*DSM-IV, ICD-09, ICD-10*	57.4	2.6	115	112; 118
2012	2008	[CDC] ADDM 2008	US	HI	14 states	337,093	8	3820	*DSM-IV*	38	4.6	113	110; 117
2012	2009	Kočovská et al.	Denmark	HI	Faroe Islands	7128	15–24	67	*DSM-IV, ICD-10, Gillberg's criteria*	–	2.7[e] (49:18)	94	73; 119
2012	2010	Lai et al.	Taiwan	HI	Nationwide	4,044,433	3–17	8072	*DSM-IV-TR, DSM-IV, DSM-III-R*	–	6.6 (7010:1062)	20	198;201[e]
2012	2010	Nygren et al.	Sweden	HI	Göteborg	5007	2	40	*DSM-IV-TR*	63[e]	4 (32:8)	80	57; 109
2013	2005–2009	Saemundsen et al.	Iceland	HI	National Database	22,229	6	267	*ICD-10*	54.7	2.8 (197:70)	120.1	106.6; 135.3
2013	2010	Davidovitch et al.	Israel	HI	Maccabi HMO Registry	423,524	1–12	2034	*DSM-IV*	–	5.2	48	45.9; 50.1
2013	2000–2010	Al-Ansari & Ahmed	Bahrain	HI	Case register whole country	1,250,000	2–27	100	*DSM-IV*	–	4.0 (80:20)	4.3	–

(continued)

Table 13.1 (continued)

Pub Year	Study Year	Authors	Country	Income level[a]	Area	Population	Age	Number Affected	Diagnostic Criteria	% with Normal IQ	Gender Ratio (M:F)	Prevalence Rate/10,000	95% CI
2013	2010	Taylor et al.	UK	HI	National Database	256,278	8	616	*DSM-IV*	–	–	24.0[e]	22.2; 26.0
2014	2009–2011	Huang et al.	China	UMI	Tianjin	8000	1.5–3	22	*DSM-IV*	–	4.5 (18:4)	27.5	16.0; 39.0
2014	2010	[CDC] ADDM 2010	US	HI	11 states	363,749	8	5338	*DSM-IV*	69	4.5	147	142.9; 150.7
2014	2010	Ouellette-Kuntz et al.	Canada	HI	Prince Edward Island & Southeastern Ontario	89,786	2–14	1173	*Clinical expertise*	–	4.8[e] (896:186)	130.6[e]	123.4; 138.3[e]
2015	–	Dekkers et al.	Ecuador	UMI	Quito	51,453	5–15	57	*DSM-III, DSM-IV*	–	4.7 (47:10)	11	8.6; 14.3[e]
2015	2009–2011 2009–2011 2009–2011 2009–2011	Atladottir et al.	Denmark Sweden Finland Australia	HI HI HI HI	Nationwide Nationwide Nationwide Western	≈ 62,000 ≈ 107,000 ≈ 60,000 ≈ 24,000	10 10 10 10	– – – –	*ICD-8, ICD-9, ICD-10* *ICD-9, ICD-10* *ICD-9, ICD-10* *ICD-9, ICD-10. DSM-IV*	– – – –	– – – –	110.3 68.1 59.8 63.6	105.7;115.0 64.6; 71.7 56.1; 63.5 58.2; 69.6
2015	–	van Bakel et al.	France	HI	Haute-Garonne, Isere, Savoy, and Upper-Savoy	307,751	7	1123	*ICD-10*	53.1	4.1 (880:213)	36.5	34.4; 38.7
2015	2011	Idring et al.	Sweden	HI	Stockholm County	735,096	0–27	11,330	*ICD-9, ICD-10, DSM-IV*	74.4	2.3	154	151; 157

2015	2014	Yang et al.	China	UMI	Longhua District, Shenzhen	15,200	3–4	398	*Clinical expertise*	–	2.1 (268:130)	260	230; 290
2016	–	Boilson et al.	Ireland	HI	Galway, Waterford, Cork	5457	6–11	63	*DSM-IV-R*	–	–	100	–
2016	2007–2008	Brugha et al.	UK	HI	England	7461	18+	107	*DSM-IV-TR, ICD-10*	–	1.43 (63:44)	110	30; 190
2016	2011–2012	Fombonne et al.	Mexico	UMI	Leon, Guanajuato	12,116	8	36	*DSM-IV-TR*	33.3	4.1 (29:7)	87	62; 110
2016	2011–2012	Poovathinal et al.	India	LMI	Shoranur, Kerala	18,480	1–30	43	*DSM-IV-TR*	–	2.1	23.3	17.3; 31.3[e]
2016	2012	[CDC] Christensen et al.	US	HI	11 states	346,978	8	5063	*DSM-IV-TR, ICD-9*	68.4	4.5 (4.2:4.8)	146	142; 150
2017	–	Raina et al.	India	LMI	Himachal Pradesh	28,070	1–10	43	*Indian Scale for Assessment of Autism*	–	(23:20)	15.3	0.15; 0.25
2017	2010–2013	Rudra et al.	India	LMI	Kolkata	11,849	3–8	6	*DSM, ICD*	–	–	23	7; 46
2017	2010–2014	Skonieczna-Zydecka et al.	Poland	HI	West Pomeranian and Pomeranian	708,029	0–16	2514	*ICD-10*	–	4.3 (2038:476)	35	34.1; 36.9[e]
2018	–	Morales-Hidalgo et al.	Spain	HI	Tarragona	5582	3–5 10–12	21 13	*DSM-5*	81.0 85.7	3.8 3.3	155 100	89; 220 48; 151
2018	2011–2012	Arora et al.	India	LMI	Palwal, Kangra, Dhenkanal, North Goa, Hyderabad	3964[f]	2–9	44	*DSM-IV-TR*	–	–	111[e]	82.8; 148.7[e]

(continued)

Table 13.1 (continued)

Pub Year	Study Year	Authors	Country	Income level[a]	Area	Population	Age	Number Affected	Diagnostic Criteria	% with Normal IQ	Gender Ratio (M:F)	Prevalence Rate/10,000	95% CI
2018	2012	Bachmann et al.	Germany	HI	Nationwide	6,400,000	0–24 6–11	21,186 –	*Claim in national Health Insurance database*	– –	2.8	38 60	–
2018	2002–2003	Golding et al.	UK	HI	Avon	13,988	11	177	*ICD-10*	–	3.66 (139:38)	127	109;146[e]
2018	2012-2016	Øzerk	Norway	HI	Oslo	108,238	1–16	337	*ICD-10*	–	4.03	31.14	28.0; 34.6[e]
2018	2014	[CDC] Baio et al.	US	HI	11 states	325,483	8	5473	*DSM-IV-TR, DSM-5*	69.0	4	168	164; 173
2018	2014–2015	Diallo et al.	Canada	HI	Quebec	16,940	1–17	207	*ICD-9/ ICD-10 Provincial insurance database*		–	122	106; 139[e]
2018	2014–2015	Jin et al.	China	UMI	Shanghai	74,252	3–12	203	*DSM-5*	7.5	3.6	8.3	6.2; 10.4
2018	2015	Ofner et al.	Canada	HI	6 Provinces	1,911,816	5–17	29,061	*Diagnosis derived from ICD/ DSM by licensed health care professional*	–	4:1	152	150; 154[e]
2018	2016	Akhter et al.	Bangladesh	LMI	Raiganj upazila, Sirajganj district	5286	1.5–3	4	*DSM-IV, MCHAT, ADOS*	–	3 (3:1)	7.6	2.1; 19.4[e]

2018	2016–2017	Narzisi et al.	Italy	HI	Pisa, Tuscany	10,138	7–9	81 46	*DSM-5*	34 –	5.23 (68:13) –	80 1.15	62; 97 83; 146
2019	–	Oner et al.	Turkey	UMI	Istanbul	6712	16–36 months	57	*DSM-5*	–	–	80	6.3; 105
2019	2008	Jariwala-Parikh et al.	US	HI	39 States	8,129,270[e]	18–65	29,745	*Client of State Medicaid database*	–	–	36.6	–
2019	2010	[CDC] Christensen et al.	US	HI	5 states	58,467	4	783	*DSM-IV-TR*	53	2.6–4.4	134	125; 144
2019	2011–2018	Al-Mamari et al.	Oman	HI	Nationwide	837,655	0–14	1705	*DSM 5*	–	3.4 (1332:373)	20.35	19.4; 21.3
2019	2012	[CDC] Christensen et al.	US	HI	5 states	59,456	4	907	*DSM-IV-TR*	56.4	3.4–4.7	153	143; 163
2019	2013–2014	Sun et al.	China	UMI	Jilin City	6240	6–10	77	*DSM-IV-TR, DSM-5*	–	–	108	87.0; 135.0
2019	2014	[CDC] Christensen et al.	US	HI	6 states	70,887	4	1208	*DSM-IV-TR, DSM-5*	53.9	3.0–5.2	170	161; 180
2019	2015	Hamad et al.	Canada	HI	Manitoba	59,438	1–5	578	*ICD-9 / ICD-10*	–	≈3.1[e]	97.2	90;105[e]
2019	2015–2018	Alshaban et al.	Qatar	HI	Whole country	133,781	6–11	1099	*DSM-5*	–	4.3	114	89; 146
2019	2016	Segev et al.	Israel	HI	Whole country	417,507	3–15	13,841	*DSM-IV, National Insurance Institute data*	–	5.0 (11,527;2314)	76[b,c]	–
2019	2016–2018	Mohammadi et al.	Iran	UMI	Nationwide	29,737	6–18	37	*DSM-IV-TR K-SADS-PL*	29.7	1.5 (22:15)	18	12; 24

(continued)

Table 13.1 (continued)

Pub Year	Study Year	Authors	Country	Income level[a]	Area	Population	Age	Number Affected	Diagnostic Criteria	% with Normal IQ	Gender Ratio (M:F)	Prevalence Rate/10,000	95% CI
2019	2017	Hoang et al.	Vietnam	LMI	Hanoi, Thai Binh, Hoa Binh	17,277	1.5–2.5	130	*DSM-IV*	–	4.7 (107:23)	75.2	62.9; 89.3
2019	2017	Pérez-Crespo et al.	Spain	HI	Catalonia	1,326,666	2–17	15,466	*ICD-9*	–	4.5 (12,647: 2819)	123	121; 125
2019	2019	Valenti et al.	Italy	HI	Register in Abruzzo region	36,938	3–18 15–17	352 89	*DSM-IV/ DSM 5 in regional referral center and disability evaluation*	–	4.1 (285:67)	95 1.19	86; 106 94; 143
2020	2014–2016	Zhou et al.	China	UMI	8 provinces	125,806	6–12	363	*DSM 5*	54.6	4.1 (292:71)	70	64; 74
2020	2015	Delobel-Ayoub et al.	Denmark	HI	Nationwide	195,293	7–9	2414	*ICD-10*	89	3.9 (1928:486)	124	119; 128
2020	2015	Delobel-Ayoub et al.	Finland	HI	Nationwide	177,193	7–9	1347	*ICD-10*	82.7	3.3 (1071:276)	76	72; 80
2020	2015	Delobel-Ayoub et al.	France	HI	Haute-Garonne county (south-west France)	15,836	7–9	115	*ICD-10*	61.1	5.4 (97:18)	73	60; 87
2020	2015	Delobel-Ayoub et al.	France	HI	Isère, Savoie and Haute-Savoie counties (south-east France)	32,342	7–9	154	*ICD-10*	76	4 (123:31)	48	40; 56
2020	2015	Delobel-Ayoub et al.	Iceland	HI	Nationwide	13,551	7–9	363	*ICD-10*	79.1	4.4(296:67)	268	241; 296
2020	2015	Rah	South	HI	Nationwide	51,529,338	All ages	5653	*ICD-10*	–	5.4	11.0	–

2020	2016	[CDC] Maenner et al.	USA	HI	11 sites	275,419	8	5108	*DSM-IV-TR, DSM-5 CDC surveillance definition*	65.8	4.3 [3.4–4.7]	185	180; 191
2020	2016	[CDC] Shaw et al.	USA	HI	6 sites	72,277	4	1125	*DSM-IV-TR, DSM-5 CDC surveillance definition*	47.4	3.5 [3.1–4.1]	156	147; 165
2020	2016	Özerk & Cardinal	Norway	HI	Nationwide	628,122	6–16	2606	*ICD-10 National Patient Register*	–	3.62 (2042:564)	41.5	39.9; 43.1[e]
2020	2016–2018	Fuentes et al.	Spain	HI	Basque country	9117	7–9	65	*DSM 5*	58.5	6.2 (56:9)	59	48; 73
2020	2019–2020	Thomaidis et al.	Greece	HI	Nationwide	182,879	10–11	2108	*Special Education Needs Autism*	–	4.36 (1715:393)	115	110; 120[e]
2020	2015–2016	Bowden et al.	New-Zealand	HI	Nationwide	317,418	5–9	3138	*3 health national databases*	70.1	4.59 (2577:561)	98.8	–

[a]According to World Bank 2020 updated classification: *LI* Low Income (28 countries), *LMI* Low Middle Income (50 countries), *UMI* Upper Middle Income (56 countries), *HI* High Income (83 countries)
[b]Estimate is cumulative incidence best reflecting that country population
[c]Data for age 8
[d][CDC] indicates a study performed by the Center for Diseases Control and Prevention (CDC) in the US as part of its Autism and Developmental Disabilities Monitoring (ADDM) network. CDC ADDM publications can be found at: https://www.cdc.gov/ncbddd/autism/articles.html
[e]Value calculated by the authors
[f]Exception made for low population because data was pulled from wide sample distribution

Otherwise Specified [PDD-NOS]; Asperger Disorder) were introduced alongside Autistic Disorder within a broader class of autism spectrum disorders denominated "pervasive developmental disorders" (PDD). The impact on survey results of using different case definitions and diagnostic criteria was well illustrated in a Finnish study where investigators evaluated how applying different diagnostic schemes to the same survey data influenced prevalence estimation. Thus, prevalence among 15–18 year olds was increased three-fold from 2.3/1000 with Kanner's criteria to 7.6/1000 with the ICD-10 broader PDD/ASD definition, everything being equal otherwise (Kielinen et al., 2000). While there was generally high interrater reliability regarding diagnosis of ASDs and high agreement between ICD-10 and DSM-IV umbrella PDD diagnosis, differences existed in their operationalization of criteria of subtypes of PDDs. For example, DSM-IV had a broad category of PDD-NOS, sometimes referred to loosely as "atypical autism" whereas ICD-10 had several corresponding diagnoses including Atypical Autism (F84.1), Other PDD (F84.8), and PDD Unspecified (F84.9). In epidemiological surveys, subtyping of PDDs proved highly inconsistent across studies. For example, the overall prevalence of ASD was similar in three independent surveys performed around the same time ranging from 0.529% to 0.674% (Baird et al., 2000; Chakrabarti & Fombonne, 2001; Bertrand et al., 2001). However, the partition of the autism spectrum led to widely inconsistent specific prevalence estimates for autistic disorder (.308%, .168%, .405%, respectively) and for other PDD subtypes (.271%, .361%, .270%, respectively). The recent DSM-5 (APA, 2013) recalibration of diagnostic criteria for ASD should improve future comparisons of findings across surveys although it may also impinge on the evaluation of time trends. A more detailed discussion of the impact of changing nosographies on surveys can be found below.

Second, case ascertainment approaches vary across studies and also over time (for a detailed discussion, see Meyers et al., 2019; MacFarlane et al., 2022). Some surveys (such as those conducted with administrative databases or registries) rely on children already diagnosed with ASD and the proportion of cases missed because they are not yet diagnosed or diagnosed with non ASD diagnoses cannot generally be estimated. The resulting loss in sensitivity varies by age, area and period of the survey. Other surveys have employed more proactive ascertainment approaches. These include screening populations for children who have diagnoses of other neurodevelopmental and psychiatric disorders (especially, language and other specific developmental disorders, global developmental delay or intellectual disability, attention deficit hyperactivity disorders) and re-evaluating them for possible misdiagnosis or for the co-occurrence of ASD, and more recently screening mainstream schools to identify children newly diagnosed (see below). Proactive survey techniques identify more cases than studies relying on passive approaches to case finding. However, they pose particular challenges for survey data analysis and interpretation of the results due to the heterogeneity of screening techniques employed, the uneven participation rates of informants (caregivers, teachers, health professionals) and participants to the various survey stages, and the variability in methods used to confirm case status. Adding to the survey variability in case definition and case finding approaches, increasing awareness over time in both the professional and lay public has resulted in improved ascertainment and increasing prevalence in more recent epidemiological surveys which complicates further the interpretation of time trends in prevalence of ASD. Thus, it is necessary to keep in mind the methodological heterogeneity of surveys listed in Table 13.1 when summarizing the results of this body of data.

The 141 surveys were conducted in 37 countries and half of them published since 2012. There was no study from any of the 28 countries classified as low-income by the World Bank (2020); seven studies originated from four of the 50 lower middle income (LMI) countries, 14 surveys from seven of the 56 upper middle income (UMI) countries, and the remaining 120 surveys from 26 of the 83 high-income (HI) countries (see Table 13.1). Some countries (e.g. Russia) have no prevalence data yet available; Africa and South America have very few published studies, most of which failed to meet our inclusion criterion due to their small sample size. The median size of populations surveyed was 62,000 (interquartile range (IQR): 15,350–284,536); there was a negative correlation between population size and prevalence (Spearman rank rho = −.213; $p = .011$) consistent with case ascertainment being less sensitive in large national databases and registries. The median age of participants was 8.0 years (IQR:6.5–9.6); only two surveys focused specifically on adults (Brugha et al., 2016; Jariwala-Parikh et al., 2019). The median number of subjects with ASD identified in studies was 177 (IQR:58-1043), the median proportion without ID was 53.4% (IQR:31.5%–65.8%), and the median male:female ratio was 4.1:1 (117 surveys; IQR:3.1–4.9). Over time, there was a marked increase in prevalence over time as indicated by a significant correlation between prevalence and year of publication (Spearman rank rho = .58; $p < .001$; Fig. 13.1a) as well as of the proportion of participants without ID as indicated by a significant correlation between proportion without ID and year of publication (67 studies; Spearman rank rho = .49; $p < .001$; Fig. 13.1b). Interested readers can also consult an interactive global autism prevalence map publicly available at: https://prevalence.spectrumnews.org/.

In order to generate summary information for current ASD prevalence, we first restricted our analysis to the 47 studies published in the last 5 years (2016–2020) given the marked time trends in prevalence (Fig. 13.1a). A comparison of prevalence in 35 high-income countries to the remaining 12 middle- and low-income countries showed that prevalence was significantly lower in the latter group (Mann-Whitney U test = 87.5; $p = .003$) that was subsequently excluded. We further restricted the analysis to surveys conducted in samples aged 4–12, excluding one survey of preschoolers (Hamad et al., 2019) and three surveys of mostly adult populations (Brugha et al., 2016; Jariwala-Parikh et al., 2019; Hong et al., 2020). The final set of 31 surveys were performed in high-income countries on a total population of 14.7 million children with a median population size of 108,000 individuals (IQR: 15,800–347,000) per survey. The median age was 8 years (range: 4–12). The median number of participants with ASD was 1,125 (IQR: 177–3138), the median male:female ratio was 4.07:1 (IQR: 3.65–4.50), and the median proportion of individuals with ID was 31.6% (IQR: 20%–44.8%). Prevalence estimates ranged from 0.2% to 2.68% (IQR: .73%–1.52%) with a median prevalence estimate of 1.14%. This figure of 1.14% can be considered a robust conservative estimate for current ASD prevalence among children ages 4–12 living in high-income countries.

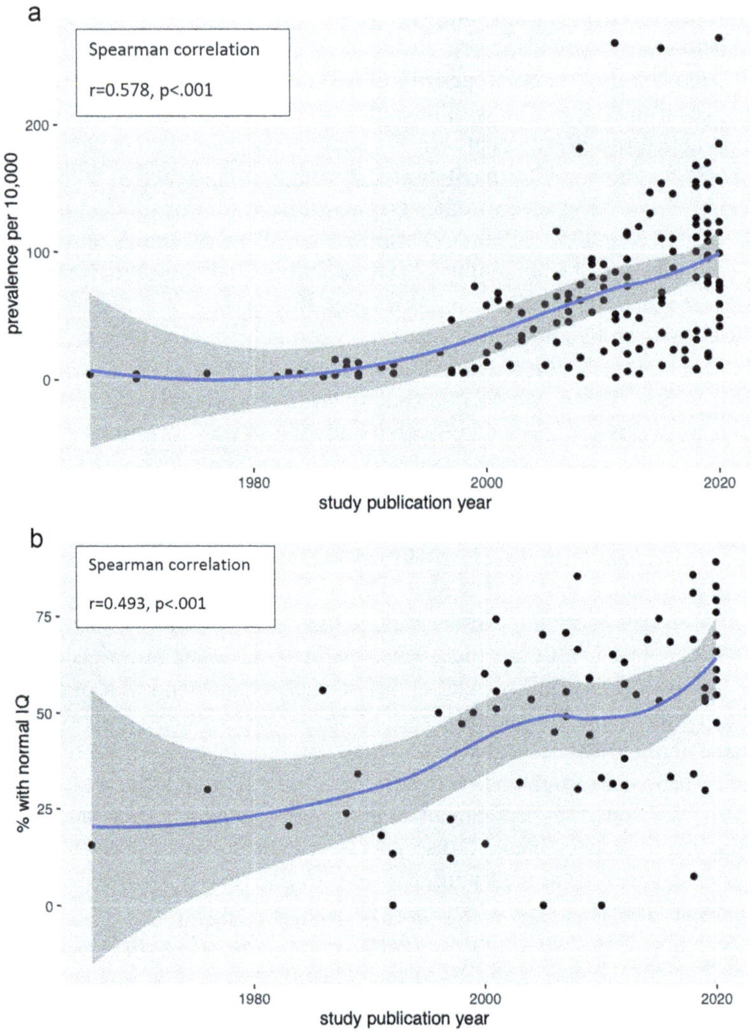

Fig. 13.1 (**a**) Prevalence of ASD over time (141 studies). (**b**) Proportion of subjects with IQ in the normal range, over time (67 studies)

13.3 Special Issues

13.3.1 Case Definition and Case Status Determination

There is no uniform approach to case definition across published studies. Some surveys simply use diagnoses from electronic medical records, some rely on an autism special education eligibility that varies across countries, areas and over time, some rely on endorsement by caregivers of a single questionnaire item while other perform in-person clinical assessments. Many, if not most, surveys use combinations of modalities. Reliance on a particular mode of defining caseness has often predictable consequences on prevalence estimation. Thus, surveys of large national registries or administrative databases usually result in downward bias in prevalence estimation since only cases already identified and diagnosed are counted. Conversely, surveys that rely on parent report in a household survey often overestimate prevalence (see below). However, in most studies, investigators have attempted to confirm directly an ASD diagnosis in participants (or a subsample of participants) by reviewing the symptomatology and developmental history and referring it to established diagnostic criteria, such as the International Classification of Diseases (ICD) or the Diagnostic and Statistical Manual (DSM).

Here, several issues need to be considered. First, the terminology of 'meeting diagnostic criteria' does not guarantee the validity of caseness unless careful attention is paid to the quality of data used to score these criteria and to how much clinical wisdom was infused into this process. The DSM/ICD algorithms for PDD or ASD are only guiding principles which can help organize the available information and provide final coherence to clinical data stemming from different data sources and informants. Yet, how data are collected, by whom, from which informants and using which methods, and how discrepancies between data sources are resolved, are essential features to consider in gauging the validity of case confirmation in a given survey.

Second, even when gold standard tools such as the Autism Diagnostic Interview-Revised (ADI-R; Rutter et al., 2003) and the Autism Diagnostic Observation Schedule (ADOS; Lord et al., 2002) are employed for in-person assessments, case status confirmation based on 'scoring above/below threshold' results are far from being sufficient. In reputable investigations (Lord et al., 2012; Lainhart et al., 2006), scoring rules and cut-offs had to be modified to maintain adequate sensitivity of both instruments in selecting participants. It is worth noting that in both studies expert clinical judgement was employed to provide final confirmation of diagnosis and inclusion in the study. Validity of case status determination does not reside solely in any instrument or its scores; rather, it requires a higher-order, interpretative, process informed by expert clinical judgement. It is important to remember that even instruments like the ADI-R and the ADOS have been developed to be used *in conjunction* and that their results must be reviewed and *interpreted* by a clinical expert (Risi et al., 2006; Lord et al., 2012; Fombonne, 2023). Mechanical translation of scores into diagnosis is unwise. Similarly, it ought to be remembered that diagnostic algorithms of the ICD and DSM have been validated against a gold standard that was precisely the clinical judgment of experts (see for example Volkmar et al., 1994). The importance of expert clinical judgement in making final decisions about caseness is generally acknowledged in epidemiological investigations even when they do not rely on in-person assessments with the ADOS and the ADI-R such as the Centers for Disease Control and Prevention (CDC) Autism and Developmental Disabilities Monitoring (ADDM) surveys.

Third, diagnostic algorithms and ADI-R/ADOS cut-offs have been calibrated against control samples that have typically included participants with either typical development or intellectual disability and developmental delays without autism. The performance of these tools may be diminished when applied to samples enriched with varied types of psychopathology (Bastiaansen et al., 2011; Matsuo et al., 2015; Grzadzinski et al., 2016; Havdahl et al., 2016; Turban & van Schalkwyk, 2018) or with other types of neurodevelopmental and genetic disorders (e.g. Garg et al., 2013; Morotti et al., 2021). For example, Grzadzinski et al. (2016) reported that 20%–30% of children with ADHD but without ASD scored over the cut-offs of standardized autism diagnostic tools (ADI-R and ADOS); likewise, in a study investigating the impact of both parent-reported and clinician-reported behavioral/emotional problems on ratings of autistic symptoms, Havdahl et al. (2016) showed that the presence of co-occurring problems increased ADOS, ADI-R and Social Responsiveness Scale (SRS; Constantino & Gruber, 2005) scores resulting in decreased specificity of ASD instruments. Moreover, epidemiological samples include school (rather than preschool) age subjects with language and intellectual skills often within the normal range. At that age, psychiatric disorders are common and frequently associated with social-communication symptoms (e.g. lack of friendships, self-centeredness, low empathy) and even restricted and repetitive behaviors (e.g. behavioral rigidity, obsessions), allowing autism symptoms to be easily 'scored' albeit wrongly endorsed. This concern is heightened for older children or adults with language and intellectual skills within the normal range when they are newly diagnosed as part of their participation in an epidemiological study, especially in the absence of a previous neurodevelopmental evaluation and/or of a developmental history suggestive of previous autistic abnormalities (Fombonne, 2023). Differentiating autism in the context of psychiatric comorbidity presents challenges both to the performance of standardized instruments and the clinical judgement.

Fourth, surveys have incorporated in their case status definition ill-defined diagnostic subtypes such as PDD-NOS whether ICD or DSM (until recently) was used. To illustrate, PDD-NOS could be diagnosed based on the presence of two diagnostic criteria (one social, one other) only, and no requirement of evidence of abnormality before age three. Therefore, contamination of cases with phenocopies of all kinds was a strong possibility. There again, false positives are more likely when mechanical rules devoid of clinical judgement are used to establish caseness.

Fifth, screening and diagnostic confirmation should rely on reasonably independent procedures. If record review is used as the main procedure to screen *and* to confirm diagnosis, the risk of circularity is very high as exemplified in one CDC diagnostic validation study (Bakian et al., 2015). For a child with a clinical ASD diagnosis or ASD special education eligibility, the documentation available in his or her medical or educational record will obviously contain descriptions in support of that classification, making it in turn difficult to truly evaluate its validity. CDC ADDM surveys are particularly vulnerable to this problem due to their specific record review methodology (Van Naarden Braun et al., 2007).

The repeated changes in nosographical systems create another source of measurement uncertainty in autism studies in general. There was relatively strong parallelism between ICD-10 and DSM-IV that was unfortunately lost with the recent changes in DSM-5. Nevertheless, the new, single, unified concept of Autism

Spectrum Disorder (ASD) that replaces the previous umbrella diagnostic class of Pervasive Developmental Disorder (PDD) has increased specificity that should benefit epidemiological research. Studies have shown that the shift from DSM-IV to DSM-5 leads to a decrease of 13% to 20% in prevalence within the same study datasets (Kim et al., 2014; Maenner et al., 2014). The decrease in prevalence is largely due to subjects with a DSM-IV PDD-NOS diagnosis no longer meeting ASD criteria in DSM-5 (−37% decrease in Kim et al., 2014). Likewise, in a recent CDC ADDM survey of children age eight (Baio et al., 2018), the prevalence of DSM-5 behavioral criteria for ASD was −18.1% lower compared to that of DSM-IV-TR (for precise calculations, see Fombonne, 2018). As it introduced a new DSM-5 based case definition, the new CDC surveillance definition provided a "grand-father" clause by which subjects with a history of a PDD diagnosis would automatically meet criteria for the new surveillance definition even though DSM-5 behavioral criteria would not necessarily be met. This practical choice was in line with DSM-5 recommendation to provide a new DSM-5 ASD diagnosis to individuals having a "well-established" DSM-IV PDD diagnosis. However, because the old and new surveillance definitions are embedded in each other, the net effect on prevalence due to the change from DSM-IV to DSM-5 cannot be properly evaluated in recent CDC ADDM surveys.

The defunct PDD-NOS diagnostic category was an ill-defined diagnostic category with poor inter-rater agreement. In a previous review (Fombonne, 2003), we noted that the proportion of PDD-NOS diagnosed in epidemiological surveys was highly variable, accounting anywhere between 20% to 70% of the spectrum diagnoses reached in surveys. In CDC-ADDM surveys, the proportion of PDD-NOS diagnoses has revolved around 40% of the caseload adding further challenges to the interpretation of CDC-ADDM surveys results (Mandell & Lecavallier, 2014; Fombonne, 2018). In an attempt to increase its specificity, the CDC surveillance case definition for PDD-NOS added the requirement of the presence of at least 1 of 19 autism discriminators (see list in Baio et al., 2018). Unfortunately, the effect on prevalence of applying or not that discriminator has not been reported. Finally, like for PDD-NOS, a poor level of reliability of Asperger disorder diagnosis was documented in epidemiological surveys (Fombonne & Tidmarsh, 2003) favoring its removal from DSM-5 as a separate diagnostic entity. Thus, findings from epidemiological studies concurred with those from other studies (e.g. Lord et al., 2012) in showing that reliability for subtypes within the autism spectrum was mediocre whereas it was excellent for differentiating spectrum and non-spectrum diagnoses.

13.3.2 The Problems of Parental Reports

In our reviews, we excluded surveys that relied solely on parental responses collected in various national health surveys due to concerns about the validity of the case definition employed and of the resulting prevalence estimates. Surveys using large nationally representative samples, such as the US National Survey of Children's

Health (NSCH), have yielded prevalence estimates relying on problematic caseness determination. Gains in sample size, participants' age range and representativeness were mitigated by reliance on simple yes/no answers by household informants to one or few survey questions ("Did a doctor or health professional ever tell you that [child's name] had autism, Asperger's disorder, pervasive developmental disorder, or autism spectrum disorder?") to establish caseness (Kogan et al., 2018). Similar unconfirmed parent reports were used in other population surveys in the US and elsewhere (Table 13.2). In these surveys, non-clinically trained interviewers recorded verbatim answers from respondents without further checking, children were not seen, and additional diagnostic evaluation reports were not collected or reviewed. In each of these surveys, the prevalence estimate was much higher than that derived from more rigorous population surveys performed at the same time in the same country. For example, the prevalence of 1.7% reported by Russell et al. (2014) in the UK compares to prevalence figures of 1.16% and .94% reported in the UK by Baird et al. (2006) and Baron-Cohen et al. (2009) at the same time. To illustrate further the limitations of this type of survey and the considerable concerns

Table 13.2 Surveys using parent reports for case definition

Author (year)	Country	Sampling method	Age range	Sample size	Prevalence %	95% CI	N of ASD
Kogan et al. (2018)	USA	National Survey of Children's Health (NSCH)	3–17	78,037	1.10	.94–1.28	913
Russell et al. (2014)	UK	Millenium Cohort Study (MCS)	6–8	13,586	1.7	1.4–2.0	209
Randall et al. (2016)	Australia	Population based Longitudinal Study of Australian Children (LSAC)	B cohort[a]: 6–7 K cohort[a]: 6–7	4239 4161	2.5 1.5	2.0–3.0 1.2–2.0	107 58
Xu et al. (2018)	USA	National Survey of Children's Health (NSCH)	3–17	43,021	2.50[b]	2.21-2.79	1133
Rydzewska et al. (2019)	UK	Scotland census 2011	8	52,325	2.4	–	1243
Zablotsky et al. (2020)	USA	National Health Interview Survey (NHIS)	3–17	33,775	2.50	2.2–2.7	856
May (2020)	Australia	Population based Longitudinal Study of Australian Children (LSAC)	B cohort[a]: 12–13 K cohort[a]: 16–17	3381 3089	4.36 2.60	3.56–5.19 2.07–3.31	145 98

[a]B cohort: Birth cohort (birth year 2003/2004); K cohort: Kinder cohort (birth year 1999/2000)
[b]data are for current ASD (not lifetime)

about what a 'case' really means, a study by Zablotsky et al. (2015) showed that changes in the wording, format and placement of the single autism question in the National Health Interview Survey resulted in a sharp prevalence increase from 1.25% in 2011 to 2.24% in 2014, a difference seen as arising purely from questionnaire design modifications. Much caution should therefore be exerted when interpreting or using these survey results.

13.3.3 Novel Approaches to Case Finding/Ascertainment

Classically, surveys identified cases by zooming in on children already diagnosed with autism or other behavioral or developmental problems. This approach to case ascertainment did not permit researchers to identify cases without a previously recognized condition and resulted in imperfect sensitivity of case ascertainment procedures (due to false negatives). The addition of a regular school survey component in recent surveys (Kim et al., 2011; Baron-Cohen et al., 2009; Fombonne et al., 2016; Alshaban et al., 2019) and in new studies in China (Zhou et al., 2020; Sun et al., 2019) has addressed this concern from a study design perspective (Table 13.3). However, new issues arose with the implementation of this approach. First, screening tools such as the Social Responsiveness Scale (SRS), the Social Communication Questionnaire (SCQ; Berument et al., 1999) and others show only mediocre specificity, especially among children with elevated levels of concurrent anxiety, attention deficit or other psychiatric symptoms (Hus et al., 2013; Grzadzinski et al., 2016; Fombonne et al., 2021a, b); moreover, their cut-offs have not been well calibrated for use in general population studies, and when both teachers and parents are used as informants, no clear rules exist for combining their often-discrepant results. As seen in Table 13.3, multiple screeners have been employed to survey school samples reflecting opportunistic rather than data-derived choices. Second, and most importantly, is the relatively low participation (30–70%) in the initial screening and other survey phases (e.g. participation in a stage-two diagnostic confirmation session). Statistical analysis of these complex survey designs was made adequate by applying a series of weights to account for different sampling fractions and participation rates at each survey phase (e.g. as in Kim et al., 2011; Fombonne et al., 2016; Alshaban et al., 2019). However, in doing so, strong, unchecked, assumptions had to be made as to whether participation was associated (or not) to caseness. In the complete absence of information about non-participants (which is the usual situation), the assumption that non-participants do not differ from participants with respect to the presence/absence of autism is a guess rather than a tested proposition. Parents of children with autism have unusually high participation in surveys (Fombonne, 2003) making it plausible that non-participants have 'less' autism than participants. Differential participation in that direction may result in upward biases in prevalence estimates. Conversely, prevalence could be underestimated if parents of children with ASD were less likely to participate.

Table 13.3 Contribution of school survey to overall prevalence estimates

Reference	Setting	Overall school target	Age	Screening tool & informant	Overall study prevalence /100	School prevalence /100	Ratio school/ overall prevalence
Baron-Cohen (2009)	England Cambridgeshire		5–9	CAST Parent	1.57	–	40.0
Kim et al. (2011)	South Korea Goyang City	36,592	7–12	ASSQ Parent & Teacher	2.64	1.89	71.6
Fombonne et al. (2016)	Mexico Leon	11,684	8	SRS Parent & Teacher	0.87	0.50	57.5
Jin (2018)	China Shanghai	84,075	3–12	SCQ Parent & Teacher	.083	.054	65.1
Narzisi (2020)	Italy Pisa	4306	7–9	Teacher Nomination (TNF) SCQ (Parent)	1.15	0.3	26.1
Morales-Hidalgo (2018)	Spain Tarragona	5555	3–5 10–12	CAST (Parent) EDUTEA (Teacher)	1.55 1.00	– –	38.1[a] 28.6[a]
Sun et al. (2019)	China Jilin	7167	6–10	CAST Parent	1.08	0.146	13.5
Alshaban et al. (2019)	Qatar Whole country	62,011	6–11	SCQ Parent	1.14	0.32	28.1
Fuentes (2020)	Spain Basque country	9177	7–9	Teacher Nomination (TN) SCQ (Parent)	0.59	–	15.4[a]
Zhou et al. (2020)	China 8 cities	125,172	6–12	ASRS Parent & Teacher	0.70	–	41.9

Abbreviations: *P* Parent, *T* Teacher, *ASSQ* Autism Spectrum Screening Questionnaire, *SRS* Social Responsiveness Scale, *SCQ* Social Communication Questionnaire, *CAST* Childhood Asperger Syndrome Test, *EDUTEA* DSM-5 derived teacher screening questionnaire, *TNF* Teacher Nomination Form, *ASRS* Autism Spectrum Rating Scale

[a]Proportion of newly diagnosed among cases

Nevertheless, important findings were obtained by adding a school survey component to the study designs. As can be seen in Table 13.3, the prevalence estimated by the school survey alone was never nil, and ranged from .054% to 1.89% confirming that screening school children allows for the identification of new cases that would otherwise have been missed by previous methodology relying on children already diagnosed with ASD or some other form of disability. Moreover, within each survey, the relative contribution of the school prevalence component to the overall population prevalence ranged from 13% to 72% (median: 33%, 10 surveys; see Table 13.3). It is probable that this wide variation reflects differences in school survey methodology across these studies although it might also reflect true differences across populations in the proportion of diagnosed/undiagnosed children. Unfortunately, there is no way to test these competing interpretations. Current preliminary evidence (Table 13.3) suggests nonetheless that up to a third of cases of autism in a population could be missed in studies that do not survey mainstream schools.

Execution of school surveys has confronted investigators with huge sample size and manpower issues, specifically due to very high numbers of screen positive, and sometimes of screen negative children to be assessed in second phases of diagnostic confirmation. Innovative techniques have been used to tackle this issue. For example, in a multisite Chinese study where 32.9% of school participants screened positive on an autism screening questionnaire, Zhou et al. (2020) implemented a second step screening procedure combining a brief semi-structured direct observation and group interviews in the school setting in order to eliminate a large proportion of false positives on the initial screener and thereby reduce the second phase sample to a manageable size. In the Qatar study, Alshaban et al. (2019) devised a brief semi-structured telephone interview allowing for rapid evaluation of a high number of screen negative children, leading to more valid and precise estimation of the prevalence estimate. The need to combine informants and data sources in efficient ways has led some European investigators (e.g. Narzisi et al., 2020; Fuentes et al., 2020) to develop a nested screening procedure whereby teachers are asked first to nominate children with suspected social communication or restricted/repetitive problems with a six-item Teacher Nomination Form (Hepburn et al., 2008). Parental screening is subsequently obtained only for the small sample of participants who first screened positively on teacher measures, allowing researchers to limit the final number of screened positive participants, hereby defined as screening positively both on teacher and parent informants. The efficiency of this approach is very attractive; however, its accuracy depends heavily on the performance and properties of the initial teacher identification, the sensitivity of which remains unknown. For example, it is possible that teacher nomination could disproportionately miss girls with autism, 'passive' autistic children in Wing's nomenclature or those without behavioral problems. The method also creates another stratification layer within the screening phase that complicates the survey data analysis.

In sum, the addition of general schools to the samples surveyed in autism epidemiology was a logical improvement that has proven to be contributory. However, the methods used to screen and confirm cases in large samples of typically developing children need to be refined and adequately tested for their performance and cost-effectiveness.

13.3.4 Worldwide Studies and Cultural Issues

It is beyond the scope of this chapter to review in detail the issues raised by the world emergence of surveys of autism and by the cross-cultural questions they pose. Two matters are addressed. First, is there evidence today that autism is either very rare or very abundant in some areas in the world? This question is important as geographical variation in incidence might provide important etiological clues either on genetic or environmental causation. The second question briefly touches upon variability across cultures of the expression of the autism phenotype and of its measurement, specifically as it applies to epidemiological surveys. Readers interested in a thorough reflection about cross-cultural issues are referred to the excellent conceptual framework proposed recently by de Leeuw et al. (2020).

The last 20 years have seen a welcome expansion of ASD epidemiological surveys of child populations, worldwide. Of the 197 world countries, prevalence estimates exist for only 37 countries (see above; and Table 13.1). Data are still lacking in many low- and middle-income countries, especially in Africa, South America, Russia, Caucasus and Central Asia. In many countries, lack of awareness and of diagnostic and intervention expertise persist alongside social stigmatization (e.g. Alshaigi et al., 2020; Yu et al., 2020). However, with the development of the internet and social media, and of advocacy organizations, it has become more difficult for governments to ignore the individual, familial and societal problems associated with autism and neurodevelopmental disorders in general. Epidemiological surveys are a natural starting point for developing clinical and research expertise on these conditions; and governments and their decision-making agencies understandably demand local, quantitative data to guide their service planning decisions.

Everywhere it has been investigated, autism has been found. Small case series appeared in the literature in 1972 for Africa and 1982 for China. Following these seminal clinical descriptions, basic surveys followed consisting of simple head counts that underestimated the prevalence as they only included diagnosed cases in areas where diagnostic services were scarce. As services expanded, prevalence increased and when more fully-developed survey methods are deployed, prevalence in the neighborhood of 1% has been reported in countries as diverse as India, Qatar, Mexico or China (Table 13.1), figures that are commensurate to those from high-income western countries. However, the variability in survey methodology from one country to the other makes it impossible to draw inferences about underlying differences, if they exist, in true population prevalence. Therefore, with today's available published data, there is no evidence that there are countries with either very low or very high autism rates, or meaningful between-country variations in prevalence. By the same token, true differences could exist and remain undetected with current methodological limitations.

Turning to the second issue, the similarity of the autism phenotype across cultural groups has been rather striking in our experience of conducting studies in varied cultural settings. Across countries, investigators have relied upon international diagnostic criteria and employed them without difficulty. Diagnostic tools such as the ADI-R and the ADOS have now been translated in multiple languages, and implemented successfully in survey diagnostic confirmation phases (e.g. Kim et al., 2011; Fombonne et al., 2016; Alshaban et al., 2019; Zhou et al., 2020). Investigators in South Korea specifically examined their cultural applicability and concluded that both DSM diagnostic criteria and scores on standardized diagnostic tools performed well in that population (Kim et al., 2016). Thus, it appears that the concept of autism has some universality even though it might be labelled and named differently in some cultures (for example, "Takiwātanga" among Maoris of New Zealand which means "in his/her own time and space"; www.tepou.co.nz and Bowden et al. (2020)).

Even though a common construct of autism is identified, differences across cultures in the expression and measurement of its manifestations may still occur. Indeed, some cultural adaptations of autism tools have been necessary here and there. For example, in China, the birthday party task of the ADOS Module 1 needed to be replaced by an equivalent task since birthday parties are not part of the familial traditions. We previously adapted a version of the Social Communication Questionnaire in Inuktitut to use as a screening tool among Inuit communities of Northern Canada, only to discover that, to mean 'No' or 'Yes', frowning the nose or raising the eyebrows are often substituted to conventional shaking and nodding the head (Fombonne et al., 2006a, b). Comparisons of Indian, English and Japanese children on the Autism Quotient showed that some items perform differently in some cultural groups (Carruthers et al., 2018). The item 'Enjoys social occasions' performed poorly with Indian parents who typically raise their children with strong expectations for social conformity. Likewise, compared to Greek and Italian counterparts, US toddlers endorsed social interaction difficulties at higher frequencies on a toddler autism screener (Matson et al., 2017). By contrast, direct observations of larger samples of Black and White children in the US evaluated with the ADOS did not find consistent differences (Fombonne & Zuckerman, 2022).

Overall, the reported differences across cultural groups are inconsistent and of small magnitude; to date, reports of cultural variation in symptom expression are best viewed as preliminary and require replication in larger samples after proper adjustment on background factors such as age, gender, language and cognitive level as well as on method of data collection. While a single item's performance might change according to cultural context, it appears that tools, in their totality, maintain measurement properties comparable to those established in Western countries where they were developed. For example, when screening tools were calibrated in local samples, the scores distributions and performances of the Social Communication Questionnaire (SCQ) or the Social Responsiveness Scale (SRS) were comparable in Qatar and Saudi Arabia (SCQ) and Mexico (SRS) to original UK and US studies (Fombonne et al., 2012; Aldosari et al., 2019). Finally, the magnitude of cultural effects on item or criterion endorsement and discriminant power does not appear to be larger than that already reported for gender, age, language or intellectual level within culturally homogeneous samples although formal comparisons of effect sizes remain to be performed.

13.3.5 Surveillance

Some countries have deployed programs aiming at monitoring autism in their population, often alongside surveillance of other developmental disabilities. A well known US surveillance program, the Autism and Developmental Disabilities Monitoring (ADDM; https://www.cdc.gov/ncbddd/autism/addm.html), was launched in 2000 by the CDC. The ADDM network comprises up to 16 sites that have estimated about every two years the prevalence of ASD among eight year-old US children. ADDM methodology relies on a systematic health and education records review that is relatively cost-effective (children are not assessed in person) and allows children without a prior diagnosis to be counted as cases if the behavioral pattern described in records meets criteria for the surveillance case definition (see Van Naarden Braun et al. (2007) and Baio et al. (2018) for details of ADDM methodology). Since 2014, the ADDM has started to track ASD among four-year old children as well. ADDM surveys have been useful in tracking over time prevalence (Fig. 13.2) and associated characteristics of ASD in the US population. Of note, ADDM surveys rely on convenience samples that are not nationally representative; in addition, the geographical repartition of ADDM sites has varied over time complicating the assessment of time trends. Other limitations of the ADDM methodology have been discussed elsewhere and include particular concerns about the

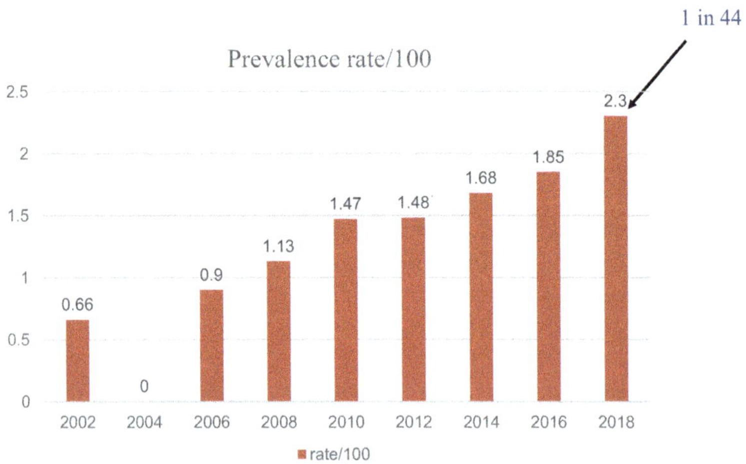

Note: most of the increase occurred in children without intellectual disability

Fig. 13.2 CDC surveys: birth cohorts 1994–2010

validity of the surveillance case definition (Mandell & Lecavallier, 2014; Fombonne, 2018).

Repeated national surveys conducted in the US (National Health Interview Survey (NHIS); National Survey of Children's Health (NSCH)) have also been used to track prevalence in the US population over time. The strengths of national surveys lie in their sampling methodology and representativeness, and their inclusion of wider age ranges, but as discussed above, they are limited by the case definition they employ (see also Table 13.2). The CDC maintains a visualization tool that allows comparisons of these different data sources in the US (www.cdc.gov/ncbddd/autism/data/index.html).

In Canada, a newly formed National Autism Surveillance System (NASS) has released its first results from 2015 concerning almost two million children ages five to 17, using administrative data from seven Provinces and Territories (Ofner et al., 2018; Table 13.1). The NASS case definition relies on ICD- or DSM-derived ASD diagnoses provided or confirmed by licensed health care professionals. The prevalence was 1.52% with a male to female ratio of 4:1 at all ages; 56% of the 29,099 cases had been diagnosed by age six, 72% by age eight, and 92% by age 12.

In Europe, 14 countries of the European Union have engaged into a large multifaceted cooperative program to develop early detection programs, validate biomarkers, train professionals, improve support for adults and propose policies (Autism Spectrum Disorders in the European Union (ASDEU); www.asdeu.eu). The other objective is to investigate the prevalence of autism in 12 countries using a methodology that focuses on school age, identifies diagnosed children as well as undiagnosed ones through school surveys, relies on common instrumentation (SCQ, ADI-R, ADOS), and uses in-person assessments. Additionally, exploration of European regional and national registries is being pursued as a complementary strategy (Delobel-Ayoub et al., 2020). Details on the methodology can be found on the web site of ASDEU and in publications (Boilson et al., 2016; Narzisi et al., 2020; Fuentes et al., 2020). Although it is too early to evaluate the efficacy and success of NASS and ASDEU as surveillance programs for Canada and the EU, the increasing interest for establishing population surveillance of ASD is noticeable in several high-income countries.

13.4 Conclusions

Epidemiological studies of ASD have expanded worldwide with a median estimate of 1.14% providing a conservative figure for ASD population prevalence. Comparisons of results across studies should be made with caution due to heterogeneity pertaining to case definition and ascertainment strategies unique to each survey. Definitions of ASD used in population surveys often do not coincide with those required in rigorous clinical research protocols such as randomized clinical trials or molecular genetic investigations. Survey definitions are influenced by the need to comprehensively capture cases (optimizing sensitivity) and to estimate service needs for developmentally impaired children (which may come at a price for specificity). The addition of a survey component where mainstream schools were surveyed has consistently proven that general school screening is required if a comprehensive picture of ASD is to be provided. However, screening schemes for school samples need to be further researched, available screeners compared, and more cost-effective approaches properly evaluated.

Several countries are now considering the implementation of national registries or surveillance programs that will help track trends in prevalence and incidence of ASD in their populations in the future. Broadening the focus of surveys to the larger realm of neurodevelopmental disorders would increase their public health relevance and would also allow examination of important questions of boundaries and overlap between the autism phenotypes and other developmental disorders and genetic syndromes. Systematic incorporation in survey protocols of standardized measures of behavioral problems and psychiatric disorders should be considered. Likewise, incorporation of biological and genetic in surveys and registries will enhance their usefulness.

References

ADDM. (2000). Autism and developmental disabilities monitoring network surveillance year 2000 principal investigators; Centers for disease control and prevention. Prevalence of autism spectrum disorders—autism and developmental disabilities monitoring network, six sites, United States, 2000. *Morbidity and Mortality Weekly Report Surveillance Summaries 2007, 56*(1), 1–11.

ADDM. (2002). Autism and developmental disabilities monitoring network surveillance year 2002 principal investigators; centers for disease control and prevention. Prevalence of autism spectrum disorders—autism and developmental disabilities monitoring network, 14 sites, United States, 2002. *Morbidity and Mortality Weekly Report. Surveillance Summaries. 2007, 56*(1), 12–28.

ADDM. (2006). Autism and developmental disabilities monitoring network surveillance year 2006 principal investigators; Centers for Disease Control and Prevention (CDC). Prevalence of autism spectrum disorders—Autism and Developmental Disabilities Monitoring Network, United States, 2006. *Morbidity and Mortality Weekly Report Surveillance Summaries 2009, 58*(10), 1–20.

ADDM. (2008). Autism and developmental disabilities monitoring network surveillance year 2008 principal investigators; centers for disease control and prevention. Prevalence of autism spectrum disorders—Autism and Developmental Disabilities Monitoring Network, 14 sites, United States, 2008. *Morbidity and Mortality Weekly Report Surveillance Summaries 2012, 61*(3), 1–19.

ADDM. (2010). Developmental disabilities monitoring network surveillance year 2010. Principal Investigators, & Centers for Disease Control and Prevention (CDC). (2014). Prevalence of autism spectrum disorder among children aged 8 years—autism and developmental disabilities monitoring network, 11 sites, United States, 2010. *Morbidity and Mortality Weekly Report. Surveillance Summaries, 63*(2), 1–21.

Akhter, S., Hussain, A. H. M. E., Shefa, J., Kundu, G. K., Rahman, F., & Biswas, A. (2018). Prevalence of Autism Spectrum Disorder (ASD) among the children aged 18-36 months in a rural community of Bangladesh: A cross sectional study. *F1000 Research, 7*, 424. https://doi.org/10.12688/f1000research.13563.1

Al-Ansari, A. M., & Ahmed, M. M. (2013). Epidemiology of autistic disorder in Bahrain: Prevalence and obstetric and familial characteristics. *Eastern Mediterranean Health Journal, 19*(9).

Aldosari, M., Fombonne, E., Aldhalaan, H., Ouda, M., Elhag, S., Alshammari, H., et al. (2019). Validation of the Arabic version of the social communication questionnaire. *Autism, 23*(7), 1655–1662.

Al-Farsi, Y. M., Al-Sharbati, M. M., Al-Farsi, O. A., Al-Shafaee, M. S., Brooks, D. R., & Waly, M. I. (2011). Brief report: Prevalence of autistic spectrum disorders in the sultanate of Oman. *Journal of Autism and Developmental Disorders, 41*(6), 821–825.

Al-Mamari, W., Idris, A. B., Dakak, S., Al-Shekaili, M., Al-Harthi, Z., Alnaamani, A. M., et al. (2019). Revisiting the prevalence of autism spectrum disorder among omani children: A multicentre study. *Sultan Qaboos University Medical Journal, 19*(4), e305.

Alshaban, F., Aldosari, M., Al-Shammari, H., El-Hag, S., Ghazal, I., Tolefat, M., et al. (2019). Prevalence and correlates of autism spectrum disorder in Qatar: A national study. *Journal of Child Psychology and Psychiatry, 60*(12), 1254–1268.

Alshaigi, K., Albraheem, R., Alsaleem, K., Zakaria, M., Jobeir, A., & Aldhalaan, H. (2020). Stigmatization among parents of autism spectrum disorder children in Riyadh, Saudi Arabia. *International Journal of Pediatrics and Adolescent Medicine, 7*(3), 140–146.

American Psychiatric Association (Ed.). (1994). *Diagnostic and statistical manual of mental disorders* (4th ed.). American Psychiatric Association.

American Psychiatric Association (Ed.). (2000). *Diagnostic and statistical manual of mental disorders: DSM-IV-TR (4th ed., text revision)*. American Psychiatric Association.

American Psychiatric Association (Ed.). (2013). *Diagnostic and statistical manual of mental disorders: DSM-5* (5th ed.). American Psychiatric Association.

Arora, N. K., Nair, M. K. C., Gulati, S., Deshmukh, V., Mohapatra, A., Mishra, D., Patel, V., Pandey, R. M., Das, B. C., Divan, G., Murthy, G. V. S., Sharma, T. D., Sapra, S., Aneja, S., Juneja, M., Reddy, S. K., Suman, P., Mukherjee, S. B., Dasgupta, R., Tudu, P., Das, M. K., Bhutani, V. K., Durkin, M. S., Pinto-Martin, J., Silberberg, D. H., Sagar, R., Ahmed, F., Babu, N., Bavdekar, S., Chandra, V., Chaudhuri, Z., Dada, T., Dass, R., Gourie-Devi, M., Remadevi, S., Gupta, J. C., Handa, K. K., Kalra, V., Karande, S., Konanki, R., Kulkarni, M., Kumar, R., Maria, A., Masoodi, M. A., Mehta, M., Mohanty, S. K., Nair, H., Natarajan, P., Niswade, A. K., Prasad, A., Rai, S. K., Russell, P. S. S., Saxena, R., Sharma, S., Singh, A. K., Singh, G. B., Sumaraj, L., Suresh, S., Thakar, A., Parthasarathy, S., Vyas, B., Panigrahi, A., Saroch, M. K., Shukla, R., Rao, K. V. R., Silveira, M. P., Singh, S., & Vajaratkar, V. (2018). Neurodevelopmental disorders in children aged 2-9 years: Population-based burden estimates across five regions in India. *Public Library of Science Medicine, 15*(7), e1002615.

Atladottir, H. O., Gyllenberg, D., Langridge, A., Sandin, S., Hansen, S. N., Leonard, H., Gissler, M., Reichenberg, A., Schendel, D. E., Bourke, J., Hultman, C. M., Grice, D. E., Buxbaum, J. D., & Parner, E. T. (2015). The increasing prevalence of reported diagnoses of childhood psychiatric disorders: A descriptive multinational comparison. *European Child & Adolescent Psychiatry, 24*(2), 173–183.

Bachmann, C. J., Gerste, B., & Hoffmann, F. (2018). Diagnoses of autism spectrum disorders in Germany: Time trends in administrative prevalence and diagnostic stability. *Autism, 22*(3), 283–290.

Baio, J., Wiggins, L., Christensen, D. L., et al. (2018). Prevalence of autism spectrum disorder among children aged 8 years—autism and developmental disabilities monitoring network, 11 sites, United States, 2014. *Morbidity and Mortality Weekly Report. Surveillance Summaries, 67*(6), 1.

Baird, G., Charman, T., Baron-Cohen, S., Cox, A., Swettenham, J., Wheelwright, S., & Drew, A. (2000). A screening instrument for autism at 18 months of age: a 6-year follow-up study. *Journal of the American Academy of Child and Adolescent Psychiatry, 39*(6), 694–702.

Baird, G., Simonoff, E., Pickles, A., Chandler, S., Loucas, T., Meldrum, D., & Charman, T. (2006). Prevalence of disorders of the autism spectrum in a population cohort of children in South Thames: The Special Needs and Autism Project (SNAP). *Lancet, 368*(9531), 210–215.

Bakian, A. V., Bilder, D. A., Carbone, P. S., Hunt, T. D., Petersen, B., & Rice, C. E. (2015). Brief report: independent validation of Autism Spectrum Disorder case status in the Utah Autism and Developmental Disabilities Monitoring (ADDM) network site. *Journal of Autism and Developmental Disorders, 45*(3), 873–880.

Baron-Cohen, S., Scott, F. J., Allison, C., Williams, J., Bolton, P., Matthews, F. E., & Brayne, C. (2009). Prevalence of autism-spectrum conditions: UK school-based population study. *The British Journal of Psychiatry, 194*(6), 500–509.

Bastiaansen, J. A., Meffert, H., Hein, S., Huizinga, P., Ketelaars, C., Pijnenborg, M., et al. (2011). Diagnosing autism spectrum disorders in adults: The use of Autism Diagnostic Observation Schedule (ADOS) module 4. *Journal of Autism and Developmental Disorders, 41*(9), 1256–1266.

Bertrand, J., Mars, A., Boyle, C., Bove, F., Yeargin-Allsopp, M., & Decoufle, P. (2001). Prevalence of Autism in a United States population: The Brick Township, New Jersey, investigation. *Pediatrics, 108*(5), 1155–1161.

Berument, S. K., Rutter, M., Lord, C., Pickles, A., & Bailey, A. (1999). Autism screening questionnaire: Diagnostic validity. *The British Journal of Psychiatry, 175*, 444–451.

Boilson, A. M., Staines, A., Ramirez, A., Posada, M., & Sweeney, M. R. (2016). Operationalisation of the European Protocol for Autism Prevalence (EPAP) for autism spectrum disorder prevalence measurement in Ireland. *Journal of Autism and Developmental Disorders, 46*(9), 3054–3067.

Bowden, N., Thabrew, H., Kokaua, J., Audas, R., Milne, B., Smiler, K., et al. (2020). Autism spectrum disorder/Takiwātanga: An Integrated Data Infrastructure-based approach to autism spectrum disorder research in New Zealand. *Autism, 24*(8), 2213–2227.

Brugha, T. S., McManus, S., Bankart, J., Scott, F., Purdon, S., Smith, J., Bebbington, P., Jenkins, R., & Meltzer, H. (2011). Epidemiology of autism spectrum disorders in adults in the community in England. *Archives of General Psychiatry, 68*(5), 459–465.

Brugha, T. S., McManus, S., Smith, J., Scott, F. J., Meltzer, H., Purdon, S., et al. (2012). Validating two survey methods for identifying cases of autism spectrum disorder among adults in the community. *Psychological Medicine, 42*(3), 647.

Brugha, T. S., Spiers, N., Bankart, J., Cooper, S. A., McManus, S., Scott, F. J., et al. (2016). Epidemiology of autism in adults across age groups and ability levels. *The British Journal of Psychiatry, 209*(6), 498–503.

Carruthers, S., Kinnaird, E., Rudra, A., Smith, P., Allison, C., Auyeung, B., et al. (2018). A cross-cultural study of autistic traits across India, Japan and the UK. *Molecular Autism, 9*(1), 52.

Chakrabarti, S., & Fombonne, E. (2001). Pervasive developmental disorders in preschool children. *Journal of the American Medical Association, 285*(24), 3093.

Chakrabarti, S., & Fombonne, E. (2005). Pervasive developmental disorders in preschool children: confirmation of high prevalence. *American Journal of Psychiatry, 162*(6), 1133–1141.

Cheon, K. A., Park, J. I., Koh, Y. J., Song, J., Hong, H. J., Kim, Y. K., et al. (2016). The social responsiveness scale in relation to DSM IV and DSM5 ASD in Korean children. *Autism Research, 9*(9), 970–980.

Chiarotti, F., & Venerosi, A. (2020). Epidemiology of autism spectrum disorders: a review of worldwide prevalence estimates Since 2014. *Brain Sciences, 10*(5), 274.

Chien, I.-C., Lin, C.-H., Chou, Y.-J., & Chou, P. (2011). Prevalence and incidence of autism spectrum disorders among national health insurance enrollees in Taiwan from 1996 to 2005. *Journal of Child Neurology, 26*(7), 830–834.

Christensen, D. L., Baio, J., Van Naarden Braun, K., Bilder, D., Charles, J., Constantino, J. N., Daniels, J., Durkin, M. S., Fitzgerald, R. T., Kurzius-Spencer, M., Lee, L. C., Pettygrove, S., Robinson, C., Schulz, E., Wells, C., Wingate, M. S., Zahorodny, W., Yeargin-Allsopp, M., & Centers for Disease Control and Prevention (CDC). (2012). Prevalence and characteristics of autism spectrum disorder among children aged 8 years—autism and developmental disabilities monitoring network, 11 Sites, United States. *Morbidity and Mortality Weekly Report Surveillance Summaries 2016, 65*(3), 1–23.

Christensen, D. L., Maenner, M. J., Bilder, D., Constantino, J. N., Daniels, J., Durkin, M. S., et al. (2019). Prevalence and characteristics of autism spectrum disorder among children aged 4 years—early autism and developmental disabilities monitoring network, seven sites, United States, 2010, 2012, and 2014. *Morbidity and Mortality Weekly Report. Surveillance Summaries, 68*(2), 1–19.

Constantino, J. N., & Gruber, C. P. (2005). *Social Responsiveness Scale (SRS)*. Western Psychological Services.

Croen, L. A., Grether, J. K., Hoogstrate, J., & Selvin, S. (2002). The changing prevalence of autism in California. *Journal of Autism and Developmental Disorders, 32*(3), 207–215.

Croen, L. A., Najjar, D. V., Fireman, B., & Grether, J. K. (2007). Maternal and paternal age and risk of autism spectrum disorders. *Archives of Pediatrics & Adolescent Medicine, 161*(4), 334–340.

Davidovitch, M., Hemo, B., Manning-Courtney, P., & Fombonne, E. (2013). Prevalence and incidence of autism spectrum disorder in an Israeli population. *Journal of Autism and Developmental Disorders, 43*(4), 785–793.

de Leeuw, A., Happé, F., & Hoekstra, R. A. (2020). A conceptual framework for understanding the cultural and contextual factors on autism across the globe. *Autism Research, 13*(7), 1029–1050.

Dekkers, L. M. S., Groot, N. A., Díaz Mosquera, E. N., Andrade Zúñiga, I. P., & Delfos, M. F. (2015a). Prevalence of autism spectrum disorders in ecuador: A pilot study in Quito. *Journal of Autism and Developmental Disorders, 45*(12), 4165–4173.

Dekkers, L. M., Groot, N. A., Mosquera, E. N. D., Zúniga, I. P. A., & Delfos, M. F. (2015b). Prevalence of autism spectrum disorders in Ecuador: A pilot study in Quito. *Journal of Autism and Developmental Disorders, 45*(12), 4165–4173.

Delobel-Ayoub, M., Saemundsen, E., Gissler, M., Ego, A., Moilanen, I., Ebeling, H., Rafnsson, V., Klapouszczak, D., Thorsteinsson, E., Arnaldsdóttir, K. M., Roge, B., Arnaud, C., & Schendel, D. (2020). Prevalence of autism spectrum disorder in 7-9-year-old children in Denmark, Finland, France and Iceland: A population-based registries approach within the ASDEU project. *Journal of Autism and Developmental Disorders, 50*(3), 949–959.

Diallo, F. B., Fombonne, É., Kisely, S., Rochette, L., Vasiliadis, H. M., Vanasse, A., Noiseux, M., Pelletier, É., Renaud, J., St-Laurent, D., & Lesage, A. (2018). Prevalence and correlates of autism spectrum disorders in Quebec: Prévalence et corrélats des troubles du spectre de l'autisme au Québec. *Canadian Journal of Psychiatry, 63*(4), 231–239.

Dietz, P. M., Rose, C. E., McArthur, D., & Maenner, M. (2020). National and state estimates of adults with autism spectrum disorder. *Journal of Autism and Developmental Disorders, 50*, 1–9.

Ellefsen, A., Kampmann, H., Billstedt, E., Gillberg, I. C., & Gillberg, C. (2007). Autism in the Faroe Islands: An epidemiological study. *Journal of Autism and Developmental Disorders, 37*(3), 437–444.

Fernell, E., & Gillberg, C. (2010). Autism spectrum disorder diagnoses in Stockholm preschoolers. *Research in Developmental Disabilities, 31*(3), 680–685.

Fombonne, E. (2003). Epidemiological surveys of autism and other pervasive developmental disorders: An update. *Journal of Autism and Developmental Disorders, 33*(4), 365–381.

Fombonne, E. (2018). The rising prevalence of autism. *Journal of Child Psychology and Psychiatry, 59*(7), 717–720.

Fombonne, É. (2019). Current issues in epidemiological studies of autism. *Psicologia: teoria e prática, 21*(3), 405–417.

Fombonne, E. (2020). Camouflage and autism. *Journal of Child Psychology and Psychiatry, 61*(7), 735–738.

Fombonne, E. (2023). Is autism overdiagnosed? *Journal of Child Psychology and Psychiatry, 64*(5), 711–714. https://doi.org/10.1111/jcpp.13806. PMID: 37055375.

Fombonne, E., & Tidmarsh, L. (2003). Epidemiologic data on Asperger disorder. *Child and Adolescent Psychiatric Clinics, 12*(1), 15–21.

Fombonne, E., & Zuckerman, K. (2022). Clinical profiles of Black and White children referred for autism diagnosis. *Journal of Autism and Developmental Disorders*, *52*(3), 1120–1130.

Fombonne, E., Simmons, H., Ford, T., Meltzer, H., & Goodman, R. (2001). Prevalence of pervasive developmental disorders in the British nationwide survey of child mental health. *Journal of the American Academy of Child and Adolescent Psychiatry, 40*(7), 820–827.

Fombonne, E., Zakarian, R., Bennett, A., Meng, L., & McLean-Heywood, D. (2006a). Pervasive developmental disorders in Montreal, Quebec, Canada: Prevalence and links with immunizations. *Pediatrics, 118*(1), e139–e150.

Fombonne, E., Morel, J., & MacArthur, J. (2006b). *No autism amongst Inuits from Northern Quebec?* International Meeting for Autism Research (IMFAR). June 1–3, 2006.

Fombonne, E., Marcin, C., Bruno, R., Tinoco, C. M., & Marquez, C. D. (2012). Screening for autism in Mexico. *Autism Research, 5*(3), 180–189.

Fombonne, E., Marcin, C., Manero, A. C., Bruno, R., Diaz, C., Villalobos, M., Ramsay, K., & Nealy, B. (2016). Prevalence of autism spectrum disorders in Guanajuato, Mexico: The Leon survey. *Journal of Autism and Developmental Disorders, 46*(5), 1669–1685.

Fombonne, E., Snyder, L. G., Daniels, A., Feliciano, P., & Chung, W. (2020). Psychiatric and medical profiles of autistic adults in the SPARK cohort. *Journal of Autism and Developmental Disorders, 50*, 1–20.

Fombonne, E., Morotti, H., Mastel, S., Keller, K., Barnard, R. A., Hall, T., & O'Roak, B. J. (2021a). Autism questionnaire scores do not only rise because of autism. *Developmental Medicine and Child Neurology, 63*(2), 235–236.

Fombonne, E., MacFarlane, H., & Salem, A. (2021b). Epidemiological surveys of autism spectrum disorder: Advances and remaining challenges. In: Special Issue: Autism in Review: 1980–2020: 40 years after DSM-III. *Journal of Autism and Developmental Disorders*. https://doi.org/10.1007/s10803-021-05005-9. Apr 17. Epub ahead of print. PMID: 33864555.

Fombonne, E., MacFarlane, H., Salem, A. C., Zuckerman, K. E. (2022). Epidemiological surveys of ASD: Current findings and new directions. In: Matson, J. L., Sturmey, P. (eds). *Handbook of Autism and Pervasive Developmental Disorder*. pp.135–184. Autism and Child Psychopathology Series. Springer, Cham. https://doi.org/10.1007/978-3-030-88538-0_6

Fuentes, J., Basurko, A., Isasa, I., Galende, I., Muguerza, M. D., García-Primo, P., García, J., Fernández-Álvarez, C. J., Canal-Bedia, R., & Posada de la Paz, M. (2020). The ASDEU autism prevalence study in northern Spain. *European Child & Adolescent Psychiatry*. 2020 May 9.

Garg, S., Green, J., Leadbitter, K., Emsley, R., Lehtonen, A., Evans, D. G., & Huson, S. M. (2013). Neuro- fibromatosis type 1 and autism spectrum disorder. *Pediatrics, 132*(6), e1642–e1648.

Gillberg, C., Cederlund, M., Lamberg, K., & Zeijlon, L. (2006). Brief report: "the autism epidemic". The registered prevalence of autism in a Swedish urban area. *Journal of Autism and Developmental Disorders, 36*(3), 429–435.

Golding, J., Rai, D., Gregory, S., Ellis, G., Emond, A., Iles-Caven, Y., Hibbeln, J., & Taylor, C. (2018). Prenatal mercury exposure and features of autism: A prospective population study. *Molecular Autism, 9*, 30.

Grzadzinski, R., Dick, C., Lord, C., & Bishop, S. (2016). Parent-reported and clinician-observed autism spectrum disorder (ASD) symptoms in children with attention deficit/hyperactivity disorder (ADHD): Implications for practice under DSM-5. *Molecular Autism, 7*(1), 7.

Gurney, J. G., Fritz, M. S., Ness, K. K., Sievers, P., Newschaffer, C. J., & Shapiro, E. G. (2003). Analysis of prevalence trends of autism spectrum disorder in Minnesota. *Archives of Pediatrics & Adolescent Medicine, 157*(7), 622–627.

Hamad, A. F., Alessi-Severini, S., Mahmud, S. M., Brownell, M., & Kuo, I. F. (2019). Annual trends in prevalence and incidence of autism spectrum disorders in Manitoba preschoolers and toddlers: 2004–2015. *Canadian Journal of Public Health, 110*(4), 476–484.

Hand, B. N., Angell, A. M., Harris, L., & Carpenter, L. A. (2020). Prevalence of physical and mental health conditions in Medicare-enrolled, autistic older adults. *Autism, 24*(3), 755–764.

Harrison, M. J., O'Hare, A. E., Campbell, H., Adamson, A., & McNeillage, J. (2006). Prevalence of autistic spectrum disorders in Lothian, Scotland: an estimate using the "capture-recapture" technique. *Archives of Disease in Childhood, 91*(1), 16–19.

Havdahl, K. A., Hus Bal, V., Huerta, M., Pickles, A., Øyen, A. S., Stoltenberg, C., Lord, C., & Bishop, S. L. (2016). Multidimensional influences on autism symptom measures: implications for use in etiological research. *Journal of the American Academy of Child and Adolescent Psychiatry, 55*(12), 1054–1063.

Hepburn, S. L., DiGuiseppi, C., Rosenberg, S., Kaparich, K., Robinson, C., & Miller, L. (2008). Use of a teacher nomination strategy to screen for autism spectrum disorders in general education classrooms: a pilot study. *Journal of Autism and Developmental Disorders, 38*(2), 373–382.

Hoang, V. M., Le, T. V., Chu, T. T. Q., Le, B. N., Duong, M. D., Thanh, N. M., Pham, V. T., et al. (2019). Prevalence of autism spectrum disorders and their relation to selected socio-demographic factors among children aged 18–30 months in northern Vietnam. *International Journal of Mental Health Systems, 13*(1), 29.

Hong, M., Lee, S. M., Park, S., Yoon, S. J., Kim, Y. E., & Oh, I. H. (2020). Prevalence and economic burden of autism spectrum disorder inSouth Korea using national health insurance data from 2008 to 2015. *Journal of Autism and Developmental Disorders, 50*(1), 333–339. https://doi.org/10.1007/s10803-019-04255-y

Hoekstra, R. A., Girma, F., Tekola, B., & Yenus, Z. (2018). Nothing about us without us: The importance of local collaboration and engagement in the global study of autism. *British Journal of Psychiatry International, 15*(2), 40–43.

Huang, J. P., Cui, S. S., Han, Y., Irva, H. P., Qi, L. H., & Zhang, X. (2014). Prevalence and early signs of autism spectrum disorder (ASD) among 18-36 month-old children of Tianjin in China. *Biomedical and Environmental Sciences, 27*(6), 453–461.

Hus, V., Bishop, S., Gotham, K., Huerta, M., & Lord, C. (2013). Factors influencing scores on the social responsiveness scale. *Journal of Child Psychology and Psychiatry, 54*(2), 216–224.

Icasiano, F., Hewson, P., Machet, P., Cooper, C., & Marshall, A. (2004). Childhood autism spectrum disorder in the Barwon region: A community based study. *Journal of Paediatrics and Child Health, 40*(12), 696–701.

Idring, S., Rai, D., Dal, H., Dalman, C., Sturm, H., Zander, E., Lee, B. K., Serlachius, E., & Magnusson, C. (2012). Autism spectrum disorders in the Stockholm Youth Cohort: Design, prevalence and validity. *Public Library of Science One, 7*(7), e41280.

Idring, S., Lundberg, M., Sturm, H., Dalman, C., Gumpert, C., Rai, D., Lee, B. K., & Magnusson, C. (2015). Changes in prevalence of autism spectrum disorders in 2001–2011: findings from the Stockholm youth cohort. *Journal of Autism and Developmental Disorders, 45*(6), 1766–1773.

Isaksen, J., Diseth, T. H., Schjølberg, S., & Skjeldal, O. H. (2012). Observed prevalence of autism spectrum disorders in two Norwegian counties. *European Journal of Paediatric Neurology, 16*(6), 592–598.

Jariwala-Parikh, K., Barnard, M., Holmes, E. R., West-Strum, D., Bentley, J. P., Banahan, B., & Khanna, R. (2019). Autism prevalence in the medicaid program and healthcare utilization and costs among adult enrollees diagnosed with autism. *Administration and Policy in Mental Health and Mental Health Services Research, 46*(6), 768–776.

Jarquin, V. G., Wiggins, L. D., Schieve, L. A., & Van Naarden-Braun, K. (2011). Racial disparities in community identification of autism spectrum disorders over time; Metropolitan Atlanta, Georgia, 2000–2006. *Journal of Developmental and Behavioral Pediatrics, 32*(3), 179–187.

Jin, Z., Yang, Y., Liu, S., Huang, H., & Jin, X. (2018). Prevalence of DSM-5 autism spectrum disorder among school-based children aged 3–12 years in Shanghai, China. *Journal of Autism and Developmental Disorders, 48*(7), 2434–2443.

Jo, H., Schieve, L. A., Rice, C. E., Yeargin-Allsopp, M., Tian, L. H., Blumberg, S. J., Kogan, M. D., & Boyle, C. A. (2015). Age at autism spectrum disorder (ASD) diagnosis by race, ethnicity, and primary household language among children with special health care needs, United States, 2009–2010. *Maternal and Child Health Journal, 19*(8), 1687–1697.

Kawamura, Y., Takahashi, O., & Ishii, T. (2008). Reevaluating the incidence of pervasive developmental disorders: Impact of elevated rates of detection through implementation of an integrated system of screening in Toyota, Japan. *Psychiatry and Clinical Neurosciences, 62*(2), 152–159.

Kielinen, M., Linna, S. L., & Moilanen, I. (2000). Autism in Northern Finland. *European Child & Adolescent Psychiatry, 9*(3), 162–167.

Kim, Y. S., Leventhal, B. L., Koh, Y. J., Fombonne, E., Laska, E., Lim, E. C., Cheon, K. A., Kim, S. J., Kim, Y. K., Lee, H. K., Song, D. H., & Grinker, R. R. (2011). Prevalence of autism spectrum disorder in a total population sample. *American Journal of Psychiatry, 168*(9), 904–912.

Kim, Y. S., Fombonne, E., Koh, Y.-J., Kim, S.-J., Cheon, K.-A., & Leventhal, B. L. (2014). A comparison of DSM-IV pervasive developmental disorder and DSM-5 autism spectrum disorder prevalence in an epidemiologic sample. *Journal of the American Academy of Child and Adolescent Psychiatry, 53*(5), 500–508.

Kim, S. H., Kim, Y. S., Koh, Y. J., Lim, E. C., Kim, S. J., & Leventhal, B. L. (2016). Often asked but rarely answered: can Asians meet DSM-5/ICD-10 autism spectrum disorder criteria? *Journal of Child and Adolescent Psychopharmacology, 26*(9), 835–842.

Kočovská, E., Biskupstø, R., Carina Gillberg, I., Ellefsen, A., Kampmann, H., Stórá, T., Billstedt, E., & Gillberg, C. (2012). The rising prevalence of autism: A prospective longitudinal study in the Faroe Islands. *Journal of Autism and Developmental Disorders, 42*(9), 1959–1966.

Kogan, M. D., Blumberg, S. J., Schieve, L. A., et al. (2009). Prevalence of parent-reported diagnosis of autism spectrum disorder among children in the US, 2007. *Pediatrics, 124*(5), 1395–1403.

Kogan, M. D., Vladutiu, C. J., Schieve, L. A., Ghandour, R. M., Blumberg, S. J., Zablotsky, B., Perrin, J. M., Shattuck, P., Kuhlthau, K. A., Harwood, R. L., & Lu, M. C. (2018). The prevalence of parent-reported autism spectrum disorder among US children. *Pediatrics, 142*(6).

Kovess-Masfety, V., Husky, M. M., Keyes, K., Hamilton, A., Pez, O., Bitfoi, A., Carta, M. G., Goelitz, D., Kuijpers, R., Otten, R., Koç, C., Lesinskiene, S., & Mihova, Z. (2016). Comparing the prevalence of mental health problems in children 6–11 across Europe. *Social Psychiatry and Psychiatric Epidemiology, 51*(8), 1093–1103.

Lai, D. C., Tseng, Y. C., Hou, Y. M., & Guo, H. R. (2012). Gender and geographic differences in the prevalence of intellectual disability in children: analysis of data from the national disability registry of Taiwan. *Research in Developmental Disabilities, 33*(6), 2301–2307.

Lainhart, J. E., Bigler, E. D., Bocian, M., Coon, H., Dinh, E., Dawson, G., Deutsch, C. K., Dunn, M., Estes, A., Tager-Flusberg, H., Folstein, S., Hepburn, S., Hyman, S., McMahon, W., Minshew, N., Munson, J., Osann, K., Ozonoff, S., Rodier, P., Rogers, S., Sigman, M., Spence, M. A., Stodgell, C. J., & Volkmar, F. (2006). Head circumference and height in autism: A study by the Collaborative Program of Excellence in Autism. *American Journal of Medical Genetics. Part A, 140*(21), 2257–2274.

Latif, A. H. A., & Williams, W. R. (2007). Diagnostic trends in autistic spectrum disorders in the South Wales valleys. *Autism, 11*(6), 479–487.

Lazoff, T., Zhong, L., Piperni, T., & Fombonne, E. (2010). Prevalence of pervasive developmental disorders among children at the English Montreal School Board. *Canadian Journal of Psychiatry, 55*(11), 715–720.

Levaot, Y., Meiri, G., Dinstein, I., Menashe, I., & Shoham-Vardi, I. (2019). Autism prevalence and severity in Bedouin-Arab and Jewish communities in Southern Israel. *Community Mental Health Journal, 55*(1), 156–160.

Lingam, R., Simmons, A., Andrews, N., Miller, E., Stowe, J., & Taylor, B. (2003). Prevalence of autism and parentally reported triggers in a north east London population. *Archives of Disease in Childhood, 88*(8), 666–670.

Loomes, R., Hull, L., & Mandy, W. P. L. (2017). What is the male-to-female ratio in autism spectrum disorder? a systematic review and meta-analysis. *Journal of the American Academy of Child and Adolescent Psychiatry, 56*(6), 466–474.

Lord, C., Rutter, M., DiLavore, P., et al. (2002). *The Autism Diagnostic Observation Scale (ADOS)*. Western Psychological Services.

Lord, C., Petkova, E., Hus, V., Gan, W., Lu, F., Martin, D. M., Ousley, O., Guy, L., Bernier, R., Gerdts, J., Algermissen, M., Whitaker, A., Sutcliffe, J. S., Warren, Z., Klin, A., Saulnier, C., Hanson, E., Hundley, R., Piggot, J., Fombonne, E., Steiman, M., Miles, J., Kanne, S. M., Goin-Kochel, R. P., Peters, S. U., Cook, E. H., Guter, S., Tjernagel, J., Green-Snyder, L. A., Bishop, S., Esler, A., Gotham, K., Luyster, R., Miller, F., Olson, J., Richler, J., & Risi, S. (2012). A multisite study of the clinical diagnosis of different autism spectrum disorders. *Archives of General Psychiatry, 69*(3), 306–313.

Lotter, V. (1966). Epidemiology of autistic conditions in young children: I. Prevalence. *Social Psychiatry, 1*, 124–137.

MacFarlane, H., Salem, A. C., Lawley, G. O., Presmanes, H. A., Zuckerman, K. E., & Fombonne, E. (2022). Epidemiology. In E. Hollander, R. Hagerman, & C. J. Ferretti (Eds.), *Autism spectrum disorders* (2nd ed., pp. 5–33). American Psychiatric Association Publishing.

Maenner, M. J., & Durkin, M. S. (2010). Trends in the prevalence of autism on the basis of special education data. *Pediatrics, 126*(5), e1018–e1025.

Maenner, M. J., Rice, C. E., Arneson, C. L., Cunniff, C., Schieve, L. A., Carpenter, L. A., Van Naarden Braun, K., Kirby, R. S., Bakian, A. V., & Durkin, M. S. (2014). Potential impact of DSM-5 criteria on autism spectrum disorder prevalence estimates. *Journal of the American Medical Association Psychiatry, 71*(3), 292–300.

Maenner, M. J., Shaw, K. A., Baio, J., et al. (2020). Prevalence of autism spectrum disorder among children aged 8 years—autism and developmental disabilities monitoring network, 11 sites, United States, 2016. *MMWR Surveillance Summaries, 69*(4), 1–12.

Maenner, M. J., Shaw, K. A., Bakian, A. V., et al. (2021). Prevalence and characteristics of autism spectrum disorder among children aged 8 years—autism and developmental disabilities monitoring network, 11 sites, United States, 2018. *MMWR Surveillance Summaries, 70*(11), 1–16.

Mandell, D., & Lecavalier, L. (2014). Should we believe the Centers for Disease Control and Prevention's autism spectrum disorder prevalence estimates? *Autism, 18*(5), 482–484.

Matson, J. L., Matheis, M., Burns, C. O., Esposito, G., Venuti, P., Pisula, E., Misiak, A., Kalyva, E., Tsakiris, V., Kamio, Y., Ishitobi, M., & Goldin, R. L. (2017). Examining cross-cultural differences in autism spectrum disorder: A multinational comparison from Greece, Italy, Japan, Poland, and the United States. *European Psychiatry, 42*, 70–76.

Matsuo, J., Kamio, Y., Takahashi, H., Ota, M., Teraishi, T., Hori, H., et al. (2015). Autistic-like traits in adult patients with mood disorders and schizophrenia. *Public Library of Science One, 10*(4), e0122711.

Mattila, M. L., Kielinen, M., Linna, S. L., Jussila, K., Ebeling, H., Bloigu, R., Joseph, R. M., & Moilanen, I. (2011). Autism spectrum disorders according to DSM-IV-TR and comparison with DSM-5 draft criteria: An epidemiological study. *Journal of the American Academy of Child and Adolescent Psychiatry, 50*(6), 583–592.e11.

May, T., Brignell, A., & Williams, K. (2020). Autism spectrum disorder prevalence in children aged 12–13 years from the longitudinal study of Australian children. *Autism Research, 13*(5), 821–827.

Meyers, J., Chavez, A., Presmanes, H. A., Zuckerman, K., & Fombonne, E. (2019). Epidemiological surveys of autism spectrum disorders. In F. Volkmar (Ed.), *Autism and pervasive developmental disorders* (2nd ed., pp. 61–88). Cambridge University Press.

Mohammadi, M. R., Ahmadi, N., Khaleghi, A., Zarafshan, H., Mostafavi, S. A., Kamali, K., et al. (2019). Prevalence of autism and its comorbidities and the relationship with maternal psychopathology: A national population-based study. *Archives of Iranian Medicine, 22*(10), 546–553.

Montiel-Nava, C., & Peña, J. A. (2008). Epidemiological findings of pervasive developmental disorders in a Venezuelan study. *Autism, 12*(2), 191–202.

Morales-Hidalgo, P., Roigé-Castellví, J., Hernández-Martínez, C., Voltas, N., & Canals, J. (2018). Prevalence and characteristics of autism spectrum disorder among Spanish school-age children. *Journal of Autism and Developmental Disorders, 48*(9), 3176–3190.

Morotti, H., Mastel, S., Keller, K., Barnard, R. A., Hall, T., O'Roak, B. J., & Fombonne, E. (2021). Autism and attention-deficit/hyperactivity disorders and symptoms in children with neurofibromatosis type 1. *Developmental Medicine and Child Neurology, 63*(2), 226–232. https://doi.org/10.1111/dmcn.14558

Narzisi, A., Posada, M., Barbieri, F., Chericoni, N., Ciuffolini, D., Pinzino, M., et al. (2020). Prevalence of Autism Spectrum Disorder in a large Italian catchment area: A school-based population study within the ASDEU project. *Epidemiology and Psychiatric Sciences, 29*, E5.

Nicholas, J. S., Carpenter, L. A., King, L. B., Jenner, W., & Charles, J. M. (2009). Autism spectrum disorders in preschool-aged children: Prevalence and comparison to a school-aged population. *Annals of Epidemiology, 19*(11), 808–814.

Nygren, G., Cederlund, M., Sandberg, E., Gillstedt, F., Arvidsson, T., Carina Gillberg, I., Westman Andersson, G., & Gillberg, C. (2012). The prevalence of autism spectrum disorders in toddlers: A population study of 2-year-old Swedish children. *Journal of Autism and Developmental Disorders, 42*(7), 1491–1497.

Ofner, M., Coles, A., Decou, M. L., Do, M., Bienek, A., Snider, J., & Ugnat, A. (2018). *Autism spectrum disorder among children and youth in Canada 2018*. Public Health Agency of Canada.

Oliveira, G., Ataíde, A., Marques, C., Miguel, T. S., Coutinho, A. M., Mota-Vieira, L., et al. (2007). Epidemiology of autism spectrum disorder in Portugal: prevalence, clinical characterization, and medical conditions. *Developmental Medicine and Child Neurology, 49*(10), 726–733.

Oner, O., & Munir, K. M. (2019). Modified Checklist for Autism in Toddlers Revised (MCHAT-- R/F) in an urban metropolitan sample of young children in Turkey. *Journal of Autism and Developmental Disorders*, 1–8.

Ouellette-Kuntz, H., Coo, H., Yu, C. T., Chudley, A. E., Noonan, A., Breitenbach, M., et al. (2006). Prevalence of pervasive developmental disorders in two Canadian provinces. *Journal of Policy and Practice in Intellectual Disabilities, 3*(3), 164–172.

Ouellette-Kuntz, H., Coo, H., Lam, M., Breitenbach, M. M., Hennessey, P. E., Jackman, P. D., Lewis, M. E., Dewey, D., Bernier, F. P., & Chung, A. M. (2014). The changing prevalence of autism in three regions of Canada. *Journal of Autism and Developmental Disorders, 44*(1), 120–136.

Özerk, K. (2018). Prevalence of autism/ASD in the capital city of Oslo, Norway. *International Electronic Journal of Elementary Education, 11*(1), 23–30.

Özerk, K., & Cardinal, D. (2020). Prevalence of autism/ASD among preschool and school-age children in Norway. *Contemporary School Psychology*, 1–10.

Pantelis, P. C., & Kennedy, D. P. (2016). Estimation of the prevalence of autism spectrum disorder in South Korea, revisited. *Autism, 20*(5), 517–527.

Parner, E. T., Schendel, D. E., & Thorsen, P. (2008). Autism prevalence trends over time in Denmark: Changes in prevalence and age at diagnosis. *Archives of Pediatrics & Adolescent Medicine, 162*(12), 1150–1156.

Parner, E. T., Thorsen, P., Dixon, G., de Klerk, N., Leonard, H., Nassar, N., Bourke, J., Bower, C., & Glasson, E. J. (2011). A comparison of autism prevalence trends in Denmark and Western Australia. *Journal of Autism and Developmental Disorders, 41*(12), 1601–1608.

Parner, E. T., Baron-Cohen, S., Lauritsen, M. B., Jørgensen, M., Schieve, L. A., Yeargin-Allsopp, M., & Obel, C. (2012). Parental age and autism spectrum disorders. *Annals of Epidemiology, 22*(3), 143–150.

Pérez-Crespo, L., Prats-Uribe, A., Tobias, A., Duran-Tauleria, E., Coronado, R., Hervás, A., & Guxens, M. (2019). Temporal and geographical variability of prevalence and incidence of Autism Spectrum Disorder diagnoses in children in Catalonia, Spain. *Autism Research, 12*(11), 1693–1705.

Poovathinal, S. A., Anitha, A., Thomas, R., Kaniamattam, M., Melempatt, N., Anilkumar, A., & Meena, M. (2016). Prevalence of autism spectrum disorders in a semiurban community in south India. *Annals of Epidemiology, 26*(9), 663–665. e8.

Posserud, M., Lundervold, A. J., Lie, S. A., & Gillberg, C. (2010). The prevalence of autism spectrum disorders: Impact of diagnostic instrument and non-response bias. *Social Psychiatry and Psychiatric Epidemiology, 45*(3), 319–327.

Powell, J. E., Edwards, A., Edwards, M., Pandit, B. S., Sungum-Paliwal, S. R., & Whitehouse, W. (2000). Changes in the incidence of childhood autism and other autistic spectrum disorders in preschool children from two areas of the West Midlands, UK. *Developmental Medicine and Child Neurology, 42*(9), 624–628.

Rah, S. S., Hong, S. B., & Yoon, J. Y. (2020). Prevalence and incidence of developmental disorders in Korea: A nationwide population-based study. *Journal of Autism and Developmental Disorders, 50*(12), 4504–4511.

Rai, D., Lewis, G., & Lundberg, M. (2012). Parental socioeconomic status and risk of offspring autism spectrum disorders in a Swedish population-based study. *Journal of the American Academy of Child and Adolescent Psychiatry, 51*(5), 467–476.

Raina, S. K., Kashyap, V., Bhardwaj, A. K., Kumar, D., & Chander, V. (2015). Prevalence of autism spectrum disorders among children (1–10 years of age)—findings of a mid-term report from Northwest India. *Journal of Postgraduate Medicine, 61*(4), 243–246.

Randall, M., Sciberras, E., Brignell, A., Ihsen, E., Efron, D., Dissanayake, C., & Williams, K. (2016). Autism spectrum disorder: Presentation and prevalence in a nationally representative Australian sample. *The Australian and New Zealand Journal of Psychiatry, 50*(3), 243–253.

Rescorla, L., Achenbach, M., Ivanova, M., Dumenci, L., Almqvist, F., Bilenberg, N., Bird, H., Chen, W., Dobrean, A., Dopfner, M., Erol, N., Fombonne, E., Fonseca, A., Frigerio, A., Grietens, H., Hannesdottir, H., Kanbayashi, Y., Lambert, M., Larsson, B., et al. (2007). Behavioral and emotional problems reported by parents of children ages 6–16 in 31 societies. *Journal of Emotional and Behavioral Disorders, 15*(3), 130–142.

Risi, S., Lord, C., Gotham, K., Corsello, C., Chrysler, C., Szatmari, P., Cook, E. H., Jr, Leventhal, B. L., & Pickles, A. (2006). Combining information from multiple sources in the diagnosis of autism spectrum disorders. *Journal of the American Academy of Child and Adolescent Psychiatry, 45*(9), 1094–1103. https://doi.org/10.1097/01.chi.0000227880.42780.0e

Rosanoff, M. J., Daniels, A. M., & Shih, A. (2015). Autism: A (key) piece of the global mental health puzzle. *Global Mental Health, (Camb), 2*, e2.

Rudra, A., Belmonte, M. K., Soni, P. K., Banerjee, S., Mukerji, S., & Chakrabarti, B. (2017). Prevalence of autism spectrum disorder and autistic symptoms in a school-based cohort of children in Kolkata, India: prevalence of ASD in India. *Autism Research, 10*(10), 1597–1605.

Russell, G., Rodgers, L. R., Ukoumunne, O. C., & Ford, T. (2014). Prevalence of parent-reported ASD and ADHD in the UK: Findings from the Millennium Cohort Study. *Journal of Autism and Developmental Disorders, 44*(1), 31–40.

Rutter, M., LeCouteur, A., & Lord, C. (2003). *The Autism Diagnostic Interview-Revised (ADI-R)*. Western Psychological Services.

Rydzewska, E., Hughes-McCormack, L. A., Gillberg, C., et al. (2019). Age at identification, prevalence and general health of children with autism: Observational study of a whole country population. *BMJ Open, 9*(7), e025904.

Saemundsen, E., Magnússon, P., Georgsdóttir, I., Egilsson, E., & Rafnsson, V. (2013). Prevalence of autism spectrum disorders in an Icelandic birth cohort. *BMJ Open, 3*(6), e002748.

Samadi, S. A., Mahmoodizadeh, A., & McConkey, R. (2012). A national study of the prevalence of autism among five-year-old children in Iran. *Autism, 16*(1), 5–14.

Scott, F. J., Baron-Cohen, S., Bolton, P., & Brayne, C. (2002). Brief report: Prevalence of autism spectrum conditions in children aged 5–11 years in Cambridgeshire, UK. *Autism, 6*(3), 231–237.

Segev, A., Weisskopf, M. G., Levine, H., Pinto, O., & Raz, R. (2019). Incidence time trends and socioeconomic factors in the observed incidence of autism spectrum disorder in Israel: A nationwide nested case–control study. *Autism Research, 12*(12), 1870–1879.

Sell, N. K., Giarelli, E., Blum, N., Hanlon, A. L., & Levy, S. E. (2012). A comparison of autism spectrum disorder DSM-IV criteria and associated features among African American and white children in Philadelphia County. *Disability and Health Journal, 5*(1), 9–17.

Shaw, K. A., Maenner, M. J., Baio, J., et al. (2016). Early identification of autism spectrum disorder among children aged 4 years—early autism and developmental disabilities monitoring network, six sites, United States. *Morbidity and Mortality Weekly Report Surveillance Summaries 2020, 69*(SS-3), 1–11.

Skonieczna-Żydecka, K., Gorzkowska, I., Pierzak-Sominka, J., & Adler, G. (2017). The prevalence of autism spectrum disorders in West Pomeranian and Pomeranian regions of Poland. *Journal of Applied Research in Intellectual Disabilities, 30*(2), 283–289.

Sun, X., Allison, C., Wei, L., Matthews, F. E., Auyeung, B., Wu, Y. Y., Griffiths, S., Zhang, J., Baron-Cohen, S., & Brayne, C. (2019). Autism prevalence in China is comparable to Western prevalence. *Molecular Autism, 10*, 7.

Taylor, B., Jick, H., & Maclaughlin, D. (2013). Prevalence and incidence rates of autism in the UK: Time trend from 2004–2010 in children aged 8 years. *BMJ Open, 3*(10), e003219.

Thomaidis, L., Mavroeidi, N., Richardson, C., Choleva, A., Damianos, G., Bolias, K., & Tsolia, M. (2020). Autism spectrum disorders in Greece: Nationwide prevalence in 10–11 year-old children and regional disparities. *Journal of Clinical Medicine, 9*(7), 2163.

Turban, J. L., & van Schalkwyk, G. I. (2018). "Gender dysphoria" and autism spectrum disorder: Is the link real? *Journal of the American Academy of Child and Adolescent Psychiatry, 57*(1), 8–9.e2.

Valenti, M., Vagnetti, R., Masedu, F., Pino, M. C., Rossi, A., Scattoni, M. L., et al. (2019). Register-based cumulative prevalence of autism spectrum disorders during childhood and adolescence in Central Italy. *Epidemiology, Biostatistics and Public Health, 16*(4).

Van Bakel, M. M., Delobel-Ayoub, M., Cans, C., Assouline, B., Jouk, P. S., Raynaud, J. P., & Arnaud, C. (2015). Low but increasing prevalence of autism spectrum disorders in a French area from register-based data. *Journal of Autism and Developmental Disorders, 45*(10), 3255–3261.

Van Balkom, I. D., Bresnahan, M., Vogtländer, M. F., van Hoeken, D., Minderaa, R. B., Susser, E., & Hoek, H. W. (2009). Prevalence of treated autism spectrum disorders in Aruba. *Journal of Neurodevelopmental Disorders, 1*(3), 197–204.

Van Naarden Braun, K., Pettygrove, S., Daniels, J., Miller, L., Nicholas, J., Baio, J., et al. (2007). Evaluation of a methodology for a collaborative multiple source surveillance network for autism spectrum disorders—Autism and developmental disabilities monitoring network, 14 sites, United States, 2002. *Morbidity and Mortality Weekly Report. Surveillance Summaries, 56*(SS1), 29–40.

Volkmar, F. R., Klin, A., Siegel, B., Szatmari, P., Lord, C., Campbell, M., et al. (1994). Field trial for autistic disorder in DSM-IV. *The American Journal of Psychiatry, 151*(9), 1361–1367.

Volkmar, F. R., Shaffer, D., & First, M. (2000). PDDNOS in DSM-IV. *Journal of Autism and Developmental Disorders, 30*(1), 74.

Wignyosumarto, S., Mukhlas, M., & Shirataki, S. (1992). Epidemiological and clinical study of autistic children in Yogyakarta, Indonesia. *The Kobe Journal of Medical Sciences, 38*(1), 1–19.

Williams, E., Thomas, K., Sidebotham, H., & Emond, A. (2008). Prevalence and characteristics of autistic spectrum disorders in the ALSPAC cohort. *Developmental Medicine and Child Neurology, 50*(9), 672–677.

Windham, G. C., Anderson, M. C., Croen, L. A., Smith, K. S., Collins, J., & Grether, J. K. (2011). Birth prevalence of autism spectrum disorders in the San Francisco Bay area by demographic and ascertainment source characteristics. *Journal of Autism and Developmental Disorders, 41*(10), 1362–1372.

Wong, V. C. N., & Hui, S. L. H. (2008). Epidemiological study of autism spectrum disorder in China. *Journal of Child Neurology, 23*(1), 67–72.

World Bank (2020) The world by income and region: https://datatopics.worldbank.org/world-development-indicators/the-world-by-income-and-region.html

World Health Organization. (1992). *International Statistical Classification of Diseases and Related Health Problems (10th Revision)*. World Health Organization.

Xu, G., Strathearn, L., Liu, B., Bao, W. (2018). Prevalence of autism spectrum disorder among US children and adolescents, 2014–2016. *JAMA, 319*(1): 81–82. https://doi.org/10.1001/jama.2017.17812. Erratum in: JAMA. 2018 Feb 6;319(5):505. Erratum in: JAMA. 2018 Feb 6;319(5):505. PMID: 29297068; PMCID: PMC5833544.

Yang, W., Xia, H., Wen, G., Liu, L., Fu, X., Lu, J., & Li, H. (2015). Epidemiological investigation of suspected autism in children and implications for healthcare system: a mainstream kindergarten-based population study in Longhua District, Shenzhen. *BioMed Central Pediatrics, 15*(1), 207.

Yeargin-Allsopp, M., Rice, C., Karapurkar, T., Doernberg, N., Boyle, C., & Murphy, C. (2003). Prevalence of autism in a US metropolitan area. *Journal of the American Medical Association, 289*(1), 49–55.

Yu, L., Stronach, S., & Harrison, A. J. (2020). Public knowledge and stigma of autism spectrum disorder: Comparing China with the United States. *Autism, 24*(6), 1531–1545.

Zablotsky, B., Black, L., Maenner, M. J., Schieve, L. A., & Blumberg, S. J. (2015). Estimated prevalence of autism and other developmental disabilities following questionnaire changes in the 2014 national health interview survey. *National Health Statistics Reports, 87*, 1–20.

Zablotsky, B., Bramlett, M. D., & Blumberg, S. J. (2020). The co-occurrence of autism spectrum disorder in children with ADHD. *Journal of Attention Disorders, 24*(1), 94–103.

Zeglam, A. M., & Maouna, A. J. (2012). Prevalence of autistic spectrum disorders in Tripoli, Libya: The need for more research and planned services. *Eastern Mediterranean Health Journal, 18*(2).

Zeidan, J., Fombonne, E., Scorah, J., Ibrahim, A., Durkin, M. S., Saxena, S., Yusuf, A., Shih, A., & Elsabbagh, M. (2022). Global prevalence of autism: A systematic review update. *Autism Research, 15*(5), 778–790.

Zhang, X., & Ji, C. Y. (2005). Autism and mental retardation of young children in China. *Biomedical and Environmental Sciences, 18*(5), 334–340.

Zhang, J., Wheeler, J. J., & Richey, D. (2006). Cultural validity in assessment instruments for children with autism from a Chinese cultural perspective. *International Journal of Special Education, 21*, 109–114.

Zhou, H., Xu, X., Yan, W., Zou, X., Wu, L., Luo, X., Li, T., Huang, Y., Guan, H., Chen, X., Mao, M., Xia, K., Zhang, L., Li, E., Ge, X., Zhang, L., Li, C., Zhang, X., Zhou, Y., Ding, D., Shih, A., Fombonne, E., Zheng, Y., Han, J., Sun, Z., Jiang, Y. H., Wang, Y., & LATENT-NHC Study Team. (2020). Prevalence of autism spectrum disorder in China: A nationwide multi-center population-based study among children aged 6 to 12 years. *Neuroscience Bulletin, 36*(9), 961–971.

Chapter 14
Metabolic Approaches to the Treatment of Autism Spectrum Disorders

Neluwa-Liyanage R. Indika, Susan C. Owens, Udara D. Senarathne, Andreas M. Grabrucker, Nelson S. K. Lam, Kerri Louati, Greer McGuinness, and Richard E. Frye

A plethora of evidence suggests that autism spectrum disorder (ASD) is associated with metabolic disturbances. While many rare inborn errors of metabolism (IEMs) are reported in patients with ASD (Žigman et al., 2021) many children with ASD have metabolic abnormalities that cannot be linked to a specific IEM. In addition, despite probable contribution by self-imposed dietary restrictions (Yule et al., 2021) and toxicities (O'Neal et al., 2020) to some metabolic abnormalities found in ASD, the origin of metabolic abnormalities is unknown in a majority.

N.-L. R. Indika (✉)
Department of Biochemistry, Faculty of Medical Sciences, University of Sri Jayewardenepura, Nugegoda, Sri Lanka
e-mail: ind.liyanage@sjp.ac.lk

S. C. Owens
Autism Oxalate Project at the Autism Research Institute, San Diego, CA, USA

U. D. Senarathne
Department of Biochemistry, Faculty of Medical Sciences, University of Sri Jayewardenepura, Nugegoda, Sri Lanka

Department of Chemical Pathology, Monash Health, Clayton, Victoria, Australia

A. M. Grabrucker
Cellular Neurobiology and Neuro-Nanotechnology Lab, Department of Biological Sciences, University of Limerick, Limerick, Ireland

Bernal Institute, University of Limerick, Limerick, Ireland

Health Research Institute (HRI), University of Limerick, Limerick, Ireland

N. S. K. Lam
Faculty of Health and Sciences, The University of Adelaide, Adelaide, SA, Australia

A. El Idrissi, D. McCloskey (eds.), *Neurobiology of Autism Spectrum Disorders*,
https://doi.org/10.1007/978-3-031-42383-3_14

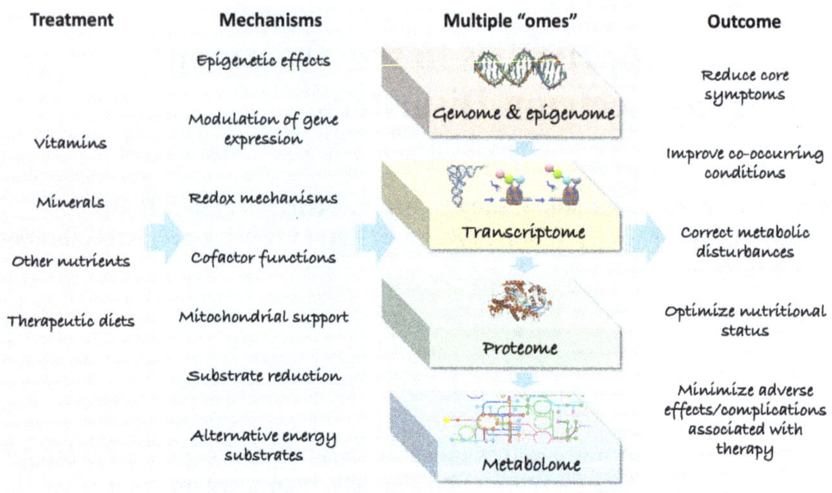

Fig. 14.1 The metabolic approaches to the treatment of autism spectrum disorder modulate multiple "omes" to bring about desirable outcomes through diverse mechanisms

Limited evidence provides some support for improving core ASD symptoms by correcting metabolic disturbances using therapeutic diets and micronutrient supplementation (Karhu et al., 2020) as well as pharmaceutical interventions such as sulforaphane or suramin (Mierau & Neumeyer, 2019). These interventions operate at different levels of multiple "omes" to bring about desirable effects in ASD (Fig. 14.1).

Some micronutrient formulations have shown beneficial effects by improving the nutritional and metabolic status of children with ASD (Adams et al., 2011a). A combination of carnitine, coenzyme Q10, and alpha-lipoic acid targeting mitochondrial function was observed to significantly improve the biomarkers related to mitochondrial function and neuro-behavioral performance in children with ASD (Legido et al., 2018). It should be noted that the dosages discussed in this chapter, may be insufficient in patients with biochemical or clinical deficiency. On the other hand, it may be impractical to attribute any clinical or biochemical improvement to a single constituent in a formulation to justify its clinical use as a monotherapy.

K. Louati
Functional Medicine Education Specialist & Clinical Lab Educator, Fort Lauderdale, FL, USA

G. McGuinness
Central Michigan University, Mount Pleasant, MI, USA

R. E. Frye
Phoenix Children's Hospital, Phoenix, AZ, USA

Rossignol Medical Center, Phoenix, AZ, USA

Therefore, given the limited evidence on specific treatments and limited biomarkers to guide their prescription, along with the heterogeneity of ASD individuals, the medical practitioner must follow a carefully personalized approach to treat individual patients based on five major goals.

1. Reduce core symptoms.
2. Improve co-occurring conditions.
3. Correct the metabolic disturbances.
4. Optimize the nutritional status.
5. Minimize adverse effects/complications associated with therapy.

A thorough assessment for metabolic therapy using the ABCDEFG (A–G) model is recommended:

A; **A**nthropometry
B; **B**iochemical investigations
C; **C**linical assessment (**C**ore symptoms as well as **C**o-morbid/**C**o-occurring conditions)
D; **D**ietary assessment
E; **E**xposure (toxins, drugs, etc.)
F; **F**unctional imaging/**Fl**uoro-deoxy-glucose (FDG) positron emission tomography-computed tomography (PET-CT)
G; **G**enetic studies

Anthropometric assessment should include weight for age, height for age, weight for height, and head circumference, which can indicate malnutrition or a genetic etiology. Biochemical assessment should include vitamin assays and metabolic screening. Clinical assessment for baseline severity of core and co-occurring symptoms is essential to evaluate the treatment response. Qualitative and quantitative assessment of diet will enable identifying nutrient inadequacies. Identifying exposure to toxins and drugs enables predicting their metabolic implications. For instance, antiepileptics may induce vitamins B9, B12, and D deficiencies (Berger et al., 2022). Metabolic (FDG PET-CT, magnetic resonance spectroscopy [MRS], resting-state functional magnetic resonance imaging [rsfMRI], functional near-infrared spectroscopy [fNIRS]) or functional (electroencephalography [EEG], quantitative EEG [qEEG], magnetoencephalography [MEG]) imaging may help identify abnormalities in the central metabolism (Żarnowska et al., 2018). Finally, genetic studies will be useful in confirming a genetic etiology, implementing a targeted therapy in the case of treatable IEMs, predicting response to treatment, and genetic counselling. This multimodal evaluation should be ideally conducted by a multidisciplinary team. Hence, this integrated approach (Fig. 14.2) ensures that the patient receives the best metabolic therapy in a personalized manner.

Below we discuss metabolic and nutritional interventions commonly used in ASD, along with evidence supporting each therapy's use. Table 14.1 summarizes the doses and adverse effects of each therapy.

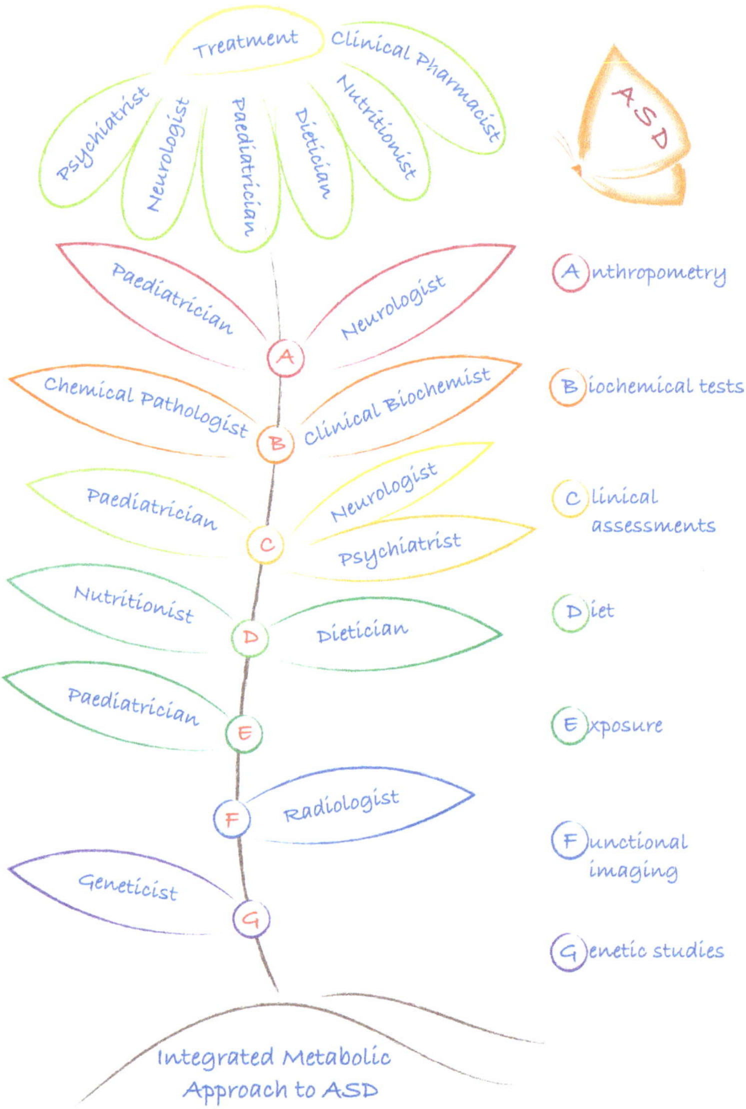

Fig. 14.2 The multidisciplinary, multimodal, integrated approach to metabolic therapy for autism spectrum disorder (ASD)

Table 14.1 Vitamins, minerals, and other nutrients used to treat autism spectrum disorder

Nutrient	Dose	Adverse effects/toxicity
Thiamin (B1)	50–100 mg daily	Not reported
Riboflavin (B2)	100–400 mg daily	Nausea at high doses
Niacin (B3)	50–100 mg daily	Flushing
Pantothenic acid (B5)	5–1200 mg/day, 1–3×/day	Diarrhea at high doses
Pyridoxine (B6)	50–250 mg/day given 1–3×/day	Headache, paresthesia, and nausea at high doses
Biotin (B7)	5–20 mg/day given 1–2×/day daily	Not reported
Folinic acid/leucovorin (B9)	400 μg/day to 4 mg/kg/day. 1–3×/day	Excitement, aggression, insomnia, tantrum, and headache
Methylcobalamin (B12)	5–2000 μg every 1–3 days	Injected methyl-B12: hyperactivity, difficulty in sleeping, and irritability Oral methyl-B12: worsening behavior/aggression
Vitamin C	100–500 mg given daily	Diarrhea at high doses
Vitamin A	10,000 units daily (in deficiency)	Rough skin, dry hair
Vitamin D	400 units daily	Abdominal pain, nausea, headache, skin reactions, laryngeal edema, hypercalcemia, and hypercalciuria
Vitamin E	200–400 IU given daily	Diarrhea and bleeding at high doses
Zinc	0.5–1 mg/kg twice daily (10–40 mg daily)	Epigastric pain, nausea, vomiting, fatigue, lethargy, and suppresses iron and copper absorption
Magnesium	Up to 20 mg/kg/day	Diarrhea
Lithium	Need therapeutic monitoring	Kidney dysfunction, gastrointestinal symptoms (nausea, diarrhea, vomiting, abdominal pain), neurological impairment (tremors, dizziness), hypothyroidism, hyperparathyroidism, and weight gain
Molybdenum	Not available	Arthralgia, gout-like symptoms, and hyperuricosuria
N-acetyl-L-cysteine (NAC)	10–70 mg/kg/day. 1–3×/day	Nausea, diarrhea, abdominal discomfort, and headache
Co-enzyme Q10 (reduced): Ubiquinol	5–30 mg/kg/day, 1–2×/day	Appetite loss, nausea, diarrhea at high doses
Co-enzyme Q10 (oxidized): Ubiquinone	10–30 mg/kg/day; 1–2×/day	Appetite loss, nausea, diarrhea at high doses
Creatine monohydrate	100 mg/kg/day; 1–3×/day	Increased urination

(continued)

Table 14.1 (continued)

Nutrient	Dose	Adverse effects/toxicity
L-carnitine or acetyl-L-carnitine	30–120 mg/kg/day, 1–3×/day	Stool loose/fishy smell
Alpha-lipoic acid	50–200 mg given daily	Headache, paresthesia, rash, and muscle cramps
Tetrahydrobiopterin	3–20 mg/kg/day	Insomnia, irritability, and mild abdominal discomfort

Derived from Niyazov et al. (2016), Robea et al. (2020), Brondino et al. (2015), and dosage information sourced from the British National Formulary for Children (2019–20) BNF

14.1 Vitamins

14.1.1 Vitamin B1

The activated form of vitamin B1 (thiamine), thiamine pyrophosphate (TPP), is a cofactor for several enzyme complexes involved in the energy metabolism (Bettendorff & Wins, 2021). It regulates antioxidants, neurotransmitters, and neurotropic enzymes (Mkrtchyan et al., 2015; Aleshin et al., 2019) and is essential for transporting folate across the blood-brain barrier into neurons. Whole blood thiamine (Adams et al., 2011b) and TPP (Anwar et al., 2016) levels have been noted to be below the average in 11% and 24% of individuals with ASD, respectively, while thiamin deficiency associated Wernicke's encephalopathy and dysautonomia have also been reported in several cases of ASD (Yule et al., 2021; Lonsdale, 2018). Limited evidence is available on the clinical application of thiamine tetrahydrofurfuryl disulfide (TTDF) supplementation in ASD with eight out of ten ASD individuals demonstrating clinical improvement in a recent study (Lonsdale, 2018).

14.1.2 Vitamin B2

Vitamin B2 (riboflavin) is involved in antioxidant functions, metabolism of other B vitamins (B3, B6, B12, and B9), and energy homeostasis (Berger et al., 2022). The available evidence on plasma B2 levels in ASD is not in agreement, as no significant difference was observed between ASD and neurotypical individuals in one study while 15% an 4% of those with ASD were found to have higher and lower vitamin B2 levels respectively compared to the controls in another study (Adams et al., 2011b). Vitamin B2 monotherapy has not been studied in ASD.

14.1.3 Vitamin B3

Vitamin B3 (niacin/nicotinamide) is the precursor for nicotinamide adenine dinucleotide (NAD^+) and nicotinamide adenine dinucleotide phosphate ($NADP^+$), the coenzymes essential for numerous metabolic reactions (Tardy et al., 2020). De novo synthesis of vitamin B3 in the human body utilizes the essential amino acid tryptophan (Tian et al., 2013). Although ASD individuals and neurotypical controls were observed to have no significant difference in blood levels of niacin (Adams et al., 2011b), abnormal niacin metabolism has been reported in relation to ASD. For example, urine nicotinamide is a biomarker that can be used to discriminate ASD patients from controls (Liang et al., 2020a), and urinary nicotinamide metabolites were found to be increased in ASD individuals, indicating increased nicotinic acid degradation (Yap et al., 2010). Furthermore, niacin is essential for several metabolic pathways important in ASD, including methylation, acetylserotonin O-methyltransferase (ASMT) activity, and mitochondrial function.

14.1.4 Vitamin B5

Vitamin B5 (pantothenic acid) is a water-soluble vitamin required for the synthesis of coenzyme A (CoA), a thiol that reacts with carboxylic acids to form thioesters (Tardy et al., 2020). Vitamin B5 deficiency is linked to ASD. Several studies report inadequate vitamin B5 intake in those with ASD (Castro et al., 2016; Lindsay et al., 2006; Stewart et al., 2015) or ASD traits (Tsujiguchi et al., 2020). Urinary vitamin B5 excretion was found to be lower in children with ASD compared to healthy controls (Liang et al., 2020b, Gevi et al., 2016), while blood vitamin B5 levels in ASD individuals was also observed to low (Adams et al., 2011b). Even though there is limited scientific evidence to support vitamin B5 deficiency in ASD, vitamin B5 may be a good candidate in multivitamin supplementation for ASD individuals.

14.1.5 Vitamin B6

Vitamin B6 (pyridoxine, pyridoxal, pyridoxamine) is found in several forms. It is intracellularly modified by the pyridoxal kinase to its active form pyridoxal 5-phosphate (PLP). Many ASD-associated symptoms overlap with symptoms of B6 deficiency, including immune abnormalities such as excessive T helper 2 (Th2) cell response (Qian et al., 2017) and transsulfuration and neurotransmitter pathway abnormalities (Clayton, 2006). Increased blood PLP levels have been observed in children with ASD not on supplementations, which has been attributed to the accumulation of PLP in the blood secondary to PLP not being transferred into the cells (Adams et al., 2006). B6 supplementation has been studied in ASD for its efficacy and safety in 18 studies with 11 of them being double-blind placebo-controlled (DBPC) trials (Adams & Holloway, 2004).

14.1.6 Vitamin B7

Biotin (B7) is critical for energy production and storage as it is a cofactor for several carboxylases involved in fatty acid synthesis, branched-chain amino acid catabolism, and gluconeogenesis (Tardy et al., 2020). Disorders of biotin metabolism have been associated with ASD (Yoldaş et al., 2021; Kiykim et al., 2016) including IEMs in biotin-dependent enzymes (Al-Owain et al., 2013; de la Bâtie et al., 2018; Cakar & Yilmazbas, 2021) and biotinidase deficiency (Frye, 2015; Inci et al., 2021). Biotin deficiency has been found in nearly 7% of ASD individuals (Adams et al., 2011b), with some cases responsive to biotin supplementation (Frye, 2015; Inci et al., 2021; Zaffanello et al., 2003; Benke et al., 2018; Colamaria et al., 1989).

14.1.7 Vitamin B9

Folate (B9) is a water-soluble essential vitamin important for neurodevelopment and is utilized as a cofactor in most metabolic systems. Abnormalities in folate metabolism have been associated with ASD, including autoantibodies to the folate receptor α (Rossignol & Frye, 2021a) and polymorphisms in genes encoding the dihydrofolate reductase, reduced folate carrier, and methylenetetrahydrofolate reductase (Frustaci et al., 2012). According to a meta-analysis, controlled trials support the use of leucovorin (aka folinic acid) as an effective treatment for neurological symptoms in ASD (Rossignol & Frye, 2021a).

14.1.8 Vitamin B12

Vitamin B12 (cobalamin) is a water-soluble essential vitamin that is key to neurodevelopment and is utilized as a cofactor in most metabolic systems. While dietary intake of B12 may be lower in children with ASD (Esteban-Figuerola et al., 2019), serum B12 has been observed to be significantly lower in ASD individuals (Chen et al., 2021) with markedly low B12 levels found in the brain tissue of ASD individuals in one study (Esteban-Figuerola et al., 2019). As outlined in a recent systematic review and meta-analysis, 17 studies, including four DBPC studies, have examined the therapeutic use of oral or injectable B12 in ASD, with most opting for injection and a majority of studies using methyl-B12 (Rossignol & Frye, 2021b).

14.1.9 Vitamin C

Vitamin C (ascorbic acid) is a potent water-soluble antioxidant that directly scavenges radicals, mitigates oxygen radical production, and recycles other antioxidants. Furthermore, it is an important cofactor/co-substrate for the biosynthesis of neurotransmitters, including serotonin (Berger et al., 2022). Vitamin C levels were higher in 29% and lower in 13% of ASD individuals compared to neurotypical controls (Adams et al., 2011b). Clinical symptoms of vitamin C deficiency (scurvy) have been reported in ASD (Yule et al., 2021; Golriz et al., 2017), and a DBPC trial in 18 children with ASD reported reduced stereotyped behaviors with vitamin C treatment. (Dolske et al., 1993).

14.1.10 Vitamin A

Vitamin A (Retinol) is a fat-soluble prohormone converted to two active metabolites: retinoic acid and retinal. It is essential for normal growth and development, vision, immunity, reproduction, and cellular differentiation in mucous membranes. Severe vitamin A deficiency has been reported in ASD (Yule et al., 2021) and has been correlated with ASD symptoms (Zhu et al., 2020; Liu et al., 2016; Xia et al., 2010; Bandini et al., 2010; Hyman et al., 2012; Sun et al., 2013). Interventional studies have reported improved emotional response, sensory perception, and communication with vitamin A therapy in individuals with ASD (Guo et al., 2018).

14.1.11 Vitamin D

Cholecalciferol (vitamin D3) and ergocalciferol (vitamin D2) are fat-soluble animal- and plant-derived forms of vitamin D, respectively. Vitamin D regulates brain development in early life and contributes to synaptic plasticity, neuroprotection, neural circuitry, immunomodulation, and turnover of the dopaminergic system in adults (Gáll & Székely, 2021). In meta-analyses, vitamin D levels were found to be lower in children with ASD (Gallardo-Carrasco et al., 2022; Wang et al., 2016) and prenatally in those that develop ASD (Wang et al., 2016; García-Serna & Morales, 2020), with vitamin D deficiency with/without rickets commonly reported among ASD individual (Yule et al., 2021). According to a meta-analysis, vitamin D supplementation was found to be beneficial for hyperactivity but not for the core symptoms or irritability associated with ASD (Li et al., 2022).

14.1.12 *Vitamin E*

Vitamin E (tocopherols, tocotrienols) is a fat-soluble vitamin linked to the other components of the redox-antioxidant cycle (vitamin C, glutathione, glutathione peroxidase, glutathione reductase, etc.) and inhibits the peroxidation of membrane lipids by scavenging highly reactive peroxyl radicals (Pangrazzi et al., 2020). Vitamin E levels are lower in individuals with ASD compared to healthy controls (Al-Gadani et al., 2009; Chen et al., 2021; El-Ansary et al., 2017) and correlate with the severity of social and cognitive impairments (Alabdali et al., 2014). In a single group assignment study involving patients with verbal apraxia, 97% reported dramatic improvements in speech, imitation, coordination, eye contact, behavior, sensory issues, and development of pain sensation following supplementation of vitamin E combined with polyunsaturated fatty acids (Morris & Agin, 2009).

14.2 Minerals

14.2.1 *Zinc*

Zinc has an important role in gastrointestinal, endocrine, immune, and nervous systems (Vela et al., 2015; Grabrucker, 2016; Maywald et al., 2017; Baltaci et al., 2019). Pre-and perinatal zinc deficiency has been consistently linked to ASD (Sayehmiri et al., 2015; Saghazadeh et al., 2017; Zhang et al., 2021). Limited studies have reported beneficial effects of zinc supplementation in children with ASD (Russo & Devito, 2011; Russo, 2011; Meguid et al., 2019).

14.2.2 *Magnesium*

Many enzymes dependent on vitamin B1 and B6 require magnesium as a cofactor (Lonsdale, 2018) and magnesium supplementation with thiamine helps absorption and utilization of thiamin (Peake et al., 2013). Although overall improvement in core ASD symptoms has been reported with vitamin B6 and magnesium therapy (Khan et al., 2021; Mousain-Bosc et al., 2006), earlier DBPC trials showed no significant effect of vitamin B6 and magnesium supplementation (Nye & Brice, 2005).

14.2.3 *Lithium*

Lithium produces a wide range of neuroprotective, neurotrophic, and neuromodulatory reactions involving various neurotransmitter systems and intracellular signalling pathways (Bojja et al., 2022). Data supporting lithium as a potential therapy for ASD is only based on *in vitro* and animal studies (Darville et al., 2016; Lutz et al., 2020).

14.2.4 Molybdenum

Molybdenum is an essential cofactor for the function of molybdoenzymes such as sulfite oxidase. Some case reports provide evidence for ASD phenotype in patients with molybdenum cofactor deficiency (Huijmans et al., 2017; Lionel et al., 2013). A randomized DBPC trial with a micronutrient containing molybdenum significantly increased the whole blood molybdenum, urinary sulfite, and sulfate levels (Adams et al., 2011a). Interestingly, sulfite is well-known to destroy thiamine (Singh et al., 2005). Therefore, ASD subjects with high sulfite excretion may benefit from a combination of thiamin and molybdenum supplementation.

14.3 Other Nutrients

14.3.1 Omega-3 Fatty Acids

Omega-3 are polyunsaturated fatty acids with more than one carbon-carbon double bond in their main chain. Previous studies have shown low levels of omega-3 fatty acids in children with ASD compared to controls (Bent et al., 2009). A recent meta-analysis has shown that supplementation improved ASD symptoms (de Andrade Wobido et al., 2021) and another study showed reduced anxiety (De Crescenzo et al., 2020).

14.3.2 N-acetyl Cysteine

N-acetyl cysteine (NAC) is the N-acetylated form of the amino acid cysteine that is widely used for various medical applications (Deepmala et al., 2015). NAC has been studied in ASD in several randomized controlled trials. Studies have found that NAC is safe and effective at reducing irritability (Hardan et al., 2012; Ghanizadeh & Moghimi-Sarani, 2013) and improving glutathione production in children with ASD (Wink et al., 2016).

14.3.3 Coenzyme Q10

Coenzyme Q10 (CoQ10) is an essential obligatory inner mitochondrial membrane co-factor for the function of the electron transport chain. Children with ASD have been shown to have depressed CoQ10 as a group (Rossignol & Frye, 2012). Two clinical studies found that CoQ10 improves gastrointestinal and sleep disorders and markers of oxidative stress in children with ASD (Mousavinejad et al., 2018) and communication, playing, sleeping, and food rejection (Gvozdjáková et al., 2014).

14.3.4 Alpha-Lipoic Acid

Alpha-lipoic acid is an essential prosthetic group of several mitochondrial enzymes. Its other functions include scavenging reactive oxygen species, heavy metal detoxification (Bjørklund et al., 2019), enhancing endogenous antioxidants like glutathione, vitamins C and E, and regulating the expression of genes associated with anti-oxidant and anti-inflammatory pathways (Park et al., 2014). Alpha-lipoic acid has been used in ASD mostly for its antioxidant and chelation properties. However, more research is warranted to support the effects of alpha-lipoic acid in ASD.

14.3.5 Creatine Monohydrate

Creatine is phosphorylated to form creatine phosphate, a high-energy phosphate buffer that stores the energy mainly in the muscle and brain (Marshall et al., 2022). Creatine monohydrate is the most widely used oral supplementation form of creatine. IEMs of creatine metabolism have been associated with ASD (Mercimek-Andrews & Salomons, 1993). However, there are no clinical trials on the use of creatine monohydrate in ASD aside from reported case studies in association to IEMs.

14.3.6 Sulforaphane

Sulforaphane is a precursor found in cruciferous vegetables such as cauliflower, broccoli, kale, cabbage, collards, and brussels sprouts (Guerrero-Beltrán et al., 2012). Recent studies have found sulforaphane to exhibit multiple biological effects, including antioxidant, anti-inflammatory, anticonvulsant, and neuroprotective effects relevant to ASD (Kim & Park, 2016). According to a systematic review, it improved the core symptoms of ASD, including irritability, lethargy, and cognitive functions (Mcguinness & Kim, 2020) while another study showed significant changes in biomarkers of glutathione redox status, mitochondrial respiration, inflammatory markers, and heat shock proteins (Zimmerman et al., 2021).

14.3.7 Melatonin

Melatonin is an indoleamine produced predominantly by the pineal gland from L-tryptophan as an endogenous neurohormone that demonstrates a circadian rhythm (Wang et al., 2002; Wu et al., 2020). Many studies have reported low levels of melatonin and melatonin metabolites during the night suggesting an abnormal melatonin physiology in ASD (Wu et al., 2020; Malow et al., 2012; Rossignol & Frye, 2014; Gagnon & Godbout, 2018; Abdelgadir et al., 2018). There is accumulating evidence of promising effects of melatonin on treating reduced sleep duration and sleep-onset latency in ASD individuals (Schroder et al., 2021; Rossignol & Frye, 2014).

14.3.8 Carnitine

Carnitine is a conditionally essential nutrient that participates in the metabolism of long-chain fatty acids and has anti-oxidative, anti-inflammatory, and neuroprotective effects (Rose et al., 2018). Multiple studies have found lower carnitine levels in individuals with ASD (Rossignol & Frye, 2012). Controlled trials have found that carnitine supplementation significantly improves core ASD symptoms (Frye & Rossignol, 2014).

14.3.9 Tetrahydrobiopterin

Tetrahydrobiopterin (BH_4) is an essential cofactor for several critical metabolic pathways, including the synthesis of monoamine neurotransmitters such as serotonin and dopamine that are focal points in ASD. BH_4 levels in cerebrospinal fluid are reported to be 42% lower in ASD children than in neurotypical children (Frye et al., 2010). DBPC studies have demonstrated improvements in social interaction, ASD-related mannerisms, hyperactivity, inappropriate speech, and social awareness following BH_4 treatment (Vargason et al., 2018).

14.4 Dietary Interventions with a Metabolic Approach

Numerous dietary interventions have been studied as potential therapies for ASD, including ketogenic dietary therapies (KDTs), gluten-free and casein-free diet (GFCF), gluten-free diet (GFD), camel milk, antioxidant-containing dietary formulations, etc. These specific diets are thought to alleviate not only core symptoms but also gastrointestinal and metabolic disturbances in children with ASD (Amadi et al., 2022; Yu et al., 2022). However, effects of some other nutritional approaches, such as the fermentable oligosaccharides, disaccharides, monosaccharides, and polyols diet (FODMAP), low oxalate diet, Feingold diet, and low-glutamate/Glutamate-Aspartate Restricted Diet (GARD) remain under-explored (Cekici & Sanlier, 2019).

14.4.1 Ketogenic Dietary Therapies

KDTs include the classic ketogenic diet (KD), the modified Atkins diet (MAD), the medium-chain triglyceride ketogenic diet (MCTKD), and the low glycemic index therapy (LGIT) diet. KD is a high-fat, adequate-protein, low-carbohydrate diet that is a well-established, nonpharmacologic treatment for drug-resistant epilepsy (Varesio et al., 2021). Abundant evidence supports that underlying metabolic dysfunction in ASD could be ameliorated by molecular mechanisms of KD (Cheng et al., 2017; Frye, 2020). A meta-analysis of clinical trials in ASD revealed that the KD had a significant impact on clinician-reported core ASD symptoms (Yu et al., 2022). Several case reports suggest that KDT appears to be a promising intervention for treating ASD associated with comorbid epilepsy, especially in those with refractory epilepsy (Arvio et al., 2010; Herbert & Buckley, 2013; Schiavoni et al., 2021; Wong et al., 2015). According to an online survey, KDT was perceived to significantly improve seizures in ASD (Frye et al., 2011). Metabolic screening for concomitant IEM is crucial before implementing KDT due to possible aggravation of certain IEMs (Kossoff et al., 2018).

14.5 Summary and Conclusions

This chapter is intended to provide a general overview of metabolic treatments for children with ASD. This is an area of medicine where a personalized approach to treatment is optimal. However, a general multivitamin supplementation can be used in most instances to treat multiple common micronutrient deficits (Trudeau et al., 2019). Several clinical trials lead by Dr. James Adam's group at Arizona State University have provided evidence for this general approach (Adams et al., 2011a, 2018; Adams & Holloway, 2004).

References

Abdelgadir, I. S., Gordon, M. A., & Akobeng, A. K. (2018). Melatonin for the management of sleep problems in children with neurodevelopmental disorders: A systematic review and meta-analysis. *Archives of Disease in Childhood, 103*, 1155–1162.

Adams, J. B., & Holloway, C. (2004). Pilot study of a moderate dose multivitamin/mineral supplement for children with autistic spectrum disorder. *Journal of Alternative and Complementary Medicine, 10*, 1033–1039.

Adams, J. B., George, F., & Audhya, T. (2006). Abnormally high plasma levels of vitamin B6 in children with autism not taking supplements compared to controls not taking supplements. *Journal of Alternative and Complementary Medicine, 12*, 59–63.

Adams, J. B., Audhya, T., Mcdonough-Means, S., Rubin, R. A., Quig, D., Geis, E., Gehn, E., Loresto, M., Mitchell, J., Atwood, S., Barnhouse, S., & Lee, W. (2011a). Effect of a vitamin/mineral supplement on children and adults with autism. *BMC Pediatrics, 11*, 111.

Adams, J. B., Audhya, T., Mcdonough-Means, S., Rubin, R. A., Quig, D., Geis, E., Gehn, E., Loresto, M., Mitchell, J., Atwood, S., Barnhouse, S., & Lee, W. (2011b). Nutritional and metabolic status of children with autism vs. neurotypical children, and the association with autism severity. *Nutrition & Metabolism (London), 8*, 34.

Adams, J. B., Audhya, T., Geis, E., Gehn, E., Fimbres, V., Pollard, E. L., Mitchell, J., Ingram, J., Hellmers, R., Laake, D., Matthews, J. S., Li, K., Naviaux, J. C., Naviaux, R. K., Adams, R. L., Coleman, D. M., & Quig, D. W. (2018). Comprehensive nutritional and dietary intervention for autism spectrum disorder-a randomized, controlled 12-month trial. *Nutrients, 10*, 369.

Alabdali, A., Al-Ayadhi, L., & El-Ansary, A. (2014). A key role for an impaired detoxification mechanism in the etiology and severity of autism spectrum disorders. *Behavioral and Brain Functions, 10*, 14.

Aleshin, V. A., Mkrtchyan, G. V., & Bunik, V. I. (2019). Mechanisms of non-coenzyme action of thiamine: Protein targets and medical significance. *Biochemistry (Mosc), 84*, 829–850.

Al-Gadani, Y., El-Ansary, A., Attas, O., & Al-Ayadhi, L. (2009). Metabolic biomarkers related to oxidative stress and antioxidant status in Saudi autistic children. *Clinical Biochemistry, 42*, 1032–1040.

Al-Owain, M., Kaya, N., Al-Shamrani, H., Al-Bakheet, A., Qari, A., Al-Muaigl, S., & Ghaziuddin, M. (2013). Autism spectrum disorder in a child with propionic acidemia. *JIMD Reports, 7*, 63–66.

Amadi, C. N., Orish, C. N., Frazzoli, C., & Orisakwe, O. E. (2022). Dietary interventions for autism spectrum disorder: An updated systematic review of human studies. *Psychiatriki, 33*, 228–242.

Anwar, A., Marini, M., Abruzzo, P. M., Bolotta, A., Ghezzo, A., Visconti, P., Thornalley, P. J., & Rabbani, N. (2016). Quantitation of plasma thiamine, related metabolites and plasma protein oxidative damage markers in children with autism spectrum disorder and healthy controls. *Free Radical Research, 50*, S85–S90.

Arvio, M., Kuisma, L., & Pöntinen, M. (2010). Modified Atkins diet brought back the joy of life to a developmentally severely disabled youth. *Duodecim, 126*, 557–560.

Baltaci, A. K., Mogulkoc, R., & Baltaci, S. B. (2019). Review: The role of zinc in the endocrine system. *Pakistan Journal of Pharmaceutical Sciences, 32*, 231–239.

Bandini, L. G., Anderson, S. E., Curtin, C., Cermak, S., Evans, E. W., Scampini, R., Maslin, M., & Must, A. (2010). Food selectivity in children with autism spectrum disorders and typically developing children. *The Journal of Pediatrics, 157*, 259–264.

Benke, P. J., Duchowny, M., & Mcknight, D. (2018). Biotin and acetazolamide for treatment of an unusual child with autism plus lack of nail and hair growth. *Pediatric Neurology, 79*, 61–64.

Bent, S., Bertoglio, K., & Hendren, R. L. (2009). Omega-3 fatty acids for autistic spectrum disorder: A systematic review. *Journal of Autism and Developmental Disorders, 39*, 1145–1154.

Berger, M. M., Shenkin, A., Amrein, K., Augsburger, M., Biesalski, H. K., Bischoff, S. C., Casaer, M. P., Gundogan, K., Lepp, H. L., De Man, A. M. E., Muscogiuri, G., Pietka, M., Pironi, L., Rezzi, S., Schweinlin, A., & Cuerda, C. (2022). ESPEN micronutrient guideline. *Clinical Nutrition, 41*, 1357–1424.

Bettendorff, L., & Wins, P. (2021). Cofactors and coenzymes | biochemistry of thiamine and thiamine phosphate compounds. In J. Jez (Ed.), *Encyclopedia of biological chemistry III* (3rd ed.). Elsevier.

Bjørklund, G., Aaseth, J., Crisponi, G., Rahman, M. M., & Chirumbolo, S. (2019). Insights on alpha lipoic and dihydrolipoic acids as promising scavengers of oxidative stress and possible chelators in mercury toxicology. *Journal of Inorganic Biochemistry, 195*, 111–119.

Bojja, S. L., Singh, N., Kolathur, K. K., & Rao, C. M. (2022). What is the role of lithium in epilepsy? *Current Neuropharmacology, 20*, 1850–1864.

Brondino, N., Fusar-Poli, L., Rocchetti, M., Provenzani, U., Barale, F., & Politi, P. (2015). Complementary and alternative therapies for autism spectrum disorder. *Evidence-based Complementary and Alternative Medicine, 2015*, 258589.

Cakar, N. E., & Yilmazbas, P. (2021). Cases of inborn errors of metabolism diagnosed in children with autism. *Ideggyógyászati Szemle, 74*, 67–72.

Castro, K., Faccioli, L. S., Baronio, D., Gottfried, C., Perry, I. S., & Riesgo, R. (2016). Feeding behavior and dietary intake of male children and adolescents with autism spectrum disorder: A case-control study. *International Journal of Developmental Neuroscience, 53*, 68–74.

Cekici, H., & Sanlier, N. (2019). Current nutritional approaches in managing autism spectrum disorder: A review. *Nutritional Neuroscience, 22*, 145–155.

Chen, L., Shi, X. J., Liu, H., Mao, X., Gui, L. N., Wang, H., & Cheng, Y. (2021). Oxidative stress marker aberrations in children with autism spectrum disorder: A systematic review and meta-analysis of 87 studies (N = 9109). *Translational Psychiatry, 11*, 15.

Cheng, N., Rho, J. M., & Masino, S. A. (2017). Metabolic dysfunction underlying autism spectrum disorder and potential treatment approaches. *Frontiers in Molecular Neuroscience, 10*, 34.

Clayton, P. T. (2006). B6-responsive disorders: A model of vitamin dependency. *Journal of Inherited Metabolic Disease, 29*, 317–326.

Colamaria, V., Burlina, A. B., Gaburro, D., Pajno-Ferrara, F., Saudubray, J. M., Merino, R. G., & Dalla Bernardina, B. (1989). Biotin-responsive infantile encephalopathy: EEG-polygraphic study of a case. *Epilepsia, 30*, 573–578.

Darville, H., Poulet, A., Rodet-Amsellem, F., Chatrousse, L., Pernelle, J., Boissart, C., Héron, D., Nava, C., Perrier, A., Jarrige, M., Cogé, F., Millan, M. J., Bourgeron, T., Peschanski, M., Delorme, R., & Benchoua, A. (2016). Human pluripotent stem cell-derived cortical neurons for high throughput medication screening in autism: A proof of concept study in SHANK3 haploinsufficiency syndrome. *eBioMedicine, 9*, 293–305.

De Andrade Wobido, K., De Sá Barreto Da Cunha, M., Miranda, S. S., Da Mota Santana, J., Da Silva, D. C. G., & Pereira, M. (2021). Non-specific effect of omega-3 fatty acid supplementation on autistic spectrum disorder: Systematic review and meta-analysis. *Nutritional Neuroscience, 25*, 1995–2007.

De Crescenzo, F., D'Alò, G. L., Morgano, G. P., Minozzi, S., Mitrova, Z., Saulle, R., Cruciani, F., Fulceri, F., Davoli, M., Scattoni, M. L., Nardocci, F., Schünemann, H. J., Amato, L., Nardocci, F., & ISACA Guideline Working Group. (2020). Impact of polyunsaturated fatty acids on patient-important outcomes in children and adolescents with autism spectrum disorder: A systematic review. *Health and Quality of Life Outcomes, 18*, 28.

De la Bâtie, C. D., Barbier, V., Roda, C., Brassier, A., Arnoux, J. B., Valayannopoulos, V., Guemann, A. S., Pontoizeau, C., Gobin, S., Habarou, F., Lacaille, F., Bonnefont, J. P., Canouï, P., Ottolenghi, C., De Lonlay, P., & Ouss, L. (2018). Autism spectrum disorders in propionic acidemia patients. *Journal of Inherited Metabolic Disease, 41*, 623–629.

Deepmala, Slattery, J., Kumar, N., Delhey, L., Berk, M., Dean, O., Spielholz, C., & Frye, R. (2015). Clinical trials of N-acetylcysteine in psychiatry and neurology: A systematic review. *Neuroscience and Biobehavioral Reviews, 55*, 294–321.

Dolske, M. C., Spollen, J., Mckay, S., Lancashire, E., & Tolbert, L. (1993). A preliminary trial of ascorbic acid as supplemental therapy for autism. *Progress in Neuro-Psychopharmacology and Biological Psychiatry, 17*, 765–774.

El-Ansary, A., Bjørklund, G., Chirumbolo, S., & Alnakhli, O. M. (2017). Predictive value of selected biomarkers related to metabolism and oxidative stress in children with autism spectrum disorder. *Metabolic Brain Disease, 32*, 1209–1221.

Esteban-Figuerola, P., Canals, J., Fernández-Cao, J. C., & Arija Val, V. (2019). Differences in food consumption and nutritional intake between children with autism spectrum disorders and typically developing children: A meta-analysis. *Autism, 23*, 1079–1095.

Frustaci, A., Neri, M., Cesario, A., Adams, J. B., Domenici, E., Dalla Bernardina, B., & Bonassi, S. (2012). Oxidative stress-related biomarkers in autism: Systematic review and meta-analyses. *Free Radical Biology & Medicine, 52*, 2128–2141.

Frye, R. E. (2015). Metabolic and mitochondrial disorders associated with epilepsy in children with autism spectrum disorder. *Epilepsy & Behavior, 47*, 147–157.

Frye, R. E. (2020). Mitochondrial dysfunction in autism spectrum disorder: Unique abnormalities and targeted treatments. *Seminars in Pediatric Neurology, 35*, 100829.

Frye, R. E., & Rossignol, D. A. (2014). Treatments for biomedical abnormalities associated with autism spectrum disorder. *Frontiers in Pediatrics, 2*, 66.

Frye, R. E., Huffman, L. C., & Elliott, G. R. (2010). Tetrahydrobiopterin as a novel therapeutic intervention for autism. *Neurotherapeutics, 7*, 241–249.

Frye, R. E., Sreenivasula, S., & Adams, J. B. (2011). Traditional and non-traditional treatments for autism spectrum disorder with seizures: An on-line survey. *BMC Pediatrics, 11*, 37.

Gagnon, K., & Godbout, R. (2018). Melatonin and comorbidities in children with autism spectrum disorder. *Current Developmental Disorders Reports, 5*, 197–206.

Gáll, Z., & Székely, O. (2021). Role of vitamin D in cognitive dysfunction: New molecular concepts and discrepancies between animal and human findings. *Nutrients, 13*, 3672.

Gallardo-Carrasco, M. C., Jiménez-Barbero, J. A., Bravo-Pastor, M. D. M., Martin-Castillo, D., & Sánchez-Muñoz, M. (2022). Serum vitamin D, folate and fatty acid levels in children with autism spectrum disorders: A systematic review and meta-analysis. *Journal of Autism and Developmental Disorders, 52*, 4708–4721.

García-Serna, A. M., & Morales, E. (2020). Neurodevelopmental effects of prenatal vitamin D in humans: Systematic review and meta-analysis. *Molecular Psychiatry, 25*, 2468–2481.

Gevi, F., Zolla, L., Gabriele, S., & Persico, A. M. (2016). Urinary metabolomics of young Italian autistic children supports abnormal tryptophan and purine metabolism. *Molecular Autism, 7*, 47.

Ghanizadeh, A., & Moghimi-Sarani, E. (2013). A randomized double blind placebo controlled clinical trial of N-acetylcysteine added to risperidone for treating autistic disorders. *BMC Psychiatry, 13*, 196–196.

Golriz, F., Donnelly, L. F., Devaraj, S., & Krishnamurthy, R. (2017). Modern American scurvy – experience with vitamin C deficiency at a large children's hospital. *Pediatric Radiology, 47*, 214–220.

Grabrucker, A. M. (2016). Zinc in the developing brain. In V. H. Moran & N. Lowe (Eds.), *Nutrition and the developing brain*. CRC Press.

Guerrero-Beltrán, C. E., Calderón-Oliver, M., Pedraza-Chaverri, J., & Chirino, Y. I. (2012). Protective effect of sulforaphane against oxidative stress: Recent advances. *Experimental and Toxicologic Pathology, 64*, 503–508.

Guo, M., Zhu, J., Yang, T., Lai, X., Liu, X., Liu, J., Chen, J., & Li, T. (2018). Vitamin A improves the symptoms of autism spectrum disorders and decreases 5-hydroxytryptamine (5-HT): A pilot study. *Brain Research Bulletin, 137*, 35–40.

Gvozdjáková, A., Kucharská, J., Ostatníková, D., Babinská, K., Nakládal, D., & Crane, F. L. (2014). Ubiquinol improves symptoms in children with autism. *Oxidative Medicine and Cellular Longevity, 2014*, 798957.

Hardan, A. Y., Fung, L. K., Libove, R. A., Obukhanych, T. V., Nair, S., Herzenberg, L. A., Frazier, T. W., & Tirouvanziam, R. (2012). A randomized controlled pilot trial of oral N-acetylcysteine in children with autism. *Biological Psychiatry, 71*, 956–961.

Herbert, M. R., & Buckley, J. A. (2013). Autism and dietary therapy: Case report and review of the literature. *Journal of Child Neurology, 28*, 975–982.

Huijmans, J. G. M., Schot, R., De Klerk, J. B. C., Williams, M., De Coo, R. F. M., Duran, M., Verheijen, F. W., Van Slegtenhorst, M., & Mancini, G. M. S. (2017). Molybdenum cofactor deficiency: Identification of a patient with homozygote mutation in the MOCS3 gene. *American Journal of Medical Genetics. Part A, 173*, 1601–1606.

Hyman, S. L., Stewart, P. A., Schmidt, B., Cain, U., Lemcke, N., Foley, J. T., Peck, R., Clemons, T., Reynolds, A., Johnson, C., Handen, B., James, S. J., Courtney, P. M., Molloy, C., & Ng, P. K. (2012). Nutrient intake from food in children with autism. *Pediatrics, 130*(Suppl 2), S145–S153.

Inci, A., Özaslan, A., Okur, I., Biberoğlu, G., Güney, E., Ezgü, F. S., Tümer, L., & Işeri, E. (2021). Autism: Screening of inborn errors of metabolism and unexpected results. *Autism Research, 14*, 887–896.

Karhu, E., Zukerman, R., Eshraghi, R. S., Mittal, J., Deth, R. C., Castejon, A. M., Trivedi, M., Mittal, R., & Eshraghi, A. A. (2020). Nutritional interventions for autism spectrum disorder. *Nutrition Reviews, 78*, 515–531.

Khan, F., Rahman, M. S., Akhter, S., Momen, A. B. I., & Raihan, S. G. (2021). Vitamin B6 and magnesium on neurobehavioral status of autism spectrum disorder: A randomized, double-blind, placebo controlled study. *Bangladesh Journal of Medicine, 32*, 12–18.

Kim, J. K., & Park, S. U. (2016). Current potential health benefits of sulforaphane. *EXCLI Journal, 15*, 571–577.

Kiykim, E., Zeybek, C. A., Zubarioglu, T., Cansever, S., Yalcinkaya, C., Soyucen, E., & Aydin, A. (2016). Inherited metabolic disorders in Turkish patients with autism spectrum disorders. *Autism Research, 9*, 217–223.

Kossoff, E. H., Zupec-Kania, B. A., Auvin, S., Ballaban-Gil, K. R., Christina Bergqvist, A. G., Blackford, R., Buchhalter, J. R., Caraballo, R. H., Cross, J. H., Dahlin, M. G., Donner, E. J., Guzel, O., Jehle, R. S., Klepper, J., Kang, H. C., Lambrechts, D. A., Liu, Y. M. C., Nathan, J. K., Nordli, D. R., Jr., Pfeifer, H. H., Rho, J. M., Scheffer, I. E., Sharma, S., Stafstrom, C. E., Thiele, E. A., Turner, Z., Vaccarezza, M. M., Van der Louw, E., Veggiotti, P., Wheless, J. W., Wirrell, E. C., Charlie, F., Matthew's, F., & Practice Committee of the Child Neurology Society. (2018). Optimal clinical management of children receiving dietary therapies for epilepsy: Updated recommendations of the International Ketogenic Diet Study Group. *Epilepsia Open, 3*, 175–192.

Legido, A., Goldenthal, M., Garvin, B., Damle, S., Corrigan, K., Connell, J., Thao, D., Valencia, I., Melvin, J., Khurana, D., Grant, M., & Newschaffer, C. (2018). Effect of a combination of carnitine, coenzyme Q10 and alpha-lipoic acid (MitoCocktail) on mitochondrial function and neurobehavioral performance in children with autism spectrum disorder (P3.313). *Neurology, 90*, P3.313.

Li, B., Xu, Y., Zhang, X., Zhang, L., Wu, Y., Wang, X., & Zhu, C. (2022). The effect of vitamin D supplementation in treatment of children with autism spectrum disorder: A systematic review and meta-analysis of randomized controlled trials. *Nutritional Neuroscience, 25*, 835–845.

Liang, Y., Ke, X., Xiao, Z., Zhang, Y., Chen, Y., Li, Y., Wang, Z., Lin, L., Yao, P., & Lu, J. (2020a). Untargeted metabolomic profiling using UHPLC-QTOF/MS reveals metabolic alterations associated with autism. *Biomed Research International, 2020*, 6105608.

Liang, Y., Xiao, Z., Ke, X., Yao, P., Chen, Y., Lin, L., & Lu, J. (2020b). Urinary metabonomic profiling discriminates between children with autism and their healthy siblings. *Medical Science Monitor, 26*, e926634.

Lindsay, R. L., Eugene Arnold, L., Aman, M. G., Vitiello, B., Posey, D. J., Mcdougle, C. J., Scahill, L., Pachler, M., Mccracken, J. T., Tierney, E., & Bozzolo, D. (2006). Dietary status and impact of risperidone on nutritional balance in children with autism: A pilot study. *Journal of Intellectual & Developmental Disability, 31*, 204–209.

Lionel, A. C., Vaags, A. K., Sato, D., Gazzellone, M. J., Mitchell, E. B., Chen, H. Y., Costain, G., Walker, S., Egger, G., Thiruvahindrapuram, B., Merico, D., Prasad, A., Anagnostou, E., Fombonne, E., Zwaigenbaum, L., Roberts, W., Szatmari, P., Fernandez, B. A., Georgieva, L., Brzustowicz, L. M., Roetzer, K., Kaschnitz, W., Vincent, J. B., Windpassinger, C., Marshall, C. R., Trifiletti, R. R., Kirmani, S., Kirov, G., Petek, E., Hodge, J. C., Bassett, A. S., & Scherer, S. W. (2013). Rare exonic deletions implicate the synaptic organizer Gephyrin (GPHN) in risk for autism, schizophrenia and seizures. *Human Molecular Genetics, 22*, 2055–2066.

Liu, X., Liu, J., Xiong, X., Yang, T., Hou, N., Liang, X., Chen, J., Cheng, Q., & Li, T. (2016). Correlation between nutrition and symptoms: Nutritional survey of children with autism spectrum disorder in Chongqing, China. *Nutrients, 8*, 294.

Lonsdale, D. (2018). Thiamin. *Advances in Food and Nutrition Research, 83*, 1–56.

Lutz, A. K., Pfaender, S., Incearap, B., Ioannidis, V., Ottonelli, I., Föhr, K. J., Cammerer, J., Zoller, M., Higelin, J., Giona, F., Stetter, M., Stoecker, N., Alami, N. O., Schön, M., Orth, M., Liebau, S., Barbi, G., Grabrucker, A. M., Delorme, R., Fauler, M., Mayer, B., Jesse, S., Roselli, F., Ludolph, A. C., Bourgeron, T., Verpelli, C., Demestre, M., & Boeckers, T. M. (2020). Autism-associated SHANK3 mutations impair maturation of neuromuscular junctions and striated muscles. *Science Translational Medicine, 12*, eaaz3267.

Malow, B., Adkins, K. W., Mcgrew, S. G., Wang, L., Goldman, S. E., Fawkes, D., & Burnette, C. (2012). Melatonin for sleep in children with autism: A controlled trial examining dose, tolerability, and outcomes. *Journal of Autism and Developmental Disorders, 42*, 1729–1737; author reply 1738.

Marshall, R. P., Droste, J. N., Giessing, J., & Kreider, R. B. (2022). Role of creatine supplementation in conditions involving mitochondrial dysfunction: A narrative review. *Nutrients, 14*, 529.

Maywald, M., Wessels, I., & Rink, L. (2017). Zinc signals and immunity. *International Journal of Molecular Sciences, 18*, 2222.

Mcguinness, G., & Kim, Y. (2020). Sulforaphane treatment for autism spectrum disorder: A systematic review. *EXCLI Journal, 19*, 892–903.

Meguid, N. A., Bjørklund, G., Gebril, O. H., Doşa, M. D., Anwar, M., Elsaeid, A., Gaber, A., & Chirumbolo, S. (2019). The role of zinc supplementation on the metallothionein system in children with autism spectrum disorder. *Acta Neurologica Belgica, 119*, 577–583.

Mercimek-Andrews, S., & Salomons, G. S. (1993). Creatine deficiency disorders. In M. P. Adam, H. H. Ardinger, R. A. Pagon, S. E. Wallace, L. J. H. Bean, K. W. Gripp, G. M. Mirzaa, & A. Amemiya (Eds.), *GeneReviews®*. University of Washington. Copyright © 1993–2022, University of Washington, Seattle. GeneReviews is a registered trademark of the University of Washington, Seattle. All rights reserved.

Mierau, S. B., & Neumeyer, A. M. (2019). Metabolic interventions in autism spectrum disorder. *Neurobiology of Disease, 132*, 104544.

Mkrtchyan, G., Aleshin, V., Parkhomenko, Y., Kaehne, T., Di Salvo, M. L., Parroni, A., Contestabile, R., Vovk, A., Bettendorff, L., & Bunik, V. (2015). Molecular mechanisms of the non-coenzyme action of thiamin in brain: Biochemical, structural and pathway analysis. *Scientific Reports, 5*, 12583.

Morris, C. R., & Agin, M. C. (2009). Syndrome of allergy, apraxia, and malabsorption: Characterization of a neurodevelopmental phenotype that responds to omega 3 and vitamin E supplementation. *Alternative Therapies in Health and Medicine, 15*, 34–43.

Mousain-Bosc, M., Roche, M., Polge, A., Pradal-Prat, D., Rapin, J., & Bali, J. P. (2006). Improvement of neurobehavioral disorders in children supplemented with magnesium-vitamin B6. II. Pervasive developmental disorder-autism. *Magnesium Research, 19*, 53–62.

Mousavinejad, E., Ghaffari, M. A., Riahi, F., Hajmohammadi, M., Tiznobeyk, Z., & Mousavinejad, M. (2018). Coenzyme Q10 supplementation reduces oxidative stress and decreases antioxidant enzyme activity in children with autism spectrum disorders. *Psychiatry Research, 265*, 62–69.

Niyazov, D. M., Kahler, S. G., & Frye, R. E. (2016). Primary mitochondrial disease and secondary mitochondrial dysfunction: Importance of distinction for diagnosis and treatment. *Molecular Syndromology, 7*, 122–137.

Nye, C., & Brice, A. (2005). Combined vitamin B6-magnesium treatment in autism spectrum disorder. *Cochrane Database of Systematic Reviews, 2005*, CD003497.

O'Neal, S., Foster, T. P., Bhatt, A., Lossius, M. N., & Dayton, K. (2020). Hypercalcemia from hypervitaminosis A in a child with autism. *Journal of Pediatric Endocrinology & Metabolism.* https://doi.org/10.1515/jpem-2020-0075

Pangrazzi, L., Balasco, L., & Bozzi, Y. (2020). Natural antioxidants: A novel therapeutic approach to autism spectrum disorders? *Antioxidants (Basel), 9*, 1186.

Park, S., Karunakaran, U., Jeoung, N. H., Jeon, J. H., & Lee, I. K. (2014). Physiological effect and therapeutic application of alpha lipoic acid. *Current Medicinal Chemistry, 21*, 3636–3645.

Peake, R. W., Godber, I. M., & Maguire, D. (2013). The effect of magnesium administration on erythrocyte transketolase activity in alcoholic patients treated with thiamine. *Scottish Medical Journal, 58*, 139–142.

Qian, B., Shen, S., Zhang, J., & Jing, P. (2017). Effects of vitamin B6 deficiency on the composition and functional potential of T cell populations. *Journal of Immunology Research, 2017*, 2197975.

Robea, M. A., Luca, A. C., & Ciobica, A. (2020). Relationship between vitamin deficiencies and co-occurring symptoms in autism spectrum disorder. *Medicina (Kaunas, Lithuania), 56*, 245.

Rose, S., Niyazov, D. M., Rossignol, D. A., Goldenthal, M., Kahler, S. G., & Frye, R. E. (2018). Clinical and molecular characteristics of mitochondrial dysfunction in autism spectrum disorder. *Molecular Diagnosis & Therapy, 22*, 571–593.

Rossignol, D. A., & Frye, R. E. (2012). Mitochondrial dysfunction in autism spectrum disorders: A systematic review and meta-analysis. *Molecular Psychiatry, 17*, 290–314.

Rossignol, D. A., & Frye, R. E. (2014). Melatonin in autism spectrum disorders. *Current Clinical Pharmacology, 9*, 326–334.

Rossignol, D. A., & Frye, R. E. (2021a). Cerebral folate deficiency, folate receptor alpha autoantibodies and leucovorin (folinic acid) treatment in autism spectrum disorders: A systematic review and meta-analysis. *Journal of Personalized Medicine, 11*, 1141.

Rossignol, D. A., & Frye, R. E. (2021b). The effectiveness of cobalamin (B12) treatment for autism spectrum disorder: A systematic review and meta-analysis. *Journal of Personalized Medicine, 11*, 784.

Russo, A. J. (2011). Increased copper in individuals with autism normalizes post zinc therapy more efficiently in individuals with concurrent GI disease. *Nutrition and Metabolic Insights, 4*, 49–54.

Russo, A. J., & Devito, R. (2011). Analysis of copper and zinc plasma concentration and the efficacy of zinc therapy in individuals with Asperger's syndrome, pervasive developmental disorder not otherwise specified (PDD-NOS) and autism. *Biomarker Insights, 6*, 127–133.

Saghazadeh, A., Ahangari, N., Hendi, K., Saleh, F., & Rezaei, N. (2017). Status of essential elements in autism spectrum disorder: Systematic review and meta-analysis. *Reviews in the Neurosciences, 28*, 783–809.

Sayehmiri, F., Babaknejad, N., Bahrami, S., Sayehmiri, K., Darabi, M., & Rezaei-Tavirani, M. (2015). Zn/Cu levels in the field of autism disorders: A systematic review and meta-analysis. *Iranian Journal of Child Neurology, 9*, 1–9.

Schiavoni, S., Spagnoli, C., Rizzi, S., Salerno, G. G., Frattini, D., Pisani, F., & Fusco, C. (2021). Beneficial effects of the ketogenic diet on drug-resistant epileptic encephalopathy associated with a de novo NBEA pathogenic variant. *Epileptic Disorders, 23*, 739–743.

Schroder, C. M., Banaschewski, T., Fuentes, J., Hill, C. M., Hvolby, A., Posserud, M. B., & Bruni, O. (2021). Pediatric prolonged-release melatonin for insomnia in children and adolescents with autism spectrum disorders. *Expert Opinion on Pharmacotherapy, 22*, 2445–2454.

Singh, M., Thompson, M., Sullivan, N., & Child, G. (2005). Thiamine deficiency in dogs due to the feeding of sulphite preserved meat. *Australian Veterinary Journal, 83*, 412–417.

Stewart, P. A., Hyman, S. L., Schmidt, B. L., Macklin, E. A., Reynolds, A., Johnson, C. R., James, S. J., & Manning-Courtney, P. (2015). Dietary supplementation in children with autism spectrum disorders: Common, insufficient, and excessive. *Journal of the Academy of Nutrition and Dietetics, 115*, 1237–1248.

Sun, C., Xia, W., Zhao, Y., Li, N., Zhao, D., & Wu, L. (2013). Nutritional status survey of children with autism and typically developing children aged 4–6 years in Heilongjiang Province, China. *Journal of Nutritional Science, 2*, e16.

Tardy, A. L., Pouteau, E., Marquez, D., Yilmaz, C., & Scholey, A. (2020). Vitamins and minerals for energy, fatigue and cognition: A narrative review of the biochemical and clinical evidence. *Nutrients, 12*, 228.

Tian, Y. J., Li, D., Ma, Q., Gu, X. Y., Guo, M., Lun, Y. Z., Sun, W. P., Wang, X. Y., Cao, Y., & Zhou, S. S. (2013). Excess nicotinamide increases plasma serotonin and histamine levels. *Sheng Li Xue Bao, 65*, 33–38.

Trudeau, M. S., Madden, R. F., Parnell, J. A., Gibbard, W. B., & Shearer, J. (2019). Dietary and supplement-based complementary and alternative medicine use in pediatric autism spectrum disorder. *Nutrients, 11*, 1783.

Tsujiguchi, H., Miyagi, S., Nguyen, T. T. T., Hara, A., Ono, Y., Kambayashi, Y., Shimizu, Y., Nakamura, H., Suzuki, K., Suzuki, F., & Nakamura, H. (2020). Relationship between autistic traits and nutrient intake among Japanese children and adolescents. *Nutrients, 12*, 2258.

Varesio, C., Grumi, S., Zanaboni, M. P., Mensi, M. M., Chiappedi, M., Pasca, L., Ferraris, C., Tagliabue, A., Borgatti, R., & De Giorgis, V. (2021). Ketogenic dietary therapies in patients with autism spectrum disorder: Facts or fads? A scoping review and a proposal for a shared protocol. *Nutrients, 13*, 2057.

Vargason, T., Kruger, U., Roth, E., Delhey, L. M., Tippett, M., Rose, S., Bennuri, S. C., Slattery, J. C., Melnyk, S., James, S. J., Frye, R. E., & Hahn, J. (2018). Comparison of three clinical trial treatments for autism spectrum disorder through multivariate analysis of changes in metabolic profiles and adaptive behavior. *Frontiers in Cellular Neuroscience, 12*, 503.

Vela, G., Stark, P., Socha, M., Sauer, A. K., Hagmeyer, S., & Grabrucker, A. M. (2015). Zinc in gut-brain interaction in autism and neurological disorders. *Neural Plasticity, 2015*, 972791.

Wang, L., Erlandsen, H., Haavik, J., Knappskog, P. M., & Stevens, R. C. (2002). Three-dimensional structure of human tryptophan hydroxylase and its implications for the biosynthesis of the neurotransmitters serotonin and melatonin. *Biochemistry, 41*, 12569–12574.

Wang, T., Shan, L., Du, L., Feng, J., Xu, Z., Staal, W. G., & Jia, F. (2016). Serum concentration of 25-hydroxyvitamin D in autism spectrum disorder: A systematic review and meta-analysis. *European Child & Adolescent Psychiatry, 25*, 341–350.

Wink, L. K., Adams, R., Wang, Z., Klaunig, J. E., Plawecki, M. H., Posey, D. J., Mcdougle, C. J., & Erickson, C. A. (2016). A randomized placebo-controlled pilot study of N-acetylcysteine in youth with autism spectrum disorder. *Molecular Autism, 7*, 26.

Wong, V. C., Fung, C. W., & Kwong, A. K. (2015). SCN2A mutation in a Chinese boy with infantile spasm – response to Modified Atkins Diet. *Brain and Development, 37*, 729–732.

Wu, Z. Y., Huang, S. D., Zou, J. J., Wang, Q. X., Naveed, M., Bao, H. N., Wang, W., Fukunaga, K., & Han, F. (2020). Autism spectrum disorder (ASD): Disturbance of the melatonin system and its implications. *Biomedicine & Pharmacotherapy, 130*, 110496.

Xia, W., Zhou, Y., Sun, C., Wang, J., & Wu, L. (2010). A preliminary study on nutritional status and intake in Chinese children with autism. *European Journal of Pediatrics, 169*, 1201–1206.

Yap, I. K., Angley, M., Veselkov, K. A., Holmes, E., Lindon, J. C., & Nicholson, J. K. (2010). Urinary metabolic phenotyping differentiates children with autism from their unaffected siblings and age-matched controls. *Journal of Proteome Research, 9*, 2996–3004.

Yoldaş, T., Gürbüz, B. B., Akar, H. T., Özmert, E. N., & Coşkun, T. (2021). Autism spectrum disorder in patients with inherited metabolic disorders-a large sample from a tertiary center. *The Turkish Journal of Pediatrics, 63*, 767–779.

Yu, Y., Huang, J., Chen, X., Fu, J., Wang, X., Pu, L., Gu, C., & Cai, C. (2022). Efficacy and safety of diet therapies in children with autism spectrum disorder: A systematic literature review and meta-analysis. *Frontiers in Neurology, 13*, 844117.

Yule, S., Wanik, J., Holm, E. M., Bruder, M. B., Shanley, E., Sherman, C. Q., Fitterman, M., Lerner, J., Marcello, M., Parenchuck, N., Roman-White, C., & Ziff, M. (2021). Nutritional deficiency disease secondary to ARFID symptoms associated with autism and the broad autism phenotype: A qualitative systematic review of case reports and case series. *Journal of the Academy of Nutrition and Dietetics, 121*, 467–492.

Zaffanello, M., Zamboni, G., Fontana, E., Zoccante, L., & Tatò, L. (2003). A case of partial biotinidase deficiency associated with autism. *Child Neuropsychology, 9*, 184–188.

Żarnowska, I., Chrapko, B., Gwizda, G., Nocuń, A., Mitosek-Szewczyk, K., & Gasior, M. (2018). Therapeutic use of carbohydrate-restricted diets in an autistic child; a case report of clinical and 18FDG PET findings. *Metabolic Brain Disease, 33*, 1187–1192.

Zhang, J., Li, X., Shen, L., Khan, N. U., Zhang, X., Chen, L., Zhao, H., & Luo, P. (2021). Trace elements in children with autism spectrum disorder: A meta-analysis based on case-control studies. *Journal of Trace Elements in Medicine and Biology, 67*, 126782.

Zhu, J., Guo, M., Yang, T., Lai, X., Tang, T., Chen, J., Li, L., & Li, T. (2020). Nutritional status and symptoms in preschool children with autism spectrum disorder: A two-center comparative study in Chongqing and Hainan Province, China. *Frontiers in Pediatrics, 8*, 469.

Žigman, T., Petković Ramadža, D., Šimić, G., & Barić, I. (2021). Inborn errors of metabolism associated with autism spectrum disorders: Approaches to intervention. *Frontiers in Neuroscience, 15*, 673600.

Zimmerman, A. W., Singh, K., Connors, S. L., Liu, H., Panjwani, A. A., Lee, L. C., Diggins, E., Foley, A., Melnyk, S., Singh, I. N., James, S. J., Frye, R. E., & Fahey, J. W. (2021). Randomized controlled trial of sulforaphane and metabolite discovery in children with autism spectrum disorder. *Molecular Autism, 12*, 38.

Chapter 15
Autism and Neurodiversity

T. A. Meridian McDonald

15.1 Heterogeneity in Autism

Autism[1] may be a single diagnosis, but the disorder is highly heterogeneous. The diagnostic criteria define autism spectrum disorder as impairments in social communication and interaction combined with narrow or repetitive behaviors and interests. Yet, the manner in which these characteristics manifest is highly variable (American Psychiatric Association, 2013). Autistic people show an array of phenotypic traits, predictive (risk[2]) factors, and biomarkers (Emberti Gialloreti et al., 2019; Lord et al., 2020a; Warrier et al., 2022).

Autism is also one of the most heritable medical conditions. The most genetic heritability occurs through polygenic clusters of common genetic variants (Grove et al., 2019). Yet, rare de novo heterozygous mutations are highly predictive of an autism diagnosis (Grove et al., 2019; Lord et al., 2020a). These two genetic pathways are also related to whether an autist has an intellectual disability. Genetic

[1] The terminology around autism is controversial (Baron-Cohen, 2017; Dwyer, 2022; Kenny et al., 2016). Controversy regarding this terminology is deeply embedded with concepts from various models of autism and disability described within this chapter. Based on the documented preferences of the autism community, (Dwyer, 2022; Kenny et al., 2016), this chapter uses the neutral term "autism" instead of autism spectrum disorder.

[2] The term "risk factor" is defined as something that increases the chance of developing a disease. Although the designation of autism as a disease or disorder is controversial, there are few neutral phrasings for the concept of "risk factors" for scientific understanding of the etiology, mechanisms, or predictive factors of autistic traits in human populations. Thus, this chapter will highlight common terms that describe scientific phenomena from a disease-centric standpoint.

T. A. M. McDonald (✉)
Vanderbilt University Medical Center, Nashville, TN, USA

A. El Idrissi, D. McCloskey (eds.), *Neurobiology of Autism Spectrum Disorders*,
https://doi.org/10.1007/978-3-031-42383-3_15

disorders have a higher association with intellectual disabilities, and common genetic variants have a higher association with a lack of intellectual disabilities and greater education attainment (Grove et al., 2019).

The heterogeneity of autism has also been linked to complex biological mechanisms and mechanistic theories of autism, such as signaling imbalance theory (Rojas et al., 2015), connectivity theory (Tang et al., 2020), serotonin regulation (Garbarino et al., 2019), executive functions (Demetriou et al., 2019), and gut microbiome (Bundgaard-Nielsen et al., 2020). To better understand these complexities, some researchers propose narrowing or redefining autism (Mottron & Bzdok, 2020). Others encourage analyzing large datasets with various techniques, such as machine learning, to identify subpopulations within autism (Hassan & Mokhtar, 2019; Jacob et al., 2019; Lombardo et al., 2019; Molloy & Gallagher, 2021). But the definition of autism, who has the authority to define it, and what research should be prioritized is controversial. This controversy is influenced by neurodiversity and movements (neurodiversity, prevention, and recovery) based on societal models of impairment, disability, and difference.

15.2 What Is Neurodiversity

The term and concept of neurodiversity was originally coined in the 1990s by the autistic sociologist, Judy Singer (Singer, 2017). Since then, the term has become so commonplace that it has its own entry in the Oxford English Dictionary (OED, 2023). According to this dictionary, neurodiversity is "a range of variation in mental or neurological functioning in a group, esp. one that includes some persons on the autism spectrum; the state or quality of being neurodiverse." This definition of differences and variation means different things to different people. As a result, proponents of the neurodiversity movement represent a wide range of ideas.

Neurodiversity is inherent in autism, largely due to the heterogeneity of autism and the vast diversity of characteristics across autistic people. This diversity influences how people view the nature of autism and how it should be handled by society (Cascio, 2012; Perry, 2014; Richman, 2020; Bumiller, 2008; Jaarsma & Welin, 2012; Lim, 2015). It also fuels controversy within the autism community and between the autism community and other communities, such as parents, clinicians, researchers, and policymakers. Autists and other stakeholders endorse concepts embedded within multiple models of disability and autism. The following section covers several of these major models.

15.3 Medical Model

15.3.1 Historical Medical Model

Throughout history, the concept of what constitutes a disability and what having a disability means for a person and for society has changed over time (Davis, 2016). For example, the prevailing medical model stems from developing and applying statistics to human development and function (Davis, 2016). In the nineteenth century, Adolphe Quetelet formed his concept of the "average man" based on the statistical average of traits within a population (Davis, 2016). These averaged traits formed the basis of an ideal, "normal," healthy person (Davis, 2016). In this way, the concept of abnormal traits that deviated from the statistical mean became synonymous with unhealthy (Davis, 2016). This connection is not coincidental—nearly all early statisticians were eugenicists (e.g., Sir. Francis Galton, Karl Peterson, R.A. Fisher) (Davis, 2016; Guthrie, 2004).

Eugenicists sought measures, including statistics, to determine the "fitness" of a population to direct the evolution of society toward what they perceived as enhanced evolutionary "fitness" (Davis, 2016; Guthrie, 2004). Although sometimes based on statistical norms of an entire population, ideas of what constituted fitness were subjective and often based on the characteristics and values of the statisticians' culture (Guthrie, 2004). In this way, women, various racial and ethnic groups, lower social-economic classes, criminals, and other groups of people were labeled with various deficiencies. For example, many Europeans believed that humans were originally white-skinned and of European culture (Guthrie, 2004). From this view, variation in skin color from white represented deviations from the cultural belief that white skin was the ideal standard. These views fueled the effort to measure and standardize human traits. For example, Aleš Hrdlička, an anthropologist at the U.S. Smithsonian Museum, measured a broad range of human traits and is known for attempting to standardize a measure of skin-color differences. Also, culturally biased "intelligence" tests that contained references to culturally specific practices and knowledge were used to demonstrate "feeblemindedness" (Davis, 2016; Guthrie, 2004).

Unfortunately, eugenicists used characteristics that systematically differed between groups, such as between people of European and African descent, to rationalize racist and eugenic beliefs and practices. For example, Lewis Terman proposed occupational stratification based on scores from culturally biased intelligence tests: "preliminary investigations indicate that an IQ below 70 rarely permits anything better than unskilled labor; that the range from 70 to 80 is preeminently that of semi-skilled labor; from 80 to 100 that of the skilled or ordinary clerical labor…" (Terman, 1920, p. 31). Terman also proposed that children's scores on intelligence tests be used to determine their course of education (Terman, 1920). More ominously, eugenicists argued that traits that departed from the idealized European man represented disabling forms of unfitness that contributed to undesirable situations in society (Davis, 2016). They believed that society would benefit if these deficiencies were eliminated from the population, which led to involuntary sterilization of people from many marginalized groups in society (Guthrie, 2004).

Disability has also been used to justify inequality in society (Baynton, 2005). Regardless of whether eugenicists' eyes fell on people from different social classes, racial and ethnic groups, women, or other groups, the focus centered on the presumed disability of these groups (Baynton, 2005). And society has used the presumed possession of a disability to justify exclusions to the rights to freedom, civil liberties, political and social equality, and access to the community (Baynton, 2005; Davis, 2016; Dolmage, 2011).

For autists, the practice of eugenics is concerning. Baker and Lang (2017) believe this concern motivates the inclusion of strengths within descriptions of autism by both Hans Asperger and Leo Kanner. Informed by the treatment of people diagnosed with intellectual disabilities, Baker and Lang assert that Kanner and Asperger sought to protect autists from the eugenics movement by emphasizing intellectual abilities and "gifts" to avoid classifying this population as "unfit." However, Baker and Lang assert that Kanner and Asperger de-emphasized, or disqualified, cases with co-occurring intellectual disabilities, which excludes important subpopulations within the autistic community.

15.3.2 Modern Medical Model

In the past decade, eugenics aims have largely been replaced by an emphasis on inequities in society that result in undesirable and preventable health outcomes (National Academies of Sciences et al., 2017; National Institute on Minority Health and Health Disparities, 2021). In other words, the modern medical model focuses on reducing suffering and mortality of people (Mission and Goals, 2014; What We Do | NIHR, n.d.), and on preventing and treating disorders and diseases. This approach is largely uncontroversial for most disorders and diseases, such as dementia, Parkinson's disease, or cancer. But it is controversial for other conditions, such as autism.

Over time, the perspectives on what constitutes a disorder or a way of being have shifted. For example, the medical community historically classified being gay men or lesbians as a disorder. Now, they consider these ways of being as part of the natural variation of sexual orientation. The rationale for classifying "homosexuality" as a disorder arose, in part, from the decreased quality of life and distress experienced by gay men and lesbians. However, these outcomes are now attributed to the discrimination and social marginalization faced by the lesbian, gay, bisexual, transgender, queer, intersex, asexual, plus (LGBTQIA+) communities.

Similar to this shift, the medical community now debates whether other conditions or ways of being, such as autism, should be considered a disorder or a different, but equally valid, way of being. Currently, the medical community classifies autism as a disorder due to differences between autistic and non-autistic people.

These differences include social, communication, and behavioral attributes (American Psychiatric Association, 2013); distress expressed by parents and other stakeholders (Silverman & Brosco, 2007); levels of independence (Bennett, 2012; Roux et al., 2015); postsecondary achievement in employment and education (Bennett et al., 2018; Milbourn et al., 2018; Roux et al., 2015); and psychological and physical health (Croen et al., 2015). The medical community primarily attributes these outcomes to the social, communication, and behavioral characteristics of autistic people (American Psychiatric Association, 2013; Lord et al., 2020b; Milbourn et al., 2018).

Some autists and other stakeholders agree with this view. They believe autism is a pathological disorder, and they desire treatments, preventions, and cures for autism and traits associated with autism (Nelson, 2021; Perry, 2014; Simplican, 2019). However, these perceptions are changing. For example, stakeholders express a need for environmental changes to better accommodate young autistic adults (Anderson et al., 2018a, b; Milbourn et al., 2018; Nicolaidis et al., 2015). Also, more and more autists are challenging the assumptions behind attributing outcomes to the characteristics of autists (Botha, 2021; Botha et al., 2020; Perry, 2014; Woods et al., 2018). Instead, these autists and other stakeholders endorse alternative models of disability, such as the social, ecological, and Indigenous models.

15.4 Social Model of Disability

According to the social model of disability, people may have different characteristics, but whether a characteristic is considered an impairment requires a level of social comparison (Davis, 2016). An impairment is only considered a disability when it is not accommodated within a society, such that the person is "isolated and excluded from full participation in society" (Union of the Physically Impaired Against Segregation & Disability Alliance, 1976). A simple example of an impairment that is not considered a disability by society is the physical ability to fly. Humans cannot fly like birds, but societies have accommodated this impairment through innovations, such as stairs, elevators, and airplanes. Of course, because all people are not able to fly, society does not view this inability as an impairment. Yet, society views an inability to walk and/or navigate stairs as a disability because the trait is relatively uncommon and is regarded as limiting.

The social model of disability asserts that excluding disabled people from society constitutes social oppression (Davis, 2016; Shakespeare, 2016). In contrast, organizations run by non-disabled people (e.g., medical communities, charitable institutions) are viewed as oppressors that (1) attribute the "problem" of societal exclusion to the traits of impaired people and (2) believe the solution to the "problem" is the prevention, cure, and/or treatment of impaired people (Shakespeare, 2016). In the social model of disability, the main approach to disability is to remove barriers and discrimination (Shakespeare, 2016). The model proposes that disabled people[3] should identify solutions appropriate for their needs and that research should include the priorities and perspectives of this population (Shakespeare, 2016).

The social model of disability has been both commended and criticized. The model is lauded for its simplicity, effectiveness for political mobilization, ability to highlight societal and environmental factors that contribute to inequities, and ability to empower disabled people through greater self-esteem and self-confidence (Shakespeare, 2016). However, the model has been criticized for its oversimplicity; lack of differentiation between impairments (or diseases and disorders) for which disabled people desire or reject prevention, cures, or treatment; lack of recognizing the relationships between impairments and societal barriers; and sole focus on oppression (and not social relationship dynamics, in general). The model has also been criticized for its view as a utopian concept that is not fully realizable in society due to limited resources and other practical constraints (Shakespeare, 2016).

Autists who endorse the neurodiversity movement include people with two differing views: some view autism as a difference in cognitive style and others view it as an impairment (Bumiller, 2008). Yet, both groups tend to endorse the social model of disability. They also both believe that society disables and marginalizes autistic people (Bumiller, 2008; Fenton & Krahn, 2007; Milton, 2012; Waltz, 2014; Woods et al., 2018) and that the challenges autistic people face fall within civil rights struggles for equality (Fenton & Krahn, 2007). These struggles for neuroequality challenge the current pathologizing of characteristics associated with neurological differences (i.e., the problem resides within the individual). They also challenge the hierarchical structure of society that privileges the needs and wants of people viewed as having normal or proper cognitive functioning (regardless of whether autists view their own traits as a difference or impairment) (Fenton & Krahn, 2007).

Both groups (cognitive differences vs impairment) also share the belief that autistic and other neurodivergent people should have access to human flourishing, such as circumstances that lead to contentment, personal achievement, self-worth, and confidence (Fredrickson & Losada, 2005; Keyes, 2002, as cited by Fenton & Krahn, 2007). However, these two views (difference vs impairment) include different ideas of what qualifies autistic people for acceptance and inclusion in society.

Some neurodiversity advocates believe autistic traits are not a deficit (i.e., disorder or disease) (Chapman, 2021; Fenton & Krahn, 2007) but, rather, an integral part of their selfhood (Chapman, 2021). They criticize deficit models of autism for ignoring the possibility that autistic people can engage successfully in social-communication activities with other autists (Chapman, 2021; Milton, 2012). For example, the double-empathy theory describes social-communication challenges as not inherent in autistic people but, rather, due to the mismatched values and

[3] Person-first versus identity-first language: Controversy exists within the disability community regarding which terminology to use when referring to people with disabilities. While it is customary to use the term(s) preferred by specific individuals involved within an interaction, the disability community does not yet unanimously endorse any specific set of terminology (Bogart & Dunn, 2019).

behaviors within interactions between autistic and non-autistic people (Milton, 2012). Simply put, the double-empathy theory asserts that non-autistic people experience as much difficulty gaining insight into autistic perceptions and culture as autistic people do with non-autistic people. However, because non-autistic people dominate culture, most do not experience social-economic pressure to understand the minds of autistic people. One exception to this pressure includes people who are socially related (e.g., family, friends, teachers, and other allies) to an autistic person.

Based on empirical support for the double-empathy theory, autistic and non-autistic groups have distinct interaction styles *within* groups that can create challenges in social communication *between* groups. For example, mixed groups of autistic and non-autistic people have less clear communication than pure groups of either autistic or non-autistic people (Crompton et al., 2020b). Also, both autistic and non-autistic people rate rapport as poorer between mixed groups than non-mixed groups. And they rate rapport as greater in pure groups of autistic people than pure groups non-autistic people. Finally, autistic people report greater ease, comfort, and understanding when engaging with other autistic people (Crompton et al., 2020a). These findings are limited by participant IQ, and more research is needed to understand social communication between pure autistic groups with people who have intellectual disabilities or use less formal language (e.g., verbal or sign language).

Some people argue that all neuro-cognitive variation cannot be viewed as natural variation in cognitive style but, rather, that some variation must be viewed as forms of impairment. For example, Jaarsma and Welin (2012) assert that environmental and biological factors that affect biological development (e.g., intellectual disabilities) can be detrimental to the individual and society. Rather than rejecting claims of neurodiversity, they assert that only a narrow definition of autism, including "high-functioning autists," is compatible with the conceptualization of neurodiversity (p. 20). They believe that including "low-functioning autists" within neurodiversity risks increasing stigma for people classified as "high-functioning" and may deny the care needed by those classified as "low-functioning." This approach does not serve either group. Nelson (2021) also rejects the idea that natural variation reflects non-disorder. Instead, he views other types of variation, such as extreme ranges of blood pressure or being blind, as disorders or impairments. He also asserts that the neurodiversity movement does not specifically define what constitutes a neurodisorder or a neurodifference. He highlights that the autism community has varied attitudes about whether autism is a disability and that the neurodiversity movement does not adequately acknowledge the experiences of autists who view autism as a disorder.

Simplican (2019) has a different—and as she describes, more ethical—view. She describes the challenging behaviors of her sibling (e.g., violence against others) as either representing disability or criminality with a disability classification. Simplican describes behavioral approaches to challenging behaviors as acknowledging and addressing power imbalances, recognizing communication and resistance, and problem-solving to increase the quality of life of disabled people. Shields and Beversdorf (2021) support her view and argue that removing the disability status would remove criminal justice protections that safeguard autists charged with a crime.

Lim (2015) rejects Jaarsma and Welin's call for focusing the neurodiversity movement on "high-functioning" autists. Instead, they focus their argument on distinguishing treatment (which aims to reduce autistic traits) from accommodation (which aims to support autistic traits) while increasing access to societal functions. Lim rejects the idea that comparative capacity (e.g., IQ) should qualify one group (e.g., those with higher IQ) for accommodation but relegate another group (e.g., those with lower IQ) for treatment. They believe that this comparative view conflates the concept of "care" with the concept of "cure" and reduces the concept of accommodation to one of acceptance. Also, the concept of relegating people with lower IQ to treatment suggests that the individual is responsible for their own (lack of) quality of life and access to societal functions, as well as the inconveniences their characteristics have on other people. In the concept of accommodation, all people deserve access to quality of life and societal functions, even if these functions inconvenience other members of society.

Another argument for a difference perspective involves identifying strengths in autism. Both Kanner (1943) and Asperger (Asperger & Frith, 1991), the earliest professionals to publish descriptions of autism, identified challenges within their original characterization of autism. However, they also included rich descriptions of the unusual strengths they saw in autistic children. Additionally, Kanner often shared detailed descriptions of similar unusual strengths in the parents of these children (Kanner, 1943). These strengths included unusual capabilities in semantic memory, attention to detail, creativity, problem-solving, the ability to sustain attentional focus for long periods of time, and more (Asperger & Frith, 1991; Kanner, 1943). Even with Kanner's and Asperger's early descriptions of these strengths, these aspects of autism have been largely ignored within research. However, some researchers highlight important strengths and their potential value to the individual and/or society. These strengths include enhanced perceptual and sensory performance (McDonald, 2021; Mottron, 2011); systemizing (Baron-Cohen, 2009; McDonald, 2021); memory, attention, object orientation, and nonconformity (McDonald, 2021); and special interests or monotropism (Murray, 2018).

Some proponents of the neurodiversity movement support a social utility model. They believe that if autism has strengths, then autism has social value and should, therefore, be considered a difference (Bumiller, 2008; Jaarsma & Welin, 2012; Van Grunsven, 2020). For example, Stenning and Rosqvist (2021) highlight how the concept of neurodiversity is altered in business/employment and programs to support social welfare, where neurodiversity is viewed from a utilitarian/social Darwinian perspective to further the interests of the neoliberal economy. Within this view, autistic strengths can be leveraged to benefit the economy. However, some organizations overlook differences that do not directly benefit them, so they are less likely to consider accommodations that would benefit the autistic person.

Other neurodiversity advocates, and neurodiversity opponents, are critical of using a social utility model (e.g., value to society, employable, independence, number of friendships,) as a marker of inherent value for autism (Bumiller, 2008; Fenton & Krahn, 2007; Stenning & Rosqvist, 2021). Some even believe that to qualify as a "proper" human citizen, one must have characteristics that enable successful living within society (Bumiller, 2008). For example, some people view autistic strengths that support employment as attributes that qualify, at least some, autistic people as proper human citizens who deserve privileges of society, such as respect, understanding, acceptance, and access to human flourishing (Stenning & Rosqvist, 2021).

In the social utility model, the implicit idea is that certain types of people deserve certain outcomes, whereas other types of people deserve different outcomes. But this idea ignores the circular problem of societal marginalization of traits that society deems as disabling (Bumiller, 2008; Stenning & Rosqvist, 2021). In other words, society determines who qualifies as having a difference or disability based on outcomes that are assigned by society. Thus, if society decides that a type of person is less than human in some way, such as through impairment or marginalization, then that person does not deserve desirable outcomes that others can attain within a society. Neurodiversity proponents believe that it is fundamentally "unfair to use successful living in such an environment as a litmus test of who properly qualifies as functional" (Stenning & Rosqvist, 2021, p. 2). Such a circular litmus test essentially *disables anyone* who is unable to achieve success through marginalization related to race and ethnicity, gender, sexual orientation, socioeconomic status, or other marginalizing situations.

Stenning and Roquivist endorse the view that neurodiversity represents problems experienced through marginalization, and they reject associating neurodivergent strengths with benefits to humanity. Specifically, they believe that what is regarded as "benefits" is deemed by societal values. They also reject the concept of "strengths" in autism because no singular autism prototype exists. Thus, they believe that strengths do not reveal "anything useful about what it is to be autistic or how to support autistic individuals" (p. 1535). Other autists do not view strengths as intrinsic to being autistic. Instead, they perceive autism as a disorder consisting of only undesirable traits and believe that their desirable traits represent aspects of their true selves (McDonald, 2016).

On the other hand, Van Grunsven (2020) asserts that for autism to be a difference, the experience of autism needs to be valued only by autists themselves. Van Grunven describes how autists engage in embodied (experiential), embedded (contextual), extended (cognition interprets the environment in some way), and enactive (shaped by and seek to shape the environment) cognitions that make them fully human (in contrast to the theory of mind that diminishes the "human" capabilities of autists). Van Grunsven's approach to neurodiversity easily extends to people of all conscious intellectual functioning.

Some neurodiversity proponents believe the combination of strengths and challenges represent a difference that is only pathologized when one group is designated as a desirable standard. Van Grunsven (2020) describes this phenomenon as "philosophical neurotypical centrism." Similarly, Kittay (2009) proposes that "when we assert that A is [or is not] like B, we take B as the template—its features are salient—and the features of A not found in B lose their salience. That is, if B is characterized by the features x, y, and z, then we come to see A only in terms of its similarity with respect to x, y, and z, even if in other contexts A's features, a, b, and c, are the salient ones.' (p. 612–13). If we consider this perspective with animals, then when a dog is held as a standard, cats may appear to have "deficits, deviances, and delays" in social-emotional development and daily living skills (McDonald, 2017; McDonald et al., 2022). This analogy shows how strengths can be ignored or, even, misattributed as evidence of deficiency. As another analogy more closely related to cognition, Bradshaw et al. (2021) describe autism as a different operating system: "just because a PlayStation cannot read an Xbox game does not mean it is broken or has a processing error – it is just a different operating system." (p. 105). They assert that autism is neither linear nor binary with high-functioning and low-functioning autism. These binary labels minimize (1) the challenges of people who appear to meet expectations of normalization and (2) the strengths of those whose characteristics are further from accepted norms (Bradshaw et al., 2021).

15.5 Ecological Model of Neurodiversity

The ecological model of neurodiversity is another critical alternative to the medical model. The ecological model also views neurodiversity as both strengths and a mix of strengths and challenges. But this model asserts that biological and evolutionary approaches largely ignore the societal systems, cultures, and contexts in which human variation has function and meaning (Chapman, 2021; Stenning & Rosqvist, 2021).

The understanding of what it means to be human changes over time (Stenning & Rosqvist, 2021). One way of understanding what it means to be human is by considering human interdependence related to the biological ecological perspective, such that biological diversity survives through interdependence (Stenning & Rosqvist, 2021). In this way, the characteristics of one entity within the ecology complement and balance the characteristics of other entities, contributing to maintaining and changing the whole ecology (Chapman, 2021; Stenning & Rosqvist, 2021).

Within human ecological systems, various forms of neurodiversity, or different neurodivergent people, represent different units. These units take on specific functions that help human ecological systems persist and thrive. In particular, the ecological model emphasizes the importance of fit between a person and the context, known as an ecological niche (Singer, 2017). In this niche, autistic differences (or strengths) may be ideal for certain contexts, and certain contexts may be ideal for autistic differences (or strengths). For example, autism is overrepresented among people in engineering and the sciences (Wheelwright & Baron-Cohen, 2001).

15.6 Indigenous Models of Neurodiversity

Indigenous people's views of autism remain understudied. According to researchers, some Indigenous cultures embrace views of autism similar to that of the neurodiversity paradigm. For example, Māori people tend to view differences that are considered disabilities in Western culture, including autism, as normal parts of life, with challenges and strengths that deserve acceptance and inclusion. However, some Māori may view autism as a punishment or consequence of family behavior, and others endorse Western medical views of autism (Lilley et al., 2020; Tupou et al., 2021). One Māori author, Keri Opai, describes autism as "Takiwātanga," which means my/their own time and space. The inspiration for this term comes from a Māori proverb about the titoki tree (*Alectryon excelsum)* (Tupou et al., 2021). This tree does not bear fruit every year, so it is difficult to cultivate. And the fruit itself is bitter. Yet, historically, Māori people value the tree and its fruit for their rare qualities, such as oil (Best, 1977). From this perspective, Māori people expect that autistic children will develop in their own time with their own, unique value.

Navajo people also express views about autism that are more accepting than Western concepts of disability. These views may be related to shared experiences of marginalization that increase acceptance within Navajo communities. For example, Kapp (2011) describes similarities in how Western society marginalizes autistic people and Navajo people, who have both experienced social classifications as "inferior," a "burden to society," and other negative social stereotypes.

The acceptance of autism within Navajo communities may involve the guiding philosophy of Hozho (Kapp, 2011). Hozho is similar to the neurodiversity paradigm in that everyone/everything has an interconnected spirit with everything else, and that everything deserves equal respect. In this way, Navajo people seek to live in harmony with people, other living beings, and the earth. Before meeting people from Western cultures, the Navajo community did not have a word for "disability." And the community fully accepts people regarded as "ill." Navajo communities also engage in cultural practices that support the needs of autistic people. For example, the Navajo believe that all people have a unique purpose in the universe, and they do not expect people to conform to a standard. They also do not judge autistic behaviors. Instead, community members lead by example and explain preferred behaviors in detail. These explanations may help autistic people better parse social behaviors and appeal to their need of rationally understanding why people behave as they do. Also, Navajo people do not emphasize a specific timeline for accomplishing "adult milestones", but they celebrate when milestones are accomplished. In this way, autistic adults can both contribute to and receive from the community.

15.7 Major Autism Movements

Since autism was identified and classified, three major movements have emerged: prevention, recovery (Perry, 2014), and neurodiversity (Perry, 2014). The prevention movement primarily involves research, which is represented by the vast majority of funded research focusing on identifying environmental factors, genes, and other mechanisms as targets for preventing autism "risk" (Cervantes et al., 2021; Perry, 2014). The recovery movement operates at all levels of society (e.g., policy, research, services, families, individuals) and focuses on reducing traits and behaviors associated with autism to help "normalize" the individual (Cascio, 2012; Perry, 2014).

Both the prevention and recovery movements are based on the medical model of disability (Cervantes et al., 2021; Perry, 2014). As a result, many organizations and research institutes operate primarily from a prevention and/or recovery approach. For example, the Autism Research Institute (creator of the now defunct Defeat Autism Now program), Cure Autism Now, Autism Speaks, the Simons foundation for Research Initiative, and the U.S. National Institutes of Health primarily view autism as a disorder or disease to be cured or reduced through treatment (Cascio, 2012; Perry, 2014). Proponents of the recovery movement assert that this approach is legitimized by science and supports the needs of caregivers by providing hope (Perry, 2014).

The neurodiversity movement is also conducted at all levels of society (Cascio, 2012). But neurodiversity is not as prevalent in research and services. However, the neurodiversity movement does not consist of a single ideology or specific set of ideologies (Bumiller, 2008). For example, some autists view autism as a naturally occurring neurological difference that is marginalized by society (Bumiller, 2008; Perry, 2014; Richman, 2020). Other autists view autism as an impairment that is disabling through marginalization in society (Bumiller, 2008; Perry, 2014; Richman, 2020).

In general, proponents of the neurodiversity movement support the social model of disability over the medical model; yet some proponents also favor other models. Other organizations, such as the Autism Self-Advocacy Network, Autism Network International, Global and Regional Autism Spectrum Partnership, AASpire, and Autistica, view autism as a neurological difference or disability that should be accommodated in society and not prevented or treated (Cascio, 2012; Perry, 2014). Proponents of the neurodiversity movement state that the recovery movement focuses on children and stigmatizes autism through bad science (Perry, 2014). They describe how anti-autism campaigns and professionals target parents before they can learn about autistic adults and neurodiversity (Cascio, 2012; Perry, 2014). Finally, proponents of the neurodiversity movement describe how the prevention and recovery movements exclude autistic adults from their cause. This exclusion occurs by ignoring autists' first-hand experiences of being autistic and being marginalized by society, and by their priorities for accommodations, services, and research (Perry, 2014).

15.8 Unifying View of Autism with Combined Models

One unifying theory of autism is based on biological and developmental models proposed from the medical, social, and ecological models of disability (McDonald, 2021). This theory is called the broader autism phenotype constellations-disability matrix paradigm (BAPCO-DMAP). The BAPCO-DMAP considers all conceptualizations of autism—as a difference in cognitive style, as an impairment, and/or as a marginalized group—to co-exist within the infinite heterogeneity of characteristics represented by autistic people (McDonald, 2021).

The BAPCO-DMAP proposes that extreme variation of any trait can impair the survival and well-being of the person, and that people, themselves, may desire treatments for recovery (e.g., from self-harm). Also, autists with co-occurring conditions may have fewer resources available to help them with communication, socialization, information and sensory/perceptual processing, executive functioning, and other facets that affect their quality of life. These people may benefit from equitable access to health care to improve these conditions. Thus, the medical model may be important for reducing undesired chronic health conditions and improving quality of life.

The BAPCO-DMAP also incorporates the ecological model. The BAPCO-DMAP acknowledges the social values and societal practices that contribute to the match between the environment and traits of any person. In this model, a greater match improves flourishing of the person and other elements within the environment, such as the local or more distal society. Further, the BAPCO-DMAP endorses the social model of disability when the traits of people are marginalized or otherwise mismatch their environment.

The BAPCO-DMAP asserts that understanding positive traits of autism is important for supporting autistic development in an empowering, holistic, and neutral stance of difference and/or disability. By understanding subtypes of autism, we can foster appropriate expectations and developmental activities that complement, not correct, the traits and developmental timing common to people within a particular subtype. Supporting these developmental activities may nurture joy and personal flourishing for the autistic individual. For example, the BAPCO-DMAP posits that when autists are not engaging in one specific, normative developmental activity, they are engaging in another activity that supports some other type of development. This paradigm posits that the perspectives of autistic adults provide the best signposts for the progression and value (personal and societal) of autistic development. Both individual and collective decisions should determine what traits comprise autistic strengths that should be valued for the benefit of the individual and society.

However, the BAPCO-DMAP does not endorse a social utility model. The BAPCO-DMAP supports that the intrinsic value of autists, regardless of the match between their traits and their environment, is not determined by their outcomes or their perceived benefit to society. Instead, the paradigm advocates that all people deserve access to a good quality of life.

15.9 Recommendations for Biological Studies of Autism

Include the autistic community The most important consideration in biological studies of autism is to include the autistic community. Historically, autism research has been fraught with stereotypes that stem from studies, service providers, and all levels of society (Botha, 2021; Gillespie-Lynch et al., 2017; Woods et al., 2018). Because of these historic—and ethical—tensions, autists should be co-creators of research and included throughout the scientific process (Barnhart & Dierickx, 2021). They need to be at the forefront of defining autism; determining what accommodations, services, and programs are needed; and weighing in on what research should be prioritized. They need to actively contribute to autism knowledge to address power dynamics of debates around autism (Waltz, 2014; Woods et al., 2018).

Enhance communication efforts Another crucial factor to address is discourse in clinical practice and research. Researchers and clinicians need to better understand how to communicate with autists and facilitate respectful communication (Barnhart & Dierickx, 2021). Autists and stakeholders also need to co-create new meanings of what constitutes illness, impairment, and disability from a patient-centered perspective (Barnhart & Dierickx, 2021). The scientific community also needs to adequately inform the public about research, ethics, and the autism community (Perry, 2011).

Expand topics of research The topics of research in autism also need to be expanded. For example, more research is needed on the intersection of autism with other identities, disabilities, ways of being, and the full range of human diversity (Woods et al., 2018). These intersections add layers of complexity on what it means to be autistic.

More research is also needed on self-diagnosed autists and their characteristics and experiences. Self-diagnosis has several benefits. For example, self-diagnosis can counteract discriminatory practices (Sarrett, 2016), such as the changing nature of autism diagnoses, or access to diagnostic services, such as those available to people with lower income, older ages, and other demographics. Also, diagnostic practices include biases affecting females, racial and ethnic groups, and people with higher IQ- essentially creating a "lost generation" of undiagnosed and underserved people (Lai & Baron-Cohen, 2015). Interestingly, both diagnosed and self-diagnosed autistic adults report similar levels of quality of life, experiences of stigma, employment rates, and autism identity (McDonald, 2020). These self-diagnosed adults tended to be older and women, and they were similar to the "lost generation" described by Lai and Baron-Cohen (2015). Self-diagnosed autists may experience similar societal marginalization as diagnosed autists, but without the support that often requires a diagnosis (McDonald, 2020).

The autism community varies in their acceptance of self-diagnosis. Some people accept self-diagnosed individuals (Woods et al., 2018). For example, some people believe that the requirement of a diagnosis from members of "out-group" authorities (e.g., scientists, clinicians, service providers) undermines the right of autists for biocitizenship (Sarrett, 2016). Thus, the autism community accepts self-diagnosis as an act of resistance against societal power dynamics that disenfranchise the self-knowledge of autistic people. Conversely, some autists reject self-diagnosis. For example, some autists are concerned about self-diagnosed (but actually non-autistic) people co-opting the representation of autistic people (Richman, 2020; Sarrett, 2016). They believe that the ability to accurately recognize and diagnose autism requires training and that self-diagnosis is too subjective. In either case, proponents of self-diagnosis describe barriers related to obtaining a diagnosis, rejection of the medical model, and the importance of self-knowledge.

Reallocate resources An important consideration in these efforts is the allocation of resources. Most money and resources in research are dedicated to understanding the cause of autism and finding preventions, cures, and treatments for autism. Little resources are allocated to research on services that improve the quality of life of autistic people (Perry, 2014; Cervantes et al., 2021). More resources are also needed for developing accommodations, programs, and services for autistic people. The different movements in autism all agree that autism is a neurological difference that puts autistic people in vulnerable situations (Richman, 2020). Thus, the autistic community needs more resources to support them in managing these situations and other challenges they face.

15.10 Conclusion

The nature of autism is hotly contested and not well understood (Perry, 2014; Richman, 2020; Sarrett, 2016). Multiple stakeholders, including autists, families of autists, and the researchers and providers that, ostensibly, serve them, differ widely in their views about the nature of autism and how autism should be treated by society. These wide-ranging views are pervasive in the neurodiversity movement. And although some autists do not endorse this movement, all groups agree that autistic people need better accommodations, services, and programs. And they are all concerned about how to improve the quality of life of autists. Researchers in the biological sciences need to be aware of these controversies, as well as the historic lessons in eugenics research with other marginalized groups. Moreover, they need to include autistic representation at every level of biological research to incorporate their perspectives and experiences and to ensure their ethical protection.

References

American Psychiatric Association. (2013). *Diagnostic and statistical manual of mental disorders: DSM-5* (5th ed.). American Psychiatric Association.

Anderson, K. A., Roux, A. M., & Link to external site, this link will open in a new window, Kuo, A., & Shattuck, P. T. (2018a). Social-ecological correlates in adult autism outcome studies: A scoping review. *Pediatrics, 141*(Suppl 4), S306–S317. http://dx.doi.org.proxy.library.vanderbilt.edu/10.1542/peds.2016-4300H

Anderson, K. A., Sosnowy, C., Kuo, A. A., & Shattuck, P. T. (2018b). Transition of individuals with autism to adulthood: A review of qualitative studies. *Pediatrics, 141*(Supplement_4), S318–S327. https://doi.org/10.1542/peds.2016-4300I

Asperger, H., & Frith, U. (1991). "Autistic psychopathy" in childhood. In U. Frith (Eds.), *Autism and Asperger syndrome* (1992-97284-002, pp. 37–92). Cambridge University Press.

Baker, J. P., & Lang, B. (2017). Eugenics and the origins of autism. *Pediatrics, 140*(2), 1–3.

Barnhart, A. J., & Dierickx, K. (2021). Cultures and cures: Neurodiversity and brain organoids. *BMC Medical Ethics, 22*, 1–6. https://doi.org/10.1186/s12910-021-00627-1

Baron-Cohen, S. (2009). The empathising-systemising theory of autism: Implications for education. *Tizard Learning Disability Review, 14*(3), 4–13. https://doi.org/10.1108/13595474200900022. Applied Social Sciences Index & Abstracts (ASSIA); Career & Technical Education Database: Health & Medicine; Career & Technical Education Database: Social Sciences; Consumer Health Database; Education Database; Health & Medical Collection; Nursing & Allied Health Database; Psychology Database; Research Library: Health & Medicine; Research Library: Social Sciences; Sociology Database.

Baron-Cohen, S. (2017). Editorial Perspective: Neurodiversity—A revolutionary concept for autism and psychiatry. *Journal of Child Psychology and Psychiatry, 58*(6), 744–747. https://doi.org/10.1111/jcpp.12703. ProQuest Central; Social Science Premium Collection.

Baynton, D. C. (2005). Disability and the justification of inequality in American history. https://doi.org/10.4324/9780203077887-8

Bennett, A. (2012). Parental involvement in early intervention programs for children with autism. *Master of Social Work Clinical Research Papers*. https://sophia.stkate.edu/msw_papers/113

Bennett, A. E., Miller, J. S., Stollon, N., Prasad, R., & Blum, N. J. (2018). Autism spectrum disorder and transition-aged youth. *Current Psychiatry Reports, 20*(11), 103. https://doi.org/10.1007/s11920-018-0967-y

Best, E. (1977). *Forest Lore of the Maori: Titoki* [Informational or Reference]. New Zealand Electronic Text Collection. https://nzetc.victoria.ac.nz/tm/scholarly/tei-BesFore-t1-body-d1-d9-d7.html

Bogart, K. R., & Dunn, D. S. (2019). Ableism special issue introduction. *Journal of Social Issues, 75*(3), 650–664. https://doi.org/10.1111/josi.12354

Botha, M. (2021). Academic, activist, or advocate? Angry, entangled, and emerging: A critical reflection on autism knowledge production. *Frontiers in Psychology, 12*. https://www.frontiersin.org/article/10.3389/fpsyg.2021.727542

Botha, M., Dibb, B., & Frost, D. M. (2020). "Autism is me": An investigation of how autistic individuals make sense of autism and stigma. *Disability & Society*. https://doi.org/10.1080/09687599.2020.1822782

Bradshaw, P., Pickett, C., van Driel, M. L., Brooker, K., & Urbanowicz, A. (2021). "Autistic" or "with autism"? *Australian Journal of General Practice, 50*(3), 104–108.

Bumiller, K. (2008). Quirky citizens: Autism, gender, and reimagining disability. *Signs, 33*(4), 967–991. https://doi.org/10.1086/528848

Bundgaard-Nielsen, C., Knudsen, J., Leutscher, P. D. C., Lauritsen, M. B., Nyegaard, M., Hagstrøm, S., & Sørensen, S. (2020). Gut microbiota profiles of autism spectrum disorder and attention deficit/hyperactivity disorder: A systematic literature review. *Gut Microbes, 11*(5), 1172–1187. https://doi.org/10.1080/19490976.2020.1748258

Cascio, M. A. (2012). Neurodiversity: Autism pride among mothers of children with autism spectrum disorder. *Intellectual and Developmental Disabilities, 50*(3), 273–283. http://dx.doi.org.proxy.library.vanderbilt.edu/10.1352/1934-9556-50.3.273

Cervantes, P. E., Matheis, M., Estabillo, J., Seag, D. E. M., Nelson, K. L., Peth-Pierce, R., Hoagwood, K. E., & Horwitz, S. M. (2021). Trends over a decade in NIH funding for autism spectrum disorder services research. *Journal of Autism and Developmental Disorders, 51*(8), 2751–2763. https://doi.org/10.1007/s10803-020-04746-3

Chapman, R. (2021). Neurodiversity and the social ecology of mental functions. *Perspectives on Psychological Science, 16*(6), 1360–1372. https://doi.org/10.1177/1745691620959833. APA PsycInfo®.

Croen, L. A., Zerbo, O., Qian, Y., Massolo, M. L., Rich, S., Sidney, S., & Kripke, C. (2015). The health status of adults on the autism spectrum. *Autism: The International Journal of Research and Practice, 19*(7), 814–823. https://doi.org/10.1177/1362361315577517

Crompton, C. J., Hallett, S., Ropar, D., Flynn, E., & Fletcher-Watson, S. (2020a). "I never realised everybody felt as happy as I do when I am around autistic people": A thematic analysis of autistic adults' relationships with autistic and neurotypical friends and family. *Autism, 24*(6), 1438–1448. https://doi.org/10.1177/1362361320908976

Crompton, C. J., Ropar, D., Evans-Williams, C. V., Flynn, E. G., & Fletcher-Watson, S. (2020b). Autistic peer-to-peer information transfer is highly effective. *Autism, 24*(7), 1704–1712. https://doi.org/10.1177/1362361320919286

Davis, L. J. (Ed.). (2016). *The disability studies reader* (5th ed.). Routledge. https://doi.org/10.4324/9781315680668

Demetriou, E. A., DeMayo, M. M., & Guastella, A. J. (2019). Executive function in autism spectrum disorder: History, theoretical models, empirical findings, and potential as an endophenotype. *Frontiers in Psychiatry, 10*, 753. https://doi.org/10.3389/fpsyt.2019.00753

Dolmage, J. (2011). Disabled upon arrival: The rhetorical construction of disability and race at Ellis Island. *Cultural Critique, 77*, 24–69.

Dwyer, P. (2022). Stigma, incommensurability, or both? Pathology-first, person-first, and identity-first aanguage and the challenges of discourse in divided autism communities. *Journal of Developmental & Behavioral Pediatrics, 43*(2), 111–113. https://doi.org/10.1097/DBP.0000000000001054

Emberti Gialloreti, L., Mazzone, L., Benvenuto, A., Fasano, A., Garcia Alcon, A., Kraneveld, A., Moavero, R., Raz, R., Riccio, M. P., Siracusano, M., Zachor, D. A., Marini, M., & Curatolo, P. (2019). Risk and protective environmental factors associated with autism spectrum disorder: Evidence-based principles and recommendations. *Journal of Clinical Medicine, 8*(2). https://doi.org/10.3390/jcm8020217

Fenton, A., & Krahn, T. (2007). Autism, neurodiversity and equality beyond the "normal". *Journal of Ethics in Mental Health, 2*(2), 1–6. APA PsycInfo®.

Garbarino, V. R., Gilman, T. L., Daws, L. C., & Gould, G. G. (2019). Extreme enhancement or depletion of serotonin transporter function and serotonin availability in autism spectrum disorder. *Pharmacological Research, 140*, 85–99. https://doi.org/10.1016/j.phrs.2018.07.010

Gillespie-Lynch, K., Kapp, S. K., Brooks, P. J., Pickens, J., & Schwartzman, B. (2017). Whose expertise is it? Evidence for autistic adults as critical autism experts. *Frontiers in Psychology, 8*. https://search.proquest.com/psycinfo/docview/1928261039/CCAA850CA4E342A3PQ/48

Grove, J., Ripke, S., Als, T. D., Mattheisen, M., Walters, R. K., Won, H., Pallesen, J., Agerbo, E., Andreassen, O. A., Anney, R., Awashti, S., Belliveau, R., Bettella, F., Buxbaum, J. D., Bybjerg-Grauholm, J., Bækvad-Hansen, M., Cerrato, F., Chambert, K., Christensen, J. H., et al. (2019). Identification of common genetic risk variants for autism spectrum disorder. *Nature Genetics, 51*(3), 431–444. https://doi.org/10.1038/s41588-019-0344-8

Guthrie, R. V. (2004). *Even the rat was white: A historical view of psychology*. Allyn and Bacon.

Hassan, M. M., & Mokhtar, H. M. O. (2019). Investigating autism etiology and heterogeneity by decision tree algorithm. *Informatics in Medicine Unlocked, 16*, 100215. https://doi.org/10.1016/j.imu.2019.100215

Jaarsma, P., & Welin, S. (2012). Autism as a natural human variation: Reflections on the claims of the neurodiversity movement. *Health Care Analysis: HCA: Journal of Health Philosophy and Policy, 20*(1), 20–30. https://doi.org/10.1007/s10728-011-0169-9

Jacob, S., Wolff, J. J., Steinbach, M. S., Doyle, C. B., Kumar, V., & Elison, J. T. (2019). Neurodevelopmental heterogeneity and computational approaches for understanding autism. *Translational Psychiatry, 9*(1), 1–12. https://doi.org/10.1038/s41398-019-0390-0

Kanner, L. (1943). Autistic disturbances of affective contact. *Nervous Child, 2*, 217–250.

Kapp, S. K. (2011). Navajo and autism: The beauty of harmony. *Disability & Society, 26*(5), 583–595. https://doi.org/10.1080/09687599.2011.589192

Kenny, L., Hattersley, C., Molins, B., Buckley, C., Povey, C., & Pellicano, E. (2016). Which terms should be used to describe autism? Perspectives from the UK autism community. *Autism: The International Journal of Research and Practice, 20*(4), 442–462.

Kittay, E. F. (2009). The personal is philosophical is political: A philosopher and mother of a cognitively disabled person sends notes from the battlefield. *Metaphilosophy, 40*(3–4), 606–627. https://doi.org/10.1111/j.1467-9973.2009.01600.x

Lai, M.-C., & Baron-Cohen, S. (2015). Identifying the lost generation of adults with autism spectrum conditions. *The Lancet Psychiatry, 2*(11), 1013–1027. https://doi.org/10.1016/s2215-0366(15)00277-1

Lilley, R., Sedgwick, M., & Pellicano, E. (2020). Inclusion, acceptance, shame and isolation: Attitudes to autism in Aboriginal and Torres Strait Islander communities in Australia. *Autism, 24*(7), 1860–1873. https://doi.org/10.1177/1362361320928830

Lim, C.-M. (2015). Accommodating autistics and treating autism: Can we have both? *Bioethics, 29*(8), 564–572. Philosopher's Index.

Lombardo, M. V., Lai, M.-C., & Baron-Cohen, S. (2019). Big data approaches to decomposing heterogeneity across the autism spectrum. *Molecular Psychiatry, 24*(10), 1435–1450. https://doi.org/10.1038/s41380-018-0321-0

Lord, C., Brugha, T. S., Charman, T., Cusack, J., Dumas, G., Frazier, T., Jones, E. J. H., Jones, R. M., Pickles, A., State, M. W., Taylor, J. L., & Veenstra-VanderWeele, J. (2020a). Autism spectrum disorder. *Nature Reviews. Disease Primers, 6*(1), 5. https://doi.org/10.1038/s41572-019-0138-4

Lord, C., McCauley, J. B., Pepa, L. A., Huerta, M., & Pickles, A. (2020b). Work, living, and the pursuit of happiness: Vocational and psychosocial outcomes for young adults with autism. *Autism, 24*(7), 1691–1703. https://doi.org/10.1177/1362361320919246

McDonald, T. A. M. (2016). *Identity as a mediator between stigma and stereotype threat on postsecondary outcomes for adults on the autism spectrum* [Ph.D., The University of Wisconsin – Madison]. http://search.proquest.com.proxy.library.vanderbilt.edu/pqdtglobal/docview/1781235653/abstract/8DEE45B70C2E4E4EPQ/1

McDonald, T. A. M. (2017). *Describing autism from a neurodiversity paradigm perspective*. Spectrum Pathways Coach Training Program.

McDonald, T. A. M. (2020). Autism identity and the "lost generation": Structural validation of the autism spectrum identity ccale and comparison of diagnosed and self-diagnosed adults on the autism spectrum. *Autism in Adulthood, 2*(1), 13–23. https://doi.org/10.1089/aut.2019.0069

McDonald, T. A. M. (2021). The broader autism phenotype constellations–disability matrix paradigm: Theoretical model for autism and the broader autism phenotype. *Medical Hypotheses, 146*, 110456. https://doi.org/10.1016/j.mehy.2020.110456

McDonald, T. A. M., Lalani, S., Chen, I., Cotton, C. M., MacDonald, L., Boursoulian, L. J., Wang, J., & Malow, B. A. (2022). Appropriateness, acceptability, and feasibility of a neurodiversity-based self-determination program for autistic adults. *Journal of Autism and Developmental Disorders*. https://doi.org/10.1007/s10803-022-05598-9

Milbourn, B., Scott, M., Black, M., Girdler, S., Falkmer, M., Bolte, S., Halladay, A., Lerner, M., & Taylor, J. L. (2018). Factors impacting employment for people with autism spectrum disorder: A scoping review. *Autism*. https://doi.org/10.1177/1362361318787789

Milton, D. E. M. (2012). On the ontological status of autism: The "double empathy problem". *Disability & Society, 27*(6), 883–887.

Mission and Goals. (2014, October 31). National Institutes of Health (NIH). https://www.nih.gov/about-nih/what-we-do/mission-goals

Molloy, C. J., & Gallagher, L. (2021). Can stratification biomarkers address the heterogeneity of autism spectrum disorder? *Irish Journal of Psychological Medicine*, 1–7. https://doi.org/10.1017/ipm.2021.73

Mottron, L. (2011). Changing perceptions: The power of autism. *Nature, 479*(7371), 33–35. https://doi.org/10.1038/479033a

Mottron, L., & Bzdok, D. (2020). Autism spectrum heterogeneity: Fact or artifact? *Molecular Psychiatry, 25*(12), 3178–3185. https://doi.org/10.1038/s41380-020-0748-y

Murray, D. (2018). Monotropism – An interest based account of autism. In F. R. Volkmar (Ed.), *Encyclopedia of autism spectrum disorders* (pp. 1–3). Springer. https://doi.org/10.1007/978-1-4614-6435-8_102269-1

National Academies of Sciences, E., Division, H. and M., Practice, B. on P. H. and P. H., States, C. on C.-B. S. to P. H. E. in the U., Baciu, A., Negussie, Y., Geller, A., & Weinstein, J. N. (2017). The state of health disparities in the United States. In *Communities in action: Pathways to health equity*. National Academies Press (US). http://www.ncbi.nlm.nih.gov/books/NBK425844/

National Institute on Minority Health and Health Disparities. (2021). *Overview [Government]*. NIMHD. http://www.nimhd.nih.gov/about/overview/

Nelson, R. H. (2021). A critique of the neurodiversity biew. *Journal of Applied Philosophy, 38*(2), 335–347. https://doi.org/10.1111/japp.12470. ProQuest Central.

Nicolaidis, C., Raymaker, D. M., Ashkenazy, E., McDonald, K. E., Dern, S., Baggs, A. E. V., Kapp, S. K., Weiner, M., & Boisclair, W. C. (2015). "Respect the way I need to communicate with you": Healthcare experiences of adults on the autism spectrum. *Autism, 19*(7), 824–831. https://doi.org/10.1177/1362361315576221

Oxford University Press. (2023, July). "neurodiversity, n". In Oxford English Dictionary. Retrieved Movember, 26, 2023.

Perry, A. (2011). Neuropluralism. *Journal of Ethics in Mental Health, 6*(1), 1–4. APA PsycInfo®.

Perry, A. (2014). Leadership, mental health ethics, and biopolitics: The case of autism. *Journal of Ethics in Mental Health, 9*, 1. Social Science Premium Collection.

Richman, K. A. (2020). Neurodiversity and autism advocacy: Who fits under the autism tent? *American Journal of Bioethics, 20*(4), 33–35. Philosopher's Index.

Rojas, D. C., Becker, K. M., & Wilson, L. B. (2015). Magnetic resonance spectroscopy studies of glutamate and GABA in autism: Implications for excitation-inhibition imbalance theory. *Current Developmental Disorders Reports, 2*(1), 46–57. https://doi.org/10.1007/s40474-014-0032-4

Roux, A. M., Shattuck, P. T., Rast, J. E., Rava, J. A., & Anderson, K. A. (2015). *National Autism Indicators report: Transition into young adulthood* (p. 68). National Autism Indicators Report. Life Course Outcomes Research Program, A.J. Drexel Autism Institute, Drexel University. http://drexel.edu/autismoutcomes/publications-and-reports/publications/National-Autism-Indicators-Report-Transition-to-Adulthood/

Sarrett, J. C. (2016). Biocertification and neurodiversity: The role and implications of self-diagnosis in autistic communities. *Neuroethics, 9*(1), 23–36. https://doi.org/10.1007/s12152-016-9247-x

Shakespeare, T. (2016). The social model of disability. In *The disability studies reader* (5th ed., pp. 195–203). Routledge.

Shields, K., & Beversdorf, D. (2021). A dilemma for neurodiversity. *Neuroethics, 14*(2), 125–141. https://doi.org/10.1007/s12152-020-09431-x. ProQuest Central.

Silverman, C., & Brosco, J. P. (2007). Understanding autism: Parents and pediatricians in historical perspective. *Archives of Pediatrics & Adolescent Medicine, 161*(4), 392–398. https://doi.org/10.1001/archpedi.161.4.392

Simplican, S. C. (2019). Behaviors that challenge disability studies. *Disability & Society, 34*(9–10), 1379–1398. https://doi.org/10.1080/09687599.2018.1552119. APA PsycInfo®.

Singer, J. (2017). *Neurodiversity: The birth of an idea*. Judy Singer.

Stenning, A., & Rosqvist, H. B. (2021). Neurodiversity studies: Mapping out possibilities of a new critical paradigm. *Disability & Society, 36*(9), 1532–1537. https://doi.org/10.1080/09687599.2021.1919503. ProQuest Central; Social Science Premium Collection.

Tang, S., Sun, N., Floris, D. L., Zhang, X., Di Martino, A., & Yeo, B. T. T. (2020). Reconciling dimensional and categorical models of autism heterogeneity: A brain connectomics and behavioral study. *Biological Psychiatry, 87*(12), 1071–1082. https://doi.org/10.1016/j.biopsych.2019.11.009

Terman, L. M. (1920). The use of intelligence tests in the grading of school children. *The Journal of Educational Research, 1*(1), 20–32.
Tupou, J., Curtis, S., Taare-Smith, D., Glasgow, A., & Waddington, H. (2021). Māori and autism: A scoping review. *Autism, 25*(7), 1844–1858. https://doi.org/10.1177/13623613211018649
Union of the Physically Impaired Against Segregation & Disability Alliance. (1976). *Fundamental principles of disability: Being a summary of the discussion held on 22nd November, 1975 and containing commentaries from each organisation*. The Union: The Alliance.
Van Grunsven, J. (2020). Perceiving "other" minds: Autism, 4E cognition, and the idea of neurodiversity. *Journal of Consciousness Studies, 27*(7–8), 115–143. Philosopher's Index.
Waltz, M. (2014). Worlds of autism: Across the spectrum of neurological difference. *Disability & Society, 29*(8), 1337–1338. https://doi.org/10.1080/09687599.2014.934064
Warrier, V., Zhang, X., Reed, P., Havdahl, A., Moore, T. M., Cliquet, F., Leblond, C. S., Rolland, T., Rosengren, A., Rowitch, D. H., Hurles, M. E., Geschwind, D. H., Børglum, A. D., Robinson, E. B., Grove, J., Martin, H. C., et al. (2022). Genetic correlates of phenotypic heterogeneity in autism. *Nature Genetics*. https://doi.org/10.1038/s41588-022-01072-5
What we do | NIHR. (n.d.). Retrieved 21 June 2022, from https://www.nihr.ac.uk/about-us/what-we-do/
Wheelwright, S., & Baron-Cohen, S. (2001). The link between autism and skills such as engineering, maths, physics and computing: A reply to Jarrold and Routh. *Autism, 5*(2), 223–227. https://doi.org/10.1177/1362361301005002010
Woods, R., Milton, D., Arnold, L., & Graby, S. (2018). Redefining critical autism studies: A more inclusive interpretation. *Disability & Society, 33*(6), 974–979. https://doi.org/10.1080/09687599.2018.1454380. APA PsycInfo®.

Chapter 16
Principal Findings of Auditory Evoked Potentials in Autism Spectrum Disorder

Carla Gentile Matas, Fernanda Cristina Leite Magliaro Aburaya, Mariana Keiko Kamita, and Rebeca Yuko Couto Kawai de Souza

16.1 Introduction

Autism is a complex condition that has been studied for decades with different approaches. The characteristic definitions of the pathology have changed since it was first described by Leo Kanner in 1943 as a central disorder in which children are unable to establish interpersonal relationships and react normally to situations in early life, with notorious social isolation (Kanner, 1943).

The latest issue of ICD-11 places autism in Chap. 6 – Mental, behavioral, and neurodevelopmental disorders. In this new version, childhood autism (F84.0), atypical autism (F84.1), childhood disintegrative disorder (F84.3), and so forth were grouped in a single classification, the autism spectrum disorder (6A02) (ASD). Hence, ASD is characterized by "persistent deficits in the ability to initiate and to sustain reciprocal social interaction and social communication, and by a range of restricted, repetitive, and inflexible patterns of behavior, interests or activities that are clearly atypical or excessive for the individual's age and sociocultural context. The onset of the disorder occurs during the developmental period, typically in early childhood, but symptoms may not become fully manifest until later, when social demands exceed limited capacities" (ICD 11; World Health Organization, 2019).

Despite the rather diversified characteristics and degrees of impairment in this pathology, there is a consensus about the occurrence of communication and language impairments. Adequate language development is known to be intrinsically related to the integrity of the auditory system. Based on this premise, various studies sought to understand auditory pathway functioning in individuals with ASD and

C. G. Matas (✉) · Fernanda Cristina Leite Magliaro Aburaya · M. K. Kamita
Rebeca Yuko Couto Kawai de Souza
Department of Physiotherapy, Speech and Language Pathology and Audiology and Occupational Therapy – FMUSP, Sao Paulo, Brazil
e-mail: cgmatas@usp.br

A. El Idrissi, D. McCloskey (eds.), *Neurobiology of Autism Spectrum Disorders*,
https://doi.org/10.1007/978-3-031-42383-3_16

investigate whether this system would be structurally or functionally impaired, impacting this population's communication skills.

Several studies have reported normal hearing thresholds in people with ASD (Rosenhall et al., 1999; Romero et al., 2014). However, they may present both discomfort to medium-intensity sounds/noises and indifference to high-intensity sounds, as they can be either hypersensitive or hyposensitive to sensory stimuli (Klin, 2006).

People with ASD can also have perceptual, attention, and memory disorders, sometimes mistaken for hearing loss because they do not consistently respond to verbal orders and environmental stimuli. Due to inconsistent responses, results obtained with routine behavioral procedures in audiological assessments may be inconclusive, requiring objective tests where results do not depend on the patient's responses.

One of the most used objective methods in audiology field are the Auditory Evoked Potentials (AEPs) evaluation, an important tool to investigate auditory integrity and functioning (Junqueira & Frizzo, 2002). These potentials can be elicited with various acoustic stimuli (e.g., pure tones, clicks, speech, sentences, etc.) and are picked up in different auditory pathway structures (from the nerve to the auditory cortex), thus identifying abnormalities in the central auditory nervous system (CANS). Some of their components have also been used in the literature as specific biomarkers of certain pathologies (Matas & Toma, 2003; Steinhoff et al., 1988).

The AEPs can be classified as short-, medium-, and long-latency responses, or respectively early, medium, and late AEPs (Ruth & Lambert, 1991). This classification is based on the time when responses occur after acoustic stimulation begins. Hence, short-latency AEPs (SLAEPs) occur between 0 and 10 milliseconds (ms) after the acoustic stimuli are presented; medium-latency AEPs (MLAEPs), between 12 and 50 ms; and longa-latency AEPs (LLAEPs), between 50 and 600 ms after the acoustic stimuli (Anias et al., 2004).

Auditory sensory alterations have been described in the last decades in studies that investigated AEPs in people with autism. One of the findings is the delayed conduction of acoustic stimuli in the Brainstem Auditory Evoked Potentials (BAEP also known as Auditory Brainstem Response – ABR one of the most clinically used SLAEPs) (Magliaro et al., 2010; Dabbous, 2012; Maziade et al., 2000; Rosenblum et al., 1980; Kwon et al., 2007; Rosenhall et al., 2003; Fujikawa-Brooks et al., 2010; Källstrand et al., 2010; Matas et al., 2009). Besides ABR, studies have also reported abnormal LLAEP results, which indicate changes in central auditory processing in this population (Magliaro et al., 2010; Gonzalez-Gadea et al., 2015; Azouz et al., 2014; Orekhova et al., 2009).

A prominent, universal, and little understood symptom in children with ASD is their atypical responsiveness to sensory stimuli. Therefore, the integrity of their auditory pathway must be investigated with objective tests to reach a precise diagnosis and provide more effective intervention.

Since AEPs analyze sensory responses to acoustic stimuli, this chapter will specifically approach the findings in the literature that investigated ABR and LLAEP in individuals with ASD to point out this population's main neuroaudiological characteristics. It is important to highlight that the literature investigating auditory responses with magnetoencephalography, behavioral assessments, fMRI, etc., will not be approached in this chapter, despite their great importance to better understand this population's auditory sensory functioning.

16.2 Evidence of Neurophysiological Abnormalities with ABR in ASD

Evoked potentials have been widely employed in neuroscience as a useful tool in functional diagnosis. Increased latencies or decreased amplitudes in responses are objective evidence of clinical and subclinical problems.

SLAEPs, or early AEPs, are obtained between 0 and 10 ms after the acoustic stimuli are presented and are picked up from the first neurons in the auditory system to the brainstem (Ruth & Lambert, 1991). ABR is generated by synchronously triggering auditory pathway structures; it is one of the most clinically used SLAEPs. ABR is mainly used to determine the acoustic signal detection threshold and infer about the functional and structural integrity of the neural components of the auditory pathway (Kraus et al., 1999).

ABR responses obtained in the time domain with click stimuli consist of seven waves generated by one or more structures throughout the auditory pathway. The classification by Möller et al. (1981) describes the following generating sites: wave I – the portion of the auditory nerve distal to the brainstem; wave II – the portion of the auditory nerve proximal to the brainstem; wave III – cochlear nucleus; wave IV – superior olivary complex; wave V – lateral lemniscus; wave VI – inferior colliculus; wave VII – medial geniculate body. Waves I, III, and V are the ones mainly analyzed to study the time of their absolute latencies and intervals.

ABR latency measures of wave peaks I, III, and V are good diagnostic parameters. The ones most used are the absolute latency of wave V, the interaural latency difference of wave V, and interpeak intervals I–III, I–V, and III–V. Moreover, the presence or absence of these components in a given latency time verifies structural and functional abnormalities that might interfere with the transmission of the acoustic stimuli throughout the auditory pathway, from the auditory nerve to the brainstem (Musiek et al., 1999). Even though ABR wave latency is commonly studied in the literature, research analyzing its amplitude is less frequent. ABR component amplitude varies widely between individuals, which hinders its standardization as a clinical measure (Hall III, 2015).

Click stimuli is the most used to record ABR waves, due to its broad frequency spectrum which stimulates more nerve fibers. However, narrow-band stimuli (e.g., tone-burst) and complex stimuli (e.g., speech) have been increasingly used in the scientific literature and clinical practice in the last decades.

The ABR is often clinically used to test auditory functioning in neonates and children who are difficult to assess with routine procedures, assess central auditory system maturation, detect acoustic nerve tumors, topographically diagnose disorders that affect the posterior cranial fossa, and so forth (Matas et al., 1998).

ABR is one of the methods most used by scientific research to study auditory pathway functioning in individuals with ASD because it is not invasive and provides valuable information on brainstem functioning. Studies indicate normal hearing in individuals with ASD, though with abnormal ABR – which demonstrates delayed conduction of acoustic stimuli in the brainstem (Magliaro et al., 2010; Dabbous, 2012; Maziade et al., 2000; Rosenblum et al., 1980; Kwon et al., 2007; Rosenhall et al., 2003; Fujikawa-Brooks et al., 2010; Källstrand et al., 2010; Talge et al., 2018; Miron et al., 2018).

The studies that analyzed ABR wave latencies and interpeak intervals found prolongations in waves III (Magliaro et al., 2010; Dabbous, 2012; Maziade et al., 2000; Rosenblum et al., 1980; Talge et al., 2018; Sersen et al., 1990; Li et al., 2020) and V (Magliaro et al., 2010; Maziade et al., 2000; Kwon et al., 2007; Rosenhall et al., 2003; Azouz et al., 2014; Miron et al., 2018; Sersen et al., 1990; Gillberg et al., 1983; ElMoazen et al., 2020; Seif et al., 2021) and in interpeak intervals I–III and I–V (Magliaro et al., 2010; Dabbous, 2012; Maziade et al., 2000; Kwon et al., 2007; Rosenhall et al., 2003; Azouz et al., 2014; Talge et al., 2018; Li et al., 2020; Taylor et al., 1982) in the population with ASD, in comparison with same-age individuals with typical development (TD).

Latency delays were also found in wave IV (Rosenblum et al., 1980) and later ABR components (III, V, VI, and VII) in sedated autistic individuals, whereas non-sedated autistic people had normal ABR latencies (Sersen et al., 1990).

These results indicate that individuals with ASD presented delay in the conduction of the nerve impulse in the brainstem regions (cochlear nucleus, superior olivary complex, lateral lemniscus, inferior colliculus). Hence, all studies pointed to a possible brainstem dysfunction in this population.

Besides latency, a smaller wave III amplitude was identified in masking paradigms in the responses of individuals with ASD in comparison with controls (Källstrand et al., 2010). Some studies identified increased latencies in response to complex stimuli, such as speech (Magliaro et al., 2010; Rosenblum et al., 1980), and at high presentation rates of acoustic stimuli (Fujikawa-Brooks et al., 2010).

Several studies describe abnormal electrophysiological responses with ABR in ASD, whereas other studies show normal wave latency results (Coutinho et al., 2002; Romero et al., 2014; Ververi et al., 2015; Russo et al., 2009b).

A study by Coutinho et al. (2002) found normal interpeak intervals and latencies of waves I, III, and V in the ASD group. On the other hand, they identified a common ABR response pattern: a prominent amplitude in wave I in relation to the amplitudes in waves III and V in all ears analyzed at 90 dB HL. Such wave I prominence had not been reported in previous studies. The authors highlighted the need for further research to establish the importance of this finding and its possible implications for these children's central auditory system. The authors hypothesized an association of this finding with hyperacusis and abnormal reaction to sound reported in children with ASD.

In their study with psychoacoustic tests, Khalfa et al. (2004) confirmed the hypothesis of increased perception of loudness in children with ASD and demonstrated a limited dynamic perception range, despite the normal hearing thresholds. Hence, the authors confirmed the existence of hyperacusis in individuals with ASD and discussed its origin (whether central or peripheral). They reported that it may be related to the abnormal behavior and auditory electrophysiological alterations perceived in autism.

The review of the literature on ABR in ASD shows abnormal results in this population. However, there is a great variety of possible results, such as prolonged wave V alone, increased interpeak intervals III–V, I–V, increased latencies of waves III and V, shortened waves, and normal latencies in part of the individuals assessed (Magliaro et al., 2010; Dabbous, 2012; Maziade et al., 2000; Rosenblum et al., 1980; Kwon et al., 2007; Rosenhall et al., 2003; Azouz et al., 2014; Talge et al., 2018; Miron et al., 2018; Sersen et al., 1990; Li et al., 2020; Gillberg et al., 1983; ElMoazen et al., 2020; Taylor et al., 1982). Thus, inconsistent responses in the literature prevent a definite conclusion so far regarding an expected result pattern in individuals with ASD.

Some hypotheses can be raised to explain the variability in the response patterns found in the studies. One of them is the diversity of methodologies used in them. Although most studies used click stimuli, the presentation rate, total number of stimuli, and intensity used in the collection were not always the same. The analysis method also changes between studies. Some of them analyzed only wave V latency, while others analyzed the latencies of all waves and interpeak intervals. Another variable in the studies is the age range of the study groups. Some studies assessed children, while others assessed adults (above 18 years old).

Another hypothesis is the variability in the characteristics presented by autistic individuals – i.e., ABR results may vary depending on each person's impairments and their severity.

It is important to point out that, despite the lack of an established ABR response pattern in this population, abnormalities such as increased wave latencies in normal hearing individuals (normal hearing thresholds) with no structural neurological impairment (such as a tumor) are strong evidence of atypical CANS functioning (Fig. 16.1).

Some authors believe that delayed ABR wave latencies are evidence of brainstem dysfunction in this population (Magliaro et al., 2010; Rosenblum et al., 1980; Gillberg et al., 1983). Furthermore, electrophysiological variability may sustain the hypothesis of perceptual inconsistency (Rosenblum et al., 1980). Brainstem lesion, cochlear dysfunction, and efferent cochlear system involvement are other possible factors that may explain ABR findings in people with ASD (Rosenhall et al., 2003).

It can be generally said that studies point to an increased stimulus transmission time in the brainstem of individuals with ASD (Magliaro et al., 2010; Maziade et al., 2000; Rosenblum et al., 1980; Rosenhall et al., 2003; Gillberg et al., 1983; Taylor et al., 1982; Tanguay et al., 1982; Gillberg et al., 1987). Hence, prolonged waves (one or more components) may be characteristic of this population and possibly a biomarker to help the therapeutic intervention process.

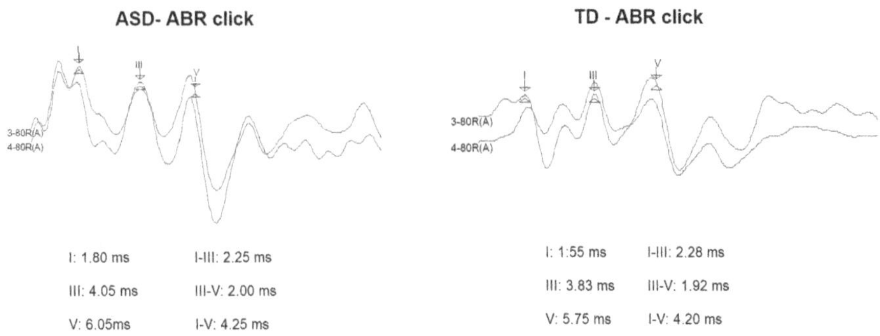

Fig. 16.1 Click ABR responses obtained at 80 dBnHL. Comparison between ASD child and TD child. Latencies of waves III and V prolonged at ASD child

Speech stimuli, which have a complex spectral-temporal structure, can also be used to trigger auditory evoked responses. This type of stimuli produces auditory evoked responses that require precisely synchronous neural activation, which makes it essential to study the neural basis of speech perception (Kraus & Nicol, 2003). Speech stimuli ABR can analyze the following time-domain components: waves V, A, C, D, E, F, and O and amplitude V–A. Waves V and A represent the beginning of the response to sound in the brainstem, in the lateral lemniscus and inferior colliculus, and comprise the transient portion (onset) of the stimulus. Waves C, D, E, and F correspond to the sound source fundamental frequency (F0) and the first formant (F1), which comprise the sustained portion (frequency following response) and involve sound coding. Wave O, the last wave analyzed in speech ABR, corresponds to the final stimulus (offset) (Kraus & Nicol, 2005).

A study by Russo et al. (2009a, b) assessed click and speech ABR in children with ASD and with TD. No differences were found between the latency values of the two groups with click stimuli. On the other hand, speech stimulus ABR revealed that the latency of waves V, A, D, and F in silence was greater in the ASD group; in noise, the amplitude of wave F decreased in the ADS group.

Other ABR changes with speech stimuli were found in patients with ASD regarding both neural synchronous response and phase locking (frequency encoding) of the speech sounds, besides a smaller amplitude and lower reproducibility rate (Russo et al., 2008, 2009a, b). The literature also shows that alterations can be found in the latencies of all ABR waves with speech stimuli in ASD groups, in comparison with groups with TD (Ramezani et al., 2019). Other scientific articles found alterations only in specific waves, such as the latency of waves C, D, E, F (El Shennawy et al., 2014) and O (El Shennawy et al., 2014; Jones et al., 2020), and changes in the amplitude of these waves in ASD groups (El Shennawy et al., 2014).

Gathering these findings helps professionals and researchers to know what results to expect in the ASD population. However, it must be pointed out that some variables are not always consistent in this population, such as the skills that individuals with ASD could or could not develop, which characteristics within the spectrum are observed in individuals with ASD, each person's age, and other issues that cannot converge to a single characteristic.

16.3 Evidence of Neurophysiological Abnormalities with LLAEP in ASD

As previously pointed out, AEPs are objective assessments, and one of their main goals is to complement behavioral assessments for whom they are difficult to be carried out, such as children with ASD (Matas & Magliaro, 2013).

Hearing assessment with LLAEP can reflect sound transmission and processing by picking up bioelectrical responses in the auditory pathways. Hence, it is possible to assess the integrity and maturation of the central auditory pathways and obtain indications of auditory information processing skills, such as neural sound decoding, auditory detection and discrimination, and temporal processing (McPherson, 1996; Hall, 2006; Garcia et al., 2013).

Like other potentials, LLAEP can be analyzed at the time-domain. Thus, the latencies and amplitudes of its positive and negative components (P1, N1, P2, N2, and P3) can be measured and analyzed. LLAEP can also be classified according to its physiological characteristics, as its components may have exogenous/sensory or endogenous/cognitive origins. The P1–N1–P2 complex is considered an exogenous/sensory component because it is generated under the greater influence of acoustic and temporal characteristics of the stimuli and CANS integrity. On the other hand, components N2 and P3 are considered endogenous/cognitive because they are generated under the greater influence of cortical stimulus processing, reflecting the activity of areas in the brain responsible for attention, discrimination, integration, and memory (McPherson, 1996; Reis & Frizzo, 2011).

Component P1 occurs approximately 50 ms after the auditory stimulation and reflects the arrival of acoustic stimuli in the auditory cortex and the activation of secondary auditory cortical areas, hippocampus, temporal plane, and lateral temporal region, which codifies the acoustic characteristics of sound, such as frequency and time (Sharma et al., 2007). Component N1 presents latency values of approximately 80–150 ms after the auditory stimulation and amplitude of 5–10 μV, generated by the primary and secondary auditory cortices, and corresponds to cortical auditory decoding and discrimination (Hall, 2007).

The latency of component P2 occurs approximately between 145 and 200 ms, with amplitude between 3 and 6 μV. This component is suggested to be influenced by the primary and secondary auditory cortices, and its amplitude values may be related to auditory discrimination (Tremblay et al., 2014). As for component N2, the literature seldom reports on its generating site, suggesting that it is related to the supratemporal auditory cortex and auditory discrimination because decreases in its amplitude are suggestive of losses in stimuli discrimination and attention (Musiek, 2006).

Component P3 is known to be actively generated, requiring a behavioral task; its response occurs at approximately 300 ms, and its amplitude may range from 5 to 15 μV. Its generation may be related to processes occurring in the auditory cortex, prefrontal and centroparietal, and hippocampus (McPherson, 1996), involving skills such as auditory attention and memory and cognition (Teixeira & Griz, 2011).

When performing LLAEP, it must be considered that some variables may influence the responses obtained in this assessment – e.g., age, sex, and the stimuli used. The CANS is known to progressively mature throughout childhood into adolescence. Hence, age may interfere with LLAEP findings (Silva et al., 2017).

Few studies are currently found in the literature correlating LLAEP with autism. Most of the research related to this topic includes a control group with individuals with TD (typical development), matched by age, to obtain an equivalent comparison (Kamita et al., 2021). Also, various studies have a greater prevalence of males with autism, which must be related to the fact that ASD is four times more prevalent in males than in females (American Psychiatric Association, 2013).

Electrophysiological assessment with LLAEP can use various types of acoustic stimuli, both verbal and nonverbal. They generate different responses in the CANS because verbal stimuli have a complex spectral-temporal structure, requiring synchronous neural activity to obtain a response to such stimuli (Kraus & Nicol, 2003). Most studies on individuals with autism were conducted with nonverbal stimuli; some of them used both verbal and nonverbal stimuli, one study used only verbal stimuli, and yet another study used biological sound stimuli (Kamita et al., 2021).

Regarding differences between verbal and nonverbal acoustic stimuli, studies found that individuals with TD had significantly different processing of both stimuli, which did not happen with individuals with ASD. Greater latencies of components P1, N1, P2, and N2 with verbal stimuli occurred only in individuals with TD (Kamita et al., 2021), and smaller latencies of components P1, N1, and P2 obtained with nonverbal stimuli likewise occurred only in individuals with TD (Swink & Stuart, 2012).

A study conducted LLAEP in children with ASD using both verbal and nonverbal stimuli and compared their results with those of a group of children with TD. It found an absence of differences in the latency and amplitude of LLAEP components between the groups, suggesting that children with ASD can process sounds like individuals with TD (Kamita et al., 2021). Another study used nonverbal stimuli to assess 10 subjects diagnosed with ASD, 10 diagnosed with Asperger syndrome, and 20 with TD, and likewise found no differences in LLAEP responses in individuals with ASD (Matas et al., 2009).

However, other findings in the literature observed differences in responses obtained in individuals with ASD, such as a greater absence of P3 responses with tone-bust stimuli (Magliaro et al., 2010), increased P3 and N4 latencies in attention activity with verbal stimuli (Whitehouse & Bishop, 2008), and decreased P3 amplitude (as observed in a study that used LLAEP with tone-burst stimuli in individuals with ASD) (Gonzalez-Gadea et al., 2015), greater latency and smaller amplitude of component N1 (Azouz et al., 2014), and decreased amplitude of components P1, N2, P3, and N4 in individuals with ASD (Whitehouse & Bishop, 2008).

Russo et al. (2009a, b) conducted a study on children with ASD and TD, using speech stimuli LLAEP (syllable /da/) to identify and analyze components P1 and N1. It was presented in two different conditions: conventional speech in silence at 80 dBSPL and in background noise at 75 dBSPL. It found statistically significant differences with increased P1 latency and decreased P1–N1 amplitude in the group with ASD, in comparison with the group with TD, in both silence and background noise. The authors observed that individuals with ASD in silence had similar responses to individuals with TD in noise, suggesting that children with ASD process complex acoustic stimuli like children with TD in noisy environments.

Sokhadze et al. (2016) conducted electrophysiological analyses with LLAEP and tone-burst stimuli and behavioral assessments before and after auditory integration training to investigate the changes that occur in the responses of children with ASD in comparison with children with TD. The ASD population, before the auditory integration training, had greater N1 and P3 latencies and smaller P3 amplitude than the TD population. After the auditory training, N1 latency in the ASD group remained greater, but without the statistically significant difference found before; the greater P3 amplitude and latency remained. These changes in LLAEP after auditory integration training demonstrated that children with ASD improved, with fewer changed results than the TD group. The behavioral assessment revealed improvements in irritability, hyperactivity, repetitive behaviors, and so forth. The authors suggest further studies to document behavioral and electrophysiological changes in this population after auditory integration training.

Research by Gomot et al. (2011) aimed to examine the neural basis of auditory changes in children with ASD with long-latency electrophysiological patterns and correlate these results with this population's intolerance to changes (using the BSE-R scale). It found results that diverge from the above mentioned ones, observing smaller latency and greater amplitude of P3 in the ASD group. The authors report that these findings may be related to this population's changed perception of new stimuli.

Recent research found different behavioral profiles between individuals with ASD and with TD. Electrophysiological response analysis revealed that when tone-burst stimuli were used, individuals with TD had smaller latency in the left ear, suggesting a greater speed to process nonverbal stimuli in the right hemisphere than in the left one. On the other hand, no difference was found between the ears in individuals with ASD, suggesting that they process sound information similarly between the two hemispheres (Kamita et al., 2021).

However, some authors analyzed hemispheric processing of verbal and nonverbal sounds in individuals with ASD and found they processed acoustic information predominantly in the left hemisphere (Sokhadze et al., 2016; Stroganova et al., 2013). On the other hand, a study in the literature found acoustic processing of electrophysiological responses predominantly in the right hemisphere of individuals with ASD (Donkers et al., 2015).

These findings may be related to differences in processing between the hemispheres, as the left hemisphere of the auditory cortex is responsible for processing verbal information, while the right hemisphere is responsible for nonverbal information (Jerger et al., 1995). Hence, the findings in the scientific literature are not homogeneous regarding auditory processing between the hemispheres, which shows the importance of individualized clinical reasoning when approaching this population (Kamita et al., 2021).

Due to the lack of consistency between studies, some authors made a meta-analysis with nine articles, analyzing LLAEP findings comparing individuals with ASD and with TD. They aimed to verify differences in auditory processing between these groups and whether such differences were observed in one or both hemispheres. They found that individuals with ASD had greater P1 latency bilaterally and greater N1 latency only in the right hemisphere. These results help us understand processing and lateralization, as the analysis demonstrated significant differences in how individuals with ASD process acoustic stimuli. Nonetheless, many gaps remain in the scientific literature, which must be addressed in future research to fully understand temporal auditory processing in ASD (Jorgensen et al., 2021) (Figs. 16.2 and 16.3).

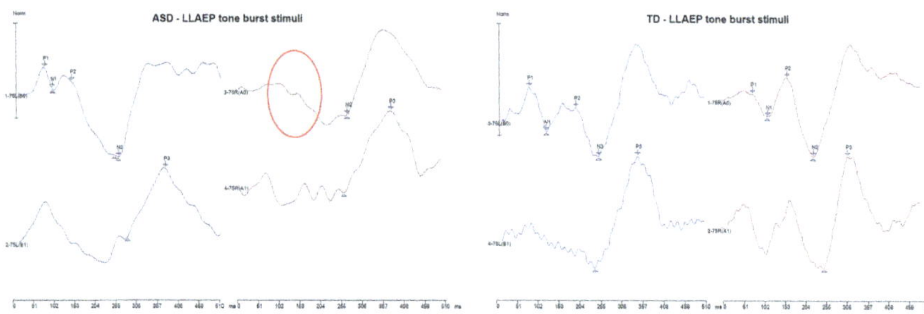

Fig. 16.2 Tone burst LLAEP responses obtained at 75 dBnHL. Odd paradigm applied to record P300 response. Frequent stimuli recorded at 1000 Hz and rare stimuli recorded at 2000 Hz. Comparison between responses for ASD and TD individuals. Note the absence of waves P1, N1 and P2 at the right ear (ipsi-lateral recording) of the ASD recording

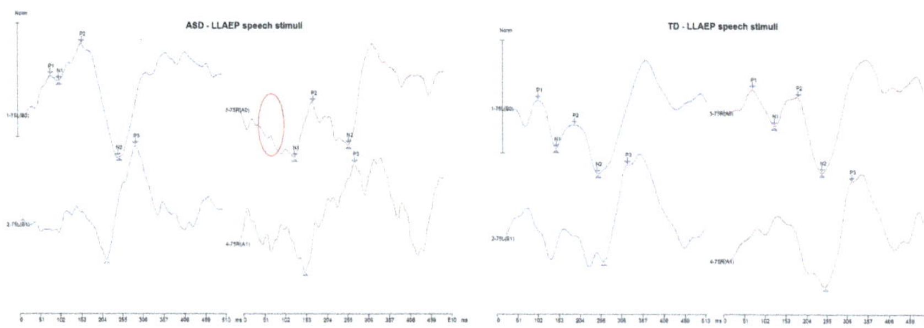

Fig. 16.3 Speech LLAEP responses obtained at 75 dBnHL. Odd paradigm applied to record P300 response. Frequent stimuli recorded with /ba/ syllable and rare stimuli recorded with /da/ syllable. Comparison between responses for ASD and TD individuals. Note the absence of wave P1 on the right ear (ipsi-lateral recording) of the ASD recording

Given the findings described above, present in the consulted literature, there is great heterogeneity in LLAEP responses in the ASD population. This diversity may be explained by a variety of intrinsic and extrinsic factors present in this population (which may influence the maturation of the central nervous system), each person's processing strategies, and consequently the electrophysiological responses. Therefore, studies with these potentials are important to better understand how individuals with ASD process sound information on the cortical level.

16.4 Conclusion

The various aspects approached in this chapter highlight the importance of electrophysiological hearing assessment with AEP in individuals with ASD to acquire further information on the functioning of the peripheral and central auditory systems, which is indispensable to characterize the auditory function in these individuals.

In clinical practice, this objective assessment aids the audiological diagnosis of a population in whom routine behavioral assessments are difficult. Hence, it investigates the integrity of the peripheral and central auditory pathways to understand how acoustic stimuli are processed in this population.

Furthermore, it helps monitor therapeutic intervention by investigating neuronal plasticity – i.e., verifying whether the CANS was structurally and/or functionally reorganized as the subject practiced a skill or was exposed to frequent stimuli. Thus, it can contribute to more effective therapeutic interventions.

It must be emphasized that further research with AEPs in the population with ASD is needed to better understand their acoustic stimulus processing. The importance of investigating the neurophysiological, cognitive, and structural basis of the wide range of electrophysiological responses in this population is also evident. Therefore, multidisciplinary/multicentric studies are the way to obtain a more comprehensive understanding of the aspects involved in sensory processing in individuals with ASD.

References

American Psychiatric Association. (2013). *Manual diagnóstico e estatístico de transtornos mentais (DSM-5)* (5th ed.). Artmed.

Anias, C. R., Lima, M. A. M., & Kós, A. O. (2004). Avaliação da influência da idade no potencial evocado auditivo de tronco encefálico. *Revista Brasileira de Oto-Rino-Laringologia, 70*(1), 84–89.

Azouz, H. G., Kozou, H., Khalil, M., Rania, M., Abdou, R. M., & Sakr, M. (2014). The correlation between central auditory processing in autistic children and their language processing abilities. *International Journal of Pediatric Otorhinolaryngology, 78*(12), 2297–2300.

Coutinho, M. B., Rocha, V., & Santos, M. C. (2002). Auditory brainstem response in two children with autism. *International Journal of Pediatric Otorhinolaryngology, 66*, 81–85.

Dabbous, A. O. (2012). Characteristics of auditory brainstem response latencies in children with autism spectrum disorders. *Audiological Medicine, 10*(3), 122–131.

Donkers, F. C., Schipul, S. E., Baranek, G. T., Cleary, K. M., Willoughby, M. T., Evans, A. M., Bulluck, J. C., Lovmo, J. E., & Belger, A. (2015). Attenuated auditory event-related potentials and associations with atypical sensory response patterns in children with autism. *Journal of Autism and Developmental Disorders, 45*(2), 506–523.

El Shennawy, A. M., El Khosht, M., Ghannoum, H., & El Meguid, N. A. (2014). Electrophysiologic assessment of auditory function in children with autism and attention-deficit and hyperactivity disorder. Ocena Elektrofizjologiczna Funkcji Słuchowych U Dzieci Z Autyzmem Oraz Zespołem Nadpobudliwości Psychoruchowej. *Journal of Hearing Science, 4*, 26–34. https://doi.org/10.17430/891185

ElMoazen, D., Sobhy, O., Abdou, R., & AbdelMotaleb, H. (2020). Binaural interaction component of the auditory brainstem response in children with autism spectrum disorder. *International Journal of Pediatric Otorhinolaryngology, 131*, 109850.

Fujikawa-Brooks, S., Isenberg, A., Osann, S., Spence, M., & Gage, N. (2010). The effect of rate stress on the auditory brainstem response in autism: A preliminary report. *International Journal of Audiology, 49*(2), 129–140.

Garcia, M., Silveira, A., & Didoné, D. (2013). Long latency auditory evoked potential in term and premature infants. *International Archives of Otorhinolaryngology, 18*(1), 16–20.

Gillberg, C., Rosenhall, U., & Johansson, E. (1983). Auditory brainstem responses in childhood psychosis. *Journal of Autism and Developmental Disorders, 13*(2), 181–195.

Gillberg, C., Steffenburg, S., & Jakobsson, G. (1987). Neurobiological findings in 20 relatively gifted children with Kanner-type autism or Asperger syndrome. *Developmental Medicine and Child Neurology, 29*, 641–649.

Gomot, M., Blanc, R., Clery, H., Roux, S., Barthelemy, C., & Bruneau, N. (2011). Candidate electrophysiological endophenotypes of hyper-reactivity to change in autism. *Journal of Autism and Developmental Disorders, 41*(6), 705–714.

Gonzalez-Gadea, M. L., Chennu, S., Bekinschtein, T. A., Rattazzi, A., Beraudi, A., Tripicchio, P., Moyano, B., Soffita, Y., Steinberg, L., Adolfi, F., Sigman, M., Marino, J., Manes, F., & Ibanez, A. (2015). Predictive coding in autism spectrum disorder and attention deficit hyperactivity disorder. *Journal of Neurophysiology, 114*, 2625–2636.

Hall, J. W. (2006). P3 response. In J. W. Hall (Ed.), *New handbook of auditory evoked responses* (2nd ed., pp. 518–547). Allyn & Bacon.

Hall, J. W. (2007). *New handbook for auditory evoked responses*. Allyn & Bacon.

Hall, J. W., III. (2015). *eHandbook of auditory evoked responses: Principles, procedures & protocols*. Kindle Direct Publishing.

Jerger, J., Alford, B., Lew, H., Rivera, V., & Chimiel, R. (1995). Dichotic listening, event-related potentials, and interhemispheric transfer in the older people. *Ear and Hearing, 16*, 482–498.

Jones, M. K., Kraus, N., Bonacina, S., Nicol, T., Otto-Meyer, S., & Roberts, M. Y. (2020). Auditory processing differences in toddlers with autism spectrum disorder. *Journal of Speech, Language, and Hearing Research, 63*, 1608–1617. https://doi.org/10.1044/2020_JSLHR-19-00061

Jorgensen, A. R., Whitehouse, A. J. O., Fox, A. M., & Maybery, M. T. (2021). Delayed cortical processing of auditory stimuli in children with autism spectrum disorder: A meta-analysis of electrophysiological studies. *Brain and Cognition, 150*, 105709. https://doi.org/10.1016/j.bandc.2021.105709

Junqueira, C. A. O., & Frizzo, A. C. F. (2002). Potenciais evocados auditivos de curta, média e longa latência. In A. M. C. M. Aquino (Ed.), *Processamento auditivo – eletrofisiologia & psicoacústica* (pp. 63–85). Lovise.

Källstrand, J., Olsson, O., Nehlstedt, S., Skold, M., & Nielzen, S. (2010). Abnormal auditory forward masking pattern in the brainstem response of individuals with Asperger syndrome. *Neuropsychiatric Disease and Treatment, 6*, 289–296.

Kamita, M. K., Silva, L. A. F., & Matas, C. G. (2021). Potenciais evocados auditivos corticais no transtorno do espectro do autismo: Revisão sistemática. *CoDAS, 33*(2), e20190207.

Kanner, L. (1943). Autistic disturbances of affective contact. *The Nervous Child, 2*, 217–250.

Khalfa, S., Bruneau, N., Rogé, B., Georgieff, N., Veuillet, E., Adrien, J.-L., et al. (2004). Increased perception of loudness in autism. *Hearing Research, 198*, 87–92.

Klin, A. (2006). Autismo e síndrome de Asperger: Uma visão geral. *Revista Brasileira de Psiquiatria, 1*, 3–11.

Kraus, N., & Nicol, T. (2003). Aggregate neural responses to speech sounds in the central auditory system. *Speech Communication, 41*, 35–47.

Kraus, N., & Nicol, T. (2005). Brainstem origins for cortical "what" and "where" pathways in the auditory system. *Trends in Neurosciences, 28*, 176–181.

Kraus, N., Kileny, P., & McGee, T. (1999). Potenciais Auditivos Evocados de Média Latência (MLR). In J. Katz (Ed.), *Tratado de Audiologia Clínica* (4th ed., pp. 384–402). Manole.

Kwon, S., Kim, J., Choe, B., Ko, C., & Park, C. (2007). Electrophysiologic assessment of central auditory processing by auditory brainstem responses in children with autism spectrum disorders. *Journal of Korean Medical Science, 22*(4), 656–659.

Li, A., Gao, G., Fu, T., Pang, W., Zhang, X., Qin, Z., & Ge, R. (2020). Continued development of auditory ability in autism spectrum disorder children: A clinical study on click-evoked auditory brainstem response. *International Journal of Pediatric Otorhinolaryngology, 138*, 110305.

Magliaro, F. C. L., Scheuer, C. I., Assumpção Júnior, F. B., & Matas, C. G. (2010). Estudo dos potenciais evocados auditivos em autismo. *Pró-Fono Revista de Atualização Científica, 22*(1), 31–36.

Matas, C. G., & Magliaro, F. C. L. (2013). Introdução aos Potenciais Evocados Auditivos e Potencial Evocado Auditivo de Tronco Encefálico. In M. C. Bevilacqua et al. (Eds.), *Tratado de Audiologia* (pp. 181–195). Santos.

Matas, C. G., & Toma, M. M. T. (2003). Audiometria de tronco encefálico (ABR): O uso do mascaramento na avaliação de indivíduos portadores de perda auditiva unilateral. *Revista Brasileira de Oto-Rino-Laringologia, 69*(3), 356–362.

Matas, C. G., Frazza, M. M., & Munhoz, M. L. L. (1998). Aplicação do potencial evocado auditivo de tronco encefálico em audiologia pediátrica. In M. C. A. Basseto, R. Brock, & R. Wajnsztejn (Eds.), *Neonatologia: Um convite à atuação fonoaudiológica* (pp. 301–310). Lovise.

Matas, C. G., Gonçalves, I. C., & Magliaro, F. C. L. (2009). Avaliação audiológica e eletrofisiológica em crianças com transtornos psiquiátricos. *Revista Brasileira de Oto-Rino-Laringologia, 75*(1), 130–138.

Maziade, M., Mérette, C., & Cayer, M. (2000). Prolongation of brainstem auditoryevoked responses in autistic probands and their unaffected relatives. *Archives of General Psychiatry, 57*(11), 1077–1083.

McPherson, D. L. (1996). *Late potentials of the auditory system (evoked potentials)*. Singular Publishing Group.

Miron, O., Beam, A. L., & Kohane, I. S. (2018). Auditory brainstem response in infants and children with autism spectrum disorder: A meta-analysis of wave V. *Autism Research, 11*(2), 355–363.

Möller, A. R., Janneta, P., Bennett, M., & Möller, M. B. (1981). Intracranially recorded responses from human auditory nerve: New insights into the origin of brainstem evoked potentials. *Electroencephalography and Clinical Neurophysiology, 52*, 18–27.

Musiek, F. E. (2006). Auditory training and CAPD. *Hearing Journal, 59*, 52.

Musiek, F. E., Borenstein, S. P., Hall, J. W., III, & Schwaber, M. K. (1999). Audiometria de tronco encefálico (ABR): Neurodiagnóstico e aplicações intra-operatórias. In J. Katz (Ed.), *Tratado de Audiologia Clínica* (4th ed., pp. 349–371). Manole.

Orekhova, E. V., Stroganova, T. A., Prokofiev, A. O., Nygren, G., Gillberg, C., & Elam, M. (2009). The right hemisphere fails to respond to temporal novelty in autism: Evidence from an ERP study. *Clinical Neurophysiology, 120*, 520–529.

Ramezani, M., Lotfi, Y., Moossavi, A., & Bakhshi, E. (2019). Auditory brainstem response to speech in children with high functional autism spectrum disorder. *Neurological Sciences, 40*, 121–125. https://doi.org/10.1007/s10072-018-3594-9

Reis, A. C. M. B., & Frizzo, A. C. F. (2011). Potencial evocado auditivo de longa latência. In C. M. Bevilacqua et al. (Eds.), *Tratado de Audiologia* (pp. 231–260). Santos.

Romero, A. C. L., Gução, A. C. B., Delecrode, C. R., Cardoso, A. C. V., Misquiatti, A. R. N., & Frizzo, A. C. F. (2014). Audiologic and electrophysiologic evaluation in the autistic spectrum disorder. *Revista CEFAC, 16*(3), 707–714.

Rosenblum, S. M., Arick, J. R., Krug, D. A., Stubbs, E. G., Young, N. B., & Pelson, R. O. (1980). Auditory brainstem evoked responses in autistic children. *Journal of Autism and Developmental Disorders, 10*(2), 215–225.

Rosenhall, U., Nordin, V., Sandström, M., Ahlsén, G., & Gillberg, C. (1999). Autism and hearing loss. *Journal of Autism and Developmental Disorders, 29*(5), 349–357.

Rosenhall, U., Nordin, V., Brantberg, K., & Gillberg, C. (2003). Autism and auditory brain stem responses. *Ear and Hearing, 24*(3), 206–214.

Russo, N., Skoe, E., Trommer, B., Nicol, T., Zecker, S., Bradlow, A., et al. (2008). Deficient brainstem encoding of pitch in children with autism spectrum disorders. *Clinical Neurophysiology, 119*, 1720–1731. https://doi.org/10.1016/j.clinph.2008.01.108

Russo, N. M., Zecker, S., Trommer, B., Chen, J., & Kraus, N. (2009a). Effects of background noise on cortical encoding of speech in autism spectrum disorders. *Journal of Autism and Developmental Disorders, 39*(8), 1185–1196.

Russo, N., Nicol, T., Trommer, B., Zecker, S., & Kraus, N. (2009b). Brainstem transcription of speech is disrupted in children with autism spectrum disorders. *Developmental Science, 12*, 557–567. https://doi.org/10.1111/j.1467-7687.2008.00790.x

Ruth, R. A., & Lambert, P. R. (1991). Auditory evoked potentials. *Otolaryngologic Clinics of North America, 24*(2), 349–370.

Seif, A., Shea, C., Schmid, S., & Stevenson, R. A. (2021). A systematic review of brainstem contributions to autism spectrum disorder. *Frontiers in Integrative Neuroscience, 15*, 760116.

Sersen, E. A., Heaney, G., Clausen, J., Belser, R., & Rainbow, S. (1990). Brainstem auditory-evoked responses with and without sedation in autism and Down's syndrome. *Biological Psychiatry, 27*(8), 834–840.

Sharma, A., Gilley, P. M., Dorman, M. F., & Baldwin, R. (2007). Deprivation-induced cortical reorganization in children with cochlear implants. *International Journal of Audiology, 46*, 494–499.

Silva, L. A., Magliaro, F. C., Carvalho, A. C., & Matas, C. G. (2017). Maturation of long latency auditory evoked potentials in hearing children: Systematic review. *Codas, 29*, e20160107.

Sokhadze, E. M., Casanova, M. F., Tasman, A., & Brockett, S. (2016). Electrophysiological and behavioral outcomes of Berard auditory integration training (AIT) in children with autism spectrum disorder. *Applied Psychophysiology and Biofeedback, 41*, 405–420.

Steinhoff, H. J., Böhnke, F., & Janssen, T. (1988). Click ABR intensity-latency characteristics in diagnosing conductive and cochlear hearing losses. *Archives of Oto-Rhino-Laryngology, 245*, 259–265.

Stroganova, T. A., Kozunov, V. V., Posikera, I. N., Galuta, I. A., Gratchev, V. V., & Orekhova, E. V. (2013). Abnormal pre-attentive arousal in young children with autism spectrum disorder contributes to their atypical auditory behavior: An ERP study. *PLoS One, 8*(7), e69100.

Swink, S., & Stuart, A. (2012). Auditory long latency responses to tonal and speech stimuli. *Journal of Speech, Language, and Hearing Research, 55*(2), 447–459.

Talge, N. M., Tudor, B. M., & Kileny, P. R. (2018). Click-evoked auditory brainstem responses and autism spectrum disorder: A meta-analytic review. *Autism Research, 11*(6), 916–927.

Tanguay, P. E., Edwards, R. M., Buchwald, J., Schwafel, J., & Allen, V. (1982). Auditory brainstem evoked responses in autistic children. *Archives of General Psychiatry, 39*(2), 174–180.

Taylor, M. J., Rosenblatt, B., & Linschoten, L. (1982). Auditory brainstem responses abnormalities in autistic children. *The Canadian Journal of Neurological Sciences, 9*(4), 429–433.

Teixeira, C. A., & Griz, S. M. S. (2011). Sistema auditivo central. In M. C. Bevilacqua et al. (Eds.), *Tratado de Audiologia* (pp. 17–27). Santos.

Tremblay, K. L., Ross, B., Inoue, K., McClannahan, K., & Colet, G. (2014). Is the auditory evoked P2 response a biomarker of learning? *Frontiers in Systems Neuroscience, 8*, 28.

Ververi, A., Vargiami, E., Papadopoulou, V., Tryfonas, D., & Zafeiriou, D. (2015). Brainstem auditory evoked potentials in boys with autism: Still searching for the hidden truth. *Iranian Journal of Child Neurology, 9*(2), 21–28.

Whitehouse, A. J., & Bishop, D. V. (2008). Do children with autism "switch off" to speech sounds an investigation using event related potentials. *Developmental Science, 11*, 516–524.

World Health Organization. (2019). 6A02 Autism spectrum disorder. In *International statistical classification of diseases and related health problems* (11th ed.). WHO. http://id.who.int/icd/entity/43781562

Chapter 17
Developmental Origins of the Structural Defects Implicated in ASD: Insights from iPSC and Post-Mortem Studies

Rana Fetit, Thomas Pratt, and David Price

17.1 Introduction

Understanding ASD aetiology requires the integration of information from multiple levels of analysis involving genetic, cellular and patient studies. The different approaches complement and augment one another, allowing us to reach a more comprehensive and systematic understanding of this complex, heterogenous and pervasive neurodevelopmental disorder. Although the majority of ASD cases are of unknown cause, commonly referred to as idiopathic or non-syndromic (Casanova et al., 2020), several genetic and environmental factors, such as maternal stress and prenatal exposure to toxins (Amaral et al., 2018; Varghese et al., 2017), play a substantial role in its onset and development. Similarly, hundreds of genomic loci and autism risk alleles have been implicated (SFARI GENE; https://gene.sfari.org). Unlike idiopathic ASD, the genetic factors known to cause syndromic ASD may be grouped into three classes. First, single gene mutations contribute to around 5% of cases, including Fragile X and Rett Syndrome (Hagerman et al., 2011; Loat et al., 2008). Secondly, large genomic copy number variants (CNVs), such as deletions and duplications of a given gene segment, account for approximately 10% of ASD cases (Ramaswami & Geschwind, 2018; Tuzun et al., 2005). Third, polygenic risk factors, due to accumulation of common de novo single-nucleotide variants, probably account for at least 20% of ASD liability (Weiner et al., 2017; Sanders et al., 2012; McKenna et al., 2018).

Post-mortem studies have greatly contributed to our understanding of ASD and several brain banks and biopsy networks exist in the UK, US and worldwide to facilitate the acquisition of post-mortem brain tissue for ASD research. Direct

R. Fetit (✉) · T. Pratt · D. Price
Centre of Discovery Brain Sciences (CDBS), University of Edinburgh, Hugh Robson Building, George Square, Edinburgh, UK
e-mail: Rana.Fetit@ed.ac.uk

A. El Idrissi, D. McCloskey (eds.), *Neurobiology of Autism Spectrum Disorders*,
https://doi.org/10.1007/978-3-031-42383-3_17

examination of post-mortem brain tissue from ASD subjects has allowed the detection of large-scale neuropathological and cytoarchitectural abnormalities, as well as impairments at the molecular, genetic and epigenetic level (Fetit et al., 2021). However, such neuropathological observations can only provide limited insights into disease mechanisms (Lewis, 2002; Amin & Pasca, 2018). ASD manifests in early developmental periods and its onset usually precedes death by decades. Thus, pathological changes due to secondary symptoms, associated comorbidities or treatment interventions confound the assessment and determination of causal pathology in end-stage post-mortem samples. Moreover, the absence of living cells in fixed post-mortem samples prevents the use of functional assays and experimental manipulation to study neural networks and test hypotheses (Habela et al., 2016; Beltrao-Braga & Muotri, 2017; Prilutsky et al., 2014; Dolmetsch & Geschwind, 2011; Acab & Muotri, 2015).

The development of induced pluripotent stem cell (iPSC) technology has provided a valuable, scalable tool that can be used to model and understand developmental programs and how they may be altered in disease. iPSCs can be derived from patient somatic cells, such as blood and skin cells, which are then reprogrammed to a pluripotent state by overexpression of specific embryonic genes (Deyle, 2015). The easy accessibility of cells from peripheral tissues and minimal invasiveness of the harvesting tools have allowed for isolating cells from ASD subjects. Indeed, iPSCs derived from individuals with ASD carry the genetic information of their donors, thus capturing the genetic diversity of the subjects. This is of great significance when it comes to studying idiopathic, non-syndromic ASD where the underlying genetic mutations remain largely unknown, making it more challenging to investigate without a personalized model to track neurodevelopment under an individual genetic background. Moreover, the availability of gene engineering techniques such as CRISPR/Cas9 has allowed for the targeted manipulation of iPSCs. Introducing genetic mutations at specific loci enabled researchers to mimic monogenic forms of ASD or ASD-related copy number variants (CNVs) through correcting or disrupting the causative gene or a specific locus. This comes with the additional advantage of generating isogenic iPSC lines that share a single genetic background and differ only at the targeted mutation site, allowing for comparisons without this confounding variable (Yuan & Shaner, 2013; Vaccarino et al., 2011; Beltrao-Braga & Muotri, 2017; Lim et al., 2015; Nestor et al., 2016).

17.1.1 2D and 3D iPSC Models

Our understanding of patterning factors that govern in vivo brain development has enabled researchers to exploit the pluripotency of iPSCs and generate an array of neural cell types in 2D-cultures and 3D-brain organoids. The different approaches established so far have been the subject of several recent reviews (Villa et al., 2021; Whiteley et al., 2022; Zhang et al., 2021; Pintacuda et al., 2021; Yang & Shcheglovitov, 2020; Wang et al., 2020; Nehme & Barrett, 2020; Lee et al., 2020; Gordon & Geschwind, 2020). 2D-neural cell cultures include glutamatergic,

GABAergic, dopaminergic and serotonergic neurons as well as astrocytic and microglial cultures that can be generated by employing pro-neural transcription factors and lineage-specifying molecules (Miskinyte et al., 2017; Vadodaria et al., 2016; Ho et al., 2015; Soubannier et al., 2020; Hasselmann & Blurton-Jones, 2020). Unlike the limited potential of neuronal samples acquired from postmortem tissue, human iPSC-derived neurons allow for detailed functional analyses of neurons generated from living subjects. Moreover, different neural cell types can be co-cultured to study specific cellular interactions. Although 2D-monolayer cultures are more scalable, facilitate imaging and offer differentiation approaches that are more direct, they are less likely to recapitulate the complexity of the brain cytoarchitecture and in vivo neuronal diversity (Paşca, 2018).

Therefore, 3D-brain organoids have been developed to provide a model system that can closely recapitulate early aspects of the emergence of neuronal subtypes, lamina formation and, to some extent, regional organization and allow for the dynamic interactions between the different cell types as well as their extracellular matrix. These are aspects of the developing human brain which are particularly relevant in ASD and cannot be properly evaluated in 2D cultures (Beltrao-Braga & Muotri, 2017, Kim et al., 2014; Lee et al., 2017b; Ben-Reuven & Reiner, 2016). Current protocols for generating human cerebral organoids fall into two main categories that utilise either free-patterning methods that rely on the default formation of the neuroepithelium (Lancaster et al., 2013), or use inductive cues like growth factors and morphogens to direct differentiation into specific brain regions that resemble the subpallium, midbrain, cerebellum or hippocampus (Chan et al., 2021). In this chapter both 2D and 3D models are considered.

17.1.2 The Need for an Integrated Approach

The difficulty in obtaining human autopsy material, let alone samples with rare and specific genetic mutations, calls for caution when interpreting and analysing data from the available studies, which are limited in number and have small sample sizes. The majority of post-mortem findings on ASD come from a small number of studies examining particular aspects of ASD brains. Most studies independently investigate different brain regions in convenience samples (collected from tissue banks close to hand owing to limited accessibility and availability) making it harder to generalise the findings or to examine their reproducibility or consistency. Differences in fixation methods, storage duration and conditions introduce additional sources of variation that need to be considered. Furthermore, the paucity of samples from younger patients or across a broad age range makes it harder to investigate the early developmental processes likely to be critical in understanding a neurodevelopmental disorder like ASD (Fetit et al., 2021). The inherent neuropathological heterogeneity of ASD can make it unclear whether any of the reported differences point to regional disruptions and desynchronization of brain development in ASD, or reflect the varying symptoms, cognitive deficits, and co-morbidities in the disorder. To collect post-mortem material for further studies that may or may not disprove these findings will take years.

The use of iPSC-based approaches is not, on its own, a panacea. The heterogeneity in iPSC-derived cultures remains an issue that can influence phenotypic read-outs. In addition to the genetic and epigenetic differences in iPSC source materials, iPSC lines tend to accumulate additional point mutations or CNVs in culture (Kanchan et al., 2020). Moreover, considerable variation between differentiation protocols and culture conditions exists and the resulting 2D and 3D-models may contain different cell types in different proportions. The technical variability can further introduce undesirable variation that may mask or influence the consequences of the inherent genetic variability within and between ASD subjects. Currently, the process of generating, maintaining, characterising, and analysing iPSC lines and their derived models is very expensive, time-consuming and labour intensive, which limits the number of cell lines an individual research lab can handle (Nestor et al., 2016; Kim et al., 2014; Beltrao-Braga & Muotri, 2017). To confidently recapitulate ASD aetiology using iPSC-models, the establishment of robust, scalable and high-throughput screening assays remains essential and additional tools, such as genome-wide expression profiling, should be routinely used to assess variability. Newly automated approaches to culture and quantify more homogenous iPSC-models are still underway (Beghin et al., 2022; Beck et al., 2022; Zhang et al., 2022; Dolmetsch & Geschwind, 2011).

It is, therefore, essential to integrate the findings from both postmortem and iPSC studies to better define and elucidate the underlying pathophysiology of ASD. The two approaches are complementary and finding similar phenotypes in both iPSC models and the intact ASD brain can provide robust evidence for alterations that arise during brain development and last postnatally. The combination of both methods can also help identify critical gaps in our understanding of ASD pathology, and direct further studies to investigate these aspects in detail. In this chapter, we will summarise the complementary findings from iPSC and post-mortem studies and highlight potential areas that require further investigation to better understand the aetiology of ASD.

17.2 Complementary Findings in iPSC and Postmortem Studies of ASD

Postmortem literature on ASD is extensive, and sufficient replicated evidence has been presented and comprehensively reviewed to implicate non-coding RNA, aberrant epigenetic profiles, GABAergic, glutamatergic and glial dysfunction in ASD pathogenesis (Fetit et al., 2021). On the other hand, only a handful of studies utilised iPSC technology to investigate ASD by generating 2D-cultures of neural progenitor cells (NPCs), neurons, astrocytes, and prefrontal cortex (PFC)-like tissue or 3D-brain organoids. Table 17.1 summarises the findings in iPSC-models. It is evident that some findings are frequently reported in both iPSC and postmortem studies and converge on certain developmental processes, which we list below and summarise in Fig. 17.1.

Table 17.1 Summary of iPSC findings in ASD

Model	Study	Condition	Findings
iPSC-derived NPCs	Zeng et al. (2013)	Monogenic autism, NRXN1 knockout	Altered expression of many genes pathways for the cell adhesion, neuronal and astrocyte differentiation
	Marchetto et al. (2017)	Non-syndromic/ idiopathic ASD with common macrencephalic phenotype	Increased cell proliferation: increased percentage of Ki67+ (cycling cells), shortening of G1 and S phases in ASD-derived NPCs owing to dysregulation of a β-catenin/BRN2 transcriptional cascade
	Sánchez-Sánchez et al. (2018)	Non-syndromic ASD with identified heterozygous missense variants in RELN gene	Abnormal interplay between Reelin–DAB1 and mTORC1 networks resulting in impaired Reelin secretion in NPCs
	Lalli et al. (2020)	CRISPRi-targeting of a set of ASD-associated genes (ADNP, ASH1L, ARID4B, CHD2, CHD8, DYRK1A, PTEN) in commercially available human NPCs	CRISPRi targeting *ASH1L* and *CHD2* resulted in decreased NPC proliferation, whereas targeting *CHD8* resulted in increased NPC proliferation, which alter neuronal differentiation.
	Connacher et al. (2022)	iPSC-derived NPCs from three idiopathic ASD subjects (only one subject exhibited macrocephalic phenotype)	Hyperproliferation observed in NPCs from macrocephalic ASD subject, hypoproliferation in NPCs from other two non-macrocephalic subjects. P-ERK is decreased in hyperproliferative but increased in hypoproliferative NPCs
iPSC-derived neurons	DeRosa et al. (2012)	Non-syndromic/ idiopathic ASD	No cellular phenotype observed between patient-derived and neurotypical subjects
	Griesi-Oliveira et al. (2015)	Non-syndromic/ idiopathic ASD with identified translocation interrupting TRPC6 gene	Morphologic and functional alterations: reduced total neurite length and dendritic arborization, lower density of dendritic spines and VGLUT1 puncta and impaired calcium dynamics.
	Marchetto et al. (2017)	Non-syndromic/ idiopathic ASD with common macrencephalic phenotype	Abnormal neurogenesis and reduced synaptogenesis leading to functional defects in neuronal networks that could be rescued by insulin growth factor 1 (IGF-1)
	Russo et al. (2018)	Non-syndromic/ idiopathic ASD	Less arborization, decrease in synaptic gene expression and protein levels, glutamate neurotransmitter release and reduced spontaneous firing rate.
	Schafer et al. (2019)	Non-syndromic/ idiopathic ASD	Aberrant neuronal maturation of cortical neurons: complex neurite branching patterns, accelerated development, abnormal temporal maturation
	Zaslavsky et al. (2019)	CRISPR-Cas9-edited control and ASD patient cohorts with SHANK2 haploinsufficiency	Increased Synapsin 1 puncta, synaptic connections, dendrite length, and branch complexity and frequency of spontaneous excitatory postsynaptic currents in patient and CRISPR-Cas9 KO excitatory neurons

(continued)

Table 17.1 (continued)

Model	Study	Condition	Findings
iPSC-derived astrocytes	Russo et al. (2018)	Non-syndromic/ idiopathic ASD	Increased ROS levels, ASD astrocytes interfered with proper neuronal development and morphology
iPSC-derived prefrontal cortex (PFC) monolayer cultures	Cederquist et al. (2020)	CRISPR-Cas9-edited control cohort of 28 high-confidence, early-developmental ASD genes	Identified two classes of ASD mutations that perturb PFC neurogenesis and are correlated to abnormal WNT/βcatenin responses: **Class 1 (ANKRD11, ASH1L, ASXL3, CUL3, DEAF1, KDM5B, KMT2C, RELN)** inhibits neurogenesis, **Class 2 (CHD8, DYRK1A, GRIN2B, KMT2A, SUV420H1)**: enhances neurogenesis.
Cerebral organoids	Mariani et al. (2015)	Non-syndromic/ idiopathic ASD with increased head circumference	Accelerated cell cycle, overproduction of GABAergic inhibitory neurons due to the upregulation of FOXG1. Increased number of synapses marked by Synapsin I puncta and VGAT puncta with no change in VGLUT. Differential expression of genes related to cytoskeletal regulation,synaptic assembly, ion transport and postsynaptic signalling, and the ASD-risk genes: NRXN1 and SCN2A.
	Wang et al. (2017)	CRISPR-Cas9-edited control cohort with *CHD8* haploinsufficiency	Alterations in DLX6-AS1 and DLX1 gene expression. Increase in TCF4 expression, enrichment of genes involved in neurogenesis, neuronal differentiation, forebrain development, Wnt/β-catenin signalling, and axonal guidance.
	Schafer et al. (2019)	Non-syndromic/ idiopathic ASD	Increased CP thickness

17.2.1 Brain Size and Neuronal Numbers

ASD is associated with a decrease in head circumference during early infancy, followed by a significant increase in head circumference after 6 months (Courchesne et al., 2003). By 2–4 years of age, there is an abrupt cessation of growth so that by adolescence and adulthood the majority of ASD cases lie within the normal range (Redcay & Courchesne, 2005). Indeed, aspects like brain size and weight can only be directly examined using brain tissue. Some postmortem studies reported that larger brain weights co-occurred with greater total prefrontal neuron counts in ASD children, suggesting a pathological increase in neuron numbers (Courchesne et al., 2011). However, individual postmortem studies have small sample sizes where the ages do not represent the entire lifespan and, therefore, do not address age-related

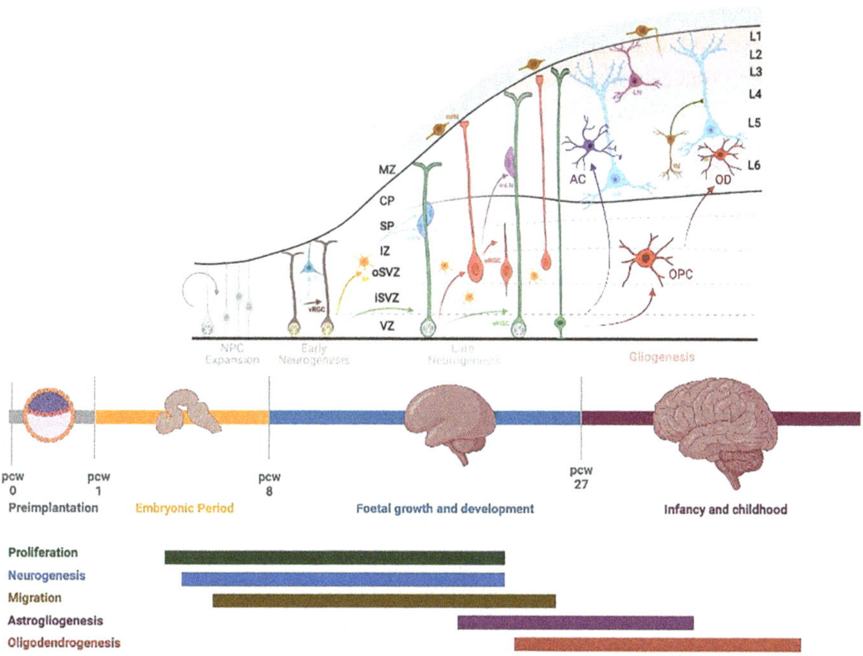

Fig. 17.1 Timeline of neurodevelopmental processes that are potentially disrupted in ASD. Illustrations show the different phases of human cortical development. Age in post conception weeks (pwc). The bars below represent the approximate timings of the different cellular processes taking place. The schematic shows the different cell types. Neural progenitor cells (NPCs) first undergo a series of symmetric cell divisions to establish an initial pool of cortical progenitors. NPCs later develop into ventricular radial glia cells (vRGCs) which undergo asymmetric cell divisions, producing another vRGC and a nascent neuron (N). The neuron migrates radially from the ventricular zone (VZ) along the basal process of a RGC to reach the cortical plate (CP). The subplate (SP) is a transient zone below the CP and above the intermediate zone (IZ). Early-born neurons (EN) settle in the deep layers (Layers 5 and 6), and later-born neurons (LN) settle in the upper layers. The subventricular zone (SVZ) is divided into the inner SVZ (iSVZ) and the outer SVZ (oSVZ). Some RGCs develop into intermediate progenitor cells (IPs) or outer radial glial cells (oRGCs) in the outer subventricular zone (oSVZ). Interneurons (INs) migrate (m) tangentially along the marginal zone (MZ) from the ventral subpallium to reach the cortical plate and integrate with mature cortical projection neurons into functional circuits. By mid-late gestation, the process of gliogenesis begins where astrocytes (AC), oligodendrocyte progenitor cells (OPCs) and oligodendrocytes (OD) are produced. (Adapted from Silbereis et al. (2016), Yang and Shcheglovitov (2020), Molnár et al. (2019) and created using Biorender.com)

changes in brain enlargement despite the evidence for a different developmental trajectory in head circumference and brain volume growth in ASD (Bedford et al., 2020).

Postmortem findings on neuronal numbers, densities and volumes are not consistent (Fetit et al., 2021), and the most severe neuronal deficits, as indicated by Wegiel et al. (2014), were found in 4–8 year old ASD children, suggesting that neuronal growth dysregulation occurs before the age of 4 years and collectively suggesting that the critical developmental period for the onset of ASD pathogenesis precedes this stage (Wegiel et al., 2014). Whether the different developmental trajectories in head circumference and brain volume growth in ASD are associated with abnormal trajectory of neuronal growth is yet to be investigated. To date, only one iPSC study used cerebral organoids to investigate the macrocephalic phenotype in non-syndromic ASD. Mariani et al. (2015) observed no significant change in organoid area between ASD and control groups, except for a transient increase in ASD-derived organoid size at day 11 (Mariani et al., 2015). However, no other iPSC-study utilised brain organoids to recapitulate the macrocephalic phenotype or investigate differences in neuronal numbers or subtypes in ASD.

17.2.2 Proliferation and Neurogenesis

In the developing brain, the earliest processes are proliferation and neurogenesis, the perturbation of which might be reflected in the brain size differences and aberrant neuronal numbers reported in a subset of ASD cases. The ability of iPSC models to recapitulate early foetal brain development has allowed us to investigate NPC proliferation and differentiation in ASD, which could not otherwise be thoroughly investigated using post-mortem tissue. NPCs are multipotent, self-renewing neural stem cells that reside in the ventricular zone (VZ) and subventricular zone (SVZ) in the developing brain (Fig. 17.1). These two proliferative regions give rise to the glial and neuronal cell types in the brain throughout the course of embryonic development (Martínez-Cerdeño & Noctor, 2018; Silbereis et al., 2016).

NPCs derived from macrocephalic ASD subjects displayed an increase in cell proliferation, as evident by the increase in the percentage of cells positive for the cell cycle marker Ki67 (Marchetto et al., 2017). Another study knocked down 13 ASD-associated genes to identify a subset of genes that converge on cell cycle control and progenitor cell proliferation and resulted in decreased or increased NPC proliferation (Lalli et al., 2020). In a similar approach, Cederquist et al. (2020) targeted 28 high-confidence, early-developmental ASD genes and generated prefrontal cortex (PFC) monolayer cultures from iPSCs revealing two subclasses that exterted opposite effects on NPC proliferations. The first subclass showed NPC enrichment and decreased neuronal output, whereas the second exhibited NPC depletion and increased neuronal output (Cederquist et al., 2020). In a recent study utilising iPSC-derived NPCs from three idiopathic ASD subjects with and without macrocephaly, Connacher et al. (2022) observed that NPCs from macrocephalic ASD subject were hyperproliferative, whereas NPCs from the other two non-macrocephalic subjects were hypoproliferative. Furthermore, cerebral organoids generated from non-syndromic ASD subjects with increased head circumference displayed an accelerated cell cycle (Mariani et al., 2015).

In the adult brain, SVZ of the lateral ventricles (LV-SVZ) is the neurogenic niche, with characteristics similar to both the embryonic ventricular zone (VZ) and SVZ (Lim & Alvarez-Buylla, 2016). Only two post-mortem studies examined the

LV-SVZ. Pearson et al. (2013) found that the mean number of Ki67+ proliferating cells in the LV-SVZ was higher in a 5-year old ASD subject (Pearson et al., 2013), and Kotagiri et al. (2014) found a higher density of proliferating cell nuclear antigen (PCNA), which is expressed at detectable levels throughout the cell cycle but peaks in S phase (Kurki et al., 1986; Kotagiri et al., 2014).

Collectively, evidence from iPSC studies, and to a lesser extent postmortem studies, implicate cell cycle dysregulation in ASD. The balance between NPC proliferation and neuronal differentiation is critical for the generation of the right numbers of functional neurons at the right time and place (Hardwick et al., 2015). The disruption of the cell cycle can either delay or accelerate the differentiation of progenitors into neurons, which may eventually alter the proportions of neuronal numbers in ASD. Rapid rates of proliferation in neural progenitors could account for the increased brain volume and neuronal numbers frequently reported in ASD, and may serve to explain the early brain overgrowth observed in the first year of life in ASD (Redcay & Courchesne, 2005).

Moreover, the WNT/βcatenin pathway has been implicated as a potential regulator of NPC proliferation in both iPSC (Marchetto et al., 2017; Cederquist et al., 2020) and postmortem (Chow et al., 2012; Garbett et al., 2008) studies of the ASD brain. During brain development, WNT signalling controls the temporal proliferation and differentiation of NPCs, whereby its stimulation in early neurogenesis induces NPC proliferation but promotes terminal neurogenesis later in development (Hirabayashi et al., 2004; Munji et al., 2011). Because NPC proliferation and neurogenesis occur very early in brain development, and altered head circumference or brain size in ASD are usually observed and reported later in postnatal development, 3D-organoid models of idiopathic ASD could be useful to provide a causative, mechanistic link between NPC proliferation, neurogenesis and brain size.

17.2.3 Differentiation and Cell-Type Comittment

In addition to perturbations in neuronal numbers or differences in brain weight and size, aberrant NPC proliferation may influence progenitor differentiation and result in changes in numbers of a specific cell type (Dalton, 2015; Hwang et al., 2021). Mariani et al. (2015), reported an overproduction of GABAergic inhibitory neurons in cerebral organoids derived from non-syndromic ASD subjects with increased head circumference (Mariani et al., 2015). However, this has not been thoroughly investigated in other iPSC-studies.

Several postmortem studies have implicated specific subpopulations of interneurons in ASD. Studies point towards a reduction in PV interneuron levels in the neocortex and possibly also the cerebellum. A decrease in the number of PV-expressing cells could result from decreased neuronal number or decreased PV protein expression (Hashemi et al., 2017; Lawrence et al., 2010; Edmonson et al., 2014; Schwede et al., 2018; Whitney et al., 2009; Soghomonian et al., 2017). Other post-mortem findings in ASD subjects revealed increased levels of calbindin (CB) interneurons in the dorsolateral pre-frontal cortex, hippocampus, posterior cingulate cortex and fusiform gyrus (Fetit et al., 2021), as well as a reduction in Reelin (RELN) interneuron levels, particularly in the dorsolateral prefrontal cortex, superior frontal, parietal, and cerebellar cortices (Fetit et al., 2021). Furthermore, four post-mortem studies consistently reported lower cell counts of Purkinje neurons, another class of GABAergic inhibitory neurons populating the cerebellum (Bauman & Kemper, 1985; Ritvo et al., 1986; Fatemi et al., 2002a; Skefos et al., 2014).

It is therefore important to further explore how the disruption of the cell cycle and, consequently, neurogenesis, may alter the ratios of different neuronal subtypes and whether a preferential differentiation of progenitors towards specific neuronal subtypes exists in ASD. Given that the reported neuronal subtype perturbations in both iPSC and postmortem studies so far are GABAergic, this should be further examined in the light of potential excitatory-inhibitory imbalances that may arise (discussed further below).

17.2.4 Apoptosis

During brain development, the removal of excess neurons is essential for formation of functional circuitry (Fricker et al., 2018). However, aberrant neuronal death may also impact neurogenesis and perturb neuronal numbers. Four postmortem studies revealed reduced anti-apoptotic proteins and/or increased pro-apoptotic proteins in ASD tissue from donors older than 4 years (Fatemi et al., 2001, Araghi-Niknam & Fatemi, 2003; Sheikh et al., 2010a, b). Thus, the trend towards increased apoptosis in ASD could align with the period where there is an abrupt cessation of growth so that by adolescence and adulthood the majority of autistic cases lie within the normal range. Such generalised processes from an early age could underlie the learning disability seen in approximately a half of people meeting diagnostic criteria for ASD. However, this should be investigated in further detail. No iPSC-studies examined apoptotic mechanisms in non-syndromic ASD so far. Moreover, due to the lack of adequate diffusion of gases and nutrients to the centre of organoids (Lancaster, 2018), cells in the organoid core usually became necrotic, which might limit the use of iPSC-derived cerebral organoids to earlier timepoints.

17.2.5 Cortical Cytoarchitecture

Following neurogenesis, newly born excitatory neurons migrate radially from the proliferative zones of the dorsal pallium to populate the cortex in an inside-out fashion and establish different cortical layers of the cortical plate (CP) (Fig. 17.1). Inhibitory interneurons, on the other hand, migrate tangentially from the ganglionic eminences in the ventral subpallium to integrate with the excitatory neurons to form functional circuits in the CP (Hu et al., 2017). In postmortem studies, Hutsler et al.

(2007) reported disturbances in cortical cell patterning in the superior temporal gyrus, dorsolateral frontal lobes and dorsal parietal lobes including dysplasias, associated disturbances in lamination and a slightly less distinct gray/white matter boundary in eight ASD subjects between 10 and 45 years (Hutsler et al., 2007). Stoner et al. (2014) identified focal regions of reduced expression or unusual patterns of markers, which they referred to as "patches" in the dorsolateral prefrontal cortex (DL-PFC) and posterior superior temporal cortex in ten autistic children aged 2–15 years. Deficits in the expression of excitatory cortical neuron markers was the most robust indicator of a patch region (Stoner et al., 2014). Two postmortem studies further examined the cytoarchitecture of the SVZ in ASD. While Pearson et al. (2013) found no differences in cortical layering and organisation in the LV-SVZ of four ASD individuals between 5 and 60 years of age (Pearson et al., 2013), Kotagiri et al. (2014) reported notable differences, specifically a lower density of total cells in the septal hypocellular gap in another cohort of four ASD cases between 10 and 42. Additionally, Casanova et al., examined minicolumns, which are the basic repeating functional unit of cortical organization consisting of radially stacked neurons, and consistently reported more minicolumns of reduced width in ASD subjects between 4 and 28 years (Casanova et al., 2002a, b, 2006a, b).

To date, only one iPSC-study using cerebral organoids from a non-syndromic ASD patient revealed an increase in CP thickness, with CP-like regions positive for TBR1, a marker of early-born neurons that occupy superficial cortical layers (Hevner et al., 2001), being significantly larger in ASD organoids (Schafer et al., 2019). Additional iPSC-studies should thoroughly investigate cortical layering in non-syndromic ASD using organoids and model any potential migratory deficits in ASD. The presence of other cell types that support neuronal migration, such as astrocytes (Cardenas et al., 2014) and oligodendrocytes (Doretto et al., 2011), as well as the possibility of fusing dorsal and ventral organoids to mimic interneuron migration make cerebral organoid a suitable model to investigate migratory mechanisms in ASD.

17.2.6 Neuropathological Alterations in Morphology and Synapses

As neurons migrate to populate the different cortical regions, they begin to mature and acquire their correct morphology by growing axons and dendrites (Sainath & Gallo, 2015) and forming synapses (Garner et al., 2002) to establish connections with one another. With the progression of neuronal maturation, the myelination of neuronal axons by oligodendrocytes occurs, a process which continues throughout adolescence and into adulthood (Williamson & Lyons, 2018).

Thinner axons with more branches were consistently reported in ASD across three postmortem studies (Zikopoulos et al., 2018; Zikopoulos & Barbas, 2010; Trutzer et al., 2019). Three additional post-mortem studies have reported increased spine densities, which are small protrusions along the dendrites of neurons representing the postsynaptic sites of most excitatory synapses (Runge et al., 2020), in ASD. This was observed in the frontal, temporal and parietal lobes as well as the amygdala (Hutsler & Zhang, 2010; Tang et al., 2014; Weir et al., 2018). Moreover, aberrant myelination patterns were also reported. A reduction in myelin thickness and lower density of myelinated axons were reported in two studies (Zikopoulos et al., 2018; Zikopoulos & Barbas, 2010), whereas a third study reported increased variability in the thickness of myelinated axons in ASD (Trutzer et al., 2019).

Morphological and synaptic perturbations have also been reported in a number of studies utilising iPSC-derived neurons. (Griesi-Oliveira et al., 2015) revealed a reduction in the total neurite length, dendritic arborization and dendritic spines density in neurons derived from non-syndromic ASD subjects carrying a mutation in TRPC6 gene (Griesi-Oliveira et al., 2015). Reduced dendritic arborization has also been observed in another study (Russo et al., 2018). Moreover, (Schafer et al., 2019) reported aberrant neuronal maturation of cortical neurons with complex neurite branching patterns and accelerated development (Schafer et al., 2019). Using iPSCs from ASD patient cohorts with SHANK2 haploinsufficiency and CRISPR/Cas9-edited lines, (Zaslavsky et al., 2019) demonstrated that both patient and CRISPR/Cas9 knockout mutant lines exhibited an increase dendrite length and branch complexity in iPSC-derived neurons.

Taken together, altered dendritic morphology is consistently implicated in ASD. However, discrepancies between iPSC and postmortem studies are present. Whilst postmortem studies report increased branching, some iPSC studies report a reduction and others observe an increase. Indeed, this could be a reflection of the heterogeneity of ASD or an effect of the differences in genetic backgrounds of the donor ASD subjects. The presence of additional mutations in specific genes like TRPC6 and SHANK2 might also contribute to the different phenotypes observed in the different studies. Conversely, this could also be due to the presence of confounding variables, such as age, sex, comorbidities or differences in differentiation protocols and sample preparation methods. More studies are needed to further elucidate the exact nature of dendritic abnormalities. Furthermore, aberrant myelination patterns are observed in postmortem ASD tissue. However, no iPSC studies to date examined disturbances in neuronal myelination. Utilising cortical organoids to model complex interactions between neurons and myelinating oligodendrocytes may be beneficial.

17.2.7 Excitatory/Inhibitory (E/I) Imbalance

The proper formation of excitatory and inhibitory synapses at a single neuron level as well as the overall ratio of excitatory to inhibitory neurons in the cortex is critical to the correct establishment of neural circuits and normal brain functioning (Sohal & Rubenstein, 2019). The imbalance between excitatory and inhibitory neurons and synapses affects how the brain processes information and regulates behaviour and is likely to be implicated in the core ASD symptoms. Indeed, perturbations to early developmental processes, such as neurogenesis, migration and synaptogenesis might contribute to this imbalance.

Several studies using iPSC-derived neurons consistently reported perturbations in glutamatergic synapses. Neurons from non-syndromic ASD with a mutation in the TRPC6 gene revealed fewer vesicular glutamate transpoter-1 (VGLUT1) puncta (Griesi-Oliveira et al., 2015). *TRPC6* is expressed in glutamatergic synapses and it is suggested that this gene is important for the regulation of excitatory synapse strength (Zhou et al., 2008). Thus, the reduction of excitatory VGLUT1 puncta density suggests that perturbations in TRPC6 expression levels may modulate glutamatergic synapse formation. Similarly, Marchetto et al. (2017) revealed a reduction in the density of both Synapsin and VGLUT1 puncta in iPSC-derived ASD neurons, suggesting deficits in the vesicular glutamate transport in ASD (Marchetto et al., 2017). Russo et al. (2018) also reported a significant decrease in synaptic gene expression and protein levels, glutamate neurotransmitter release and, consequently, reduced spontaneous firing rate in iPSC-derived neurons from non-syndromic ASD subjects (Russo et al., 2018). Increased Synapsin 1 puncta, synaptic connections and frequency of spontaneous excitatory postsynaptic currents were also identified in excitatory neurons from ASD subjects with SHANK2 haploinsufficiency and CRISPR/Cas9 knockouts (Zaslavsky et al., 2019). In addition, Zaslavsky et al. (2019) reported an upregulation of glutamate receptor signalling genes, which suggested that higher expression of glutamate receptors could explain the observed neuronal hyperconnectivity (Zaslavsky et al., 2019).

Whilst the majority of iPSC studies focused on synaptic receptors, postmortem studies covered other aspects of the glutamatergic system such as glutamatergic enzymes, transporters and receptors. Nicolini et al. (2015) found a decrease in the protein levels of postsynaptic density protein 95 (PSD-95) in the fusiform gyrus of 11 subjects with idiopathic autism (Nicolini et al., 2015). Further, Chandley et al. (2015) investigated gene expression levels of 16 synaptic genes relevant to glutamatergic neurotransmission, revealing a robust decrease in NTRK2 expression, which encodes the neurotrophic receptor kinase B (TrkB) and modulates glutamate receptor function (Minichiello, 2009). Trends towards low expression levels of genes encoding a metabotropic glutamate receptor (GRM8), an ionotropic glutamate receptor subunit (GRIN1), a glutamate transporter (SLC1A1), and a glutamate receptor anchoring protein (GRIP1) were also found (Chandley et al., 2015). Other glutamatergic dysfunctions reported in postmortem studies included enzymatic dysfunctions in the glutamate-glutamine cycle (Shimmura et al., 2013) and increased glutamate receptors and transporters reported by three other postmortem studies (Purcell et al., 2001; Fatemi et al., 2011; Chana et al., 2015).

Several aspects of GABAergic abnormalities in ASD have been thoroughly investigated in numerous postmortem studies. Reductions in Parvalbumin (PV)+ and Reelin (RELN)+ interneurons as well as increases in calbindin (CB)+ interneurons have been reported (Hashemi et al., 2017; Lawrence et al., 2010; Edmonson et al., 2014; Schwede et al., 2018; Whitney et al., 2009; Soghomonian et al., 2017). Consistent reductions in GABA receptors have also been reported (Fatemi et al.,

2009, 2011; Oblak et al., 2009; Crider et al., 2014). There is evidence, albeit weaker, for reductions in GAD levels (Fatemi et al., 2002b; Yip et al., 2009). Other studies that reported the presence of minicolumnopathies in ASD also reported a defect in GABAergic fibres (Casanova et al., 2003).

On the other hand, only one iPSC study reported an overproduction of inhibitory GABAergic neurons in cerebral organoids derived from non-syndromic ASD cases with increased head circumference (Mariani et al., 2015). In addition, an increase in the number of synapses marked by Synapsin1 puncta was found. This was due to an increase of inhibitory pre-synaptic puncta marked by the vesicular GABA transporter VGAT and not due to the increase in excitatory VGLUT1 puncta (Mariani et al., 2015), suggesting an increase in inhibitory synapse numbers in ASD. Transcriptomic analysis of organoids with *CHD8* haploinsufficiency revealed alterations in long non-coding antisense RNA DLX6-AS1, which is regulator of GABAergic interneuron development, as well as perturbations in the expression of DLX1, a transcription factor essential for GABAergic interneuron differentiation (Wang et al., 2017).

iPSC-derived cerebral organoids can be further exploited to explore the molecular processes which may contribute to E/I imbalance in ASD. To date, no study utilised region-specific organoids to generate ventral telencephalic organoids and specifically investigate GABAergic impairments in non-syndromic ASD or specifically looked into perturbations of interneuron subtypes. In addition to investigating neurogenesis, the fusion of region-specific organoids can be used to recapitulate interneuron migration and investigate their functional integration into circuits (Birey et al., 2017). Organoids can also be used to model neural oscillations, which are regular network activity events that emerge early in development and depend on the balance between glutamatergic and GABAergic signalling (Trujillo et al., 2019; Buzsaki & Draguhn, 2004; Culotta & Penzes, 2020).

17.2.8 Reelin (RELN) Dysfunction

RELN is a large, extracellular matrix glycoprotein, that is essential for regulating neuronal migration and brain lamination in the developing neocortex (Tissir & Goffinet, 2003). Cajal-Retzius cells are the early embryonic source of RELN, but shortly after birth inhibitory interneurons become the main source of Reelin (Alcántara et al., 1998; Pesold et al., 1998). In addition, RELN plays a role in regulating neuronal development and maintenance, where its loss has been shown to affect dendrite extension (Arikkath, 2012). Thus, RELN dysfunction may play a role in ASD pathology through regulating the processes of migration and dendritic branching. Failure of neurons, both excitatory and inhibitory, to reach their intended cortical destinations, might manifest in abnormal cortical layering and consequently impact their integration into functional circuitry.

To date, only one iPSC study implicated RELN dysfunction in ASD. Using iPSC-derived NPCs from non-syndromic ASD individuals with identified heterozygous missense variants in RELN gene, (Sánchez-Sánchez et al., 2018) revealed impaired RELN secretion in NPCs (Sánchez-Sánchez et al., 2018). Similarly, two postmortem studies point towards reduced RELN levels (Fatemi et al., 2001, 2005) and two implicate epigenetic dysregulation of RELN gene expression in the ASD cerebral cortex and cerebellum (Zhubi et al., 2014; Lintas et al., 2016). However, no clear link is established between epigenetic dysregulation and lower RELN expression.

Collectively, very few iPSC and postmortem studies implicated RELN dysfunction in ASD, and further studies are required to confirm and expand on the nature of these RELN-associated perturbations from gene expression to epigenetic regulation and signalling.

17.2.9 Gliogenesis

Following neurogenesis in the developing brain, the generation of astrocytes and oligodendrocytes begin around mid-gestation (Silbereis et al., 2016). Within the central nervous system, astrocytes play critical roles in regulating cell-cell interactions, preventing neuronal oxidative stress and homeostasis whereas oligodendrocytes undertake the process of myelination. Through their cross-talk, astrocytes and oligodendrocytes also participate in several immune processes (Nutma et al., 2020) and are essential for maintaining neural circuits in the brain (Robertson, 2013). Thus, their disruption can potentially contribute to ASD pathology.

To date, only one study generated iPSC-derived astrocytes from non-syndromic ASD subjects. (Russo et al., 2018) revealed a significant increase of reactive oxygen species (ROS) levels in ASD-derived astrocytes. Moreover, ASD-derived astrocytes interfered with proper neuronal development, in that neurons co-cultured with ASD-astrocytes revealed altered neuronal morphology. Russo et al. (2018) also identified interleukin-6 (IL6) secretions from ASD-astrocytes as a possible causative agent for neural defects, and were able to reverse synaptogenesis deficits by blocking IL6 levels (Russo et al., 2018). In another study using hiPSC-NPCs with a functional knockdown of Neurexin-1 (NRNX1), a presynaptic neuronal adhesion molecule essential for synapse formation and maintenance (Südhof, 2008) and its mutation is highly associated with ASD pathogenesis (Kim et al., 2008), Zeng et al. (2013) observed a reduction in astrocyte differentiation potential, and a reduced GFAP expression, suggesting an impairment in astrocyte differentiation (Zeng et al., 2013).

Postmortem literature generally focuses on microglial impairments, with a few studies investigating astrocytes. Microglia, a third population of glial cells, are generated in the yolk sac and migrate to the developing brain to establish contacts with neurons and astrocytes (Gomez Perdiguero et al., 2015; Menassa & Gomez-Nicola, 2018). The elevation of glial markers, specifically GFAP, has been consistently reported in post-mortem ASD tissue, especially in the prefrontal cortex and cerebellum (Laurence & Fatemi, 2005; Fatemi et al., 2011; Crawford et al., 2015; Edmonson et al., 2014). One study reported no increase in the fusiform gyrus (Nicolini et al., 2015), but the expression of glial markers in this region has not been further investigated. Fatemi et al. (2008) further investigated the expression of two astrocytic markers revealing an increase in Connexin 43 expression, a major component protein in astrocytic gap junctions, in the superior frontal cortex and a decrease in Aquaporin 4 expression, a transmembrane water channel protein, in the cerebellum. One study examined microglial and neuronal organization to find increased short-distance microglia–neuron interaction (Morgan et al., 2012) and three other studies consistently reported an increase in microglial densities (Tetreault et al., 2012, Menassa et al., 2017; Lee et al., 2017a), but none of these findings has been replicated in the same brain region. Previous studies have reported an increase in IL6 levels in the postmortem ASD brain (Vargas et al., 2005; Wei et al., 2011; Li et al., 2009). Transcriptomic analysis of cortical postmortem ASD tissue also implicated IL6 as part of aberrant immune response pathways (Guan et al., 2019). However, no postmortem study has specifically investigated IL6 in association to astrocytes or microglia.

Very little data is presented to elucidate the exact nature of astrocytic impairments in non-syndromic ASD. Although it is possible to efficiently generate iPSC-derived microglia (Haenseler et al., 2017; Douvaras et al., 2017; Abud et al., 2017), no studies to date have attempted to generate iPSC-derived microglia from non-syndromic ASD subjects. Similarly, there are no reports of using organoids to model the interactions between neurons and oligodendrocytes.

17.2.10 Transcriptional Landscape of ASD from iPSC and Post-Mortem Studies

The neurodevelopmental processes implicated above involve hundreds of genes within the human genome. Large sequencing studies have identified a myriad of genes that are associated with increased ASD risk (Satterstrom et al., 2020). Transcriptomic analysis on postmortem and iPSC-derived samples from ASD subjects converge on a number of pathways, mainly those regulating neuronal differentiation, the cytoskeleton, cell adhesion and synaptic function. In addition, postmortem studies further implicated metabolic, cell cycle, apoptosis and immune pathways in ASD (Fetit et al., 2021).

Compared to the large number of transcriptomic studies performed on postmortem ASD tissue (Garbett et al., 2008; Chow et al., 2012; Guan et al., 2016; Fournier et al., 2010; Schwede et al., 2018; Broek et al., 2014; Ander et al., 2015; Stamova et al., 2015; Schumann et al., 2017; Anitha et al., 2012), only a few studies performed transcriptomic analysis of iPSC-derived ASD organoids (Mariani et al., 2015; Wang et al., 2017) and NPCs (Zeng et al., 2013). The three studies revealed a significant overlap in the differentially expressed genes, despite differences in genetic background. The first study used iPSCs from a non-syndromic ASD subject with increased head circumference (Mariani et al., 2015), the second utilised a

CRISPR/Cas9-edited control cohort with *CHD8* haploinsufficiency (Wang et al., 2017) while the third used iPSCs carrying a functional knockdown of NRNX1. In addition to the common pathways listed above, Mariani et al. (2015) revealed the upregulation of genes relevant to ion transport and postsynaptic signaling, as well as the ASD-risk genes NRXN1 and SCN2A. Wang et al. (2017) further demonstrated the upregulation of genes related to forebrain development, Wnt/β-catenin signaling and axonal guidance. Both studies also revealed alterations in long non-coding antisense RNA DLX6-AS1 and DLX1 gene expression.

17.3 Conclusions and Future Directions

It is inevitable that the dysregulation of the tightly controlled, spatiotemporal developmental processes renders the brain vulnerable and may individually, or collectively, contribute to the pathogenesis of ASD. The application of iPSC technology in ASD research has clearly shown that deviations from neurotypical brain development may occur at stages as early as progenitor proliferation and differentiation into neurons. Coupled with the analysis of postmortem ASD tissue, altered neurogenesis, and possibly apoptotic mechanisms, may result in aberrant neuronal numbers and subtype specification.

This neural progeny potentially exhibits morphological alterations, such as aberrant dendritic branching, that are described by several post-mortem and iPSC studies, and may fail to migrate to correct cortical positions. Reelin dysfunction is a potential contributor to aberrant migration in ASD, although the evidence from both iPSC and postmortem studies is minimal. Eventually, aberrant migration manifests as irregular cortical cytoarchitecture observed in postmortem tissue. This could be explored further using iPSC-organoids.

Dysregulation of these processes may result in an imbalance between excitation and inhibition, which is supported by a wealth of evidence from both postmortem and iPSC-studies. The field can further benefit from utilising region-specific organoids to specifically investigate early interneuron development and migration. Eventually, the E/I imbalance may contribute to dysfunctional neural circuits and cerebral organoids could be used to model neural oscillations as a way for assessing glutamatergic and GABAergic signalling.

While postmortem ASD studies investigated other neuronal populations in several brain regions such as cerebellar Purkinje cells and von-economo neurons as well as neurons in the amygdala and brain stem, iPSC-studies to date have only focused on excitatory and inhibitory neurons in the cortex. It is therefore essential to further exploit iPSC-derived region-specific organoids to examine the nature and extent of early neuropathological impairments in non-cortical brain regions, such as the hippocampus, thalamus, midbrain and cerebellum (Chan et al., 2021).

Perturbations of glial cell types in the developing brain can further contribute to ASD pathology. However, it remains unclear the exact nature of astrocytic impairments in non-syndromic ASD and future iPSC-studies should consider investigating iPSC-derived microglia from non-syndromic ASD subjects. Similarly, organoids can be used to model the interactions between neurons and oligodendrocytes. Finally, transcriptomic findings from post-mortem and iPSC-studies converged on common neuronal differentiation, cytoskeletal, cell adhesion and synaptic pathways. The postmortem literature also describes several perturbations in nonprotein-coding RNA and epigenetic alterations in ASD, which are yet to be thoroughly investigated using iPSC-technology. Single-cell RNA sequencing is yet to be exploited to provide a detailed assessment of the observed cellular phenotypes implicated here in both postmortem and iPSC-studies of the ASD brain.

Despite the limitations of both methodologies, postmortem and iPSC-studies offer complementary insights on the perturbations that may occur during proliferation, neurogenesis, migration, neuronal morphogenesis and synaptogenesis as well as gliogenesis and myelination. Indeed, there are several remaining gaps in our knowledge, some of which we have highlighted here, and which may further require the integration of additional layers of analysis from genetic studies, animal models, and functional neuroimaging studies in living individuals. Such a multi-system approach can leverage the advantages of the individual methods to reach a more comprehensive understanding of the neuropathology of ASD.

References

Abud, E. M., Ramirez, R. N., Martinez, E. S., Healy, L. M., Nguyen, C. H., Newman, S. A., Yeromin, A. V., Scarfone, V. M., Marsh, S. E., & Fimbres, C. (2017). iPSC-derived human microglia-like cells to study neurological diseases. *Neuron, 94*, 278–293.e9.

Acab, A., & Muotri, A. R. (2015). The use of induced pluripotent stem cell technology to advance autism research and treatment. *Neurotherapeutics, 12*, 534–545.

Alcántara, S., Ruiz, M., D'Arcangelo, G., Ezan, F., De Lecea, L., Curran, T., Sotelo, C., & Soriano, E. (1998). Regional and cellular patterns of reelin mRNA expression in the forebrain of the developing and adult mouse. *The Journal of Neuroscience, 18*, 7779–7799.

Amaral, D. G., Anderson, M. P., Ansorge, O., Chance, S., Hare, C., Hof, P. R., Miller, M., Nagakura, I., Pickett, J., & Schumann, C. (2018). Autism BrainNet: A network of postmortem brain banks established to facilitate autism research. *Handbook of Clinical Neurology, 150*, 31–39.

Amin, N. D., & Pasca, S. P. (2018). Building models of brain disorders with three-dimensional organoids. *Neuron, 100*, 389–405.

Ander, B. P., Barger, N., Stamova, B., Sharp, F. R., & Schumann, C. M. (2015). Atypical miRNA expression in temporal cortex associated with dysregulation of immune, cell cycle, and other pathways in autism spectrum disorders. *Molecular Autism, 6*, 1–13.

Anitha, A., Nakamura, K., Thanseem, I., Yamada, K., Iwayama, Y., Toyota, T., Matsuzaki, H., Miyachi, T., Yamada, S., & Tsujii, M. (2012). Brain region-specific altered expression and association of mitochondria-related genes in autism. *Molecular Autism, 3*, 1–12.

Araghi-Niknam, M., & Fatemi, S. H. (2003). Levels of Bcl-2 and P53 are altered in superior frontal and cerebellar cortices of autistic subjects. *Cellular and Molecular Neurobiology, 23*, 945–952.

Arikkath, J. (2012). Molecular mechanisms of dendrite morphogenesis. *Frontiers in Cellular Neuroscience, 6*, 61.

Bauman, M., & Kemper, T. L. (1985). Histoanatomic observations of the brain in early infantile autism. *Neurology, 35*, 866–866.

Beck, L. E., Lee, J., Coté, C., Dunagin, M. C., Lukonin, I., Salla, N., Chang, M. K., Hughes, A. J., Mornin, J. D., & Gartner, Z. J. (2022). Systematically quantifying morphological features reveals constraints on organoid phenotypes. *Cell Systems, 13*, 547–560.e3.

Bedford, S. A., Park, M. T. M., Devenyi, G. A., Tullo, S., Germann, J., Patel, R., Anagnostou, E., Baron-Cohen, S., Bullmore, E. T., & Chura, L. R. (2020). Large-scale analyses of the relationship between sex, age and intelligence quotient heterogeneity and cortical morphometry in autism spectrum disorder. *Molecular Psychiatry, 25*, 614–628.

Beghin, A., Grenci, G., Sahni, G., Guo, S., Rajendiran, H., Delaire, T., Mohamad Raffi, S. B., Blanc, D., De Mets, R., & Ong, H. T. (2022). Automated high-speed 3D imaging of organoid cultures with multi-scale phenotypic quantification. *Nature Methods, 19*, 1–12.

Beltrao-Braga, P. C. B., & Muotri, A. R. (2017). Modeling autism spectrum disorders with human neurons. *Brain Research, 1656*, 49–54.

Ben-Reuven, L., & Reiner, O. (2016). Modeling the autistic cell: iPSCs recapitulate developmental principles of syndromic and nonsyndromic ASD. *Development, Growth & Differentiation, 58*, 481–491.

Birey, F., Andersen, J., Makinson, C. D., Islam, S., Wei, W., Huber, N., Fan, H. C., Metzler, K. R. C., Panagiotakos, G., & Thom, N. (2017). Assembly of functionally integrated human forebrain spheroids. *Nature, 545*, 54–59.

Broek, J. A., Guest, P. C., Rahmoune, H., & Bahn, S. (2014). Proteomic analysis of post mortem brain tissue from autism patients: Evidence for opposite changes in prefrontal cortex and cerebellum in synaptic connectivity-related proteins. *Molecular Autism, 5*, 1–8.

Buzsaki, G., & Draguhn, A. (2004). Neuronal oscillations in cortical networks. *Science, 304*, 1926–1929.

Cardenas, A., Kong, M., Alvarez, A., Maldonado, H., & Leyton, L. (2014). Signaling pathways involved in neuron-astrocyte adhesion and migration. *Current Molecular Medicine, 14*, 275–290.

Casanova, M. F., Buxhoeveden, D. P., Switala, A. E., & Roy, E. (2002a). Minicolumnar pathology in autism. *Neurology, 58*, 428–432.

Casanova, M. F., Buxhoeveden, D. P., Switala, A. E., & Roy, E. (2002b). Neuronal density and architecture (gray level index) in the brains of autistic patients. *Journal of Child Neurology, 17*, 515–521.

Casanova, M. F., Buxhoeveden, D., & Gomez, J. (2003). Disruption in the inhibitory architecture of the cell minicolumn: Implications for autisim. *The Neuroscientist, 9*, 496–507.

Casanova, M., Van Kooten, I., Switala, A., Van Engeland, H., Heinsen, H., Steinbusch, H., Hof, P., & Schmitz, C. (2006a). Abnormalities of cortical minicolumnar organization in the prefrontal lobes of autistic patients. *Clinical Neuroscience Research, 6*, 127–133.

Casanova, M. F., Van Kooten, I. A., Switala, A. E., Van Engeland, H., Heinsen, H., Steinbusch, H. W., Hof, P. R., Trippe, J., Stone, J., & Schmitz, C. (2006b). Minicolumnar abnormalities in autism. *Acta Neuropathologica, 112*, 287–303.

Casanova, M. F., Casanova, E. L., Frye, R. E., Baeza-Velasco, C., Lasalle, J. M., Hagerman, R. J., Scherer, S. W., & Natowicz, M. R. (2020). Secondary vs. idiopathic autism. *Frontiers in Psychiatry, 11*, 297.

Cederquist, G. Y., Tchieu, J., Callahan, S. J., Ramnarine, K., Ryan, S., Zhang, C., Rittenhouse, C., Zeltner, N., Chung, S. Y., & Zhou, T. (2020). A multiplex human pluripotent stem cell platform defines molecular and functional subclasses of autism-related genes. *Cell Stem Cell, 27*, 35–49.e6.

Chan, W. K., Fetit, R., Griffiths, R., Marshall, H., Mason, J. O., & Price, D. J. (2021). Using organoids to study human brain development and evolution. *Developmental Neurobiology, 81*, 608–622.

Chana, G., Laskaris, L., Pantelis, C., Gillett, P., Testa, R., Zantomio, D., Burrows, E. L., Hannan, A. J., Everall, I. P., & Skafidas, E. (2015). Decreased expression of mGluR5 within the dorsolateral prefrontal cortex in autism and increased microglial number in mGluR5 knockout mice: Pathophysiological and neurobehavioral implications. *Brain, Behavior, and Immunity, 49*, 197–205.

Chandley, M. J., Crawford, J. D., Szebeni, A., Szebeni, K., & Ordway, G. A. (2015). NTRK2 expression levels are reduced in laser captured pyramidal neurons from the anterior cingulate cortex in males with autism spectrum disorder. *Molecular Autism, 6*, 1–12.

Chow, M. L., Pramparo, T., Winn, M. E., Barnes, C. C., Li, H.-R., Weiss, L., Fan, J.-B., Murray, S., April, C., & Belinson, H. (2012). Age-dependent brain gene expression and copy number anomalies in autism suggest distinct pathological processes at young versus mature ages. *PLoS Genetics, 8*, e1002592.

Connacher, R., Williams, M., Prem, S., Yeung, P. L., Matteson, P., Mehta, M., Markov, A., Peng, C., Zhou, X., & Mcdermott, C. R. (2022). Autism NPCs from both idiopathic and CNV 16p11.2 deletion patients exhibit dysregulation of proliferation and mitogenic responses. *Stem Cell Reports, 17*, 1380–1394.

Courchesne, E., Carper, R., & Akshoomoff, N. (2003). Evidence of brain overgrowth in the first year of life in autism. *JAMA, 290*, 337–344.

Courchesne, E., Mouton, P. R., Calhoun, M. E., Semendeferi, K., Ahrens-Barbeau, C., Hallet, M. J., Barnes, C. C., & Pierce, K. (2011). Neuron number and size in prefrontal cortex of children with autism. *JAMA, 306*, 2001–2010.

Crawford, J. D., Chandley, M. J., Szebeni, K., Szebeni, A., Waters, B., & Ordway, G. A. (2015). Elevated GFAP protein in anterior cingulate cortical white matter in males with autism spectrum disorder. *Autism Research, 8*, 649–657.

Crider, A., Pandya, C. D., Peter, D., Ahmed, A. O., & Pillai, A. (2014). Ubiquitin-proteasome dependent degradation of GABAAα1 in autism spectrum disorder. *Molecular Autism, 5*, 1–10.

Culotta, L., & Penzes, P. (2020). Exploring the mechanisms underlying excitation/inhibition imbalance in human iPSC-derived models of ASD. *Molecular Autism, 11*, 32.

Dalton, S. (2015). Linking the cell cycle to cell fate decisions. *Trends in Cell Biology, 25*, 592–600.

Derosa, B. A., Van Baaren, J. M., Dubey, G. K., Lee, J. M., Cuccaro, M. L., Vance, J. M., Pericak-Vance, M. A., & Dykxhoorn, D. M. (2012). Derivation of autism spectrum disorder-specific induced pluripotent stem cells from peripheral blood mononuclear cells. *Neuroscience Letters, 516*, 9–14.

Deyle, D. R. (2015). Generation of induced pluripotent stem cells. In *Osteoporosis and osteoarthritis* (pp. 43–58). Springer.

Dolmetsch, R., & Geschwind, D. H. (2011). The human brain in a dish: The promise of iPSC-derived neurons. *Cell, 145*, 831–834.

Doretto, S., Malerba, M., Ramos, M., Ikrar, T., Kinoshita, C., De Mei, C., Tirotta, E., Xu, X., & Borrelli, E. (2011). Oligodendrocytes as regulators of neuronal networks during early postnatal development. *PLoS One, 6*, e19849.

Douvaras, P., Sun, B., Wang, M., Kruglikov, I., Lallos, G., Zimmer, M., Terrenoire, C., Zhang, B., Gandy, S., & Schadt, E. (2017). Directed differentiation of human pluripotent stem cells to microglia. *Stem Cell Reports, 8*, 1516–1524.

Edmonson, C., Ziats, M. N., & Rennert, O. M. (2014). Altered glial marker expression in autistic post-mortem prefrontal cortex and cerebellum. *Molecular Autism, 5*, 1–9.

Fatemi, S. H., Stary, J. M., Halt, A. R., & Realmuto, G. R. (2001). Dysregulation of Reelin and Bcl-2 proteins in autistic cerebellum. *Journal of Autism and Developmental Disorders, 31*, 529–535.

Fatemi, S. H., Halt, A. R., Realmuto, G., Earle, J., Kist, D. A., Thuras, P., & Merz, A. (2002a). Purkinje cell size is reduced in cerebellum of patients with autism. *Cellular and Molecular Neurobiology, 22*, 171–175.

Fatemi, S. H., Halt, A. R., Stary, J. M., Kanodia, R., Schulz, S. C., & Realmuto, G. R. (2002b). Glutamic acid decarboxylase 65 and 67 kDa proteins are reduced in autistic parietal and cerebellar cortices. *Biological Psychiatry, 52*, 805–810.

Fatemi, S. H., Snow, A. V., Stary, J. M., Araghi-Niknam, M., Reutiman, T. J., Lee, S., Brooks, A. I., & Pearce, D. A. (2005). Reelin signaling is impaired in autism. *Biological Psychiatry, 57*, 777–787.

Fatemi, S. H., Folsom, T. D., Reutiman, T. J., & Lee, S. (2008). Expression of astrocytic markers aquaporin 4 and connexin 43 is altered in brains of subjects with autism. *Synapse, 62*, 501–507.

Fatemi, S. H., Reutiman, T. J., Folsom, T. D., & Thuras, P. D. (2009). GABAA receptor down-regulation in brains of subjects with autism. *Journal of Autism and Developmental Disorders, 39*, 223–230.

Fatemi, S. H., Folsom, T. D., Kneeland, R. E., & Liesch, S. B. (2011). Metabotropic glutamate receptor 5 upregulation in children with autism is associated with underexpression of both Fragile X mental retardation protein and GABAA receptor beta 3 in adults with autism. *The Anatomical Record, 294*, 1635–1645.

Fetit, R., Hillary, R. F., Price, D. J., & Lawrie, S. M. (2021). The neuropathology of autism: A systematic review of post-mortem studies of autism and related disorders. *Neuroscience and Biobehavioral Reviews, 129*, 35–62.

Fournier, K. A., Hass, C. J., Naik, S. K., Lodha, N., & Cauraugh, J. H. (2010). Motor coordination in autism spectrum disorders: A synthesis and meta-analysis. *Journal of Autism and Developmental Disorders, 40*, 1227–1240.

Fricker, M., Tolkovsky, A. M., Borutaite, V., Coleman, M., & Brown, G. C. (2018). Neuronal cell death. *Physiological Reviews, 98*, 813–880.

Garbett, K., Ebert, P. J., Mitchell, A., Lintas, C., Manzi, B., Mirnics, K., & Persico, A. M. (2008). Immune transcriptome alterations in the temporal cortex of subjects with autism. *Neurobiology of Disease, 30*, 303–311.

Garner, C. C., Zhai, R. G., Gundelfinger, E. D., & Ziv, N. E. (2002). Molecular mechanisms of CNS synaptogenesis. *Trends in Neurosciences, 25*, 243–250.

Gomez Perdiguero, E., Klapproth, K., Schulz, C., Busch, K., Azzoni, E., Crozet, L., Garner, H., Trouillet, C., De Bruijn, M. F., & Geissmann, F. (2015). Tissue-resident macrophages originate from yolk-sac-derived erythro-myeloid progenitors. *Nature, 518*, 547–551.

Gordon, A., & Geschwind, D. H. (2020). Human in vitro models for understanding mechanisms of autism spectrum disorder. *Molecular Autism, 11*, 26.

Griesi-Oliveira, K., Acab, A., Gupta, A. R., Sunaga, D. Y., Chailangkarn, T., Nicol, X., Nunez, Y., Walker, M. F., Murdoch, J. D., & Sanders, S. J. (2015). Modeling non-syndromic autism and the impact of TRPC6 disruption in human neurons. *Molecular Psychiatry, 20*, 1350–1365.

Guan, J., Yang, E., Yang, J., Zeng, Y., Ji, G., & Cai, J. J. (2016). Exploiting aberrant mRNA expression in autism for gene discovery and diagnosis. *Human Genetics, 135*, 797–811.

Guan, J., Cai, J. J., Ji, G., & Sham, P. C. (2019). Commonality in dysregulated expression of gene sets in cortical brains of individuals with autism, schizophrenia, and bipolar disorder. *Translational Psychiatry, 9*, 1–15.

Habela, C. W., Song, H. J., & Ming, G. L. (2016). Modeling synaptogenesis in schizophrenia and autism using human iPSC derived neurons. *Molecular and Cellular Neurosciences, 73*, 52–62.

Haenseler, W., Sansom, S. N., Buchrieser, J., Newey, S. E., Moore, C. S., Nicholls, F. J., Chintawar, S., Schnell, C., Antel, J. P., & Allen, N. D. (2017). A highly efficient human pluripotent stem cell microglia model displays a neuronal-co-culture-specific expression profile and inflammatory response. *Stem Cell Reports, 8*, 1727–1742.

Hagerman, R., Au, J., & Hagerman, P. (2011). FMR1 premutation and full mutation molecular mechanisms related to autism. *Journal of Neurodevelopmental Disorders, 3*, 211–224.

Hardwick, L. J., Ali, F. R., Azzarelli, R., & Philpott, A. (2015). Cell cycle regulation of proliferation versus differentiation in the central nervous system. *Cell and Tissue Research, 359*, 187–200.

Hashemi, E., Ariza, J., Rogers, H., Noctor, S. C., & Martínez-Cerdeño, V. (2017). The number of parvalbumin-expressing interneurons is decreased in the prefrontal cortex in autism. *Cerebral Cortex, 27*, 1931–1943.

Hasselmann, J., & Blurton-Jones, M. (2020). Human iPSC-derived microglia: A growing toolset to study the brain's innate immune cells. *Glia, 68*, 721–739.

Hevner, R. F., Shi, L., Justice, N., Hsueh, Y.-P., Sheng, M., Smiga, S., Bulfone, A., Goffinet, A. M., Campagnoni, A. T., & Rubenstein, J. L. (2001). Tbr1 regulates differentiation of the preplate and layer 6. *Neuron, 29*, 353–366.

Hirabayashi, Y., Itoh, Y., Tabata, H., Nakajima, K., Akiyama, T., Masuyama, N., & Gotoh, Y. (2004). The Wnt/β-catenin pathway directs neuronal differentiation of cortical neural precursor cells. *Development, 131*, 2791–2801.

Ho, S.-M., Topol, A., & Brennand, K. J. (2015). From "directed differentiation" to "neuronal induction": Modeling neuropsychiatric disease: Supplementary issue: Stem cell biology. *Biomarker Insights, 10*, 31–41.

Hu, J. S., Vogt, D., Sandberg, M., & Rubenstein, J. L. (2017). Cortical interneuron development: A tale of time and space. *Development, 144*, 3867–3878.

Hutsler, J. J., & Zhang, H. (2010). Increased dendritic spine densities on cortical projection neurons in autism spectrum disorders. *Brain Research, 1309*, 83–94.

Hutsler, J. J., Love, T., & Zhang, H. (2007). Histological and magnetic resonance imaging assessment of cortical layering and thickness in autism spectrum disorders. *Biological Psychiatry, 61*, 449–457.

Hwang, Y., Hidalgo, D., & Socolovsky, M. (2021). The shifting shape and functional specializations of the cell cycle during lineage development. *WIREs Mechanisms of Disease, 13*, e1504.

Kanchan, K., Iyer, K., Yanek, L. R., Carcamo-Orive, I., Taub, M. A., Malley, C., Baldwin, K., Becker, L. C., Broeckel, U., & Cheng, L. (2020). Genomic integrity of human induced pluripotent stem cells across nine studies in the NHLBI NextGen program. *Stem Cell Research, 46*, 101803.

Kurki, P., Vanderlaan, M., Dolbeare, F. E. T. A. L., Gray, J., & Tan, E. M. (1986). Expression of proliferating cell nuclear antigen (PCNA)/cyclin during the cell cycle. *Experimental cell research, 166*(1), 209–219.

Kim, H.-G., Kishikawa, S., Higgins, A. W., Seong, I.-S., Donovan, D. J., Shen, Y., Lally, E., Weiss, L. A., Najm, J., & Kutsche, K. (2008). Disruption of neurexin 1 associated with autism spectrum disorder. *American Journal of Human Genetics, 82*, 199–207.

Kim, D. S., Ross, P. J., Zaslavsky, K., & Ellis, J. (2014). Optimizing neuronal differentiation from induced pluripotent stem cells to model ASD. *Frontiers in Cellular Neuroscience, 8*, 109.

Kotagiri, P., Chance, S. A., Szele, F. G., & Esiri, M. M. (2014). Subventricular zone cytoarchitecture changes in autism. *Developmental Neurobiology, 74*, 25–41.

Lalli, M. A., Avey, D., Dougherty, J. D., Milbrandt, J., & Mitra, R. D. (2020). High-throughput single-cell functional elucidation of neurodevelopmental disease–associated genes reveals convergent mechanisms altering neuronal differentiation. *Genome Research, 30*, 1317–1331.

Lancaster, M. A., Renner, M., Martin, C. A., Wenzel, D., Bicknell, L. S., Hurles, M. E., ... & Knoblich, J. A. (2013). Cerebral organoids model human brain development and microcephaly. *Nature, 501*(7467), 373–379.

Lancaster, M. A. (2018). Brain organoids get vascularized. *Nature Biotechnology, 36*, 407–408.

Laurence, J., & Fatemi, S. (2005). Glial fibrillary acidic protein is elevated in superior frontal, parietal and cerebellar cortices of autistic subjects. *Cerebellum, 4*, 206–210.

Lawrence, Y., Kemper, T., Bauman, M., & Blatt, G. (2010). Parvalbumin-, calbindin-, and calretinin-immunoreactive hippocampal interneuron density in autism. *Acta Neurologica Scandinavica, 121*, 99–108.

Lee, A. S., Azmitia, E. C., & Whitaker-Azmitia, P. M. (2017a). Developmental microglial priming in postmortem autism spectrum disorder temporal cortex. *Brain, Behavior, and Immunity, 62*, 193–202.

Lee, C. T., Bendriem, R. M., Wu, W. W., & Shen, R. F. (2017b). 3D brain organoids derived from pluripotent stem cells: Promising experimental models for brain development and neurodegenerative disorders. *Journal of Biomedical Science, 24*, 59.

Lee, K. M., Hawi, Z. H., Parkington, H. C., Parish, C. L., Kumar, P. V., Polo, J. M., Bellgrove, M. A., & Tong, J. (2020). The application of human pluripotent stem cells to model the neuronal and glial components of neurodevelopmental disorders. *Molecular Psychiatry, 25*, 368–378.

Lewis, D. A. (2002). The human brain revisited: Opportunities and challenges in postmortem studies of psychiatric disorders. *Neuropsychopharmacology, 26*, 143–154.

Li, X., Chauhan, A., Sheikh, A. M., Patil, S., Chauhan, V., Li, X.-M., Ji, L., Brown, T., & Malik, M. (2009). Elevated immune response in the brain of autistic patients. *Journal of Neuroimmunology, 207*, 111–116.

Lim, D. A., & Alvarez-Buylla, A. (2016). The adult ventricular–subventricular zone (V-SVZ) and olfactory bulb (OB) neurogenesis. *Cold Spring Harbor Perspectives in Biology, 8*, a018820.

Lim, C. S., Yang, J. E., Lee, Y. K., Lee, K., Lee, J. A., & Kaang, B. K. (2015). Understanding the molecular basis of autism in a dish using hiPSCs-derived neurons from ASD patients. *Molecular Brain, 8*, 57.

Lintas, C., Sacco, R., & Persico, A. M. (2016). Differential methylation at the RELN gene promoter in temporal cortex from autistic and typically developing post-puberal subjects. *Journal of Neurodevelopmental Disorders, 8*, 1–12.

Loat, C., Curran, S., Lewis, C., Duvall, J., Geschwind, D., Bolton, P., & Craig, I. (2008). Methyl-CpG-binding protein 2 polymorphisms and vulnerability to autism. *Genes, Brain, and Behavior, 7*, 754–760.

Marchetto, M. C., Belinson, H., Tian, Y., Freitas, B. C., Fu, C., Vadodaria, K., Beltrao-Braga, P., Trujillo, C. A., Mendes, A. P., & Padmanabhan, K. (2017). Altered proliferation and networks in neural cells derived from idiopathic autistic individuals. *Molecular Psychiatry, 22*, 820–835.

Mariani, J., Coppola, G., Zhang, P., Abyzov, A., Provini, L., Tomasini, L., Amenduni, M., Szekely, A., Palejev, D., & Wilson, M. (2015). FOXG1-dependent dysregulation of GABA/glutamate neuron differentiation in autism spectrum disorders. *Cell, 162*, 375–390.

Martínez-Cerdeño, V., & Noctor, S. C. (2018). Neural progenitor cell terminology. *Frontiers in Neuroanatomy, 12*, 104.

Mckenna, B., Koomar, T., Vervier, K., Kremsreiter, J., & Michaelson, J. J. (2018). Whole-genome sequencing in a family with twin boys with autism and intellectual disability suggests multimodal polygenic risk. *Molecular Case Studies, 4*, a003285.

Menassa, D. A., & Gomez-Nicola, D. (2018). Microglial dynamics during human brain development. *Frontiers in Immunology, 9*, 1014.

Menassa, D. A., Sloan, C., & Chance, S. A. (2017). Primary olfactory cortex in autism and epilepsy: Increased glial cells in autism. *Brain Pathology, 27*, 437–448.

Minichiello, L. (2009). TrkB signalling pathways in LTP and learning. *Nature Reviews. Neuroscience, 10*, 850–860.

Miskinyte, G., Devaraju, K., Grønning Hansen, M., Monni, E., Tornero, D., Woods, N. B., Bengzon, J., Ahlenius, H., Lindvall, O., & Kokaia, Z. (2017). Direct conversion of human fibroblasts to functional excitatory cortical neurons integrating into human neural networks. *Stem Cell Research & Therapy, 8*, 1–18.

Molnár, Z., Clowry, G. J., Šestan, N., Alzu'Bi, A., Bakken, T., Hevner, R. F., Hüppi, P. S., Kostović, I., Rakic, P., & Anton, E. (2019). New insights into the development of the human cerebral cortex. *Journal of Anatomy, 235*, 432–451.

Morgan, J. T., Chana, G., Abramson, I., Semendeferi, K., Courchesne, E., & Everall, I. P. (2012). Abnormal microglial–neuronal spatial organization in the dorsolateral prefrontal cortex in autism. *Brain Research, 1456*, 72–81.

Munji, R. N., Choe, Y., Li, G., Siegenthaler, J. A., & Pleasure, S. J. (2011). Wnt signaling regulates neuronal differentiation of cortical intermediate progenitors. *The Journal of Neuroscience, 31*, 1676–1687.

Nehme, R., & Barrett, L. E. (2020). Using human pluripotent stem cell models to study autism in the era of big data. *Molecular Autism, 11*, 21.

Nestor, M. W., Phillips, A. W., Artimovich, E., Nestor, J. E., Hussman, J. P., & Blatt, G. J. (2016). Human inducible pluripotent stem cells and autism spectrum disorder: Emerging technologies. *Autism Research, 9*, 513–535.

Nicolini, C., Ahn, Y., Michalski, B., Rho, J. M., & Fahnestock, M. (2015). Decreased mTOR signaling pathway in human idiopathic autism and in rats exposed to valproic acid. *Acta Neuropathologica Communications, 3*, 1–13.

Nutma, E., Van Gent, D., Amor, S., & Peferoen, L. A. (2020). Astrocyte and oligodendrocyte cross-talk in the central nervous system. *Cell, 9*, 600.

Oblak, A., Gibbs, T., & Blatt, G. (2009). Decreased GABAA receptors and benzodiazepine binding sites in the anterior cingulate cortex in autism. *Autism Research, 2*, 205–219.

Paşca, S. P. (2018). The rise of three-dimensional human brain cultures. *Nature, 553*, 437–445.

Pearson, B. L., Corley, M. J., Vasconcellos, A., Blanchard, D. C., & Blanchard, R. J. (2013). Heparan sulfate deficiency in autistic postmortem brain tissue from the subventricular zone of the lateral ventricles. *Behavioural Brain Research, 243*, 138–145.

Pesold, C., Impagnatiello, F., Pisu, M., Uzunov, D., Costa, E., Guidotti, A., & Caruncho, H. (1998). Reelin is preferentially expressed in neurons synthesizing γ-aminobutyric acid in cortex and hippocampus of adult rats. *Proceedings of the National Academy of Sciences, 95*, 3221–3226.

Pintacuda, G., Martin, J. M., & Eggan, K. C. (2021). Mind the translational gap: Using iPS cell models to bridge from genetic discoveries to perturbed pathways and therapeutic targets. *Molecular Autism, 12*, 10.

Prilutsky, D., Palmer, N. P., Smedemark-Margulies, N., Schlaeger, T. M., Margulies, D. M., & Kohane, I. S. (2014). iPSC-derived neurons as a higher-throughput readout for autism: Promises and pitfalls. *Trends in Molecular Medicine, 20*, 91–104.

Purcell, A., Jeon, O., Zimmerman, A., Blue, M., & Pevsner, J. (2001). Postmortem brain abnormalities of the glutamate neurotransmitter system in autism. *Neurology, 57*, 1618–1628.

Ramaswami, G., & Geschwind, D. H. (2018). Genetics of autism spectrum disorder. *Handbook of Clinical Neurology, 147*, 321–329.

Redcay, E., & Courchesne, E. (2005). When is the brain enlarged in autism? A meta-analysis of all brain size reports. *Biological Psychiatry, 58*, 1–9.

Ritvo, E. R., Freeman, B., Scheibel, A. B., Duong, T., Robinson, H., Guthrie, D., & Ritvo, A. (1986). Lower Purkinje cell counts in the cerebella of four autistic subjects: Initial findings of the UCLA-NSAC autopsy research report. *The American Journal of Psychiatry, 143*, 862–866.

Robertson, J. M. (2013). Astrocyte domains and the three-dimensional and seamless expression of consciousness and explicit memories. *Medical Hypotheses, 81*, 1017–1024.

Runge, K., Cardoso, C., & De Chevigny, A. (2020). Dendritic spine plasticity: Function and mechanisms. *Frontiers in Synaptic Neuroscience, 12*, 36.

Russo, F. B., Freitas, B. C., Pignatari, G. C., Fernandes, I. R., Sebat, J., Muotri, A. R., & Beltrão-Braga, P. C. B. (2018). Modeling the interplay between neurons and astrocytes in autism using human induced pluripotent stem cells. *Biological Psychiatry, 83*, 569–578.

Sainath, R., & Gallo, G. (2015). Cytoskeletal and signaling mechanisms of neurite formation. *Cell and Tissue Research, 359*, 267–278.

Sánchez-Sánchez, S. M., Magdalon, J., Griesi-Oliveira, K., Yamamoto, G. L., Santacruz-Perez, C., Fogo, M., Passos-Bueno, M. R., & Sertié, A. L. (2018). Rare RELN variants affect Reelin–DAB1 signal transduction in autism spectrum disorder. *Human Mutation, 39*, 1372–1383.

Sanders, S. J., Murtha, M. T., Gupta, A. R., Murdoch, J. D., Raubeson, M. J., Willsey, A. J., Ercan-Sencicek, A. G., DiIullo, N. M., Parikshak, N. N., & Stein, J. L. (2012). De novo mutations revealed by whole-exome sequencing are strongly associated with autism. *Nature, 485*, 237–241.

Satterstrom, F. K., Kosmicki, J. A., Wang, J., Breen, M. S., De Rubeis, S., An, J.-Y., Peng, M., Collins, R., Grove, J., & Klei, L. (2020). Large-scale exome sequencing study implicates both developmental and functional changes in the neurobiology of autism. *Cell, 180*, 568–584.e23.

Schafer, S. T., Paquola, A., Stern, S., Gosselin, D., Ku, M., Pena, M., Kuret, T. J., Liyanage, M., Mansour, A. A., & Jaeger, B. N. (2019). Pathological priming causes developmental gene network heterochronicity in autistic subject-derived neurons. *Nature Neuroscience, 22*, 243–255.

Schumann, C. M., Sharp, F. R., Ander, B. P., & Stamova, B. (2017). Possible sexually dimorphic role of miRNA and other sncRNA in ASD brain. *Molecular Autism, 8*, 1–10.

Schwede, M., Nagpal, S., Gandal, M. J., Parikshak, N. N., Mirnics, K., Geschwind, D. H., & Morrow, E. M. (2018). Strong correlation of downregulated genes related to synaptic transmission and mitochondria in post-mortem autism cerebral cortex. *Journal of Neurodevelopmental Disorders, 10*, 1–9.

Sheikh, A., Li, X., Wen, G., Tauqeer, Z., Brown, W., & Malik, M. (2010a). Cathepsin D and apoptosis related proteins are elevated in the brain of autistic subjects. *Neuroscience, 165*, 363–370.

Sheikh, A. M., Malik, M., Wen, G., Chauhan, A., Chauhan, V., Gong, C. X., Liu, F., Brown, W. T., & Li, X. (2010b). BDNF-Akt-Bcl2 antiapoptotic signaling pathway is compromised in the brain of autistic subjects. *Journal of Neuroscience Research, 88*, 2641–2647.

Shimmura, C., Suzuki, K., Iwata, Y., Tsuchiya, K. J., Ohno, K., Matsuzaki, H., Iwata, K., Kameno, Y., Takahashi, T., & Wakuda, T. (2013). Enzymes in the glutamate-glutamine cycle in the anterior cingulate cortex in postmortem brain of subjects with autism. *Molecular Autism, 4*, 1–7.

Silbereis, J. C., Pochareddy, S., Zhu, Y., Li, M., & Sestan, N. (2016). The cellular and molecular landscapes of the developing human central nervous system. *Neuron, 89*, 248–268.

Skefos, J., Cummings, C., Enzer, K., Holiday, J., Weed, K., Levy, E., Yuce, T., Kemper, T., & Bauman, M. (2014). Regional alterations in purkinje cell density in patients with autism. *PLoS One, 9*, e81255.

Soghomonian, J. J., Zhang, K., Reprakash, S., & Blatt, G. J. (2017). Decreased parvalbumin mRNA levels in cerebellar Purkinje cells in autism. *Autism Research, 10*, 1787–1796.

Sohal, V. S., & Rubenstein, J. L. (2019). Excitation-inhibition balance as a framework for investigating mechanisms in neuropsychiatric disorders. *Molecular Psychiatry, 24*, 1248–1257.

Soubannier, V., Maussion, G., Chaineau, M., Sigutova, V., Rouleau, G., Durcan, T. M., & Stifani, S. (2020). Characterization of human iPSC-derived astrocytes with potential for disease modeling and drug discovery. *Neuroscience Letters, 731*, 135028.

Stamova, B., Ander, B. P., Barger, N., Sharp, F. R., & Schumann, C. M. (2015). Specific regional and age-related small noncoding RNA expression patterns within superior temporal gyrus of typical human brains are less distinct in autism brains. *Journal of Child Neurology, 30*, 1930–1946.

Stoner, R., Chow, M. L., Boyle, M. P., Sunkin, S. M., Mouton, P. R., Roy, S., Wynshaw-Boris, A., Colamarino, S. A., Lein, E. S., & Courchesne, E. (2014). Patches of disorganization in the neocortex of children with autism. *The New England Journal of Medicine, 370*, 1209–1219.

Südhof, T. C. (2008). Neuroligins and neurexins link synaptic function to cognitive disease. *Nature, 455*, 903–911.

Tang, G., Gudsnuk, K., Kuo, S.-H., Cotrina, M. L., Rosoklija, G., Sosunov, A., Sonders, M. S., Kanter, E., Castagna, C., & Yamamoto, A. (2014). Loss of mTOR-dependent macroautophagy causes autistic-like synaptic pruning deficits. *Neuron, 83*, 1131–1143.

Tetreault, N. A., Hakeem, A. Y., Jiang, S., Williams, B. A., Allman, E., Wold, B. J., & Allman, J. M. (2012). Microglia in the cerebral cortex in autism. *Journal of Autism and Developmental Disorders, 42*, 2569–2584.

Tissir, F., & Goffinet, A. M. (2003). Reelin and brain development. *Nature Reviews. Neuroscience, 4*, 496–505.

Trujillo, C. A., Gao, R., Negraes, P. D., Gu, J., Buchanan, J., Preissl, S., Wang, A., Wu, W., Haddad, G. G., & Chaim, I. A. (2019). Complex oscillatory waves emerging from cortical organoids model early human brain network development. *Cell Stem Cell, 25*, 558–569.e7.

Trutzer, I. M., García-Cabezas, M. Á., & Zikopoulos, B. (2019). Postnatal development and maturation of layer 1 in the lateral prefrontal cortex and its disruption in autism. *Acta Neuropathologica Communications, 7*, 1–23.

Tuzun, E., Sharp, A. J., Bailey, J. A., Kaul, R., Morrison, V. A., Pertz, L. M., Haugen, E., Hayden, H., Albertson, D., & Pinkel, D. (2005). Fine-scale structural variation of the human genome. *Nature Genetics, 37*, 727–732.

Vaccarino, F. M., Stevens, H. E., Kocabas, A., Palejev, D., Szekely, A., Grigorenko, E. L., & Weissman, S. (2011). Induced pluripotent stem cells: A new tool to confront the challenge of neuropsychiatric disorders. *Neuropharmacology, 60*, 1355–1363.

Vadodaria, K. C., Marchetto, M. C., Mertens, J., & Gage, F. H. (2016). Generating human serotonergic neurons in vitro: Methodological advances. *BioEssays, 38*, 1123–1129.

Vargas, D. L., Nascimbene, C., Krishnan, C., Zimmerman, A. W., & Pardo, C. A. (2005). Neuroglial activation and neuroinflammation in the brain of patients with autism. *Annals of Neurology, 57*, 67–81.

Varghese, M., Keshav, N., Jacot-Descombes, S., Warda, T., Wicinski, B., Dickstein, D. L., Harony-Nicolas, H., De Rubeis, S., Drapeau, E., & Buxbaum, J. D. (2017). Autism spectrum disorder: Neuropathology and animal models. *Acta Neuropathologica, 134*, 537–566.

Villa, C., Combi, R., Conconi, D., & Lavitrano, M. (2021). Patient-derived induced pluripotent stem cells (iPSCs) and cerebral organoids for drug screening and development in autism spectrum disorder: Opportunities and challenges. *Pharmaceutics, 13*, 280.

Wang, P., Mokhtari, R., Pedrosa, E., Kirschenbaum, M., Bayrak, C., Zheng, D., & Lachman, H. M. (2017). CRISPR/Cas9-mediated heterozygous knockout of the autism gene CHD8 and

characterization of its transcriptional networks in cerebral organoids derived from iPS cells. *Molecular Autism, 8*, 1–17.
Wang, M. Y., Zhang, L., & Gage, F. H. (2020). Modeling neuropsychiatric disorders using human induced pluripotent stem cells. *Protein & Cell, 11*, 45–59.
Wegiel, J., Flory, M., Kuchna, I., Nowicki, K., Ma, S. Y., Imaki, H., Wegiel, J., Cohen, I. L., London, E., & Brown, W. T. (2014). Brain-region–specific alterations of the trajectories of neuronal volume growth throughout the lifespan in autism. *Acta Neuropathologica Communications, 2*, 1–18.
Wei, H., Zou, H., Sheikh, A. M., Malik, M., Dobkin, C., Brown, W. T., & Li, X. (2011). IL-6 is increased in the cerebellum of autistic brain and alters neural cell adhesion, migration and synaptic formation. *Journal of Neuroinflammation, 8*, 1–10.
Weiner, D. J., Wigdor, E. M., Ripke, S., Walters, R. K., Kosmicki, J. A., Grove, J., Samocha, K. E., Goldstein, J. I., Okbay, A., & Bybjerg-Grauholm, J. (2017). Polygenic transmission disequilibrium confirms that common and rare variation act additively to create risk for autism spectrum disorders. *Nature Genetics, 49*, 978–985.
Weir, R., Bauman, M., Jacobs, B., & Schumann, C. (2018). Protracted dendritic growth in the typically developing human amygdala and increased spine density in young ASD brains. *The Journal of Comparative Neurology, 526*, 262–274.
Whiteley, J. T., Fernandes, S., Sharma, A., Mendes, A. P. D., Racha, V., Benassi, S. K., & Marchetto, M. C. (2022). Reaching into the toolbox: Stem cell models to study neuropsychiatric disorders. *Stem Cell Reports, 17*, 187–210.
Whitney, E. R., Kemper, T. L., Rosene, D. L., Bauman, M. L., & Blatt, G. J. (2009). Density of cerebellar basket and stellate cells in autism: Evidence for a late developmental loss of Purkinje cells. *Journal of Neuroscience Research, 87*, 2245–2254.
Williamson, J. M., & Lyons, D. A. (2018). Myelin dynamics throughout life: An ever-changing landscape? *Frontiers in Cellular Neuroscience, 12*, 424.
Yang, G., & Shcheglovitov, A. (2020). Probing disrupted neurodevelopment in autism using human stem cell-derived neurons and organoids: An outlook into future diagnostics and drug development. *Developmental Dynamics, 249*, 6–33.
Yip, J., Soghomonian, J. J., & Blatt, G. J. (2009). Decreased GAD65 mRNA levels in select subpopulations of neurons in the cerebellar dentate nuclei in autism: An in situ hybridization study. *Autism Research, 2*, 50–59.
Yuan, S. H., & Shaner, M. (2013). Bioengineered stem cells in neural development and neurodegeneration research. *Ageing Research Reviews, 12*, 739–748.
Zaslavsky, K., Zhang, W.-B., Mccready, F. P., Rodrigues, D. C., Deneault, E., Loo, C., Zhao, M., Ross, P. J., El Hajjar, J., & Romm, A. (2019). SHANK2 mutations associated with autism spectrum disorder cause hyperconnectivity of human neurons. *Nature Neuroscience, 22*, 556–564.
Zeng, L., Zhang, P., Shi, L., Yamamoto, V., Lu, W., & Wang, K. (2013). Functional impacts of Nrxn1 knockdown on neurodevelopment in stem cell models. *PLoS One, 8*, e59685.
Zhang, X., Li, Z. L., Liu, Y., & Gai, Z. T. (2021). Great expectations: Induced pluripotent stem cell technologies in neurodevelopmental impairments. *International Journal of Medical Sciences, 18*, 459–473.
Zhang, J., Xin, W., Qin, Y., Hong, Y., Xiahou, Z., Zhang, K., Fu, P., & Yin, J. (2022). "All-in-one" zwitterionic granular hydrogel bioink for stem cell spheroids production and 3D bioprinting. *Chemical Engineering Journal, 430*, 132713.
Zhou, J., Du, W., Zhou, K., Tai, Y., Yao, H., Jia, Y., Ding, Y., & Wang, Y. (2008). Critical role of Trpc6 channels in the formation of excitatory synapses. *Nature Neuroscience, 11*, 741–743.
Zhubi, A., Chen, Y., Dong, E., Cook, E., Guidotti, A., & Grayson, D. (2014). Increased binding of MeCP2 to the GAD1 and RELN promoters may be mediated by an enrichment of 5-hmC in autism spectrum disorder (Asd) cerebellum. *Translational Psychiatry, 4*, e349.
Zikopoulos, B., & Barbas, H. (2010). Changes in prefrontal axons may disrupt the network in autism. *The Journal of Neuroscience, 30*, 14595–14609.
Zikopoulos, B., Liu, X., Tepe, J., Trutzer, I., John, Y. J., & Barbas, H. (2018). Opposite development of short-and long-range anterior cingulate pathways in autism. *Acta Neuropathologica, 136*, 759–778.

Chapter 18
Genes and their Involvement in the Pathogenesis of Autism Spectrum Disorder: Insights from Earlier Genetic Studies

Rishabh Chaudhary and Emma Steinson

18.1 Introduction

Autism spectrum disorder (ASD) is a highly complex and heterogeneous neurodevelopmental condition, which is clinically categorized by impairments in social interaction, communication along with irregular patterns of repetitive, restrictive and stereotypical activities (Fig. 18.1, Table 18.1) (Chaudhary et al., 2021). Dr. Leo kanner, an American psychiatrist, in 1943 first issued a report on eleven of his patients at the Baltimore clinic titled *"Autistic Disturbances of Affective Contact"* (Kanner, 1968). Though the theory of autism was first devised in 1911 by Eugen Bleulerto, a Swiss psychiatrist, who characterized the strange symptoms exhibited by a schizophrenic patient (Asperger, 2009). The prevalence rate of ASD is about 1 in 44 (23.0 per 1000) children (Maenner et al., 2021). Previous studies have shown that about 31% of ASD individuals exhibits intellectual disabilities (Baio et al., 2018) and about 20–37% of them have epileptic conditions (Canitano, 2007; Yasuhara, 2010). In addition, ASD is often associated with several psychological illnesses, such as anxiety, depression, attention deficit hyperactivity disorder (ADHD), sleep disorders as well as gastrointestinal problems (Valicenti-McDermott et al., 2006; MacNeil et al., 2009; Richdale & Schreck, 2009). Till now lots of system models and theories concerning ASD pathogenesis were put forward. But it is clearly asserted to be significantly linked to genomic and environmental factors (De Rubeis et al., 2014; Ng et al., 2017). As per the Diagnostic and Statistical Manual of Mental Disorders (DSM) – 4, there are total five different subtypes of ASD,

R. Chaudhary (✉)
Department of Pharmaceutical Sciences, Babasaheb Bhimrao Ambedkar University, Lucknow, India

E. Steinson
School of Chemistry and Molecular Biosciences, The University of Queensland, St Lucia, QLD, Australia

A. El Idrissi, D. McCloskey (eds.), *Neurobiology of Autism Spectrum Disorders*,
https://doi.org/10.1007/978-3-031-42383-3_18

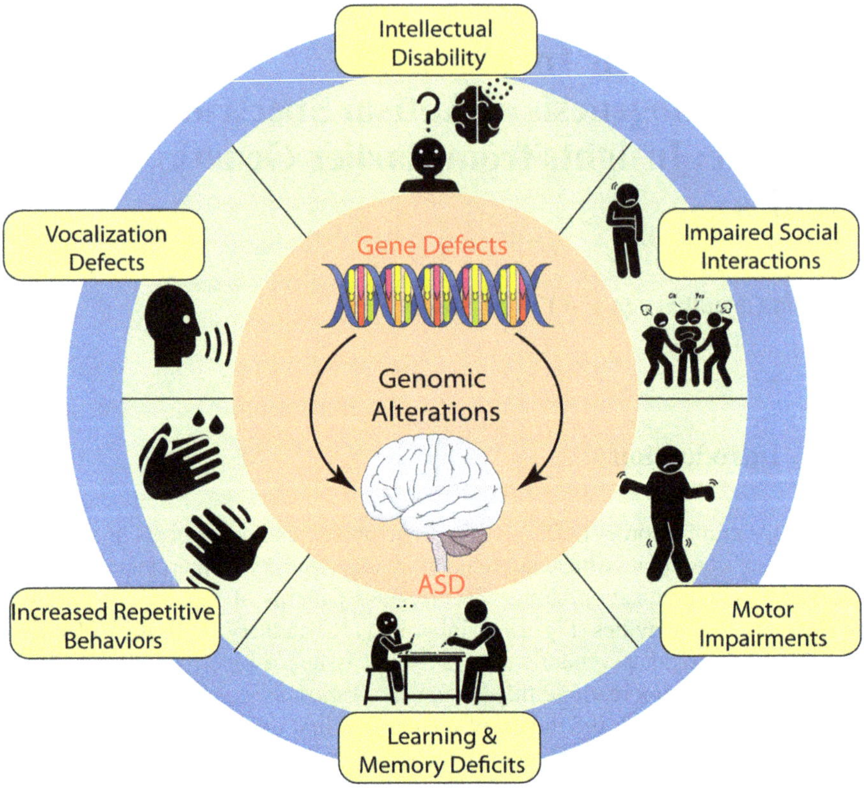

Fig. 18.1 ASD and its clinical characteristics. Individuals with ASD can experience difficulties with attention, language, learning, behavior, social interactions, motor abilities, and various other cognitive functions

including autism, Asperger syndrome, childhood disintegrative disorder (CDD), pervasive developmental disorder – not otherwise specified (PDD-NOS) and Rett syndrome (Muhle et al., 2004; Walsh et al., 2008; Geschwind, 2011). However, the Diagnostic and Statistical Manual of Mental Disorders, 5th Edition (DSM-5) and the International Classification of Diseases 11th Revision (ICD-11), has replaced four subtypes (i.e., Asperger, autism, CDD and PDD-NOS) with one specific diagnostic term, 'autism spectrum disorder' (ASD) to describe and differentiate individuals. The DSM category no longer includes Rett syndrome (Del Barrio, 2016; Lord et al., 2020). Recent advancements in brain imaging technologies have allowed researchers to investigate the brains of autistic patients and to discern various physical alterations, such as atypical neural networking, abnormal brain growth, neuropathology, electrophysiology, neurochemistry and neurotransmission. Individuals with autism, for example, have far more white and gray matter than healthy individuals. Eventually, the gray matter declines, but the white matter continues to rise. This gives autistic individuals more white connective tissues than healthy

Table 18.1 Clinical characteristics of autism spectrum disorder as defined in DSM-5

Impairments in social interaction
Deficits in social communication because of lack of understanding of socio-emotional features.
Impairment in the development of social relationships.
Weakened communicative and social behaviors, including eye contact and facial expression.
Impairments in communication
Impairment in extemporaneous imaginary and social imitative play.
Absence of or delay in language development together with a lack of social norm of whatsoever language skills are present.
Reduced flexibility and lack of imagination in language communication.
Absence of accompanying gesture signals to accentuate or support the meaning of communication.
Poor cognitive reactions to other people's verbal and nonverbal overtures.
Repetitive and stereotyped behaviors, interests, and activities
Might have stereotyped obsessions with energetic interests, such as timing, schedules, date or routes.
Abnormal stereotypic motor activities, such as hand flapping, finger twisting and spinning around.
The propensity to enforce rigidity and ritual in multiple areas of daily functioning.
Hypo-or hyperactivity of sensory stimuli. Has an unusual response to the way things taste, sound, appear, smell, feel or look.

individuals throughout their lifespan. Usually, the structure of the hippocampus and the amygdala of the autistic patient are found to have smaller proportions than the average, which might play a very important part in multiple cognitive difficulties later seen in an autistic individual's life (Ha et al., 2015). Studies stated that autism occurs in all ethnic groups and that males are diagnosed with autism 2–3 times more than females, suggesting biological sex variations in the disorder or an under diagnosis of females, or a combination of both (Baron-Cohen et al., 2005). Similar to various other complex neurodevelopmental disorders, autism was first not acknowledged as having a profound genetic link until the 1980s. Extensive family-based experiments have reported that ASD is caused by the combined actions of more than one gene with epistatic associations. In an earlier study, Rutter et al. asserted that interaction of at least 3–12 genes is involved in the susceptibility of autism (Rutter, 2005). Therefore, researchers have focused their efforts primarily more on genetic backgrounds in an attempt to find genes that increases the risk of ASD.

18.2 Genetic Studies of ASD

Previous studies strongly suggests that genes actively play a significant part in the development and progression of ASD. In these studies, researchers generally focused on the concordance rate, or the likelihood that two individuals with identical genes would acquire the same disease or condition. Because identical twins share almost all of their genetic information with one another, it is plausible to hypothesize that both of the twins will exhibit autistic-like behaviors if there is a genetic basis for the condition. Sixty percent to 90% of the time, when one identical twin is on the autistic spectrum, the other twin is also on the spectrum. This percentage is significantly higher than the rate of concordance among fraternal twins and the rate of ASD recurrence among sisters and brothers, both of which are about 20%. Currently, genetic factors may be identified in around 20% of these situations. These genetic factors may include chromosomal abnormalities, alterations in single genes, or copy number variations (CNVs) including deletions and duplications (Wiśniowiecka-Kowalnik & Nowakowska, 2019). The significant degree of heterogeneity found in ASD cases has presented a foremost task to decipher the genetic lineage influencing the ASDs. An estimated 10–12% of ASD cases are attributable to defined mutations, genetic disorders, and de novo copy number variations (Devlin & Scherer, 2012; Ronemus et al., 2014). Overall, there are approximately 103 genes and 44 genetic loci were reported in individuals with ASD or autistic behaviors. To isolate the suspected genes which are involved in ASD, several experiments were conducted, including linkage studies, low-scale candidate gene association studies, and the widely used genome-wide copy number variation (CNV) as well as genome-wide association studies (GWAS) studies.

18.2.1 Twin Studies

Four twin studies on autism disorder between 1977 and 1995 recorded a range of concordance rates of about 36–91% and 0–21% for monozygotic (MZ) twins and dizygotic (DZ) twins, respectively (Folstein & Rutter, 1977; Ritvo et al., 1985; Steffenburg et al., 1989; Le Couteur et al., 1995). Folstein and Rutter did the first twin research on autism in 1977 and documented a cohort of 11 MZ twins and 10 DZ twins (Folstein & Rutter, 1977). According to the findings of their research, MZ twins had a higher rate of concordance for autism when compared to DZ twins. Additionally, three other twin studies conducted between 2005 and 2009 with quite large groups of twins (>3000 twin pairs) have revealed a higher concordance rate of approximately 77–95% for MZ twins compared to 31% for DZ twins (Ronald et al., 2005; Skuse et al., 2005; Ronald et al., 2006; Rosenberg et al., 2009). In sum, all these different twin studies when taken into account reports higher concordance rates for ASD in MZ relative to DZ twins. One of the largest population-based twin studies was performed by Hallmayer and his colleagues in 2011, which reported that 55% of the variance in liability to autism could possibly be clarified by the environmental factors shared by twins; where the genetic impact is somewhat moderate (Hallmayer et al., 2011). Another study indicated that the genetic liability to autism was higher than the other causes, which is based on a study of twins from UK population and the genetic heritability varied from 56% to 95% (Colvert et al., 2015). Frazier et al. reported that higher rates of autism with social, repetitive and stereotyped behavioral symptoms display high heritability confirming significant genetic susceptibility to autism, whereas the lower level of autism symptoms show low heredity (Frazier et al., 2014). A meta-analysis study by Tick and colleagues on >6000 twins supported all these accumulating evidences and indicated a strong genetic associations with significant shared environmental effects for ASD (Tick et al., 2016).

18.2.2 Family Studies

Parents and other close relatives are also known to have some low degree of symptoms that place them within the broad autism phenotype (BAP). The term "BAP" is used to describe non-autistic relatives who display milder form of autistic characteristics (Cook, 1998). It has been reported that the recurrence risk of autism among parents with autistic probands is roughly 5% and decreases significantly to <1% among second and third degree relatives (Ritvo et al., 1989). The study by Gronborg et al. showed differences in the probability of recurrence risk between half and full siblings suggesting a genetic correlation in autism (Grønborg et al., 2013). However, the increased risk of recurrence in maternal half-siblings as compared to paternal half-siblings shows an environmental influence associated with pregnancy and the intrauterine ASD environment (Risch et al., 2014). One of the largest population based study to evaluate the familial risk of ASD was performed by Sandin and colleagues (Sandin et al., 2014). In the sample of 2,049,899 children, 14,516 achieved an ASD diagnosis. The relative recurrence risk rate of ASD was assessed to 153.0 (95% confidence interval (CI): 56.7–412.8; 27 vs. 6273 per 100,000 person-years) for MZ twins, 8.2 (95% CI 3.7–18.1; 55 vs. 805 per 100,000 person-years) for DZ twins, 10.3 (95% CI 9.4–11.2; 49 vs. 829 per 100,000 person-years) for full-siblings, 3.3 (95% CI 2.6–4.2; 94 vs. 492 per 100,000 person-years) for maternal half siblings, 2.9 (95% CI: 2.2–3.7; 85 vs. 371 per 100,000 person-years) for paternal half siblings, and 2.0 (95% CI: 1.8–2.2; 49 vs. 155 per 100,000 person-years) for cousins (Sandin et al., 2014).

18.2.3 Linkage Studies

Linkage studies are primary method for the discovery of broad genomic regions co-inherited with the phenotype. It is a powerful tool for detecting rare high-risk disease associated genes. These studies have found several chromosomal loci, including 2q region that harbors *SLC25A12* (Solute Carrier Family 25 Member 13), *HOXD1* (Homeobox D1), *NRXN1* (Neurexin-1), *PAX3* (Paired Box Gene 3), *DLX1* (Distal-Less Homeobox 1), *DLX2* and *CENTG2* (Centaurin, Gamma 2) genes (Buxbaum et al., 2001; Shao, Raiford, et al., 2002b), 7q region that harbors *MET*, *NRCAM* (Neuronal Cell Adhesion Molecule), *RELN* (Reelin), *EN2* and *FOXP2* (Forkhead box protein P2) genes (Ashley-Koch et al., 1999), 15q region consisting of the *GABA* (gamma-aminobutyric acid) receptor genes cluster (Sutcliffe et al., 2003),

17q region that harbors *SLC6A4* gene (Cantor et al., 2005), and X chromosome region *MeCP2* (methyl CpG binding protein 2) and *NLGN* (neuroligin) genes (Liu et al., 2001; Auranen et al., 2002; Shao et al., 2002a). Lowe et al. conducted linkage analyses using Social Responsiveness Scale (SRS) scores in 590 families from the Autism Genetic Resource Exchange (AGRE), a very complex and large ASD cohort (Lowe et al., 2015). The scores from the SRS measures heritable, quantitative variation in autistic-like social impairments (Constantino, 2013). In correlated analyses, using genome-wide significant linkage for SRS score from teacher/parent respondents they identified two novel loci on chromosome 8, chr8p21.3 and chr8q24.22 with LOD (logarithm of odd) score 4.11 and 2.63, respectively. Furthermore, using parent respondents only, they identified genome wide significant evidence for linkage at chr8p21.3 (LOD = 3.65) as well as at chr8q24.22 (LOD = 4.54) (Lowe et al., 2015).

18.2.4 Genome-Wide Association Studies

Vast numbers of genetic variations linked to complex and diverse characteristics and illnesses, including ASD, have been discovered by GWAS analysis (Anney et al., 2017; Grove et al., 2019). This method utilizes single nucleotide polymorphisms (SNPs) and/or microsatellite markers/tags to ascertain chromosomal locations that co-segregate with the phenotype within the family members. A number of experiments have been performed to identify regions harboring one or more genetic variants that are responsible for ASD susceptibility in families (Ashley-Koch et al., 1999; Shao et al., 2003; Yonan et al., 2003; Ylisaukko-Oja et al., 2004; Schellenberg et al., 2006; Szatmari et al., 2007; Anney et al., 2010). Moreover, two other GWAS studies were conducted that include Japan (Liu et al., 2016) and Taiwan (Kuo et al., 2015). The findings of these research led to the discovery of several particular chromosomal regions as a potential location for autism susceptibility genes, including significant peaks on chromosomes 1p, 2q, 5q, 7q, 15q, 16p, 17q, 19p and X (Veenstra-VanderWeele & Cook, 2004; Weiss et al., 2009; Li et al., 2011). Though, GWAS studies have failed to distinguish many common susceptibility variants that are deviated from the common variant – common disorder hypothesis for the underlying genetic causes of ASD. Nonetheless, incorporating GWAS datasets with regulatory variants seemed to have the ability and potential to divulge new evidence about the molecular mechanisms of ASD.

18.2.5 Cytogenetic Studies

Data from genome scan analysis allowed the identification of broad chromosomal regions suggesting the location of susceptibility loci, but not the gene of interest. The study of chromosomal aberrations is an alternate methodology that has typically shown to be extremely efficient in discovering and mapping genes for illnesses. It has been found that certain people with ASD have chromosomal abnormalities at a much greater rate than those typically developing individuals (Castermans et al., 2004; Vorstman et al., 2006). Such impairments were shown to be present on every chromosome. An evidence of weak linkage were observed on chromosome 2 and 7 for specific language impairment (Bartlett et al., 2004). Multiple genes, such as *RAY1*, *AUTS2*, *MMP16*, *PAX3*, *A2BP1*, *GRPR*, *NBEA*, and *GABARG1* have been found as physically damaged by chromosomal rearrangements (Muhle et al., 2004). The most prominent ASD candidate loci disrupted by chromosomal abnormalities includes *SHANK3* on chromosome location 22q13 (Durand et al., 2007; Moessner et al., 2007), *NLGN3* and *NLGN4* on X chromosome (Jamain et al., 2003), and *neurexin* gene on chromosome location 2p16 (Chubykin et al., 2005). Microarray-based comparative genomic hybridization (CGH) is an advanced technique that detects DNA gain or loss in CNV at a resolution of 5 Mb in the genome and thus has more advantage over conventional karyotyping. Recent data on CGH analysis have shown that *de novo* CNVs possibly are common risk factors in ASD pathophysiology (Glessner et al., 2009).

18.2.6 Copy Number Variation (CNV) Analysis

CNVs in chromosomes are primarily submicroscopic structural variations which encompass deletions, duplications, inversions, and translocations that may span many kilobases in length (Marshall et al., 2008). It's a much more recent molecular cytogenetic technique designed for determining the deletion or insertion of DNA segments typically greater than 50 kilobyte (kb) in length (Piggot et al., 2009). Population-based studies have identified thousands of CNVs involved in ASD that are supposed to modulate cellular phenotypes by functionally influencing the gene expression directly or indirectly. Generally, CNVs are generated through diverse mutational mechanisms that occur as a consequence of abnormal meiotic recombination, errors in DNA repair and replication processes. Simons Foundation Autism Research Initiative (SFARI) has indicated that roughly 2145 CNVs are implicated in ASD (Abrahams et al., 2013). The most common ASD-related CNVs are found at chromosome location 1q21 containing *BCL9* gene, 7q11.23 containing *AUTS2* (autism susceptibility candidate 2) gene, 15q11-q13 harboring *UBE3A* (Ubiquitin protein ligase E3A), *GABRB3* (Gamma-aminobutyric acid receptor subunit beta-3), *GABRG3* (Gamma-Aminobutyric Acid Type A Receptor Subunit Gamma3) and *GABRA5* (Gamma-Aminobutyric Acid Type A Receptor Subunit Alpha5) genes, 15q13.3 harboring *CHRNA7* (Cholinergic Receptor Nicotinic Alpha 7 Subunit) gene, 16p11.2 containing *MVP* (Major Vault Protein) and *GDPD3* (Glycerophosphodiester Phosphodiesterase Domain Containing 3) genes, 22q11.2 including *PI4KA* (Phosphatidylinositol 4-kinase alpha), *SNAP29* (Synaptosome Associated Protein 29), *TBX1* (T-Box Transcription Factor 1), *SLC7A4* (Solute Carrier Family 7 Member 4), *SLC25A1*, *HIC2*, *UBE2L3*, *YPEL1* and *MAPK1* genes,

22q13.33 containing *SHANK3* gene and Xp27–28 containing *SPANXC* and *MAGE8* genes are revealed by Fluorescence *in situ* hybridization (FISH), SNP arrays and CGH methodologies (Christian et al., 2008; Glessner et al., 2009; Pinto et al., 2010; Bremer et al., 2011; Matsunami et al., 2013).

18.3 Genes Involved in ASD Pathogenesis

The genetic makeup of ASD remains obscure, and the exact variables that produce the disorder have yet to be determined. ASD has a wide range of manifestations, which might be related to the large number of genes (Table 18.2) or environmental factors involved, resulting in various genomic discrepancies that transpire in each individual. Since autism is a neurodevelopmental condition, researchers have been studying its genetic element as well as other variables that influence neuronal development. As a result, studies have been undertaken to identify genes associated to ASD pathogenesis and to determine if the genes are potentially implicated in each ASD symptom. Owing to multiple ASD signs, ASD-linked genes are also potentially associated to other neurodevelopmental diseases, for example Rett syndrome and fragile X syndrome (FXS) (Amir et al., 1999). Thus, the complex interrelationship amongst different ASD-linked genes makes it very complicated to understand the aetiology of ASD.

18.3.1 RELN (Reelin) Gene

Reelin is a large extra-cellular matrix glycoprotein produced by the Cajal Retzius (CR) neurons of the developing brain. It is synthesized as 3460 amino acids precursor protein with a molecular weight of 410 kDa. The protein consists of eight reelin repeat domains. Each reelin repeat domain is made up of two subsequent domains (A and B), which are connected by an epidermal growth factor (EGF)–like domain (D'Arcangelo et al., 1995; Sato et al., 2015). In 2001, the International Molecular Genetic Study of Autism Consortium (IMGSAC) identified the chromosome 7q region as a crucial linkage and the first autism susceptible point (Bailey et al., 1998). *RELN* gene maps to the AUTS1 (Autism, Susceptibility To, 1) locus on chromosome location 7q22 (Scherer et al., 2003). At an early stage of embryonic development, reelin performs crucial role in the placement of neuronal cells in inferior olivery complex, cerebellum, and cerebral cortex regions (Goffinet, 1983a, b, 1984). It is also responsible for the neuronal proliferation and directs neurons to establish a proper laminar structure within the cortex. Impairment of the reelin signaling pathway during development may cause multiple neurodevelopmental disorders, including autism (Fatemi et al., 2005).

Fatemi et al. first described that the deregulation of *RELN* gene could be responsible for structural and behavioral abnormalities in autism. In their study, they

Table 18.2 Profoundly documented ASD related genes

Gene	Chromosome location	Aliases	Interacting proteins	Protein function	Human diseases
RELN	7q22.1	Reelin, LIS2, PRO1598	VLDLR, LRP8, DAB1, PCDHA6	Regulation of corticogenesis, neuronal cell positioning and cellular maturation	ASD, schizophrenia, arthritis, lissencephaly, epilepsy, bipolar disorder
SHANK1	19q13.33	SPANK1, SSTRIP, synamon, SH3 and multiple ankyrin repeat domains 1	HOMER1, DLG4, DLGAP1, NEK4, TANC1, STAT1, SPTAN1	Structural and functional organization of dendritic spine and synaptic junctions	ASD, Phelan-McDermid syndrome, pitt-hopkins like syndrome 1, chromosomal deletion syndrome
SHANK2	11q13.3 – q13.4	AUTS17, CORTBP1, CTTNBP1, ProSAP1, SPANK3	ARHGEF7, SHARPIN, DLG4, DLGAP1, DNM2, SRC	As molecular scaffolds in the post-synaptic density	ASD, Patellofemoral pain syndrome, pervasive developmental disorder
SHANK3	22q13.33	ProSAP2, SPANK2, SCZD15, DEL22q13.3	ARHGEF7, DLGAP1, GRB2, PLCG1, NCK1, FLNB	Synapse formation and dendritic spine maturation	Schizophrenia, Phelan McDermid syndrome, mutism, ASD
NLGN1	3q26.31	Neuroligin 1, EC3.1.1, NL, KIAA1070	NRXN1, NRXN2, NRXN3, DLG4, DLG3	Synaptic signaling transmission, formation and remodeling of synapses	Asperger syndrome, post-traumatic stress disorder, childhood disintegrative disorder, autism
NLGN2	17p13.1	Neuroligin 2, KIAA1366, NL2	NRXN1, NRXN2, NRXN3, DLG4, MDGA1	Synapse function and synaptic signal transmission	Childhood disintegrative disorder, autism, pervasive developmental disorder
NLGN3	Xq13.1	Neuroligin 3, Gliotactin, KIAA1480, HNL3, NL3	NRXN1, NRXN2, NRXN3, PIK3R3, PIK3R1	Synapse function and synaptic signal transmission	Asperger syndrome, autism x-linked, pervasive developmental disorder

(continued)

Table 18.2 (continued)

Gene	Chromosome location	Aliases	Interacting proteins	Protein function	Human diseases
NLGN4	Xp22.32 – p22.31	Neuroligin 4, neuroligin X, KIAA1260, AUTSX2	NLGN3, DLG4, DLGAP2, BRK1	Formation of functional synapses, cell-cell interactions	Asperger syndrome, autism x-linked, pervasive developmental disorder
NLGN4Y	Yq11.221	Neuroligin 4 Y-linked, neuroligin Y, KIAA0951, HNLAY	NLGN3, PRIM1, PRIM2, HNRNPL	Formation of functional synapses, cell-cell interactions	ASD
OXTR	3p25.3	Oxytocin receptor, OT-R	OXT, AVP, THOC5, CYFIP2, ASH1L, GRK5	Receptor for oxytocin	Postpartum depression, conduct disorder, adenomyosis, ASD
GABRB3	15q12	ECA5, EIEE43	AKAP5, PPP2CA, GABRA6, ASIC4, MCOLN3, MEOX2	Formation of functional inhibitory GABAergic synapses	Epilepsy, childhood absence, lennox-gastuat syndrome, ASD
GABRA5	15q12	EIEE79	GABRA3, GABRG2, GPHN, GABRB1, GABRB2	Involved in GABA-A receptor assembly, immobilization and accumulation at the synapses	Angelman syndrome, childhood absence epilepsy, prader-willi syndrome, epileptic encephalopathy
GABRG3	15q12	–	GABRAP, GABRB1, GABRB3, GPHN, GABRB2	Mediates neuronal inhibition by binding to the GABA receptor and opening an integral chloride channel	Asperger syndrome, alcohol dependence, ASD, albinism
GABRD	1p36.33	EIG10, EJM7, GEESP5	CCR8, UBQLN1, ATF6, GABRA1, GABRA2, GABRA3	Mediates neuronal inhibition by binding to the GABA receptor and opening an integral chloride channel	Chromosome 1p36 deletion syndrome, febrile seizures, ASD

(continued)

Table 18.2 (continued)

Gene	Chromosome location	Aliases	Interacting proteins	Protein function	Human diseases
GABRB1	4p12	EIEE45	FKBP5, UBQLN1, PIK3CA, AKAP5, NOD1, GABRA1, GABRA2	Mediates cellular responses	Schizophrenia, alcohol dependence, epileptic encephalopathy
GABRA4	4p12	–	PRKCG, DGAT1, APP, GABRB2, GABRA1, GABRA2, GABRA3, GABRA5	Mediates neuronal inhibition by binding to the GABA receptor and opening an integral chloride channel	Autism, schizophrenia, restless leg syndrome, alcohol dependence
MET	7q31.2	HGF/SF receptor, proto-oncogene C-Met, AUTS9, RCCP2, HGFR, DFNB97	CBLC, CBL, HGF, PTPN1, INPPL1	Gastrulation development, migration of muscles, neuronal precursors, angiogenesis and kidney formation	Renal cell carcinoma, hepatocellular carcinoma, osteofibrous dysplasia, autism
SLC6A4	17q11.2	5HTT, SERT, OCD1, 5-HTTLPR	STX1A, PICK1, SNCA, SNCG, GPM6B	Regulation of serotonergic signaling	Obsessive compulsive disorder, anxiety, borderline personality disorder, mood disorder, depression
SLC25A12	2q31.1	ARALAR1, EIEE39, AGC1	IMMT, GNAS, CHCHD6, TDRD1, SLC25A13	Solute: Cation symporter activity	Asperger syndrome, autism, epileptic encephalopathy
ITGB3	17q21.32	Antigen CD61, GPIIIa, GT, BDPLT16, Platelet glycoprotein IIIa	SHC1, FN1, ITGA2B, ITGAV, FLNA	Synaptic transmission and plasticity, controls excitatory synaptic length	ASD, glanzmann thrombasthenia, myocardial infarction, thrombocytopenia

(continued)

Table 18.2 (continued)

Gene	Chromosome location	Aliases	Interacting proteins	Protein function	Human diseases
MAOA	Xp11.3	Amine oxidase (Flavin-containing) A, EC1.4.3, BRNRS	MAOB, COQ9, MIEF1, ATP6V1B12	Catalyzes the oxidative deamination of biogenic and xenobiotic amines, metabolism of neuroactive and vasoactive amines in the CNS and peripheral tissues	Brunner syndrome, antisocial personality disorder, social phobia, conduct disorder, endogenous depression
MAOB	Xp11.3	Amine oxidase (Flavin-containing) B, Adrenaline oxidase	MAOA, ATP2B2, SRPK1, NFYB	Catalyzes the oxidative deamination of biogenic and xenobiotic amines, metabolism of neuroactive and vasoactive amines in the CNS and peripheral tissues	Norrie disease, pathological gambling, post-traumatic stress disorder, antisocial personality disorder, psychotic disorder

measured the level of *RELN* expression in the cerebellar cortex of the brains of autistic and control subjects and observed a 44% decrease in RELN in the cerebellum of five subjects with autism relative to eight normal control individuals. Moreover, they also observed that the levels of reelin protein and mRNA in the frontal cortex region were significantly decreased, that the mRNA for adaptor Disabled-1 (Dab-1) was drastically reduced, but the level of mRNA for the reelin receptor very-low density lipoprotein receptor (VLDLR) was significantly elevated in the superior frontal lobe and cerebellar regions of the brains in autistic individuals (Fatemi et al., 2001). The research was further supported by another study, which showed that the expression of reelin was reduced in anatomical brain regions like parietal cortex, superior frontal cortex, and cerebellum in ASD patients (Lammert & Howell, 2016). Further research has shown that abnormalities in expression of reelin during the developmental phase leads to atypical anatomical forms, including damage of the cerebellar, cerebral, and hippocampal-cortical layers, formation of improper cortical laminar development, and impaired synaptic connectivity (Boyle et al., 2011). Since *RELN* is found highly sensitive to epigenetic alterations caused by the environment, it is possible that a single genetic mutation that reduces its expression, along with environmental downregulation of reelin synthesis, may keep reelin protein levels in the brain well below the critical range. *RELN* mutations are found in around four times the number of men as women. Furthermore, testosterone and oestrogen levels affect the expression of *RELN* and HRM phenotypes in distinct manners (Absil et al., 2003; Biamonte et al., 2009; Macrì et al., 2010). Nevertheless, much further work is needed to obtain a thorough picture of reelin and its signaling processes and to translate such findings into clinical settings. As a result, animal studies of *RELN* mutations are required to clarify the relationship between *RELN* and ASD and other complicated neurodevelopmental etiologies.

18.3.2 SHANK (SH3 and Multiple Ankyrin Repeat Domains Protein) *Gene*

There is total three human *SHANK* genes: *SHANK 1*, *2*, and *3*. They enciphered a family of post-synaptic scaffolding proteins whose key function is to organize protein structures at the postsynaptic densities (PSDs) of excitatory glutamatergic synapses in the central nervous system (CNS) (Boeckers et al., 1999; Naisbitt et al., 1999). The *SHANK1* gene is found on chromosome location 19q13.33, containing 24 exons and two alternative promoters resulting to two different isoforms. It is exceedingly expressed in the hippocampal and cerebral cortex regions (Lim et al., 1999). The *SHANK2* (also known as *ProSAP1*) gene is present on chromosome location 11q13.2, containing 25 exons, three alternative promoters and one alternative stop codon, which leads to the generation of four different isoforms. It has high levels of expression in the brain, particularly in cerebellar purkinje cells (Böckers et al., 2004; Schmeisser et al., 2012). The *SHANK3* (also known as ProSAP2) gene is present on chromosome location 22q13.3, consisting of 22 exons and one alternative stop codon located in the exon 21b, resulting in three different isoform generation. It is found strongly expressed in thalamus, striatum, and in cerebellar granular cells (Peça et al., 2011; Wang et al., 2014). The *SHANK3* gene encodes a multi-domain protein containing ankyrin repeats (ANK), an Src homology 3 (SH3) domain, a polysynaptic density 95/discs large/zone occludens-1 (PDZ) domain, a proline rich region, a homer-binding area, a cortactin-binding region and a sterile alpha motif (SAM). SHANK3 is a scaffolding protein and is mainly expressed in neurons, especially in synapses. It interacts with numerous synaptic molecules in the brain, including with the N-methyl-D-aspartate receptor (NMDAR) via the post-synaptic density-95 (PSD-95)/guanylate kinase-associated protein (GKAP) complex, the metabotropic glutamate receptor (mGlur) via Homer and the GluR1 α-amino-3-hydroxy-5-methyl-4-isoxazole propionic acid receptor (AMPA-R) (Naisbitt et al., 1999; Uchino et al., 2006). Many studies have shown that mutations in these three *SHANK* genes were associated to idiopathic and syndromic ASD development, in addition to other psychiatric disorders such as intellectual disability (ID) and schizophrenia (Berkel et al., 2010; Jiang & Ehlers, 2013; Sala et al., 2015).

Because of the notable genetic link between ASD and *SHANK3* mutations, a lot of scientific studies have centered on this particular gene.

In humans, the 22q13 deletion syndrome, now known as Phelan McDermid syndrome (PMS), was found in people with *SHANK* gene mutations. PMS is a variation of ID caused by *SHANK3* haploinsufficiency and is usually linked with ASD (Phelan & McDermid, 2012). Several other finding demonstrated that not only *SHANK3* mutation but also *SHANK3* duplication is also linked with ASD development (Durand et al., 2007; Moessner et al., 2007) indicating that *SHANK3* is extremely vital for the proper brain functioning. Zhu et al. have found higher concentrations of *SHANK3* methylation in a cohort of ASD individuals, which demonstrated that epigenetic deregulation of *SHANK3* may also be correlated with ASD (Zhu et al., 2014). In addition, Leblond et al. have shown that there is a substantial link between cognitive impairments and various different shank proteins. Individuals with *SHANK3* mutations are affected more significantly than individuals with *SHANK1* and *2* mutations (Leblond et al., 2014).

Apart from the *SHANK3's* recurring role in ASD, multiple investigations have also shown detrimental mutations in the *SHANK1* and *2* genes. Sato et al. first reported the role of *SHANK1* in ASD. Six members of the same family with high-functioning autism or BAP have a hereditary deletion of 63.8 kb in length, encompassing *SHANK1* and the *CLEC11A (C-Type Lectin Domain Containing 11A)* gene. Males are carriers but not females (Sato et al., 2012). Berkel et al. identified three ASD and ID patients with *de novo* heterozygous *SHANK2* deletion (Berkel et al., 2010). This was further supported by two different studies, which found out that *SHANK2* is critical for the regulation and modulation of cerebellar excitatory synaptic density and cellular functions, signifying that the lack of *SHANK2* in purkinje cells is accountable for disruption in motor learning and coordination activities, as well as for repetitive and anxiety-like behaviors (Ha et al., 2016; Peter et al., 2016). In sum, all *SHANK* gene variants have been linked to ASD, although further studies are needed to thoroughly comprehend their prevalence and specific clinical backgrounds.

18.3.3 NLGN (Neuroligin) Gene

Neuroligins are a broad family of alternatively spliced transmembrane molecules found primarily in post-synaptic membranes, where they are believed to interact with pre-synaptic neurexin proteins to mediate heterophilic adhesion (Ichtchenko et al., 1996; Nguyen & Südhof, 1997; Song et al., 1999). There are five classes of *NLGN* genes: *NLGN1, NLGN2, NLGN3, NLGN4* and *NLGN4Y* that are localized on chromosome location 3q26, 17p13, Xq13, Xp22.3 and Yq11.2, respectively. They transcribe trans-synaptic cell adhesion molecules called neuroligins that are important for brain plasticity, synaptic integration and maintenance (Jamain et al., 2003). *NLGN1* and *NLGN2* are primarily found at glutamatergic and GABAergic synapses, respectively (Song et al., 1999; Varoqueaux et al., 2004). While *NLGN3* and *NLGN4* are located post-synaptically at both inhibitory and excitatory synapses (Felix-Ortiz & Tye, 2014; Banerjee et al., 2016; Haar et al., 2016). Various cytogenetic and genome sequencing studies have reported that *neuroligins (NLGNs)* are one of the subsets of ASD-associated synaptic protein encoding genes (Jamain et al., 2003; Südhof, 2008). Jamain et al. first described the mutations in two X-linked genes that encode *NLGN3* and *NLGN4* in ASD. They hypothesized that defect in these two genes could suppress the development and stabilization of specific synapses, which are indispensable for the transmission and communication processes that are deficient in ASD subjects (Jamain et al., 2003). Following this study, several other researchers studied these two genes and found positive genetic association for *NLGN3* and *NLGN4* (Yan et al., 2005; Lawson-Yuen et al., 2008; Rothwell et al., 2014). However, studies have also stated negative genetic association for *NLGN3* and *4* (Talebizadeh et al., 2004; Gauthier et al., 2005; Wermter et al., 2008).

To study the function of *NLGN* mutations in ASD, various *NLGN* mutant mice models have been designed. The *NLGN3* R451C knock-in mouse model corresponding to the human non-synonymous SNP (R451C) in *NLGN3* was found similar to ASD individuals. It showed deficits in socialization, enhanced spatial memory, as well as an electrophysiological phenotype in the somatosensory cortex area with increased inhibitory synaptic transmission activities (Tabuchi et al., 2007; Etherton et al., 2011a). In addition, Foldy et al. have demonstrated that both *NLGN3* R451C-knock-in and *NLGN3* knock-out mutations in mice displayed disrupted tonic but not phasic endo-cannabinoid signaling, indicating the possibility that alterations in endo-cannabinoid signaling mechanism could lead to ASD development (Földy et al., 2013). In another study, Etherton et al. presented the *NLGN3* R704C mutation into the mouse model by homologous recombination method and investigated its effect on synapses *in vivo* and *in vitro*. Electrophysiological and morphological studies have demonstrated that the mutation of *NLGN3* R704C did not substantially affect the synapse development. However, the functions of synapses are significantly hampered. Furthermore, the R704C mutation reduced AMPA receptor-mediated synaptic transmission in pyramidal neurons of the hippocampus significantly and selectively, without affecting GABA or NMDA receptor-mediated synaptic transmission or presynaptic neurotransmitter release (Etherton et al., 2011b).

Mice with a loss-of-function (LOF) mutation in *NLGN4* ortholog exhibited extremely substantial dysregulations in reciprocal social interactions and impaired social communication, both of which are selective ASD traits. In addition, there were decrease in the brain size of adult *NLGN4* knock-outs which substantiate findings of decreased brain size in adult ASD individuals. Thus, in terms of the ASD specificity, the *NLGN4* knock-out model is a perfect tool for further ASD-related studies. However, in *NLGN4* knock-out mice models researchers did not detected any abnormal repetitive behavioral patterns which are typically observed in ASD individuals (Jamain et al., 2008). Mutations within *NLGN1* have also been found to be linked to neurodevelopment. Glessner et al. observed duplication of copy number variants within *NLGN1* in ASD individuals relative to controls, whereas Ylisaukko-oja et al. identified mutation (rs1488545) within the Finnish autistic cohort (Ylisaukko-oja et al., 2005; Glessner et al., 2009). Furthermore, Millson et al. found an inherited 2.1 Mb deletion (3q26.31 – 3q26.32) that involved NAALADL2 and the 3′ end of *NLGN1* in a 6.5 year old female with microcephaly, seizures, and severe intellectual disability (Millson et al., 2012). This was further confirmed by Blundell et al. who demonstrated that NLGN1 knockout mice exhibited alterations in spatial memory and learning in the water maze test, which was theorized to be linked to altered hippocampal long-term potentiation. Such mice displayed a significant rise in repetitive and stereotyped grooming behaviors, as well as impaired nest building activities, characterized as ASD-relevant abnormalities (Blundell et al., 2010).

Alterations in *NLGN2* are also linked to neurodevelopmental activities. Belligni et al. first reported that an individual with afebrile seizures, mental retardation and dysmorphic features had a *de novo* 790–830 kb duplication on chromosomal band 17p13.1, of which one particular gene affected was *NLGN2* (Belligni et al., 2012). The *NLGN2* knockout mice exhibited an anxiety-related phenotype, with decreased time spent in the middle of an open area and anxiety-like behaviors in the light/dark box, while their socializing activities were found totally normal (Blundell et al., 2009). In sum, all these studies propose that functional mutations in the *NLGN* genes and malformation in synaptogenesis may predispose an individual to autism spectrum.

18.3.4 OXTR (Oxytocin Receptor) Gene

Oxytocin is a well-known nonapeptide that is synthesized in the hypothalamic regions and released into the circulation via posterior pituitary gland (Lerer et al., 2008). The *OXTR* gene encodes 389 amino acids and is located on chromosome location 3p25–3p26.2 in the human genome as a single copy (Inoue et al., 1994; Simmons et al., 1995). The exact position of the *OXTR* gene is chr3:8,750,408–8,769,617 and is approximately 19.2 kb in length (Kent et al., 2002). It contains three introns and four exon regions. Other than modulating and directing lactation and uterine contractions, it also works as a neuromodulator within the CNS, regulating various social and emotional activities, such as normal cognitive mechanisms, dealing with social stress, modulation of social memories, and advancement of social interactions (Meyer-Lindenberg et al., 2011). Animal studies and clinical experiments have both proven the significance of oxytocin in social and repetitive behaviors (Green & Hollander, 2010). As a result, oxytocin may have a role in the aetiology of ASD, and the OXTR gene has been identified as a prime target for further investigation in this area.

The *OXTR* gene contains several SNPs that are significantly associated with functional and structural differences in the limbic circuitry like hypothalamus, amygdala and cingulate gyrus that are related to the neurobiology of social behaviors, and to the ASD pathology (Inoue et al., 2010; Furman et al., 2011; Tost et al., 2011). Several research have examined the relationships between the ASD diagnosis and variation of *OXTR* SNPs (Jacob et al., 2007; Liu et al., 2010; Tansey et al.,

2010; Wermter et al., 2010; Francis et al., 2016). After collecting the data of published association studies between *OXTR* SNPs and ASD by LoPero and Waldman, a meta-analysis study concluded that certain alleles of *OXTR* gene polymorphisms reduce the probability of autistic development: rs7632287 A, rs237887 A, rs2268491 T, and rs2254298 A (LoParo & Waldman, 2015). After this, Kranz et al. performed a new meta-analysis and reported a link between rs237889 A and ASD. (Kranz et al., 2016). In addition, another analysis recently analyzed three different *OXTR* gene SNPs: rs2228485, rs4686302 and rs237902. It showed that the *OXTR* rs237902 polymorphism is related with ASD severity (Ocakoğlu et al., 2018). The *OXTR* gene contains a synonymous variation called Rs237902, which is located at the exon 3 position (Kent et al., 2002). As a result, this modification has no effect on the sequence of amino acids in the protein assembly. However, synonymous mutations may influence expression, conformation and function of a protein by affecting mRNA splicing and its stability, deregulating the ability of binding of miRNAs and translation processes (Sauna & Kimchi-Sarfaty, 2011). Moreover, these mutations have been linked with different neurodevelopmental disorders, including ASD.

The majority of studies examining epigenetics concentrate on DNA methylation, which is a chemical modification of the DNA. The *OXTR* gene comprises CpG island, and methylation within this region has been shown to have a negative impact on *OXTR* transcription processes in a variety of tissues (Kusui et al., 2001; Gregory et al., 2009; Gouin et al., 2017). Gregory et al. first demonstrated a link between *OXTR* methylation and autism. They reported that CpG island of *OXTR* results in aberrant gene silencing, which is mediated by significant increase in the DNA methylation status of *OXTR* in temporal cortex and peripheral blood cells of autism cases, indicating that epigenetic deregulation could be linked with ASD development (Gregory et al., 2009). An additional piece of evidence came from a report exploring the influence of *OXTR* methylation on variations in neural brain responses that occur during the processing of emotional facial expressions (Puglia et al., 2015). Methylation of the *OXTR* site-934 was discovered to have a positive association with activity in regions of the brain associated with the processing of emotions. These regions include the amygdala, insular cortex, and dorsal anterior cingulate cortex (dACC), among others (pSTS). However, the functional pairing between areas of social cognition, such as cingulate cortex, insular cortex and orbitofrontal cortex were negatively associated with *OXTR* methylation, i.e. higher pairing with lower methylation rates (Puglia et al., 2015). Furthermore, via resequencing the full-length receptor gene, Liu et al. identified 28 novel variants in the intronic region and one rare missense variant (R150S) in ASD. They hypothesized that the rare variants of *OXTR* might be another source that contributes to the likelihood of ASD development (Liu et al., 2015b). In conclusion, the effects of DNA methylation and SNPs have been extensively connected to the cognitive deficits that characterize ASD. Future study that uses a multimodal approach will further define the regulatory mechanisms and might unravels the complex nature of this and other neurodevelopmental diseases. This is because the oxytocinergic system tends to be a cornerstone in such conditions.

18.3.5 GABR (Gamma-Aminobutyric Acid Receptor) Gene

GABA, also known as gamma-aminobutyric acid, is a major inhibitory neurotransmitter found in the brain system. It exerts its effect by binding to the GABA receptor. During the initial stages of neuronal development, GABAergic neurotransmission is significantly vital for the construction of synapses and neural circuits, as well as proper brain circuitry activities. GABA is primarily released by GABAergic interneurons, which is responsible for much of the inhibition in the hippocampus, cerebral cortex, amygdala and striatum regions. Impaired inter-neuronal function has been recognized as the underlying aetiology of various neurodevelopmental diseases, including ASD (Powell, 2013). The GABA receptor is a multimeric transmembrane protein that is composed of five subunits. These subunits are arranged in a cluster around a central pore in the receptor's structure. The GABA receptor subunits are similar to one another, but they differ from one another in terms of their structures and their functions (Menold et al., 2001). Three GABA receptor subunits (GABRB3, GABRA5, and GABRG3) are encoded by chromosome 15 genes, which are clustered at the 15q11–q13 region of the genome that is implicated in the expression of genes, genetic imprinting, genomic instability and recombination processes (Cook et al., 1998; Martin et al., 2000). The chromosomal region 15q11–q13 was first linked to ASD based on a number of different studies that found frequent duplication of this region in ASD individuals (Bolton et al., 2004; Kwasnicka-Crawford et al., 2007; Cai et al., 2008; Depienne et al., 2009). Nakatinin et al. duplicated chromosome 7 mirroring the human chromosomal region 15q11 – q13, which harbors the GABR cluster in mice through chromosome engineering, and the generated mouse model displayed the ASD-like phenotypes, including deficits in social interaction, abnormal ultrasonic vocalizations and behaviors (Nakatani et al., 2009). In another animal model study, mice deficient of PX-RICS, an N-terminal splicing variant of β-catenin interacting protein with GAP activity toward Cdc42 and Rac1 (RICS), displayed ASD-like behavioral abnormalities and comorbidities. Such mice have decreased membrane-bound GABA A receptor and GABA A receptor-mediated social behavior deficits that can be rescued by GABAA receptor agonist (Nakamura et al., 2016). Furthermore, a study looking for CNVs in autistic patients reported that 15q duplication is the second most common duplication with a frequency of 1 in 500 ASD cases, which supports the likelihood of a GABR cluster as a possible ASD risk gene (Moreno-De-Luca et al., 2013).

In several regions of the ASD individual's brain, including the motor cortex, auditory cortex, and frontal lobe lower GABA and/or glutamate concentrations have been found (Harada et al., 2011; Gaetz et al., 2014; Rojas & Wilson, 2014). ASD individuals also exhibit substantial reductions in GABA A receptor levels in the superior frontal cortex, parietal cortex, cingulate cortex, cerebellum, hippocampus and fusiform gyrus (Blatt et al., 2001; Fatemi et al., 2009; Oblak et al., 2009), as well as lowered levels of GABA B receptors in the superior frontal cortex, parietal cortex, cerebellum, fusiform gyrus, and anterior and posterior cingulate gyrus (Buxbaum et al., 2002; Oblak et al., 2010). Moreover, mutations in genes producing

GABA receptor subunits like GABRB1, GABRB3, GABRG3, GABRA4, and GABRD have also been linked to the development of ASD (Buxbaum et al., 2002; Ma et al., 2005; Ashley-Koch et al., 2006; Delahanty et al., 2011; Sesarini et al., 2015).

Parvalbumin (PV) neurons, a subset of GABAergic interneurons, are crucial for adequate excitatory/inhibitory regulation of synaptic activities in cortex and hippocampal regions (Mao et al., 2015). These PV neurons may be implicated in GABA dysregulation in certain cases of ASD. According to Hashemi et al. alteration in the density and absolute numbers of PV neurons may result in ASD and could be an underlying cause of behavioral and cognitive symptoms (Hashemi et al., 2018). PV interneurons are important in the production of gamma oscillations (Vreugdenhil et al., 2003; Sohal, 2012). ASDs are frequently associated with gamma-band abnormalities. Such disruptions in rhythmic synchronizations indicate dysfunctional brain processing mechanisms, which can lead to perceptual and cognitive impairments seen in some ASD individuals (Simon & Wallace, 2016).

18.3.6 MET (Mesenchymal Epithelial Transition) Gene

The *MET* gene expresses a transmembrane receptor tyrosine kinase (RTK) of the hepatocyte growth factor/scatter factor (HGF/SF), which has pleiotropic function in embryonic development and influences a multitude of neurological processes, including neural activation, neurotropic and chemotropic properties of developing axons, and mutagenic properties of the specialized central neuronal populations (Streit et al., 1995; Ebens et al., 1996; Maina et al., 1997; Powell et al., 2001). The *MET* gene and expression of proteins are significantly regulated in excitatory projection neurons during synapse formation in the forebrain (Judson et al., 2009; Eagleson et al., 2011; Judson et al., 2011). *MET* is highly expressed in the neocortex of mouse (Judson et al., 2009). Although in humans it is restricted to the medial parietal cortex, temporal and occipital regions that comprise neuronal circuits underpinning the processing of socially relevant information (Mukame et al., 2011). The human *MET* gene is located on chromosome band 7q31.2, spans around >120 kb in length and consists of total 21 exons. The MET receptor is a heterodimer composed of an extracellular α-subunit (50 kDa) and an N-terminal and a β-subunit (140 kDa) linked by disulfide bonds (Liu, 1998). Campbell et al. first analyzed the *MET* gene in a family-based ASD study. They revealed a strong genetic association of the non-coding promoter variant of the human *MET* gene, the rs1858830 “C” allele, which results in a significant decrease in the promoter activity of MET and dysregulated binding of transcription factor complexes (Campbell et al., 2006). In another study, they discovered considerably lower MET protein levels and higher mRNA expression for proteins implicated in MET signaling activity regulation (Campbell et al., 2007). Moreover, they also examined the exons and 5′ promoter regions for variants in the five genes encoding the proteins which regulates MET expression, and discovered that genetic predisposition affects several components of the MET signaling mechanism, which increases the risk of ASD development (Campbell et al., 2008). Individuals with ASD who have rs1858830 "C" allele exhibit more severe cognitive and social impairments than those who do not (Campbell et al., 2010). The relationship between the *MET* gene and the risk of ASD has been further validated by a number of other genetic investigations conducted in a variety of populations. In two additional ASD cohorts, Sousa et al. observed a positive connection between the variant rs38845 in the *MET* gene and autism. On the other hand, rs1858830 "C" allele variations and autism were shown to have no association at all (Sousa et al., 2009). Another independent study conducted by Jackson et al. screened two additional cohorts and showed that the rs1858830 "C" mutation was connected with the autism cohort but was not correlated with the PDD cohort (Jackson et al., 2009). Another group conducted two association studies of *MET* with ASD in the Japanese population around the same period, and they discovered a new SNP in intron 1 called rs38841 that is associated with an increased risk of ASD (Thanseem et al., 2010). Thus, regardless of the source of genomic variance, all of these investigations have shown a significant relationship between *MET* and ASD.

The research conducted by Rudie and colleagues using functional imaging has provided data on the ways in which *MET* might influence the anatomical and physiological interconnectivity of brain circuits. They did this by assessing the correlation between *MET* risk genotype and functional activation patterns in response to social stimuli. This allowed them to investigate the impact of a functional ASD risk variation known as rs1858830 "CC" on network functioning in ASD and control participants. The *MET* risk genotype, also known as the "CC allele," is capable of predicting atypical fMRI (functional magnetic resonance imaging) activation and deactivation patterns in response to social stimuli. This genotype is also associated with decreased functional and structural integration at the temporoparietal junction, which is a region that is thought to have elevated rates of *MET* expression. In addition, the *MET* rs1858830 "CC" risk allele is the one that shows the significant modifications in structural and functional endophenotypes in those who have ASD condition (Rudie et al., 2012). Human *MET* gene transcription can be regulated by MeCP2 and FOXP2 (Forkhead box protein P2) transcriptional factors. Mukamel et al. first investigated the expression and transcriptional regulatory activities of *MET* by FOXP2, a transcription factor involved in the modulation of cognitive and language abilities that is disrupted in ASD. They discovered that FOXP2 binds to regulatory sequences in *MET* genomic DNA both *in vitro* and *in vivo*, and that forced expression of FOXP2 contributes to the downregulation of *MET*, suggesting that FOXP2 can independently modulate the expression activities of *MET* (Mukame et al., 2011). In addition, Plummer et al. proposed that MeCP2 is a potential candidate as an auxiliary transcriptional modulator of *MET*. They also discovered sexuality-based decline in *MET* expression in male ASD individuals but not in female ASD cases. The results of their experiments concluded that MeCP2 regulates *MET* transcription in the human neocortex in an allele-specific manner (Plummer et al., 2013). All of these studies have shown that the *MET* gene's expression in certain human neocortical areas, as well as its regulation, play an important role in autism susceptibility.

18.3.7 SLC6A4 (Solute Carrier Family 6 Member 4) Gene

The *SLC6A4* gene is found on chromosome location 17q11.1–q12. It contains 15 exons encompassing around 35 kb in length, with the first two exons (1a and 1b) alternately synthesized. The *SLC6A4* gene produces serotonin transporter protein (5-HTT), a 630-amino-acid integral membrane protein with 12 transmembrane domains that controls the levels of 5-HT in synapses and influences synaptic excitation activities (Ramamoorthy et al., 1993). The serotonergic pathway modulates the expression of glutamate and GABA neurotransmitters in presynaptic terminals, as well as neurogenesis, cell migration and axonal guidance (Heils et al., 1996; Ciranna, 2006; Cai et al., 2013; Belmer et al., 2017). Various linkage and populations studies provide strong evidence on the role of *SLC6A4* gene in imprinting risk to ASD. It has been considered as one of the best susceptibility genes for ASD development as there is well-studied evidence revealing high blood platelet serotonin level in one-third of autistic probands (Cook et al., 1997; McCauley et al., 2004; Sutcliffe et al., 2005; Muller et al., 2016; Garbarino et al., 2019). Polymorphisms in the *SLC6A4* gene have been linked to varying degrees of transcriptional regulation and, as a result, varied levels of protein synthesis (Kenna et al., 2012). The 5-hydroxytryptamine transporter linked polymorphic region (5-HTTLPR), located at the proximal 5′ promoter region, is by far the most thoroughly investigated polymorphism in the SLC6A4 gene. Due to the presence or lack of repeated sequence segments, the 5-HTTLPR locus has two common alleles, a short (S) and a long (L) allele, as well as an additional atypical allele (Lesch et al., 1996; Coutinho et al., 2004). Coutinho et al. proposed that variation in the 5-HTTLPR in the promoter region of SLC6A4, and variable number of tandem repeats (VNTRs) in intron 2 of the *SLC6A4* gene influence the amount of serotonin concentration in the circulation (Coutinho et al., 2004). A family-based association research in a Korean population demonstrated a genomic link between the long allele of 5-HTTLPR and *SLC6A4* gene polymorphisms with ASD (Cho et al., 2007). Various other studies by utilizing transmission disequilibrium test (TDT) further confirmed the positive correlation of 5-HTTLPR polymorphism in the promoter region of the *SLC6A4* gene with autism (Cook et al., 1997; McCauley et al., 2004; Sutcliffe et al., 2005; Anderson et al., 2009). Despite these positive outcomes, nine family-based studies and one case-control study found no sign of a linkage between *SLC6A4* polymorphism and autism (Klauck et al., 1997; Maestrini et al., 1999; Zhong et al., 1999; Persico et al., 2000; Tordjman et al., 2001; Wu et al., 2005; Koishi et al., 2006; Ramoz et al., 2006). Furthermore, an Indian group of researchers also carried out number of investigations but uncovered no conclusive proven link between *SLC6A4* polymorphism and autism (Guhathakurta et al., 2006, 2008, 2009).

Serotonin-reuptake transporter (SERT) is transiently expressed and controls serotonin signaling in specific sets of glutamatergic neurons and uptakes extra-synaptic serotonin during the perinatal CNS development. In 1998, Bengel et al. first created an *SLC6A4* knockout mouse model and, in their study, they recognized that levels of 5-HT decreased by 60–80% in the hippocampus, brainstem, striatum and frontal cortex regions, possibly due to the reduced synthesis and recycling of 5-HT (Bengel et al., 1998). In addition, such mice displayed a 50% decrease in number of serotonergic cells in the dorsal raphe nucleus (DRN) (Lira et al., 2003). Moreover, *SLC6A4* knockout mice exhibits amplified abnormal anxiety-like behaviors, exaggerated stress responses, and reduced aggression and motor strength (Holmes et al., 2003). The *SLC6A4* knock-in mouse, on the other hand, was shown to be hyperserotonemic, with disrupted CNS 5-HT network functioning and social and repetitive behavioral abnormalities (Veenstra-VanderWeele et al., 2012). Ellegood et al. have recently reported that an increase or decrease in SERT expression has a major influence on neuroanatomy in 5-HT related areas, notably the raphe nuclei region (Ellegood et al., 2018). To fully comprehend the relationship between 5-HTTLPR polymorphism and ASD risk, additional study with a larger sample size and homogenous autistic individuals is needed.

18.3.8 *SLC25A12* (Solute Carrier Family 25 Member 12) Gene

The *SLC25A12* gene is found on chromosome location 2q31.1, has 18 exons, and spans approximately 110 kb in length. The *SLC25A12* gene encodes mitochondrial aspartate/glutamate carrier isoform 1 (AGC1) protein, a calcium-binding solute carrier found in the inner mitochondrial membrane and mostly produced in the brain, heart and skeletal muscles (Del Arco & Satrústegui, 1998). AGC1 is a key protein transporter actively involved in the malate/aspartate shuttle and oxidative phosphorylation for the processing of ATP (Palmieri et al., 2001). Since the physiological functioning of neurons is highly dependent upon energy supply, any disruption in mitochondrial functioning or ATP production may trigger alterations in the neurons (Del Arco & Satrústegui, 1998). Several linkage-directed association studies and replications led to the discovery of *SLC25A12* as an autism susceptibility gene (Ramoz et al., 2004, 2008; Segurado et al., 2005; Turunen et al., 2008). In 2004, Ramoz et al. conducted mutation analysis of nine potential genes in the 2q31 chromosomal region. They discovered a significant link between autism and two SNPs in the third and sixteenth introns of the *SLC25A12* gene. These SNPs are described as rs2056202 (I3–21 C/T) and rs2292813 (I16 + 70 C/T), respectively (Ramoz et al., 2004). An independent study sample of 167 autistic trios provided further support for the above findings (Segurado et al., 2005) with the data demonstrating that such variations may contribute to the diversity of autism-related rituals and ceremonies linked with *SLC25A12* polymorphism (Silverman et al., 2008). Following that, multiple similar studies were directed to investigate the correlation between *SLC25A12* and the risk of ASD (Blasi et al., 2006; Correia et al., 2006;

Rabionet et al., 2006; Chien et al., 2010; Palmieri et al., 2010; Prandini et al., 2012; Durdiaková et al., 2014). The results, however, were contentious and ambiguous and such dramatic outcomes might be attributable to a combination of factors, including lack of resources, false positive results, and demographic stratification (Liu et al., 2015a; Aoki & Cortese, 2016).

The *SLC25A12* gene is transcribed in the developing brain, and its expression in the dorsolateral frontal cortex is 1.5 times greater in those with ASDs (Lepagnol-Bestel et al., 2008). In a postmortem study, Palmieri et al. reported an elevated activity of the mitochondrial aspartate/glutamate transporter protein as well as an increased expression of proteins in the superior temporal region of autistic persons relative to controls (Palmieri et al., 2010). In another study, Lepagnol-Bestel et al. also showed elevated *SLC25A12* transcript expression in the prefrontal cortex of autistic individuals; however, not in patients with schizophrenic or depressive-like conditions (Lepagnol-Bestel et al., 2008). Taken together, all these findings indicate that aberrant *SLC25A12* gene expression and activities may play a pivotal role in the pathogenesis of autism. Thus, to substantiate the role and importance of the *SLC25A12* gene in autism pathology as well as to define the precise predisposing allele variants from this particular gene sequence, present studies should include searching for risk allelic variants in neighboring regions using denser genetic markers, including all tagging SNPs and dense resequencing.

18.3.9 MAO (Monoamine Oxidase) Gene

The enzyme MAO was first discovered as tyramine oxidase by Hare in 1928. It belongs to a family of flavin monoamine oxidase, which catalyzes the oxidation of monoamines, such as dopamine, adrenaline, and tryptamine, and is bound to the outermost mitochondrial membrane in most cell types, especially in the histocyte of the liver, brain and kidney (Binda et al., 2011; Nunes et al., 2011). There are two types of monoamine oxidase i.e. MAO-A and MAO-B. *MAO-A* and *B* genes are located together on the short arm of the X chromosome, between chromosome bands Xp11.23 and Xp11.4, and consist of 15 exons with identical intron-exon organization (Huang et al., 2004; Shih & Chen, 2004). In brain, both MAO-A and B regulate the concentrations of important neurotransmitters, such as dopamine, noradrenaline and adrenaline, related to movement, emotion and cognition. Since abnormal levels of monoamine neurotransmitter are observed in ASD individuals, one of the causes that generate such a situation includes dysfunction of its metabolic pathway, thereby leading to the accumulation of monoamine neurotransmitters (Hranilović et al., 2008). There are five case-control correlation studies, indicating a positive effect of the MAO gene variants on ASD phenotypes (Yoo et al., 2009; Tassone et al., 2011; Salem et al., 2013; Verma et al., 2014; Chakraborti et al., 2016). However, two other family based studies were unable to show any substantial genetic effects (Philippe et al., 2002; Yirmiya et al., 2002).

Maternal genotypes have been found to act as modifiers that define the frequency of ASD symptoms such as IQ level in probands (Cohen et al., 2003; Jones et al., 2004). The upstream variable number tandem repeat (uVNTR) polymorphism within the MAO A promoter region have been studied in numerous psychiatric phenotypes, including autism (Cohen, 2004), and it has shown multiple structural brain abnormalities, including increased occurrence of macroencephaly, increased overall brain volume and enlargement of cerebral cortical white and gray matter (Piven et al., 1992; Meyer-Lindenberg & Weinberger, 2006). Multiple studies have shown that decreased *MAO-A* allele activity and increased cortical densities in children's brain may have been linked to lower IQ and more intense autistic manifestations (Davis et al., 2008; Cohen et al., 2011). Thus, *MAO-A* promoter polymorphism tends to reflect a broad influence on cerebral cortical regions in children with autism and various other neurodevelopmental conditions.

Behavioral anomalies in people with MAO-A and MAO-B combined impairment have been labeled as "autistic-like" on several instances (Lenders et al., 1996; Whibley et al., 2010). In a study, Bortolao et al. reported increased monoamine levels in the brain of MAO knockout mice. And both *MAO-A* and *MAO-B* knockout mice exhibited neuropathological changes similar to those seen in typical ASD individuals, such as decreased thickness of corpus callosum, excessive dendritic arborization of pyramidal neurons in the prefrontal cortex, and distorted cerebellar microarchitecture (Bortolato et al., 2013). Furthermore, by using array comparative genomic hybridization analysis Saito et al. demonstrated the deletions in the chromosomal regions of the monoamine oxidase genes. Loss of *MAO* resulted in autistic traits and symptoms indicating that both genes were candidate susceptibility genes for ASD (Saito et al., 2014). Recently, Chen et al. demonstrated that treatment of *MAO-A* knockout mice with the serotonin synthesis inhibitor (parachlorophenylalanine; pCPA) from postnatal day 1 to 7 restored serotonin levels and corrected repetitive behaviors (Chen et al., 2017). This indicates that modulation of *MAO* genes by increased serotonin level may result in neurodevelopmental alterations and also cause autistic-like behaviors. Further studies on MAO's function in brain development and behavioral activities may clear its involvement and significance in ASD.

18.3.10 ITGB3 (Integrin-β 3) Gene

Elevated serotonin (5-HT) blood levels were the first biomarker discovered in autism research, and they have been seen consistently in 20–30% of people with ASD (Schain & Freedman, 1961; Anderson et al., 1987; Cook & Leventhal, 1996; Veenstra-Vanderweele & Blakely, 2012). Genes encoding proteins involved in 5-HT neurotransmission and metabolism include the ITGB3 (Integrin-β 3) gene. The *ITGB3* gene is located on human chromosome 17q21.32, approximately 20 centimorgans (cM) distal to the *SLC6A4* gene. It is approximately 60 kilobases in length and has 15 exons. The encoded β-integrin is a component of the heterodimeric fibrinogen receptor that is exclusive to megakaryocytes and platelets, in addition to being a part of the heterodimeric vitronectin receptor that is widely expressed. In addition to this, it was discovered that integrin receptors participate in signal transduction activities, which have the potential to influence the processes of transcription and/or translation (Riikonen et al., 1995; Hapke et al., 2001). Through at least two separate gene variations positioned at the 5′ and at the 3′ ends of the gene, respectively as tagged by rs2317385 and rs12603582, *ITGB3* plays a vital role in both hyperserotonemia and in the risk of autism. This is because *ITGB3* is tagged by both of these variants (Napolioni et al., 2011). According to Napolioni et al. the presence of SNP at the 5′ terminal of the *ITGB3* gene is significantly correlated with elevated levels of 5-HT in plasma, one of the most prevalent characteristics in autistic individuals (275). In certain investigations, allelic variations of the ITGB3 gene have been discovered, either alone or interacting with allelic variants of the gene encoding for 5-HT transporter (SLC6A4) (Napolioni et al., 2011). In addition to that, Schuch et al. studied five different single nucleotide polymorphisms that were present in the *ITGB3* gene. They found a feeble link between the rs15908 and rs12603582 polymorphisms with epileptic symptoms, echolalia, and aggression, all of which are possible characteristics of individuals with ASD (Schuch et al., 2014). Recent research conducted by Gabriele and colleagues has shown that the presence of the "C" allele at the rs55827077 locus results in increased activity of the *ITGB3* gene promoter. In addition, this same allele is also strongly linked with increased 5-HT blood levels in autistic individuals (Gabriele et al., 2019). This suggested that the rs55827077 variation might be a functioning *ITGB3* gene promoter leading to elevated 5-HT blood levels in individuals with ASD condition. Therefore, identification of different *ITGB3* gene variants should now be the main objective of future studies to understand the *ITGB3* effects and their impacts and risks on such neurodevelopmental disorders.

18.4 Concluding Remarks

While it is true that our understanding of ASD has grown remarkably over the last few years, but, the genetic architecture of autism is still a challenge. And the data of various studies remains inconclusive for most of the candidate genes tested so far. From these studies, however, we can conclude that autism is not a monogenic condition with Mendelian-like inheritance, but a highly complex genetic disorder that results from genomic variations in hundreds and thousands of genes. Thus, it might possible that multiple factors and elements are involved in the regulation and modification of ASD dynamics, including genetic, epigenetic and environmental. Now the investigations have been carried out especially via using linkage, cytogenetic and GWAS studies to discover novel genes, it turns out to be more challenging and demanding to delineate the number of genes related with ASD and their molecular mechanisms and signaling pathways behind the contribution to ASD pathogenesis. Gene discovery methods have been dynamic in the discovery of rare *de novo* and inherited mutations in novel candidate genes, and next generation sequencing techniques along with gene expression profile studies are methodically adding to the existing autism knowledge and understanding. On this note, segregation of the spectrum into groups based on therapeutic effectiveness should also be explored. There is a potential possibility that genes and their protein products, responsible for distinct subsets of autism, could become better candidates for targeted clinical interventions. Furthermore, stratifying autism into genetic, molecular, clinical and statistical elements for analysis and the assessment of their collective role in influencing the disorder is extensive but necessary to resolve the gap between its genotypic origins and the phenotypic outcomes, and to establish diagnostic and therapeutic perspectives. Finally, genetic studies will also facilitate other fields of autism research, including neuroimaging and biochemical studies, which will, in addition, offer evidence and constructive inklings for future studies.

Data Availability Statement No data was used for the research described in the article.

Conflicts of Interest The authors declare no conflict of interest.

Funding This research did not receive any specific grant from funding agencies in the public, commercial, or not-for-profit sectors.

References

Abrahams, B. S., et al. (2013). SFARI Gene 2.0: A community-driven knowledgebase for the autism spectrum disorders (ASDs). *Molecular Autism, 4*(1), 36. https://doi.org/10.1186/2040-2392-4-36

Absil, P., et al. (2003). Effects of testosterone on Reelin expression in the brain of male European starlings. *Cell and Tissue Research, 312*(1), 81–93. https://doi.org/10.1007/s00441-003-0701-9

Amir, R. E., et al. (1999). Rett syndrome is caused by mutations in X-linked MECP2, encoding methyl- CpG-binding protein 2. *Nature Genetics, 23*(2), 185–188. https://doi.org/10.1038/13810

Anderson, G. M., et al. (1987). Whole blood serotonin in autistic and normal subjects. *Journal of Child Psychology and Psychiatry, 28*(6), 885–900. https://doi.org/10.1111/j.1469-7610.1987.tb00677.x

Anderson, B. M., et al. (2009). Examination of association of genes in the serotonin system to autism. *Neurogenetics, 10*(3), 209–216. https://doi.org/10.1007/s10048-009-0171-7

Anney, R., et al. (2010). A genome-wide scan for common alleles affecting risk for autism. *Human Molecular Genetics, 19*(20), 4072–4082. https://doi.org/10.1093/hmg/ddq307

Anney, R. J. L., Ripke, S., Anttila, V., & Jakob Grove, P. H. (2017). Meta-analysis of GWAS of over 16,000 individuals with autism spectrum disorder highlights a novel locus at 10q24.32 and a significant overlap with schizophrenia. *Molecular Autism, 8*, 21. https://doi.org/10.1186/s13229-017-0137-9

Aoki, Y., & Cortese, S. (2016). Mitochondrial aspartate/glutamate carrier SLC25A12 and autism spectrum disorder: A meta-analysis. *Molecular Neurobiology, 53*(3), 1579–1588. https://doi.org/10.1007/s12035-015-9116-3

Ashley-Koch, A., et al. (1999). Genetic studies of autistic disorder and chromosome 7. *Genomics, 61*(3), 227–236. https://doi.org/10.1006/geno.1999.5968

Ashley-Koch, A. E., et al. (2006). An analysis paradigm for investigating multi-locus effects in complex disease: Examination of three GABAA receptor subunit genes on 15q11-q13 as risk factors for autistic disorder. *Annals of Human Genetics, 70*(3), 281–292. https://doi.org/10.1111/j.1469-1809.2006.00253.x

Asperger, H. (2009). *'Autistic psychopathy' in childhood, Autism and Asperger syndrome*. Edited by U. Frith. Cambridge University Press. https://doi.org/10.1017/cbo9780511526770.002

Auranen, M., et al. (2002). A genomewide screen for autism-spectrum disorders: Evidence for a major susceptibility locus on chromosome 3q25-27. *American Journal of Human Genetics, 71*(4), 777–790. https://doi.org/10.1086/342720

Bailey, A., et al. (1998). A full genome screen for autism with evidence for linkage to a region on chromosome 7q. *Human Molecular Genetics, 7*(3), 571–578. https://doi.org/10.1093/hmg/7.3.571

Baio, J., et al. (2018). Prevalence of autism spectrum disorder among children aged 8 years - autism and developmental disabilities monitoring network, 11 sites, United States, 2014. *MMWR Surveillance Summaries, 67*(6), 1–23. https://doi.org/10.15585/mmwr.ss6706a1

Banerjee, A., et al. (2016). Overexpression of Homer1a in the basal and lateral amygdala impairs fear conditioning and induces an autism-like social impairment. *Molecular Autism, 7*(1), 16. https://doi.org/10.1186/s13229-016-0077-9

Baron-Cohen, S., Knickmeyer, R. C., & Belmonte, M. K. (2005). Sex differences in the brain: Implications for explaining autism. *Science, 310*, 819–823. https://doi.org/10.1126/science.1115455

Bartlett, C. W., et al. (2004). Examination of potential overlap in autism and language loci on chromosomes 2, 7, and 13 in two independent samples ascertained for specific language impairment. *Human Heredity, 57*(1), 10–20. https://doi.org/10.1159/000077385

Belligni, E. F., et al. (2012). 790 Kb microduplication in chromosome band 17p13.1 associated with intellectual disability, afebrile seizures, dysmorphic features, diabetes, and hypothyroidism. *European Journal of Medical Genetics, 55*(3), 222–224. https://doi.org/10.1016/j.ejmg.2012.01.016

Belmer, A., et al. (2017). Mapping the connectivity of serotonin transporter immunoreactive axons to excitatory and inhibitory neurochemical synapses in the mouse limbic brain. *Brain Structure and Function, 222*(3), 1297–1314. https://doi.org/10.1007/s00429-016-1278-x

Bengel, D., et al. (1998). Altered brain serotonin homeostasis and locomotor insensitivity to 3,4- methylenedioxymethamphetamine ('ecstasy') in serotonin transporter-deficient mice. *Molecular Pharmacology, 53*(4), 649–655. https://doi.org/10.1124/mol.53.4.649

Berkel, S., et al. (2010). Mutations in the SHANK2 synaptic scaffolding gene in autism spectrum disorder and mental retardation. *Nature Genetics, 42*(6), 489–491. https://doi.org/10.1038/ng.589

Biamonte, F., et al. (2009). Interactions between neuroactive steroids and reelin haploinsufficiency in Purkinje cell survival. *Neurobiology of Disease, 36*(1), 103–115. https://doi.org/10.1016/j.nbd.2009.07.001

Binda, C., Mattevi, A., & Edmondson, D. E. (2011). Structural properties of human monoamine oxidases A and B. *International Review of Neurobiology, 100*, 1–11. https://doi.org/10.1016/B978-0-12-386467-3.00001-7

Blasi, F., et al. (2006). SLC25A12 and CMYA3 gene variants are not associated with autism in the IMGSAC multiplex family sample. *European Journal of Human Genetics, 14*(1), 123–126. https://doi.org/10.1038/sj.ejhg.5201444

Blatt, G. J., et al. (2001). Density and distribution of hippocampal neurotransmitter receptors in autism: An autoradiographic study. *Journal of Autism and Developmental Disorders, 31*(6), 537–543. https://doi.org/10.1023/A:1013238809666

Blundell, J., et al. (2009). Increased anxiety-like behavior in mice lacking the inhibitory synapse cell adhesion molecule neuroligin 2. *Genes, Brain and Behavior, 8*(1), 114–126. https://doi.org/10.1111/j.1601-183X.2008.00455.x

Blundell, J., et al. (2010). Neuroligin-1 deletion results in impaired spatial memory and increased repetitive behavior. *Journal of Neuroscience, 30*(6), 2115–2129. https://doi.org/10.1523/JNEUROSCI.4517-09.2010

Böckers, T. M., et al. (2004). Differential expression and dendritic transcript localization of shank family members: Identification of a dendritic targeting element in the 3′ untranslated region of Shank1 mRNA. *Molecular and Cellular Neuroscience, 26*(1), 182–190. https://doi.org/10.1016/j.mcn.2004.01.009

Boeckers, T. M., et al. (1999). Proline-rich synapse-associated protein-1/cortactin binding protein 1 (ProSAP1/CortBP1) is a PDZ-domain protein highly enriched in the postsynaptic density. *Journal of Neuroscience, 19*(15), 6506–6518. https://doi.org/10.1523/jneurosci.19-15-06506.1999

Bolton, P. F., et al. (2004). Chromosome 15q11-13 abnormalities and other medical conditions in individuals with autism spectrum disorders. *Psychiatric Genetics, 14*(3), 131–137. https://doi.org/10.1097/00041444-200409000-00002

Bortolato, M., et al. (2013). Monoamine oxidase A and A/B knockout mice display autistic-like features. *International Journal of Neuropsychopharmacology, 16*(4), 869–888. https://doi.org/10.1017/S1461145712000715

Boyle, M. P., et al. (2011). Cell-type-specific consequences of reelin deficiency in the mouse neocortex, hippocampus, and amygdala. *Journal of Comparative Neurology, 519*(11), 2061–2089. https://doi.org/10.1002/cne.22655

Bremer, A., et al. (2011). Copy number variation characteristics in subpopulations of patients with autism spectrum disorders. *American Journal of Medical Genetics, Part B: Neuropsychiatric Genetics, 156*(2), 115–124. https://doi.org/10.1002/ajmg.b.31142

Buxbaum, J. D., et al. (2001). Evidence for a susceptibility gene for autism on chromosome 2 and for genetic heterogeneity. *American Journal of Human Genetics, 68*(6), 1514–1520. https://doi.org/10.1086/320588

Buxbaum, J. D., et al. (2002). Association between a GABRB3 polymorphism and autism. *Molecular Psychiatry, 7*(3), 311–316. https://doi.org/10.1038/sj.mp.4001011

Cai, G., et al. (2008). Multiplex ligation-dependent probe amplification for genetic screening in autism spectrum disorders: Efficient identification of known microduplications and identification of a novel microduplication in ASMT. *BMC Medical Genomics, 1*(1). https://doi.org/10.1186/1755-8794-1-50

Cai, X., et al. (2013). Local potentiation of excitatory synapses by serotonin and its alteration in rodent models of depression. *Nature Neuroscience, 16*(4), 464–472. https://doi.org/10.1038/nn.3355

Campbell, D. B., et al. (2006). A genetic variant that disrupts MET transcription is associated with autism. *Proceedings of the National Academy of Sciences of the United States of America, 103*(45), 16834–16839. https://doi.org/10.1073/pnas.0605296103

Campbell, D. B., et al. (2007). Disruption of cerebral cortex MET signaling in autism spectrum disorder. *Annals of Neurology, 62*(3), 243–250. https://doi.org/10.1002/ana.21180

Campbell, D. B., et al. (2008). Genetic evidence implicating multiple genes in the met receptor tyrosine kinase pathway in autism spectrum disorder. *Autism Research, 1*(3), 159–168. https://doi.org/10.1002/aur.27

Campbell, D. B., et al. (2010). Association of MεT with social and communication phenotypes in individuals with autism spectrum disorder. *American Journal of Medical Genetics, Part B: Neuropsychiatric Genetics, 153*(2), 438–446. https://doi.org/10.1002/ajmg.b.30998

Canitano, R. (2007). Epilepsy in autism spectrum disorders. *European Child and Adolescent Psychiatry, 16*(1), 61–66. https://doi.org/10.1007/s00787-006-0563-2

Cantor, R. M., et al. (2005). Replication of autism linkage: Fine-mapping peak at 17q21. *American Journal of Human Genetics, 76*(6), 1050–1056. https://doi.org/10.1086/430278

Castermans, D., et al. (2004). Chromosomal anomalies in individuals with autism: A strategy towards the identification of genes involved in autism. *Autism, 8*(2), 141–161. https://doi.org/10.1177/1362361304042719

Chakraborti, B., et al. (2016). Genetic variants of MAOB affect serotonin level and specific behavioral attributes to increase autism spectrum disorder (ASD) susceptibility in males. *Progress in*

Neuro-Psychopharmacology and Biological Psychiatry, 71, 123–136. https://doi.org/10.1016/j.pnpbp.2016.07.001

Chaudhary, R., et al. (2021). Involvement of myocyte enhancer factor 2c in the pathogenesis of autism spectrum disorder. *Heliyon, 7*(4), e06854. https://doi.org/10.1016/j.heliyon.2021.e06854

Chen, K., et al. (2017). Altered gene expression in early postnatal monoamine oxidase A knockout mice. *Brain Research, 1669*, 18–26. https://doi.org/10.1016/j.brainres.2017.05.017

Chien, W. H., et al. (2010). Association study of the SLC25A12 gene and autism in Han Chinese in Taiwan. *Progress in Neuro-Psychopharmacology and Biological Psychiatry, 34*(1), 189–192. https://doi.org/10.1016/j.pnpbp.2009.11.004

Cho, I. H., et al. (2007). Family-based association study of 5-HTTLPR and the 5-HT2A receptor gene polymorphisms with autism spectrum disorder in Korean trios. *Brain Research, 1139*(1), 34–41. https://doi.org/10.1016/j.brainres.2007.01.002

Christian, S. L., et al. (2008). Novel submicroscopic chromosomal abnormalities detected in autism spectrum disorder. *Biological Psychiatry, 63*(12), 1111–1117. https://doi.org/10.1016/j.biopsych.2008.01.009

Chubykin, A. A., et al. (2005). Dissection of synapse induction by neuroligins: Effect of a neuroligin mutation associated with autism. *Journal of Biological Chemistry, 280*(23), 22365–22374. https://doi.org/10.1074/jbc.M410723200

Ciranna, L. (2006). Serotonin as a modulator of glutamate- and GABA-mediated neurotransmission: Implications in physiological functions and in pathology. *Current Neuropharmacology, 4*(2), 101–114. https://doi.org/10.2174/157015906776359540

Cohen, B. I. (2004). Rationale for further investigation of chromosome 16p 13.3, a region implicated for autism. *Autism, 8*(4), 445–447. https://doi.org/10.1177/1362361304047226

Cohen, I. L., et al. (2003). Association of autism severity with a monoamine oxidase a functional polymorphism. *Clinical Genetics, 64*(3), 190–197. https://doi.org/10.1034/j.1399-0004.2003.00115.x

Cohen, I. L., et al. (2011). Autism severity is associated with child and maternal MAOA genotypes. *Clinical Genetics, 79*(4), 355–362. https://doi.org/10.1111/j.1399-0004.2010.01471.x

Colvert, E., et al. (2015). Heritability of autism spectrum disorder in a UK population-based twin sample. *JAMA Psychiatry, 72*(5), 415–423. https://doi.org/10.1001/jamapsychiatry.2014.3028

Constantino, J. N. (2013). Social responsiveness scale. In F. R. Volkmar (Ed.), *Encyclopedia of autism spectrum disorders* (pp. 2919–2929). New York, NY. https://doi.org/10.1007/978-1-4419-1698-3_296

Cook, E. H. (1998). Genetics of autism. *Mental Retardation and Developmental Disabilities Research Reviews, 4*(2), 113–120. https://doi.org/10.1002/(SICI)1098-2779(1998)4:2<113::AID-MRDD8>3.0.CO;2-Q

Cook, E. H., & Leventhal, B. L. (1996). The serotonin system in autism. *Current Opinion in Pediatrics, 8*(4), 348–354. https://doi.org/10.1097/00008480-199608000-00008

Cook, E. H., et al. (1997). Evidence of linkage between the serotonin transporter and autistic disorder. *Molecular Psychiatry, 2*(3), 247–250. https://doi.org/10.1038/sj.mp.4000266

Cook, E. H., et al. (1998). Linkage-disequilibrium mapping of autistic disorder, with 15q11-13 markers. *American Journal of Human Genetics, 62*(5), 1077–1083. https://doi.org/10.1086/301832

Correia, C., et al. (2006). Brief report: High frequency of biochemical markers for mitochondrial dysfunction in autism: No association with the mitochondrial aspartate/glutamate carrier SLC25A12 gene. *Journal of Autism and Developmental Disorders, 36*(8), 1137–1140. https://doi.org/10.1007/s10803-006-0138-6

Coutinho, A. M., et al. (2004). Variants of the serotonin transporter gene (SLC6A4) significantly contribute to hyperserotonemia in autism. *Molecular Psychiatry, 9*(3), 264–271. https://doi.org/10.1038/sj.mp.4001409

D'Arcangelo, G., et al. (1995). A protein related to extracellular matrix proteins deleted in the mouse mutant reeler. *Nature, 374*(6524), 719–723. https://doi.org/10.1038/374719a0

Davis, L. K., et al. (2008). Cortical enlargement in autism is associated with a functional VNTR in the monoamine oxidase a gene. *American Journal of Medical Genetics, Part B: Neuropsychiatric Genetics, 147*(7), 1145–1151. https://doi.org/10.1002/ajmg.b.30738

De Rubeis, S., et al. (2014). Synaptic, transcriptional and chromatin genes disrupted in autism. *Nature, 515*(7526), 209–215. https://doi.org/10.1038/nature13772

Del Arco, A., & Satrústegui, J. (1998). Molecular cloning of aralar, a new member of the mitochondrial carrier superfamily that binds calcium and is present in human muscle and brain. *Journal of Biological Chemistry, 273*(36), 23327–23334. https://doi.org/10.1074/jbc.273.36.23327

Del Barrio, V. (2016). *Diagnostic and statistical manual of mental disorders, the curated reference collection in neuroscience and biobehavioral psychology*. American Psychiatric Publishing. https://doi.org/10.1016/B978-0-12-809324-5.05530-9

Delahanty, R. J., et al. (2011). Maternal transmission of a rare GABRB3 signal peptide variant is associated with autism. *Molecular Psychiatry, 16*(1), 86–96. https://doi.org/10.1038/mp.2009.118

Depienne, C., et al. (2009). Screening for genomic rearrangements and methylation abnormalities of the 15q11-q13 region in autism spectrum disorders. *Biological Psychiatry, 66*(4), 349–359. https://doi.org/10.1016/j.biopsych.2009.01.025

Devlin, B., & Scherer, S. W. (2012). Genetic architecture in autism spectrum disorder. *Current Opinion in Genetics and Development, 22*, 229–237. https://doi.org/10.1016/j.gde.2012.03.002

Durand, C. M., et al. (2007). Mutations in the gene encoding the synaptic scaffolding protein SHANK3 are associated with autism spectrum disorders. *Nature Genetics, 39*(1), 25–27. https://doi.org/10.1038/ng1933

Durdiaková, J., et al. (2014). Single nucleotide polymorphism rs6716901 in SLC25A12 gene is associated with Asperger syndrome. *Molecular Autism, 5*(1). https://doi.org/10.1186/2040-2392-5-25

Eagleson, K. L., et al. (2011). The autism risk genes MET and PLAUR differentially impact cortical development. *Autism Research, 4*(1), 68–83. https://doi.org/10.1002/aur.172

Ebens, A., et al. (1996). Hepatocyte growth factor/scatter factor is an axonal chemoattractant and a neurotrophic factor for spinal motor neurons. *Neuron, 17*(6), 1157–1172. https://doi.org/10.1016/S0896-6273(00)80247-0

Ellegood, J., et al. (2018). Analysis of neuroanatomical differences in mice with genetically modified serotonin transporters assessed by structural magnetic resonance imaging. *Molecular Autism, 9*(1). https://doi.org/10.1186/s13229-018-0210-z

Etherton, M., et al. (2011a). Autism-linked neuroligin-3 R451C mutation differentially alters hippocampal and cortical synaptic function. *Proceedings of the National Academy of Sciences of the United States of America, 108*(33), 13764–13769. https://doi.org/10.1073/pnas.1111093108

Etherton, M. R., et al. (2011b). An autism-associated point mutation in the neuroligin cytoplasmic tail selectively impairs AMPA receptor-mediated synaptic transmission in hippocampus. *EMBO Journal, 30*(14), 2908–2919. https://doi.org/10.1038/emboj.2011.182

Fatemi, S. H., et al. (2001). Dysregulation of reelin and Bcl-2 proteins in autistic cerebellum. *Journal of Autism and Developmental Disorders, 31*(6), 529–535. https://doi.org/10.1023/A:1013234708757

Fatemi, S. H., et al. (2005). Reelin signaling is impaired in autism. *Biological Psychiatry, 57*(7), 777–787. https://doi.org/10.1016/j.biopsych.2004.12.018

Fatemi, S. H., et al. (2009). GABAA receptor downregulation in brains of subjects with autism. *Journal of Autism and Developmental Disorders, 39*(2), 223–230. https://doi.org/10.1007/s10803-008-0646-7

Felix-Ortiz, A. C., & Tye, K. M. (2014). Amygdala inputs to the ventral hippocampus bidirectionally modulate social behavior. *Journal of Neuroscience, 34*(2), 586–595. https://doi.org/10.1523/JNEUROSCI.4257-13.2014

Földy, C., Malenka, R. C., & Südhof, T. C. (2013). Autism-associated neuroligin-3 mutations commonly disrupt tonic endocannabinoid signaling. *Neuron, 78*(3), 498–509. https://doi.org/10.1016/j.neuron.2013.02.036

Folstein, S., & Rutter, M. (1977). Infantile autism: A genetic study of 21 twin pairs. *Journal of Child Psychology and Psychiatry, 18*(4), 297–321. https://doi.org/10.1111/j.1469-7610.1977.tb00443.x

Francis, S. M., et al. (2016). ASD and genetic associations with receptors for oxytocin and vasopressin-AVPR1A, AVPR1B, and OXTR. *Frontiers in Neuroscience, 10*(NOV). https://doi.org/10.3389/fnins.2016.00516

Frazier, T. W., et al. (2014). A twin study of heritable and shared environmental contributions to autism. *Journal of Autism and Developmental Disorders, 44*(8), 2013–2025. https://doi.org/10.1007/s10803-014-2081-2

Furman, D. J., Chen, M. C., & Gotlib, I. H. (2011). Variant in oxytocin receptor gene is associated with amygdala volume. *Psychoneuroendocrinology, 36*(6), 891–897. https://doi.org/10.1016/j.psyneuen.2010.12.004

Gabriele, S., et al. (2019). Evidence that ITGB3 promoter variants increase serotonin blood levels by regulating platelet serotonin transporter trafficking. *Human Molecular Genetics, 28*(7), 1153–1161. https://doi.org/10.1093/hmg/ddy421

Gaetz, W., et al. (2014). GABA estimation in the brains of children on the autism spectrum: Measurement precision and regional cortical variation. *NeuroImage, 86*, 1–9. https://doi.org/10.1016/j.neuroimage.2013.05.068

Garbarino, V. R., et al. (2019). Extreme enhancement or depletion of serotonin transporter function and serotonin availability in autism spectrum disorder. *Pharmacological Research, 140*, 85–99. https://doi.org/10.1016/j.phrs.2018.07.010

Gauthier, J., et al. (2005). NLGN3/NLGN4 gene mutations are not responsible for autism in the Quebec population. *American Journal of Medical Genetics - Neuropsychiatric Genetics, 132 B*(1), 74–75. https://doi.org/10.1002/ajmg.b.30066

Geschwind, D. H. (2011). Genetics of autism spectrum disorders. *Trends in Cognitive Sciences, 15*, 409–416. https://doi.org/10.1016/j.tics.2011.07.003

Glessner, J. T., et al. (2009). Autism genome-wide copy number variation reveals ubiquitin and neuronal genes. *Nature, 459*(7246), 569–572. https://doi.org/10.1038/nature07953

Goffinet, A. M. (1983a). The embryonic development of the cerebellum in normal and reeler mutant mice. *Anatomy and Embryology, 168*(1), 73–86. https://doi.org/10.1007/BF00305400

Goffinet, A. M. (1983b). The embryonic development of the inferior olivary complex in normal and reeler (rlORL) mutant mice. *Journal of Comparative Neurology, 219*(1), 10–24. https://doi.org/10.1002/cne.902190103

Goffinet, A. M. (1984). Events governing organization of postmigratory neurons: Studies on brain development in normal and reeler mice. *Brain Research Reviews, 7*, 261–296. https://doi.org/10.1016/0165-0173(84)90013-4

Gouin, J. P., et al. (2017). Associations among oxytocin receptor gene (OXTR) DNA methylation in adulthood, exposure to early life adversity, and childhood trajectories of anxiousness. *Scientific Reports, 7*(1), 7446. https://doi.org/10.1038/s41598-017-07950-x

Green, J. J., & Hollander, E. (2010). Autism and oxytocin: New developments in translational approaches to therapeutics. *Neurotherapeutics, 7*(3), 250–257. https://doi.org/10.1016/j.nurt.2010.05.006

Gregory, S. G., et al. (2009). Genomic and epigenetic evidence for oxytocin receptor deficiency in autism. *BMC Medicine, 7*. https://doi.org/10.1186/1741-7015-7-62

Grønborg, T. K., Schendel, D. E., & Parner, E. T. (2013). Recurrence of autism spectrum disorders in full- and half-siblings and trends over time: A population-based cohort study. *JAMA Pediatrics, 167*(10), 947–953. https://doi.org/10.1001/jamapediatrics.2013.2259

Grove, J., et al. (2019). Identification of common genetic risk variants for autism spectrum disorder. *Nature Genetics, 51*(3), 431–444. https://doi.org/10.1038/s41588-019-0344-8

Guhathakurta, S., et al. (2006). Serotonin transporter promoter variants: Analysis in Indian autistic and control population. *Brain Research, 1092*(1), 28–35. https://doi.org/10.1016/j.brainres.2006.03.078

Guhathakurta, S., et al. (2008). Population-based association study and contrasting linkage disequilibrium pattern reveal genetic association of SLC6A4 with autism in the Indian population from West Bengal. *Brain Research, 1240*, 12–21. https://doi.org/10.1016/j.brainres.2008.08.063

Guhathakurta, S., et al. (2009). Analysis of serotonin receptor 2A gene (HTR2A): Association study with autism spectrum disorder in the Indian population and investigation of the gene expression in peripheral blood leukocytes. *Neurochemistry International, 55*(8), 754–759. https://doi.org/10.1016/j.neuint.2009.07.008

Ha, S., et al. (2015). Characteristics of brains in autism spectrum disorder: Structure, function and connectivity across the lifespan. *Experimental Neurobiology, 24*(4), 273–284. https://doi.org/10.5607/en.2015.24.4.273

Ha, S., et al. (2016). Cerebellar shank2 regulates excitatory synapse density, motor coordination, and specific repetitive and anxiety-like behaviors. *Journal of Neuroscience, 36*(48), 12129–12143. https://doi.org/10.1523/JNEUROSCI.1849-16.2016

Haar, S., et al. (2016). Anatomical abnormalities in autism? *Cerebral Cortex, 26*(4), 1440–1452. https://doi.org/10.1093/cercor/bhu242

Hallmayer, J., et al. (2011). Genetic heritability and shared environmental factors among twin pairs with autism. *Archives of General Psychiatry, 68*(11), 1095–1102. https://doi.org/10.1001/archgenpsychiatry.2011.76

Hapke, S., et al. (2001). beta(3)A-integrin downregulates the urokinase-type plasminogen activator receptor (u-PAR) through a PEA3/ets transcriptional silencing element in the u-PAR promoter. *Molecular and Cellular Biology, 21*(6), 2118–2132. https://doi.org/10.1128/MCB.21.6.2118-2132.2001

Harada, M., et al. (2011). Non-invasive evaluation of the GABAergic/glutamatergic system in autistic patients observed by MEGA-editing proton MR spectroscopy using a clinical 3 tesla instrument. *Journal of Autism and Developmental Disorders, 41*(4), 447–454. https://doi.org/10.1007/s10803-010-1065-0

Hashemi, E., et al. (2018). Erratum: The number of parvalbumin-expressing interneurons is decreased in the prefrontal cortex in autism (Cerebral Cortex (2017) 27 (1931–1943) DOI: 10.1093/cercor/bhw021). *Cerebral Cortex, 28*(2), 690. https://doi.org/10.1093/cercor/bhx063

Heils, A., et al. (1996). Allelic variation of human serotonin transporter gene expression. *Journal of Neurochemistry, 66*(6), 2621–2624. https://doi.org/10.1046/j.1471-4159.1996.66062621.x

Holmes, A., et al. (2003). Abnormal anxiety-related behavior in serotonin transporter null mutant mice: The influence of genetic background. *Genes, Brain and Behavior, 2*(6), 365–380. https://doi.org/10.1046/j.1601-1848.2003.00050.x

Hranilović, D., et al. (2008). Hyperserotonemia in autism: The potential role of 5HT-related gene variants. *Collegium Antropologicum, 32*(SUPPL. 1), 75–80.

Huang, Y. Y., et al. (2004). An association between a functional polymorphism in the monoamine oxidase a gene promoter, impulsive traits and early abuse experiences. *Neuropsychopharmacology, 29*(8), 1498–1505. https://doi.org/10.1038/sj.npp.1300455

Ichtchenko, K., Nguyen, T., & Südhof, T. C. (1996). Structures, alternative splicing, and neurexin binding of multiple neuroligins. *Journal of Biological Chemistry, 271*(5), 2676–2682. https://doi.org/10.1074/jbc.271.5.2676

Inoue, T., et al. (1994). Structural organization of the human oxytocin receptor gene. *Journal of Biological Chemistry, 269*(51), 32451–32456. https://doi.org/10.1016/s0021-9258(18)31656-9

Inoue, H., et al. (2010). Association between the oxytocin receptor gene and amygdalar volume in healthy adults. *Biological Psychiatry, 68*(11), 1066–1072. https://doi.org/10.1016/j.biopsych.2010.07.019

Jackson, P. B., et al. (2009). Further evidence that the rs1858830 C variant in the promoter region of the MET gene is associated with autistic disorder. *Autism Research, 2*(4), 232–236. https://doi.org/10.1002/aur.87

Jacob, S., et al. (2007). Association of the oxytocin receptor gene (OXTR) in Caucasian children and adolescents with autism. *Neuroscience Letters, 417*(1), 6–9. https://doi.org/10.1016/j.neulet.2007.02.001

Jamain, S., et al. (2003). Mutations of the X-linked genes encoding neuroligins NLGN3 and NLGN4 are associated with autism. *Nature Genetics, 34*(1), 27–29. https://doi.org/10.1038/ng1136

Jamain, S., et al. (2008). Reduced social interaction and ultrasonic communication in a mouse model of monogenic heritable autism. *Proceedings of the National Academy of Sciences of the United States of America, 105*(5), 1710–1715. https://doi.org/10.1073/pnas.0711555105

Jiang, Y. h., & Ehlers, M. D. (2013). Modeling autism by SHANK gene mutations in mice. *Neuron, 78*, 8–27. https://doi.org/10.1016/j.neuron.2013.03.016

Jones, M. B., et al. (2004). Modifier effects in autism at the MAO-A and DBH loci. *American Journal of Medical Genetics – Neuropsychiatric Genetics, 126 B*(1), 58–65. https://doi.org/10.1002/ajmg.b.20172

Judson, M. C., et al. (2009). Dynamic gene and protein expression patterns of the autism-associated met receptor tyrosine kinase in the developing mouse forebrain. *Journal of Comparative Neurology, 513*(5), 511–531. https://doi.org/10.1002/cne.21969

Judson, M. C., Amaral, D. G., & Levitt, P. (2011). Conserved subcortical and divergent cortical expression of proteins encoded by orthologs of the autism risk gene MET. *Cerebral Cortex, 21*(7), 1613–1626. https://doi.org/10.1093/cercor/bhq223

Kanner, L. (1968). Autistic disturbances of affective contact. *Acta Paedopsychiatrica, 35*(4), 100–136.

Kenna, G. A., et al. (2012). Association of the 5-HTT gene-linked promoter region (5-HTTLPR) polymorphism with psychiatric disorders: Review of psychopathology and pharmacotherapy. *Pharmacogenomics and Personalized Medicine, 5*, 19–35. https://doi.org/10.2147/PGPM.S23462

Kent, W. J., et al. (2002). The human genome browser at UCSC. *Genome Research, 12*(6), 996–1006. https://doi.org/10.1101/gr.229102

Klauck, S. M., et al. (1997). Serotonin transporter (5-HTT) gene variants associated with autism? *Human Molecular Genetics, 6*(13), 2233–2238. https://doi.org/10.1093/hmg/6.13.2233

Koishi, S., et al. (2006). Serotonin transporter gene promoter polymorphism and autism: A family-based genetic association study in Japanese population. *Brain and Development, 28*(4), 257–260. https://doi.org/10.1016/j.braindev.2005.09.003

Kranz, T. M., et al. (2016). Meta-analysis and association of two common polymorphisms of the human oxytocin receptor gene in autism spectrum disorder. *Autism Research, 9*(10), 1036–1045. https://doi.org/10.1002/aur.1597

Kuo, P. H., et al. (2015). Genome-wide association study for autism spectrum disorder in taiwanese han population. *PLoS One, 10*(9), 1–15. https://doi.org/10.1371/journal.pone.0138695

Kusui, C., et al. (2001). DNA methylation of the human oxytocin receptor gene promoter regulates tissue-specific gene suppression. *Biochemical and Biophysical Research Communications, 289*(3), 681–686. https://doi.org/10.1006/bbrc.2001.6024

Kwasnicka-Crawford, D. A., Roberts, W., & Scherer, S. W. (2007). Characterization of an autism-associated segmental maternal heterodisomy of the chromosome 15q11-13 region. *Journal of Autism and Developmental Disorders, 37*(4), 694–702. https://doi.org/10.1007/s10803-006-0225-8

Lammert, D. B., & Howell, B. W. (2016). RELN mutations in autism spectrum disorder. *Frontiers in Cellular Neuroscience, 10*. https://doi.org/10.3389/fncel.2016.00084

Lawson-Yuen, A., et al. (2008). Familial deletion within NLGN4 associated with autism and Tourette syndrome. *European Journal of Human Genetics, 16*(5), 614–618. https://doi.org/10.1038/sj.ejhg.5202006

Le Couteur, A. L., et al. (1995). Autism as a strongly genetic disorder evidence from a british twin study. *Psychological Medicine, 25*(1), 63–77. https://doi.org/10.1017/S0033291700028099

Leblond, C. S., et al. (2014). Meta-analysis of SHANK mutations in autism spectrum disorders: A gradient of severity in cognitive impairments. *PLoS Genetics, 10*(9), e1004580. https://doi.org/10.1371/journal.pgen.1004580

Lenders, J. W. M., et al. (1996). Specific genetic deficiencies of the A and B isoenzymes of monoamine oxidase are characterized by distinct neurochemical and clinical phenotypes. *Journal of Clinical Investigation, 97*(4), 1010–1019. https://doi.org/10.1172/JCI118492
Lepagnol-Bestel, A. M., et al. (2008). SLC25A12 expression is associated with neurite outgrowth and is upregulated in the prefrontal cortex of autistic subjects. *Molecular Psychiatry, 13*(4), 385–397. https://doi.org/10.1038/sj.mp.4002120
Lerer, E., et al. (2008). Association between the oxytocin receptor (OXTR) gene and autism: Relationship to Vineland Adaptive Behavior Scales and cognition. *Molecular Psychiatry, 13*(10), 980–988. https://doi.org/10.1038/sj.mp.4002087
Lesch, K. P., et al. (1996). Association of anxiety-related traits with a polymorphism in the serotonin transporter gene regulatory region. *Science, 274*(5292), 1527–1531. https://doi.org/10.1126/science.274.5292.1527
Li, N., et al. (2011). Prevalence of autism-caused disability among Chinese children: A national population-based survey. *Epilepsy and Behavior, 22*(4), 786–789. https://doi.org/10.1016/j.yebeh.2011.10.002
Lim, S., et al. (1999). Characterization of the Shank family of synaptic proteins. Multiple genes, alternative splicing, and differential expression in brain and development. *Journal of Biological Chemistry, 274*(41), 29510–29518. https://doi.org/10.1074/jbc.274.41.29510
Lira, A., et al. (2003). Altered depression-related behaviors and functional changes in the dorsal raphe nucleus of serotonin transporter-deficient mice. *Biological Psychiatry, 54*(10), 960–971. https://doi.org/10.1016/S0006-3223(03)00696-6
Liu, Y. (1998). The human hepatocyte growth factor receptor gene: Complete structural organization and promoter characterization. *Gene, 215*(1), 159–169. https://doi.org/10.1016/S0378-1119(98)00264-9
Liu, J., et al. (2001). A genomewide screen for autism susceptibility loci. *American Journal of Human Genetics, 69*(2), 327–340. https://doi.org/10.1086/321980
Liu, X., et al. (2010). Association of the oxytocin receptor (OXTR) gene polymorphisms with autism spectrum disorder (ASD) in the Japanese population. *Journal of Human Genetics, 55*(3), 137–141. https://doi.org/10.1038/jhg.2009.140
Liu, J., et al. (2015a). Association between genetic variants in SLC25A12 and risk of autism spectrum disorders: An integrated meta-analysis. *American Journal of Medical Genetics, Part B: Neuropsychiatric Genetics, 168*(4), 236–246. https://doi.org/10.1002/ajmg.b.32304
Liu, X., et al. (2015b). Novel rare variations of the oxytocin receptor (OXTR) gene in autism spectrum disorder individuals. *Human Genome Variation, 2*(1), 15024. https://doi.org/10.1038/hgv.2015.24
Liu, W., Li, M., & Yi, L. (2016). Identifying children with autism spectrum disorder based on their face processing abnormality: A machine learning framework. *Autism Research, 9*(8), 888–898. https://doi.org/10.1002/aur.1615
LoParo, D., & Waldman, I. D. (2015). The oxytocin receptor gene (OXTR) is associated with autism spectrum disorder: A meta-analysis. *Molecular Psychiatry, 20*(5), 640–646. https://doi.org/10.1038/mp.2014.77
Lord, C., et al. (2020). Autism spectrum disorder. *Nature Reviews Disease Primers, 6*(1), 5. https://doi.org/10.1038/s41572-019-0138-4
Lowe, J. K., et al. (2015). Social responsiveness, an autism endophenotype: Genomewide significant linkage to two regions on chromosome 8. *American Journal of Psychiatry, 172*(3), 266–275. https://doi.org/10.1176/appi.ajp.2014.14050576
Ma, D. Q., et al. (2005). Identification of significant association and gene-gene interaction of GABA receptor subunit genes in autism. *American Journal of Human Genetics, 77*(3), 377–388. https://doi.org/10.1086/433195
MacNeil, B. M., Lopes, V. A., & Minnes, P. M. (2009). Anxiety in children and adolescents with Autism Spectrum Disorders. *Research in Autism Spectrum Disorders, 3*(1), 1–21. https://doi.org/10.1016/j.rasd.2008.06.001

Macrì, S., et al. (2010). Perseverative responding and neuroanatomical alterations in adult heterozygous reeler mice are mitigated by neonatal estrogen administration. *Psychoneuroendocrinology, 35*(9), 1374–1387. https://doi.org/10.1016/j.psyneuen.2010.03.012

Maenner, M. J., et al. (2021). Prevalence and characteristics of autism spectrum disorder among children aged 8 years - autism and developmental disabilities monitoring network, 11 sites, United States, 2018. *Morbidity and Mortality Weekly Report. Surveillance Summaries (Washington, D.C. : 2002), 70*(11), 1–16. https://doi.org/10.15585/mmwr.ss7011a1

Maestrini, E., et al. (1999). Serotonin transporter (5-HTT) and γ-aminobutyric acid receptor subunit β3 (GABRB3) gene polymorphisms are not associated with autism in the IMGSA families. *American Journal of Medical Genetics - Neuropsychiatric Genetics, 88*(5), 492–496. https://doi.org/10.1002/(SICI)1096-8628(19991015)88:5<492::AID-AJMG11>3.0.CO;2-X

Maina, F., et al. (1997). Met receptor signaling is required for sensory nerve development and HGF promotes axonal growth and survival of sensory neurons. *Genes and Development, 11*(24), 3341–3350. https://doi.org/10.1101/gad.11.24.3341

Mao, W., et al. (2015). Shank1 regulates excitatory synaptic transmission in mouse hippocampal parvalbumin-expressing inhibitory interneurons. *European Journal of Neuroscience, 41*(8), 1025–1035. https://doi.org/10.1111/ejn.12877

Marshall, C. R., et al. (2008). Structural variation of chromosomes in autism spectrum disorder. *American Journal of Human Genetics, 82*(2), 477–488. https://doi.org/10.1016/j.ajhg.2007.12.009

Martin, E. R., et al. (2000). Analysis of linkage disequilibrium in γ-aminobutyric acid receptor subunit genes in autistic disorder. *American Journal of Medical Genetics - Neuropsychiatric Genetics, 96*(1), 43–48. https://doi.org/10.1002/(SICI)1096-8628(20000207)96:1<43::AID-AJMG9>3.0.CO;2-3

Matsunami, N., et al. (2013). Identification of rare recurrent copy number variants in high-risk autism families and their prevalence in a large ASD population. *PLoS One, 8*(1), e52239. https://doi.org/10.1371/journal.pone.0052239

McCauley, J. L., et al. (2004). Linkage and association analysis at the serotonin transporter (SLC6A4) locus in a rigid-compulsive subset of autism. *American Journal of Medical Genetics - Neuropsychiatric Genetics, 127 B*(1), 104–112. https://doi.org/10.1002/ajmg.b.20151

Menold, M. M., et al. (2001). Association analysis of chromosome 15 GABAA receptor subunit genes in autistic disorder. *Journal of Neurogenetics, 15*(42067), 245–259. https://doi.org/10.3109/01677060109167380

Meyer-Lindenberg, A., & Weinberger, D. R. (2006). Intermediate phenotypes and genetic mechanisms of psychiatric disorders. *Nature Reviews Neuroscience, 7*(10), 818–827. https://doi.org/10.1038/nrn1993

Meyer-Lindenberg, A., et al. (2011). Oxytocin and vasopressin in the human brain: Social neuropeptides for translational medicine. *Nature Reviews Neuroscience, 12*, 524–538. https://doi.org/10.1038/nrn3044

Millson, A., et al. (2012). Chromosomal loss of 3q26.3-3q26.32, involving a partial neuroligin 1 deletion, identified by genomic microarray in a child with microcephaly, seizure disorder, and severe intellectual disability. *American Journal of Medical Genetics, Part A, 158 A*(1), 159–165. https://doi.org/10.1002/ajmg.a.34349

Moessner, R., et al. (2007). Contribution of SHANK3 mutations to autism spectrum disorder. *American Journal of Human Genetics, 81*(6), 1289–1297. https://doi.org/10.1086/522590

Moreno-De-Luca, D., et al. (2013). Using large clinical data sets to infer pathogenicity for rare copy number variants in autism cohorts. *Molecular Psychiatry, 18*(10), 1090–1095. https://doi.org/10.1038/mp.2012.138

Muhle, R., Trentacoste, S. V., & Rapin, I. (2004). The genetics of autism. *Pediatrics, 113*, e472–e486. https://doi.org/10.1542/peds.113.5.e472

Mukame, Z., et al. (2011). Regulation of MET by FOXP2, genes implicated in higher cognitive dysfunction and autism risk. *Journal of Neuroscience, 31*(32), 11437–11442. https://doi.org/10.1523/JNEUROSCI.0181-11.2011

Muller, C. L., Anacker, A. M. J., & Veenstra-VanderWeele, J. (2016). The serotonin system in autism spectrum disorder: From biomarker to animal models. *Neuroscience, 321*, 24–41. https://doi.org/10.1016/j.neuroscience.2015.11.010
Naisbitt, S., et al. (1999). Shank, a novel family of postsynaptic density proteins that binds to the NMDA receptor/PSD-95/GKAP complex and cortactin. *Neuron, 23*(3), 569–582. https://doi.org/10.1016/S0896-6273(00)80809-0
Nakamura, T., et al. (2016). PX-RICS-deficient mice mimic autism spectrum disorder in Jacobsen syndrome through impaired GABA A receptor trafficking. *Nature Communications, 7*, 10861. https://doi.org/10.1038/ncomms10861
Nakatani, J., et al. (2009). Abnormal behavior in a chromosome- engineered mouse model for human 15q11-13 duplication seen in autism. *Cell, 137*(7), 1235–1246. https://doi.org/10.1016/j.cell.2009.04.024
Napolioni, V., et al. (2011). Family-based association study of ITGB3 in autism spectrum disorder and its endophenotypes. *European Journal of Human Genetics, 19*(3), 353–359. https://doi.org/10.1038/ejhg.2010.180
Ng, M., et al. (2017). Environmental factors associated with autism spectrum disorder: A scoping review for the years 2003–2013. In *Health promotion and chronic disease prevention in Canada* (pp. 1–23). https://doi.org/10.24095/hpcdp.37.1.01
Nguyen, T., & Südhof, T. C. (1997). Binding properties of neuroligin 1 and neurexin 1β reveal function as heterophilic cell adhesion molecules. *Journal of Biological Chemistry, 272*(41), 26032–26039. https://doi.org/10.1074/jbc.272.41.26032
Nunes, S. F., et al. (2011). Monoamine oxidase and semicarbazide-sensitive amine oxidase kinetic analysis in mesenteric arteries of patients with type 2 diabetes. *Physiological Research, 60*(2), 309–315. https://doi.org/10.33549/physiolres.931982
Oblak, A., Gibbs, T. T., & Blatt, G. J. (2009). Decreased GABAa receptors and benzodiazepine binding sites in the anterior cingulate cortex in autism. *Autism Research, 2*(4), 205–219. https://doi.org/10.1002/aur.88
Oblak, A. L., Gibbs, T. T., & Blatt, G. J. (2010). Decreased GABAB receptors in the cingulate cortex and fusiform gyrus in autism. *Journal of Neurochemistry, 114*(5), 1414–1423. https://doi.org/10.1111/j.1471-4159.2010.06858.x
Ocakoğlu, F. T., et al. (2018). The oxytocin receptor gene polymorphism -rs237902- is associated with the severity of autism spectrum disorder: A pilot study. *Asian Journal of Psychiatry, 31*, 142–149. https://doi.org/10.1016/j.ajp.2018.01.002
Palmieri, L., et al. (2001). Citrin and aralar1 are Ca2+–stimulated aspartate/glutamate transporters in mitochondria. *EMBO Journal, 20*(18), 5060–5069. https://doi.org/10.1093/emboj/20.18.5060
Palmieri, L., et al. (2010). Altered calcium homeostasis in autism-spectrum disorders: Evidence from biochemical and genetic studies of the mitochondrial aspartate/glutamate carrier AGC1. *Molecular Psychiatry, 15*(1), 38–52. https://doi.org/10.1038/mp.2008.63
Peça, J., et al. (2011). Shank3 mutant mice display autistic-like behaviours and striatal dysfunction. *Nature, 472*(7344), 437–442. https://doi.org/10.1038/nature09965
Persico, A. M., et al. (2000). Lack of association between serotonin transporter gene promoter variants and autistic disorder in two ethnically distinct samples. *American Journal of Medical Genetics - Neuropsychiatric Genetics, 96*(1), 123–127. https://doi.org/10.1002/(SICI)1096-8628(20000207)96:1<123::AID-AJMG24>3.0.CO;2-N
Peter, S., et al. (2016). Dysfunctional cerebellar Purkinje cells contribute to autism-like behaviour in Shank2-deficient mice. *Nature Communications, 7*. https://doi.org/10.1038/ncomms12627
Phelan, K., & McDermid, H. E. (2012). The 22q13.3 deletion syndrome (Phelan-McDermid syndrome). *Molecular Syndromology, 2*(3–5), 186–201. https://doi.org/10.1159/000334260
Philippe, A., et al. (2002). Analysis of ten candidate genes in autism by association and linkage. *American Journal of Medical Genetics - Neuropsychiatric Genetics, 114*(2), 125–128. https://doi.org/10.1002/ajmg.10041

Piggot, J., et al. (2009). Neural systems approaches to the neurogenetics of autism spectrum disorders. *Neuroscience, 164*(1), 247–256. https://doi.org/10.1016/j.neuroscience.2009.05.054

Pinto, D., et al. (2010). Functional impact of global rare copy number variation in autism spectrum disorders. *Nature, 466*(7304), 368–372. https://doi.org/10.1038/nature09146

Piven, J., et al. (1992). Magnetic resonance imaging in autism: Measurement of the cerebellum, pons, and fourth ventricle. *Biological Psychiatry, 31*(5), 491–504. https://doi.org/10.1016/0006-3223(92)90260-7

Plummer, J. T., et al. (2013). Transcriptional regulation of the MET receptor tyrosine kinase gene by MeCP2 and sex-specific expression in autism and Rett syndrome. *Translational Psychiatry, 3*, e316. https://doi.org/10.1038/tp.2013.91

Powell, E. M. (2013). Interneuron development and epilepsy: Early genetic defects cause long-term consequences in seizures and susceptibility. *Epilepsy Currents, 13*(4), 172–176. https://doi.org/10.5698/1535-7597-13.4.172

Powell, E. M., Mars, W. M., & Levitt, P. (2001). Hepatocyte growth factor/scatter factor is a motogen for interneurons migrating from the ventral to dorsal telencephalon. *Neuron, 30*(1), 79–89. https://doi.org/10.1016/S0896-6273(01)00264-1

Prandini, P., et al. (2012). The association of rs4307059 and rs35678 markers with autism spectrum disorders is replicated in Italian families. *Psychiatric Genetics, 22*(4), 177–181. https://doi.org/10.1097/YPG.0b013e32835185c9

Puglia, M. H., et al. (2015). Epigenetic modification of the oxytocin receptor gene influences the perception of anger and fear in the human brain. *Proceedings of the National Academy of Sciences of the United States of America, 112*(11), 3308–3313. https://doi.org/10.1073/pnas.1422096112

Rabionet, R., et al. (2006). Lack of association between autism and SLC25A12. *American Journal of Psychiatry, 163*(5), 929–931. https://doi.org/10.1176/ajp.2006.163.5.929

Ramamoorthy, S., et al. (1993). Antidepressant- and cocaine-sensitive human serotonin transporter: Molecular cloning, expression, and chromosomal localization. *Proceedings of the National Academy of Sciences of the United States of America, 90*(6), 2542–2546. https://doi.org/10.1073/pnas.90.6.2542

Ramoz, N., et al. (2004). Linkage and association of the mitochondrial aspartate/glutamate carrier SLC25A12 gene with autism. *American Journal of Psychiatry, 161*(4), 662–669. https://doi.org/10.1176/appi.ajp.161.4.662

Ramoz, N., et al. (2006). Lack of evidence for association of the serotonin transporter gene SLC6A4 with autism. *Biological Psychiatry, 60*(2), 186–191. https://doi.org/10.1016/j.biopsych.2006.01.009

Ramoz, N., et al. (2008). An analysis of candidate autism loci on chromosome 2q24-q33: Evidence for association to the STK39 gene. *American Journal of Medical Genetics, Part B: Neuropsychiatric Genetics, 147*(7), 1152–1158. https://doi.org/10.1002/ajmg.b.30739

Richdale, A. L., & Schreck, K. A. (2009). Sleep problems in autism spectrum disorders: Prevalence, nature, & possible biopsychosocial aetiologies. *Sleep Medicine Reviews, 13*, 403–411. https://doi.org/10.1016/j.smrv.2009.02.003

Riikonen, T., et al. (1995). Integrin α2β1 is a positive regulator of collagenase (MMP-1) and collagen α1(I) gene expression. *Journal of Biological Chemistry, 270*(22), 13548–13552. https://doi.org/10.1074/jbc.270.22.13548

Risch, N., et al. (2014). Familial recurrence of autism spectrum disorder: Evaluating genetic and environmental contributions. *American Journal of Psychiatry, 171*(11), 1206–1213. https://doi.org/10.1176/appi.ajp.2014.13101359

Ritvo, E. R., et al. (1985). Concordance for the syndrome of autism in 40 pairs of afflicted twins. *American Journal of Psychiatry, 142*(1), 74–77. https://doi.org/10.1176/ajp.142.1.74

Ritvo, E. R., et al. (1989). The UCLA-University of Utah epidemiologic survey of autism: Prevalence. *American Journal of Psychiatry, 146*(2), 194–199. https://doi.org/10.1176/ajp.146.2.194

Rojas, D. C., & Wilson, L. B. (2014). Γ-band abnormalities as markers of autism spectrum disorders. *Biomarkers in Medicine, 8*, 353–368. https://doi.org/10.2217/bmm.14.15

Ronald, A., Happé, F., & Plomin, R. (2005). The genetic relationship between individual differences in social and nonsocial behaviours characteristic of autism. *Developmental Science, 8*(5), 444–458. https://doi.org/10.1111/j.1467-7687.2005.00433.x

Ronald, A., et al. (2006). Genetic heterogeneity between the three components of the autism spectrum: A twin study. *Journal of the American Academy of Child and Adolescent Psychiatry, 45*(6), 691–699. https://doi.org/10.1097/01.chi.0000215325.13058.9d

Ronemus, M., et al. (2014). The role of de novo mutations in the genetics of autism spectrum disorders. *Nature Reviews Genetics, 15*, 133–141. https://doi.org/10.1038/nrg3585

Rosenberg, R. E., et al. (2009). Characteristics and concordance of autism spectrum disorders among 277 twin pairs. *Archives of Pediatrics and Adolescent Medicine, 163*(10), 907–914. https://doi.org/10.1001/archpediatrics.2009.98

Rothwell, P. E., et al. (2014). Autism-associated neuroligin-3 mutations commonly impair striatal circuits to boost repetitive behaviors. *Cell, 158*(1), 198–212. https://doi.org/10.1016/j.cell.2014.04.045

Rudie, J. D., et al. (2012). Autism-associated promoter variant in MET impacts functional and structural brain networks. *Neuron, 75*(5), 904–915. https://doi.org/10.1016/j.neuron.2012.07.010

Rutter, M. (2005). Aetiology of autism: Findings and questions. *Journal of Intellectual Disability Research, 49*(4), 231–238. https://doi.org/10.1111/j.1365-2788.2005.00676.x

Saito, M., et al. (2014). MAOA/B deletion syndrome in male siblings with severe developmental delay and sudden loss of muscle tonus. *Brain and Development, 36*(1), 64–69. https://doi.org/10.1016/j.braindev.2013.01.004

Sala, C., et al. (2015). Shank synaptic scaffold proteins: Keys to understanding the pathogenesis of autism and other synaptic disorders. *Journal of Neurochemistry, 135*, 849–858. https://doi.org/10.1111/jnc.13232

Salem, A. M., et al. (2013). Genetic variants of neurotransmitter-related genes and miRNAs in egyptian autistic patients. *The Scientific World Journal*. Edited by S. Mastana, J. L. Vilotte, and K. Csiszar, *2013*, 670621. https://doi.org/10.1155/2013/670621

Sandin, S., et al. (2014). The familial risk of autism. *JAMA - Journal of the American Medical Association, 311*(17), 1770–1777. https://doi.org/10.1001/jama.2014.4144

Sato, D., et al. (2012). SHANK1 deletions in males with autism spectrum disorder. *American Journal of Human Genetics, 90*(5), 879–887. https://doi.org/10.1016/j.ajhg.2012.03.017

Sato, Y., et al. (2015). Determination of cleavage site of Reelin between its sixth and seventh repeat and contribution of meprin metalloproteases to the cleavage. *Journal of Biochemistry, 159*(3), 305–312. https://doi.org/10.1093/jb/mvv102

Sauna, Z. E., & Kimchi-Sarfaty, C. (2011). Understanding the contribution of synonymous mutations to human disease. *Nature Reviews Genetics, 12*, 683–691. https://doi.org/10.1038/nrg3051

Schain, R. J., & Freedman, D. X. (1961). Studies on 5-hydroxyindole metabolism in autistic and other mentally retarded children. *The Journal of Pediatrics, 58*(3), 315–320. https://doi.org/10.1016/S0022-3476(61)80261-8

Schellenberg, G. D., et al. (2006). Evidence for multiple loci from a genome scan of autism kindreds. *Molecular Psychiatry, 11*(11), 1049–1060. https://doi.org/10.1038/sj.mp.4001874

Scherer, S. W., et al. (2003). Human chromosome 7: DNA sequence and biology. *Science, 300*(5620), 767–772. https://doi.org/10.1126/science.1083423

Schmeisser, M. J., et al. (2012). Autistic-like behaviours and hyperactivity in mice lacking ProSAP1/Shank2. *Nature, 486*(7402), 256–260. https://doi.org/10.1038/nature11015

Schuch, J. B., et al. (2014). The role of β3 integrin gene variants in autism spectrum disorders - diagnosis and symptomatology. *Gene, 553*(1), 24–30. https://doi.org/10.1016/j.gene.2014.09.058

Segurado, R., et al. (2005). Confirmation of association between autism and the mitochondrial aspartate/glutamate carrier SLC25A12 gene on chromosome 2q31. *American Journal of Psychiatry, 162*(11), 2182–2184. https://doi.org/10.1176/appi.ajp.162.11.2182

Sesarini, C. V., et al. (2015). Association between GABA(A) receptor subunit polymorphisms and autism spectrum disorder (ASD). *Psychiatry Research, 229*(1–2), 580–582. https://doi.org/10.1016/j.psychres.2015.07.077

Shao, Y., Wolpert, C. M., et al. (2002a). Genomic screen and follow-up analysis for autistic disorder. *American Journal of Medical Genetics, Part B: Neuropsychiatric Genetics, 114*(1), 99–105. https://doi.org/10.1002/ajmg.10153

Shao, Y., Raiford, K. L., et al. (2002b). Phenotypic homogeneity provides increased support for linkage on chromosome 2 in autistic disorder. *American Journal of Human Genetics, 70*(4), 1058–1061. https://doi.org/10.1086/339765

Shao, Y., et al. (2003). Fine mapping of autistic disorder to chromosome 15q11-q13 by use of phenotypic subtypes. *American Journal of Human Genetics, 72*(3), 539–548. https://doi.org/10.1086/367846

Shih, J. C., & Chen, K. (2004). Regulation of MAO-A and MAO-B gene expression. *Current Medicinal Chemistry, 11*(15), 1995–2005.

Silverman, J. M., et al. (2008). Autism-related routines and rituals associated with a mitochondrial aspartate/glutamate carrier SLC25A12 polymorphism. *American Journal of Medical Genetics, Part B: Neuropsychiatric Genetics, 147*(3), 408–410. https://doi.org/10.1002/ajmg.b.30614

Simmons, C. F., et al. (1995). The oxytocin receptor gene (OXTR) localizes to human chromosome 3p25 by fluorescence in situ hybridization and PCR analysis of somatic cell hybrids. *Genomics, 26*(3), 623–625. https://doi.org/10.1016/0888-7543(95)80188-R

Simon, D. M., & Wallace, M. T. (2016). Dysfunction of sensory oscillations in Autism Spectrum Disorder. *Neuroscience and Biobehavioral Reviews, 68*, 848–861. https://doi.org/10.1016/j.neubiorev.2016.07.016

Skuse, D. H., Mandy, W. P. L., & Scourfield, J. (2005). Measuring autistic traits: Heritability, reliability and validity of the social and communication disorders checklist. *British Journal of Psychiatry, 187*(DEC.), 568–572. https://doi.org/10.1192/bjp.187.6.568

Sohal, V. S. (2012). Insights into cortical oscillations arising from optogenetic studies. *Biological Psychiatry, 71*, 1039–1045. https://doi.org/10.1016/j.biopsych.2012.01.024

Song, J. Y., et al. (1999). Neuroligin 1 is a postsynaptic cell-adhesion molecule of excitatory synapses. *Proceedings of the National Academy of Sciences of the United States of America, 96*(3), 1100–1105. https://doi.org/10.1073/pnas.96.3.1100

Sousa, I., et al. (2009). MET and autism susceptibility: Family and case-control studies. *European Journal of Human Genetics, 17*(6), 749–758. https://doi.org/10.1038/ejhg.2008.215

Steffenburg, S., et al. (1989). A twin study of autism in Denmark, Finland, Iceland, Norway and Sweden. *Journal of Child Psychology and Psychiatry, 30*(3), 405–416. https://doi.org/10.1111/j.1469-7610.1989.tb00254.x

Streit, A., et al. (1995). A role for HGF/SF in neural induction and its expression in Hensen's node during gastrulation. *Development, 121*(3), 813–824. https://doi.org/10.1242/dev.121.3.813

Südhof, T. C. (2008). Neuroligins and neurexins link synaptic function to cognitive disease. *Nature, 455*, 903–911. https://doi.org/10.1038/nature07456

Sutcliffe, J. S., et al. (2003). Partial duplication of the APBA2 gene in chromosome 15q13 corresponds to duplicon structures. *BMC Genomics, 4*(1), 15. https://doi.org/10.1186/1471-2164-4-15

Sutcliffe, J. S., et al. (2005). Allelic heterogeneity at the serotonin transporter locus (SLC6A4) confers susceptibility to autism and rigid-compulsive behaviors. *American Journal of Human Genetics, 77*(2), 265–279. https://doi.org/10.1086/432648

Szatmari, P., et al. (2007). Mapping autism risk loci using genetic linkage and chromosomal rearrangements. *Nature Genetics, 39*(3), 319–328. https://doi.org/10.1038/ng1985

Tabuchi, K., et al. (2007). A neuroligin-3 mutation implicated in autism increases inhibitory synaptic transmission in mice. *Science, 318*(5847), 71–76. https://doi.org/10.1126/science.1146221

Talebizadeh, Z., et al. (2004). Letter to the editor: Do known mutations in neuroligin genes (NLGN3 and NLGN4) cause autism? *Journal of Autism and Developmental Disorders, 34*, 735–736. https://doi.org/10.1007/s10803-004-5295-x

Tansey, K. E., et al. (2010). Oxytocin receptor (OXTR) does not play a major role in the aetiology of autism: Genetic and molecular studies. *Neuroscience Letters, 474*(3), 163–167. https://doi.org/10.1016/j.neulet.2010.03.035

Tassone, F., et al. (2011). MAOA, DBH, and SLC6A4 variants in CHARGE: A case-control study of autism spectrum disorders. *Autism Research, 4*(4), 250–261. https://doi.org/10.1002/aur.196

Thanseem, I., et al. (2010). Further evidence for the role of MET in autism susceptibility. *Neuroscience Research, 68*(2), 137–141. https://doi.org/10.1016/j.neures.2010.06.014

Tick, B., et al. (2016). Heritability of autism spectrum disorders: A meta-analysis of twin studies. *Journal of Child Psychology and Psychiatry and Allied Disciplines, 57*(5), 585–595. https://doi.org/10.1111/jcpp.12499

Tordjman, S., et al. (2001). Role of the serotonin transporter gene in the behavioral expression of autism. *Molecular Psychiatry, 6*(4), 434–439. https://doi.org/10.1038/sj.mp.4000873

Tost, H., et al. (2011). Neurogenetic effects of OXTR rs2254298 in the extended limbic system of healthy caucasian adults. *Biological Psychiatry, 70*, e37–e39. https://doi.org/10.1016/j.biopsych.2011.06.034

Turunen, J. A., et al. (2008). Mitochondrial aspartate/glutamate carrier SLC25A12 gene is associated with autism. *Autism Research, 1*(3), 189–192. https://doi.org/10.1002/aur.25

Uchino, S., et al. (2006). Direct interaction of post-synaptic density-95/Dlg/ZO-1 domain-containing synaptic molecule Shank3 with GluR1 α-amino-3-hydroxy-5-methyl-4-isoxazole propionic acid receptor. *Journal of Neurochemistry, 97*(4), 1203–1214. https://doi.org/10.1111/j.1471-4159.2006.03831.x

Valicenti-McDermott, M., et al. (2006). Frequency of gastrointestinal symptoms in children with autistic spectrum disorders and association with family history of autoimmune disease. *Journal of Developmental and Behavioral Pediatrics, 27*(2 SUPPL. 2), S128–S136. https://doi.org/10.1097/00004703-200604002-00011

Varoqueaux, F., Jamain, S., & Brose, N. (2004). Neuroligin 2 is exclusively localized to inhibitory synapses. *European Journal of Cell Biology, 83*(9), 449–456. https://doi.org/10.1078/0171-9335-00410

Veenstra-Vanderweele, J., & Blakely, R. D. (2012). Networking in autism: Leveraging genetic, biomarker and model system findings in the search for new treatments. *Neuropsychopharmacology, 37*(1), 196–212. https://doi.org/10.1038/npp.2011.185

Veenstra-VanderWeele, J., & Cook, E. H. (2004). Molecular genetics of autism spectrum disorder. *Molecular Psychiatry, 9*(9), 819–832. https://doi.org/10.1038/sj.mp.4001505

Veenstra-VanderWeele, J., et al. (2012). Autism gene variant causes hyperserotonemia, serotonin receptor hypersensitivity, social impairment and repetitive behavior. *Proceedings of the National Academy of Sciences of the United States of America, 109*(14), 5469–5474. https://doi.org/10.1073/pnas.1112345109

Verma, D., et al. (2014). Sexual dimorphic effect in the genetic association of monoamine oxidase A (MAOA) markers with autism spectrum disorder. *Progress in Neuro-Psychopharmacology and Biological Psychiatry, 50*, 11–20. https://doi.org/10.1016/j.pnpbp.2013.11.010

Vorstman, J. A. S., et al. (2006). Overview of cytogenetic regions of interest (CROIs) associated with the autism phenotype across the human genome. *Molecular Psychiatry, 11*(1), 1–1, 28. https://doi.org/10.1038/sj.mp.4001781

Vreugdenhil, M., et al. (2003). Parvalbumin-deficiency facilitates repetitive IPSCs and gamma oscillations in the hippocampus. *Journal of Neurophysiology, 89*(3), 1414–1422. https://doi.org/10.1152/jn.00576.2002

Walsh, C. A., Morrow, E. M., & Rubenstein, J. L. R. (2008). Autism and brain development. *Cell, 135*, 396–400. https://doi.org/10.1016/j.cell.2008.10.015

Wang, X., et al. (2014). Transcriptional and functional complexity of Shank3 provides a molecular framework to understand the phenotypic heterogeneity of SHANK3 causing autism and Shank3 mutant mice. *Molecular Autism, 5*(1). https://doi.org/10.1186/2040-2392-5-30

Weiss, L. A., et al. (2009). A genome-wide linkage and association scan reveals novel loci for autism. *Nature, 461*(7265), 802–808. https://doi.org/10.1038/nature08490

Wermter, A. K., et al. (2008). No evidence for involvement of genetic variants in the x-linked neuroligin genes NLGN3 and NLGN4X in probands with autism spectrum disorder on high functioning level. *American Journal of Medical Genetics, Part B: Neuropsychiatric Genetics, 147*(4), 535–537. https://doi.org/10.1002/ajmg.b.30618

Wermter, A. K., et al. (2010). Evidence for the involvement of genetic variation in the oxytocin receptor gene (OXTR) in the etiology of autistic disorders on high-functioning level. *American Journal of Medical Genetics, Part B: Neuropsychiatric Genetics, 153*(2), 629–639. https://doi.org/10.1002/ajmg.b.31032

Whibley, A., et al. (2010). Deletion of MAOA and MAOB in a male patient causes severe developmental delay, intermittent hypotonia and stereotypical hand movements. *European Journal of Human Genetics, 18*(10), 1095–1099. https://doi.org/10.1038/ejhg.2010.41

Wiśniowiecka-Kowalnik, B., & Nowakowska, B. A. (2019). Genetics and epigenetics of autism spectrum disorder—Current evidence in the field. *Journal of Applied Genetics, 60*(1), 37–47. https://doi.org/10.1007/s13353-018-00480-w

Wu, S., et al. (2005). Lack of evidence for association between the serotonin transporter gene (SLC6A4) polymorphisms and autism in the Chinese trios. *Neuroscience Letters, 381*(1–2), 1–5. https://doi.org/10.1016/j.neulet.2005.01.073

Yan, J., et al. (2005). Analysis of the neuroligin 3 and 4 genes in autism and other neuropsychiatric patients [1]. *Molecular Psychiatry, 10*, 329–332. https://doi.org/10.1038/sj.mp.4001629

Yasuhara, A. (2010). Correlation between EEG abnormalities and symptoms of autism spectrum disorder (ASD). *Brain and Development, 32*(10), 791–798. https://doi.org/10.1016/j.braindev.2010.08.010

Yirmiya, N., et al. (2002). Brief research communication: Family-based and population study of a functional promoter-region monoamine oxidase A polymorphism in autism: Possible association with IQ. *American Journal of Medical Genetics - Neuropsychiatric Genetics, 114*(3), 284–287. https://doi.org/10.1002/ajmg.10189

Ylisaukko-Oja, T., et al. (2004). Genome-wide scan for loci of Asperger syndrome. *Molecular Psychiatry, 9*(2), 161–168. https://doi.org/10.1038/sj.mp.4001385

Ylisaukko-oja, T., et al. (2005). Analysis of four neuroligin genes as candidates for autism. *European Journal of Human Genetics, 13*(12), 1285–1292. https://doi.org/10.1038/sj.ejhg.5201474

Yonan, A. L., et al. (2003). Bioinformatic analysis of autism positional candidate genes using biological databases and computational gene network prediction. *Genes, Brain and Behavior, 2*(5), 303–320. https://doi.org/10.1034/j.1601-183X.2003.00041.x

Yoo, H. J., et al. (2009). Family- and population-based association studies of monoamine oxidase A and autism spectrum disorders in Korean. *Neuroscience Research, 63*(3), 172–176. https://doi.org/10.1016/j.neures.2008.11.007

Zhong, N., et al. (1999). 5-HTTLPR variants not associated with autistic spectrum disorders. *Neurogenetics, 2*(2), 129–131. https://doi.org/10.1007/s100480050064

Zhu, L., et al. (2014). Epigenetic dysregulation of SHANK3 in brain tissues from individuals with autism spectrum disorders. *Human Molecular Genetics, 23*(6), 1563–1578. https://doi.org/10.1093/hmg/ddt547

Chapter 19
Electrophysiology of Semantic Processing in ASD

Mirella Manfredi and Emily Coderre

19.1 Electrophysiology of Semantic Processing in Healthy Population

Semantic processing refers to the ability to receive a stimulus, whether a spoken or written word, picture, or sound, and apply meaning to it based on previously stored knowledge (Kelley et al., 2006; Yang & Small, 2015). Since the mid-late twentieth century, psycholinguistic research on semantic processing has used electroencephalography (EEG), a technique which measures the electrical activity of the brain via electrodes that are placed on the scalp. There are two primary ways of analyzing EEG data in cognitive neuroscience research. One is to derive *event-related potentials (ERPs)*, which are time-locked changes in the EEG signal in response to a stimulus. The particular polarity and latency of various components in the ERP signal are interpreted as reflecting various cognitive processes in the processing stream. The second analysis method is to break the overall EEG signal into its component frequency bands and look at oscillations. In this method, EEG *power* refers to the amplitude of oscillations in a certain frequency band, while EEG *coherence* refers to the synchrony of oscillations in a certain frequency band between two electrodes. Below we will elaborate on the specific ERP components and oscillatory bands that are representative of semantic processing, both in neurotypical populations and in individuals with ASD.

M. Manfredi (✉)
Department of Psychology, University of Zurich, Zurich, Switzerland

Jacobs Center for Productive Youth Development, University of Zurich, Zurich, Switzerland
e-mail: m.manfredi@psychologie.uzh.ch

E. Coderre
Department of Communication Sciences and Disorders, University of Vermont, Burlington, VT, USA
e-mail: Emily.Coderre@med.uvm.edu

A. El Idrissi, D. McCloskey (eds.), *Neurobiology of Autism Spectrum Disorders*,
https://doi.org/10.1007/978-3-031-42383-3_19

19.2 Classic ERPs Components Associated with Semantic Processing

19.2.1 N400

The classic ERP component related to semantic processing is called the N400, so named because it is a negativity that occurs from approximately 300–500 milliseconds (ms), but peaking at approximately 400 ms, post-stimulus. First described by Kutas and Hillyard in 1980, the N400 has been studied extensively across many different modalities, populations, and experimental paradigms. Although the N400 ERP component is elicited by any semantic processing, in neurotypical populations its amplitude is modulated by the difficulty or ease of semantic processing. In the linguistic domain, a larger N400 amplitude is elicited by semantically incongruous words in a sentence (e.g., "He spread the warm bread with SOCKS" (Kutas & Hillyard, 1980) compared to semantically congruous words (e.g., "It was his first day at WORK"). This enhanced amplitude is thought to reflect increased difficulty with integrating the target word with the preceding semantic context (Marta Kutas & Federmeier, 2011).

The N400 is also modulated by the predictability of a given word in a sentence, called *cloze probability*. "High cloze" words, which are highly predictable given the semantic constraints of the preceding sentence (e.g., "He mailed the letter without a STAMP"), elicit smaller N400 amplitudes compared to "low cloze" words which are less predictable in a sentence (e.g., "He was soothed by the gentle WIND") (Marta Kutas & Hillyard, 1984). The reduced amplitude for high-cloze words is thought to reflect facilitation of semantic processing by a highly constrained semantic context (Marta Kutas & Federmeier, 2011). This effect is also continuous, such that cloze rating is inversely correlated with N400 negative amplitude (with larger negative amplitudes occurring for lower-cloze words) (Van Petten & Luka, 2012).

It should be noted that there is still debate and discussion around the specific functions of semantic processing that the N400 component reflects, such as semantic integration or retrieval of lexical information (Kutas & Hillyard, 1980; Brouwer et al., 2012; Lau et al., 2008; see Kutas & Federmeier, 2011, for a broader discussion). There is also ample discussion surrounding the exact onset and end time of this effect. Some researchers argue that the N400 effect begins immediately following early lexical access processes, such as the N170 effect, which represents the identification of letter strings and occurs 170 ms after the presentation of a stimulus (Sereno & Rayner, 2003). Other researchers argue that the N400 has an onset of strictly 400 ms after stimulus presentation, while still others argue that the N400 effect encompasses even higher-order processes such as syntactic processing (Brouwer et al., 2017). In a recent model attempting to integrate theoretical semantics with empirical psycholinguistic research, Baggio (2018) proposes that these functional interpretations of the N400 may not necessarily be mutually exclusive. This model proposes that within ~250 ms, information flows from sensory input regions (in the visual or temporal cortices, depending on whether input is visual or auditory) to the posterior middle superior temporal gyrus (pMSTG) in the brain, which mediates retrieval of lexical knowledge, and the anterior temporal lobe (ATL), which houses amodal semantic concepts. These regions are involved in activating lexico-semantic *types*, which are general, abstract semantic concepts. Information then flows forward to the left inferior frontal gyrus (LIFG), which binds the representation of lexical types to *token* representations, which are more specific instances of a concept. This forward flow of information is reflected as the ascending phase of the N400 component; information reaches the LIFG at ~400 ms, reflected as the N400 peak. Information then flows back to the pMSTG to enable top-down binding, reflected as the descending phase of the N400. This model thus suggests that the N400 may reflect both feed-forward lexical access as well as backward-flowing integrative processes. Regardless of the specific semantic function(s) that the N400 reflects, it has become well-established as a marker of semantic processing in the brain.

Importantly, the N400 ERP component is not elicited only by words in sentences. It is also elicited in *semantic priming* paradigms, in which the processing of a word is affected by the semantic relatedness of a preceding word. For example, the N400 amplitude is reduced in response to a target word that is semantically related to a prime word (e.g., dog-CAT) compared to an unrelated prime word (e.g., rock-CAT). In a semantic priming paradigm with single words, the prime can act as the "context" for the target word, leading to a reduction in the N400 amplitude when the two words are related compared to when they are unrelated.

Furthermore, although the N400 is *sensitive* to language, it is not *specific* to language and can be elicited in other domains. N400 effects have been reported in response to static pictures (Marta Kutas & Federmeier, 2011), visual event cognition (Amoruso et al., 2013; Sitnikova et al., 2008), scene perception (Võ, 2021), and even music (Zhang et al., 2016). Its ubiquity across modalities suggests that the N400 is a domain-general reflection of semantic processing mechanisms in the brain.

There are, however, some differences in the N400 effect between modalities. For instance, the N400 effect in response to pictures typically has a more frontal distribution than the more parietal N400 elicited by language (Ganis et al., 1996; Hamm et al., 2002; McPherson & Holcomb, 1999). This effect also generally begins slightly earlier than for words, leading some researchers to suggest that this frontal negativity is a separate component from the N400 termed the 'N300' (Cohn et al., 2012; Hamm et al., 2002; McPherson & Holcomb, 1999; West & Holcomb, 2002). The N300 is also related to semantic processing, but is thought to reflect access to the semantic features of objects in visual images (Cohn et al., 2012; Sitnikova et al., 2006) as opposed to the semantic features of words.

19.2.2 Later Positivities

While the N400 is the ERP component most commonly associated with semantic processing, several other later components, occurring after the N400, have also been implicated in this cognitive function.

The P600 ERP component is a slow, positive-going deflection occurring after the N400, with a midpoint of approximately 600 ms post-stimulus. First described by Osterhout and Holcomb in 1992, the P600 was initially thought to be reflective of grammatical, or *syntactic*, processing, since its amplitude is enhanced for words that are incongruous with the expected sentence structure (e.g., "The broker persuaded to sell the stock") compared to words in syntactically expected sentences (e.g., "The broker hoped to sell the stock"). However, more recently there has been debate about the functional significance of this ERP component, since P600 effects have also been reported for critical words in sentences that violate the expected semantic-thematic relationships (e.g., "Every morning at breakfast, the eggs would EAT…"): the so-called "semantic P600" (Kuperberg, 2007). Some researchers propose that the N400 indexes the retrieval of word meaning while the P600 is indicative of semantic integration (Brouwer et al., 2012, 2017; Brouwer & Hoeks, 2013).

The P600 is part of a broader category of sustained positivities occurring after the N400 effect, generally called *late positive components (LPCs)*. An LPC is often (but not always) reported to occur after the N400 effect and has been proposed to reflect a more elaborate monitoring process that triggers re-analysis and re-interpretation of semantic information when initial integration fails (Pijnacker et al., 2010; van de Meerendonk et al., 2010).

19.2.3 Later Sustained Negativity

In addition to the late positivity, some studies have observed a post-N400 sustained fronto-central negativity (500–700 ms time window) in response to incongruent versus congruent information. Similar sustained negativities have been observed in response to both linguistic and visual abnormalities (Cohn et al., 2012; Manfredi et al., 2018; West & Holcomb, 2002). However, the functional meaning of this late electrophysiological response is not univocal.

According to some researchers, in continuity with the earlier N400, this sustained negativity might reflect a general cost of incongruent information processing (Manfredi et al., 2018). From this perspective, this ERP response might reflect higher cognitive demands. However, other evidence has observed sustained negativity in response to nonsense information that was not necessarily more cognitively demanding or involved complex inference processing (Kröger et al., 2013).

Finally, a sustained anterior negativity has also been linked to an increased working memory load to resolve inconsistency during frame-shifting processes, which is necessary to re-establish semantic coherence (Manfredi et al., 2021).

In general, we could argue that while the late positivity might reflect model revision or updating, the sustained negativity might be a general index of sustained interpretive processing within a mental model (Baggio et al., 2008).

19.3 Electrophysiology of Semantic Processing in ASD

Autism spectrum disorder (ASD) is characterized by strong impairments in communication and social behaviour (APA, 2000). Furthermore, individuals with ASD are known to exhibit abnormalities in various aspects of language, ranging from semantic-pragmatic deficits to the total or partial absence of language (Coderre, 2020).

In the next section, we will review the studies that have analyzed the electrophysiological responses (i.e., ERPs) during the semantic processing of verbal information in individuals with autism. In particular, using the N400 effect as an index of the semantic processing (Kutas & Hillyard, 1980; Kutas & Federmeier, 2000), these studies have revealed abnormalities in the neural processing of verbal information in the ASD population.

19.3.1 Semantic Processing of Verbal Information

Several ERP studies on adults with ASD revealed deficits in neural processing of verbal semantic information (Braeutigam et al., 2008; Fishman et al., 2011; Kujala et al., 2013; O'Connor, 2012; Pijnacker et al., 2010). Pijnacker et al. (2010) performed an ERP study to examine context sensitivity in high-functioning adults with autistic disorder (HFA) and Asperger syndrome at two levels: sentence processing and solving reasoning problems. The results of this study showed that sentence context and reasoning context had an immediate ERP effect in adults with Asperger syndrome, as in matched controls. Indeed, both groups showed a typical N400 effect, a late positive component for the sentence conditions, and a sustained negativity for the reasoning conditions. In contrast, the HFA group demonstrated neither an N400 effect nor a sustained negativity. However, the HFA group showed a late positive component which was larger for semantically anomalous sentences than congruent sentences. The authors concluded that because sentence context had a modulating effect in a later phase, semantic processing is perhaps less automatic in HFA, and presumably, more elaborate processes are needed to arrive at a sentence interpretation.

Similar results were observed in a MEG study performed by Braeutigam et al. (2008). In this study, adult participants read sentences that ended with semantically incongruous words. Spatial evoked signals at 400 ms (N4) and 750 ms (LPC) and synchronized gamma oscillations were analyzed. The results showed a distinct difference in sentence processing among individuals with autism compared to typically developing ones. In particular, compared to typically developing individuals, N4 responses following incongruous words were more attenuated in the left temporal cortex. In addition, individuals with ASD exhibit a long-lasting gamma-oscillations above 40 Hz, suggesting that this group uses different strategies to resolve semantic ambiguity.

In 2011, Fishman and colleagues contrasted brain activity in two neurodevelopmental disorders, Williams syndrome (WS) and autism. These two disorders present opposing social profiles: While autism is characterized by social avoidance, Williams syndrome exhibits hypersociality. In this study, the authors analyzed contextual integration ability and semantic processing in the two groups. In particular, they presented a classic paradigm in which final word of an auditory sentence was semantically congruent or incongruent with the meaning of the rest of the sentence. The results revealed that individuals with WS showed a significantly larger N400 effect than typical controls and individuals with autism, while the latter group showed the smallest N400 effect. The results demonstrate completely opposite profiles of neural correlates of language processing in WS and Autism, reflecting their contrasting social phenotypes.

Finally, a recent study conducted by Coderre et al. (2018) compared the N400 response in verbal and visual narratives in a group of adults with ASD and a group of neurotypical individuals. Specifically, the authors presented two tasks: In the first task, they were presented with short verbal narratives in which the final word was congruent or incongruent; In the second task, short visual narratives (comics) with semantically congruent or incongruent final images were presented. They found that the N400 effects were attenuated by both the linguistic and the visual narratives for the ASD population compared to the controls.

In conclusion, the results of these studies show that adult individuals with ASD process the meaning of verbal information qualitatively and quantitatively differently from neurotypical individuals, in line with the deficits in their socio-communicative function.

Other studies have analyzed the comprehension of verbal information in children with ASD. In this section, we will present the findings of studies that, as in the case of previous studies conducted on adult populations, analyzed the electrophysiological responses during the semantic processing of verbal information.

Dunn and colleagues have conducted two studies aimed at examining the processing of semantic categories in children with autism and controls with typical development (Dunn et al., 1999; Dunn & Bates, 2005). In the first study (Dunn et al., 1999), children were presented with animal names or other categories while their EEG and behavioural responses (i.e., accuracy and response time) were recorded. The study's results revealed that children with ASD were slower in classifying animal target words and made more errors than neurotypical children. Concerning electrophysiological responses, as in the studies conducted on adults, a deficit in semantic processing was also revealed in the group of children with autism. In fact, no difference was observed in the N400 response to non-targets compared to targets in the normal control group. In another study conducted by the same authors (Dunn & Bates, 2005), neurotypical children and children with ASD were instructed to listen to lists, one of the animal names and another of names that belonged to a different semantic category. The results revealed that semantically incongruent non-animal words elicited a significant N400 effect in control children. However, this N400 effect was negligible in children with ASD and completely absent in the youngest group (8–9-year-old children). In both studies, the authors interpreted these data as evidence of altered linguistic-semantic processing in autism that improves slightly with age.

More recently, Manfredi et al. (2020) compared the semantic processing of auditory sentences and visual narratives in children and adolescents with ASD and in typically developing (TD) children and adolescents. Although the authors of this study employed a very similar experimental paradigm to that used by Coderre and colleagues in adults with ASD, the results of the present investigation are slightly different from those of Coderre's study. As in Coderre's study, the results of Manfredi's study revealed a central N400 focal effect to incongruent versus congruent words, which was slightly attenuated in ASD adolescents compared to TD adolescents. Furthermore, at later latency, we observed sustained negativity to incongruent versus congruent words, but only for TD adolescents, probably reflecting a cost for further processing of incongruent auditory information, which was absent in ASD adolescents. However, unlike Coderre's study, in this work incongruent critical panels in visual narratives evoked a greater fronto-central N400 than congruent panels in visual narratives in both groups. We will discuss these findings related to visual processing in the next section.

In conclusion, these results highlight that the deficit related to linguistic-semantic processing is present in individuals with autism from childhood through adulthood.

19.3.2 Semantic Processing of Visual Information

Semantic processing of visual stimuli involves accessing the meaning of pictures. This has been assessed in ASD at the level of processing pairs of pictures as well as sequential images. Contrary to much of the literature on linguistic semantic processing, there is some evidence that semantic processing of non-verbal stimuli, such as pictures and auditory sounds, is intact in individuals with ASD.

In a recent study using pairs of matching or mismatching pictures (visual condition) or written words (linguistic condition), Coderre et al. (2017) found that autistic adults showed similar N400 responses as non-autistic adults during visuo-semantic processing, suggesting again that visuo-semantic processing of static pictures is intact in ASD. In a follow-up study, O'Rourke and Coderre (2021) used similar stimuli but with an implicit task, in which the semantic relatedness of the stimuli was not emphasized, and found interesting interactions of levels of autistic traits and age. Across both linguistic and visual conditions, lower levels of autistic traits were associated with increased sensitivity to semantic relatedness with age. However, this effect was less pronounced for those with higher levels of autistic traits in the word condition, which the authors speculate could reflect the development of compensatory mechanisms as participants get older. In the picture condition, this effect was flipped: those with higher levels of autistic traits showed *decreasing* sensitivity to semantic relatedness with age. This could suggest that visuo-semantic processing becomes harder as autistic individuals get older, which the authors speculate may be due to the focus on literacy in education.

Although this evidence suggests that visuo-semantic processing is intact at the level of single pictures, at more complex levels of cognition, such as sequential images, the evidence tells a different story. Several studies have examined processing of sequential images, or visual narratives, in individuals with ASD and have found evidence of impaired comprehension, similar to that found in linguistic narrative comprehension. For example, as discussed above, Coderre et al. (2018) contrasted visual and linguistic narrative comprehension in adults with and without ASD. Visual narratives consisted of comic strips that ended in a panel that was either semantically congruent or incongruent with the preceding narrative. Linguistic narratives consisted of short written sentences, and the final word in the narrative was either congruent or incongruent with the preceding story. Coderre et al. (2018) observed that the ASD group showed smaller N400 effects (i.e., smaller differences between the congruent and incongruent stimuli) in both visual and linguistic modalities, suggesting that difficulties with narrative comprehension occurred in both modalities. In a similar paradigm comparing short visual narratives with short spoken sentences, Manfredi et al. (2020) also found a smaller N400 effect in response to linguistic stimuli for autistic children and adolescents. Incongruent pictures, however, elicited an increased N400 in both groups (Manfredi et al., 2020).

19.3.3 Semantic Processing of Cross-Modal Information

Finally, several studies have focused on the semantic processing of cross-modal information in individuals with ASD (i.e., children and adolescents). For example, in 2013, Ribeiro and colleagues performed an ERP study using multimodal stimuli. Specifically, they used visual targets preceded by musical and linguistic cues to explore cross-modal semantic integration. ~~The participants were ASD and neurotypical children.~~ Their neuroelectric activity was recorded in response to visual targets that were related or unrelated to a preceding spoken sentence or musical excerpt. The data showed that, only in a verbal context, the N400 was sensitive to semantic incongruence in neurotypical but not in autistic children, whereas the LPC showed a flipped pattern. Given these results, the authors pointed out a functional decoupling between the generators of the N400 and LPC, which, according to them, may indicate a general delay in semantic processing in autism.

In another study, McCleery et al. (2010) recorded EEG while children were presented with matching or mismatching pairs of pictures and spoken words (a linguistic condition, such as a picture of a ball paired with the spoken word "ball") or pictures and environmental sounds (a non-linguistic condition, such as a picture of a ball paired with the sound of a ball bouncing). The results showed that children with ASD showed reduced N400 effects (i.e., smaller differences between matching and mismatching conditions) than TD peers in the linguistic condition but comparable N400 effects for the non-linguistic condition. According to the authors, these results provide evidence for a deficit in the automatic activation of semantic representations in children with ASD and suggest that this deficit is somehow more selective or more severe in the verbal domain than in the nonverbal domain.

Russo et al. (2012) analyzed audiovisual semantic integration in adolescents with ASD during a forced choice multisensory semantic congruence task. The results of this study indicate that people with ASD are able to process incongruence, but with different latency and by recruiting different areas of the cortex than TD people. For TD participants, the expected N400 effect occurred in the expected latency range (300–500 ms) and was detected in the frontal regions, while for people with ASD, the modulation was observed at an earlier latency (between 150 and 300 ms) and was more pronounced posterior and centro-temporal regions.

In 2016, Cantiani and colleagues analyzed the electrophysiological responses associated with visual and auditory sensory perception and lexical-semantic processing in nonverbal (NV) or minimally verbal (MV) children with autism spectrum disorder (ASD). Children's ERPs responses were recorded during the presentation of an image-word matching paradigm. The results showed that NT children showed an N400 effect in responses to mismatched verbal labels, but minimally/nonverbal children with ASD did not show it statistically. However, visual inspection of individual data revealed that five of the 10 children with ASD showed negativity consistent with the N400 effect, suggesting that some children in the ASD group were capable of processing stimuli at the semantic-linguistic level. However, the authors concluded that in NV/MV children with ASD, higher levels of lexical-semantic processing are impaired.

Using a similar paradigm, DiStefano et al. (2019) analyzed the electrophysiological responses during semantic processing in TD, verbal ASD, and minimally verbal ASD children. They presented an image-word matching paradigm in which participants consisted of presenting pairs of basic linguistic or non-linguistic nouns. Consistent with some previous studies (Ribeiro et al., 2013; McCleery et al., 2010) the data of this work showed that the N400 effect was present in all groups, with a shorter latency in the TD group than in the ASD children in both groups. The authors of this study concluded that children with ASD in both groups are characterized by delayed semantic processing speed.

In a recent work, Manfredi et al. (2021) analyzed ERP responses associated with auditory descriptive words and sounds associated with visual narratives in children and adolescents with ASD. The words and sounds were semantically congruent or incongruent with the culminating visual event. A similar N400 effect was observed in both neurotypical and ASD children and adolescents when they accessed different types of auditory information (i.e., words and sounds) in a visual narrative.

Overall, these studies on semantic processing in crossmodal contexts suggest that although difficulties with processing verbal material persist, verbal information processing in adolescents with ASD might be facilitated by direct association with meaningful visual information. In addition, differences emerged in the distribution of scalp and latency of brain responses between the groups of ASD and neurotypical children (Manfredi et al., 2021; Ribeiro et al., 2013; McCleery et al., 2010; Russo et al., 2012). This might suggests that children and adolescents with ASD differ from neurotypical children and adolescents in processing the multimodal combination of visual information at different temporal stages of semantic processing.

19.4 Oscillatory Activation

As discussed above, an alternative way of analyzing EEG data is to perform time-frequency decomposition and calculate the power (i.e., amplitude of oscillations) and coherence (i.e., synchrony of oscillations) in specific frequency bands.

Curl and Coderre (2022) used EEG coherence to examine neural connectivity during semantic processing in linguistic and visual modalities in adults with and without ASD. This study used the same semantic priming paradigm of related and unrelated pictures and words as was employed in Coderre et al. (2017). Curl and Coderre (2022) found that coherence in the theta frequency band, which is typically associated with semantic processing, was reduced for word, but not picture, stimuli in autistic individuals, suggesting impaired lexico-semantic connectivity. Even more intriguingly, this study found that these differences in neural connectivity during language processing were localized to extremely early time windows (within 100–300 ms), which implies fundamental differences in the neural connections of semantic processing in autism.

Ortiz-Mantilla et al. (2019) investigated the neural mechanisms underlying visual processing of common objects in MV-ASD children and control children. Children in both groups were exposed to an image-word matching paradigm. Compared with controls, the ASD group showed lower power in theta bands in occipital regions and gamma bands in frontal regions when processing semantically related visual information. In addition, reduced gamma power was found at the level of the anterior cingulate cortex, suggesting fewer attentional resources for processing semantically related visual information. The authors concluded that language difficulties in minimally verbal children with ASD may result from impaired representations of object labels in the cortex and reduced attentional specificity.

19.4.1 Future Directions

19.4.1.1 The Visual Ease Assumption

In the clinical and educational literature in autism, there is a widespread assumption (often implicit, sometimes explicit) that visual narratives provide a non-linguistic way of assessing cognition which may bypass any verbal processing bottlenecks: a proposal that Coderre (2020) has termed the Visual Ease Assumption.

This assumption has led to the creation of many tests and assessments that use visual stimuli to assess cognition. One such task is a narrative elicitation task, which ask participants to tell a story from a wordless picture book. Transcript analyses are then used to examine narrative and discourse production abilities (Berman, 2009; Liles, 1993). This type of task is included in many diagnostic assessments for clinical populations, including the ADOS, which includes a section on Telling a Story from a Book. Another task is a picture arrangement task, in which participants are asked to arrange randomly presented pictures in a temporal order to construct a coherent story (Allen et al., 1991). This type of task is included in many intelligence tests, including the Wechsler intelligence scales (Wechsler, 2008, 2014).

Visual narratives are also used to assess other aspects of cognition such as theory of mind (Hutchins et al., 2008; Sivaratnam et al., 2012) and are often used to compensate for deficits in other cognitive areas such as attention, auditory processing, sequencing, and organization (Dettmer et al., 2000), and language. For instance, various forms of augmentative and alternative communication (AAC) use picture stimuli, such as the Picture Exchange Communication System (PECS), used often for individuals with ASD who have limited functional language (Bondy & Frost, 1994; Hutchins & Prelock, 2014; Preston & Carter, 2009).

The VEA has also led to interventions that use visual narratives to train narrative comprehension abilities (Davis et al., 2007), social cognition (e.g., Social Stories (Gray & Garand, 1993; Hutchins & Prelock, 2006, 2013; Sansosti et al., 2004) and Comic Strip Conversations (Glaeser et al., 2003; Gray, 1994; Hutchins & Prelock, 2006)), and reading comprehension (e.g., the Visualizing and Verbalizing (V/V) program in the Lindamood-Bell Learning Processes reading intervention program (Bell, 1991; Bell & Lindamood, 1991)).

Despite the pervasive use of visual stimuli in cognitive tasks and interventions for ASD, as we have seen from the literature reviewed above, there is mixed evidence on the notion that visuo-semantic processing is "easier" for individuals with ASD. Thus far, the literature seems to suggest that at levels of basic, perceptual processing, as well as at the level of single pictures, visual processing is intact in ASD and does not drastically differ from NT peers. However, at higher levels of semantic processing with more complex visual stimuli, such as sequential images or visual narratives, challenges with visuo-semantic processing emerge.

This may suggest that it is not the *modality* that determines whether individuals with ASD will experience challenges, but the *level of complexity* of semantic processing. This proposal would also fit in with patterns seen in the linguistic realm: even in language, deficits are not universal in ASD. Individuals with ASD often demonstrate intact word decoding abilities, and some evidence reviewed above has even found similar neural responses for semantic processing of single words or pairs of words. However, when moving to more complex levels of semantic processing such as the level of sentences or narratives, difficulties are more pronounced and group differences more often occur. Taken together, what this evidence seems to be pointing to is not that semantic processing of language is impaired while semantic processing of visual stimuli is intact; that is, it is not the *modality* that matters. Rather, lower-level semantic processing seems to be relatively intact in ASD but higher-level semantic processing is impaired; that is, it is the *level of complexity* that matters.

At this point, however, this conclusion remains speculative because the Visual Ease Assumption remains to be extensively tested. As discussed in Coderre (2020), an empirical test of this assumption would require that an ASD group be directly compared against a TD group and that linguistic and visual modalities be directly

Table 19.1 Summary of findings for verbal, visual and cross-modal semantic processing in ASD individuals

	Main findings	Electrophysiological responses
Verbal processing	Deficit in the linguistic-semantic domain is present in individuals with autism from childhood through adulthood	N400: attenuated or absent Sustained negativity: attenuated or absent LP: attenuated or absent Reduced theta band
Visual processing	Mixed findings: evidence of impaired comprehension, similar to that found for linguistic material, and evidence of intact processing.	N400: attenuated or similar to that observed in neurotypical individuals. Sustained negativity: absent LP: absent Reduced theta and gamma bands
Cross-modal processing	Verbal information processing in children and adolescents with ASD might be facilitated by direct association with meaningful visual information.	N400, LP, Sustained negativity: attenuated or similar to that observed in neurotypical individuals.

compared. This second requirement in particular is rarely included in experimental designs, but is a crucial comparison for any claim that visual processing is "easier" than linguistic processing. Ultimately, more research is needed before any claims of "visual strengths" in individuals with ASD can be substantiated (Table 19.1).

References

Allen, M. H., Lincoln, A. J., & Kaufman, A. S. (1991). Sequential and simultaneous processing abilities of high-functioning autistic and language-impaired children. *Journal of Autism and Developmental Disorders, 21*(4), 483–502.

American Psychiatric Association (2000). Diagnostic and Statistical Manual of Mental Disorders (4th edn, text revision), Washington, DC. American Psychiatric Association.

Amoruso, L., Gelormini, C., Aboitiz, F., Alvarez González, M., Manes, F., Cardona, J., & Ibanez, A. (2013). N400 ERPs for actions: Building meaning in context. *Frontiers in Human Neuroscience, 7*, 1–16. https://doi.org/10.3389/fnhum.2013.00057

Baggio, G. (2018). *Meaning in the brain*. MIT Press.

Baggio, G., van Lambalgen, M., & Hagoort, P. (2008). Computing and recomputing discourse models: An ERP study. *Journal of Memory and Language, 52*(1), 36–53. https://doi.org/10.1016/j.jml.2008.02.005

Bell, N. (1991). Gestalt imagery: A critical factor in language comprehension. *Annals of Dyslexia, 41*, 246–260.

Bell, N., & Lindamood, P. (1991). *Visualizing and verbalizing for language comprehension and thinking*. Academy of Reading Publications.

Berman, R. A. (2009). Language development in narrative contexts. In E. L. Bavin (Ed.), *Cambridge handbook of child language* (pp. 356–375). Cambridge University Press.

Bondy, A. S., & Frost, L. A. (1994). The picture exchange communication system. *Focus on Autistic Behavior, 9*(3), 1–19.

Braeutigam, S., Swithenby, S. J., & Bailey, A. J. (2008). Contextual integration the unusual way: A magnetoencephalographic study of responses to semantic violation in individuals with autism spectrum disorders. *European Journal of Neuroscience, 27*(4), 1026–1036. https://doi.org/10.1111/j.1460-9568.2008.06064.x

Brouwer, H., & Hoeks, J. C. J. (2013). A time and place for language comprehension: Mapping the N400 and the P600 to a minimal cortical network. *Frontiers in Human Neuroscience, 7*, 1–12. https://doi.org/10.3389/fnhum.2013.00758

Brouwer, H., Fitz, H., & Hoeks, J. (2012). Getting real about Semantic Illusions: Rethinking the functional role of the P600 in language comprehension. *Brain Research, 1446*, 127–143.

Brouwer, H., Crocker, M. W., Venhuizen, N. J., & Hoeks, J. C. J. (2017). A neurocomputational model of the N400 and the P600 in language processing. *Cognitive Science, 41*, 1318–1352. https://doi.org/10.1111/cogs.12461

Cantiani, C., Choudhury, N. A., Yu, Y. H., Shafer, V. L., Schwartz, R. G., & Benasich, A. A. (2016). From sensory perception to lexical-semantic processing: An ERP study in non-verbal children with autism. *PLoS One, 11*(8), 1–31. https://doi.org/10.1371/journal.pone.0161637

Coderre, E. L. (2020). Dismantling the "visual ease assumption": A review of visual narrative processing in clinical populations. *Topics in Cognitive Science, 12*(1), 224–225.

Coderre, E. L., Chernenok, M., Gordon, B., & Ledoux, K. (2017). Linguistic and non-linguistic semantic processing in individuals with autism spectrum disorders: An ERP study. *Journal of Autism and Developmental Disorders, 47*(3), 795–812. https://doi.org/10.1007/s10803-016-2985-0

Coderre, E. L., Cohn, N., Slipher, S. K., Chernenok, M., Ledoux, K., & Gordon, B. (2018). Brain and Language Visual and linguistic narrative comprehension in autism spectrum disorders: Neural evidence for modality-independent impairments. *Brain and Language, 186*(June), 44–59. https://doi.org/10.1016/j.bandl.2018.09.001

Cohn, N., Paczynski, M., Jackendoff, R., Holcomb, P. J., & Kuperberg, G. R. (2012). (pea)nuts and bolts of visual narrative: Structure and meaning in sequential image comprehension. *Cognitive Psychology, 65*(1), 1–38. https://doi.org/10.1016/j.cogpsych.2012.01.003

Curl, A., & Coderre, E. L. (2022). The time-locked neurodynamics of semantic processing in autism spectrum disorder: An EEG study. *Cognitive Neurodynamics, 16*(1), 43–72. https://doi.org/10.1007/s11571-021-09697-8

Davis, M., Dautenhahn, K., Nehaniv, C. L., & Powell, S. D. (2007). The narrative construction of our (social) world: Steps towards an interactive learning environment for children with autism. *Universal Access in the Information Society, 6*(2), 145–157.

Dettmer, S., Simpson, R. L., Smith Myles, B., & Ganz, J. B. (2000). The use of visual supports to facilitate transition of students with autism. *Focus on Autism and Other Developmental Disabilities, 15*(3), 163–169.

DiStefano, C., Senturk, D., & Jeste, S. S. (2019). ERP evidence of semantic processing in children with ASD. *Developmental Cognitive Neuroscience, 36*(March), 100640. https://doi.org/10.1016/j.dcn.2019.100640

Dunn, M. A., & Bates, J. C. (2005). Developmental change in neutral processing of words by children with autism. *Journal of Autism and Developmental Disorders, 35*(3), 361–376. https://doi.org/10.1007/s10803-005-3304-3

Dunn, M., Vaughan, H., Jr., Kreuzer, J., & Kurtzberg, D. (1999). Electrophysiologic correlates of semantic classification in autistic and normal children. *Developmental Neuropsychology, 16*(1), 79–99. https://doi.org/10.1207/S15326942DN160105

Fishman, I., Yam, A., Bellugi, U., Lincoln, A., & Mills, D. (2011). Contrasting patterns of language-associated brain activity in autism and Williams syndrome. *Social Cognitive and Affective Neuroscience, 6*(5), 630–638. https://doi.org/10.1093/scan/nsq075

Ganis, G., Kutas, M., & Sereno, M. I. (1996). The search for "common sense": An electrophysiological study of the comprehension of words and pictures in reading. *Journal of Cognitive Neuroscience, 8*(2), 89–106.

Glaeser, B. C., Pierson, M. R., & Fritschmann, N. (2003). Comic strip conversations: A positive behavioral support strategy. *Teaching Exceptional Children, 36*(2), 14–19.

Gray, C. A. (1994). *Comic strip conversations: Illustrated interactions that teach conversation skills to students with autism and related disorders*. Future Horizons.
Gray, C. A., & Garand, J. D. (1993). Social stories: Improving responses of students with autism with accurate social information. *Focus on Autistic Behavior, 8*(1), 1–10.
Hamm, J. P., Johnson, B. W., & Kirk, I. J. (2002). Comparison of the N300 and N400 ERPs to picture stimuli in congruent and incongruent contexts. *Clinical Neurophysiology, 113*, 1339–1350.
Hutchins, T. L., & Prelock, P. A. (2006). Using social stories and comic strip conversations to promote socially valid outcomes for children with autism. *Seminars in Speech and Language, 27*(1), 47–59.
Hutchins, T. L., & Prelock, P. A. (2013). The social validity of Social Stories™ for supporting the behavioural and communicative functioning of children with autism spectrum disorder. *International Journal of Speech-Language Pathology, 15*(4), 383–395.
Hutchins, T. L., & Prelock, P. A. (2014). Using communication to reduce challenging behaviors in individuals with autism spectrum disorders and intellectual disability. *Child and Adolescent Psychiatric Clinics of North America, 23*, 41–55.
Hutchins, T. L., Prelock, P. A., & Chace, W. (2008). Test retest reliability of a theory of mind task battery for children with autism spectrum disorders. *Focus on Autism and Other Developmental Disabilities, 23*(4), 195–206.
Kelley, E., Paul, J. J., Fein, D., & Naigles, L. R. (2006). Residual language deficits in optimal outcome children with a history of autism. *Journal of Autism and Developmental Disorders, 36*(6), 807–828. https://doi.org/10.1007/s10803-006-0111-4
Kröger, S., Rutter, B., Hill, H., Windmann, S., Hermann, C., & Abraham, A. (2013). An ERP study of passive creative conceptual expansion using a modified alternate uses task. *Brain Research, 1527*, 189–198. https://doi.org/10.1016/j.brainres.2013.07.007
Kujala, T., Lepistö, T., & Näätänen, R. (2013). The neural basis of aberrant speech in autism spectrum disorders. *Neuroscience & Biobehavioral Reviews, 37*, 1–8. http://www.sciencedirect.com/science/article/pii/S0149763413000079
Kuperberg, G. R. (2007). Neural mechanisms of language comprehension: Challenges to syntax. *Brain Research, 1146*, 23–49. https://doi.org/10.1016/j.brainres.2006.12.063
Kutas, M., & Federmeier, K. D. (2000). Electrophysiology reveals semantic memory use in language comprehension. *Trends in Cognitive Sciences, 4*(2), 463–470.
Kutas, M., & Federmeier, K. D. (2011). Thirty years and counting: Finding meaning in the N400 component of the event-related brain potential (ERP). *Annual Review of Psychology, 62*, 621–647. https://doi.org/10.1146/annurev.psych.093008.131123
Kutas, M., & Hillyard, S. A. (1980). Reading senseless sentences: Brain potentials reflect semantic incongruity. *Science, 207*(4427), 203–205. https://doi.org/10.1126/science.7350657
Kutas, M., & Hillyard, S. (1984). Brain potentials during reading reflect word expectancy and semantic association. *Nature, 307*(12), 161–163.
Lau, E. F., Phillips, C., & Poeppel, D. (2008). A cortical network for semantics: (de)constructing the N400. *Nature Reviews. Neuroscience, 9*(12), 920–933. https://doi.org/10.1038/nrn2532
Liles, B. Z. (1993). Narrative discourse in children with language disorders and children with normal language: A critical review of the literature. *Journal of Speech and Hearing Research, 36*(5), 868–882.
Manfredi, M., Cohn, N., De Araújo Andreoli, M., & Boggio, P. S. (2018). Listening beyond seeing: Event-related potentials to audiovisual processing in visual narrative. *Brain and Language, 185*, 1–8. https://doi.org/10.1016/j.bandl.2018.06.008
Manfredi, M., Cohn, N., Sanchez Mello, P., Fernandez, E., & Boggio, P. S. (2020). Visual and verbal narrative comprehension in children and adolescents with autism spectrum disorders: An ERP study. *Journal of Autism and Developmental Disorders, 50*, 0123456789–0123452672. https://doi.org/10.1007/s10803-020-04374-x
Manfredi, M., Cohn, N., Ribeiro, B., Sanchez, P., Fernandes, E., Pereira, R., & Sergio, P. (2021). The electrophysiology of audiovisual processing in visual narratives in adolescents with autism spectrum disorder. *Brain and Cognition, 151*(February), 105730.

McCleery, J. P., Ceponiene, R., Burner, K. M., Townsend, J., Kinnear, M., & Schreibman, L. (2010). Neural correlates of verbal and nonverbal semantic integration in children with autism spectrum disorders. *Journal of Child Psychology and Psychiatry and Allied Disciplines, 51*(3), 277–286. https://doi.org/10.1111/j.1469-7610.2009.02157.x

McPherson, W. B., & Holcomb, P. J. (1999). An electrophysiological investigation of semantic priming with pictures of real objects. *Psychophysiology, 36*(1), 53–65. https://doi.org/10.1017/S0048577299971196

O'Connor, K. (2012). Auditory processing in autism spectrum disorder: A review. *Neuroscience and Biobehavioral Reviews, 36*(2), 836–854. https://doi.org/10.1016/j.neubiorev.2011.11.008

O'Rourke, E., & Coderre, E. L. (2021). Implicit semantic processing of linguistic and non-linguistic stimuli in adults with autism spectrum disorder. *Journal of Autism and Developmental Disorders, 51*(8), 2611–2630. https://doi.org/10.1007/s10803-020-04736-5

Ortiz-Mantilla, S., Cantiani, C., Shafer, V. L., & Benasich, A. A. (2019). Minimally-verbal children with autism show deficits in theta and gamma oscillations during processing of semantically-related visual information. *Scientific Reports, 9*(1), 1–12.

Osterhout, L., & Holcomb, P. J. (1992). Event-related brain potentials elicited by syntactic anomaly. *Journal of Memory and Language, 31*, 785–806.

Pijnacker, J., Geurts, B., van Lambalgen, M., Buitelaar, J., & Hagoort, P. (2010). Exceptions and anomalies: An ERP study on context sensitivity in autism. *Neuropsychologia, 48*(10), 2940–2951. https://doi.org/10.1016/j.neuropsychologia.2010.06.003

Preston, D., & Carter, M. (2009). A review of the efficacy of the picture exchange communication system intervention. *Journal of Autism and Developmental Disorders, 39*(10), 1471–1486.

Ribeiro, T. C., Valasek, C. A., Minati, L., & Boggio, P. S. (2013). Altered semantic integration in autism beyond language: A cross-modal event-related potentials study. *Neuroreport, 24*(8), 414–418. https://doi.org/10.1097/WNR.0b013e328361315e

Russo, N., Mottron, L., Burack, J. A., & Jemel, B. (2012). Parameters of semantic multisensory integration depend on timing and modality order among people on the autism spectrum: Evidence from event-related potentials. *Neuropsychologia, 50*(9), 2131–2141. https://doi.org/10.1016/j.neuropsychologia.2012.05.003

Sansosti, F. J., Powell-Smith, K. A., & Kincaid, D. (2004). A research synthesis of social story interventions for children with autism Spectrum disorders. *Focus on Autism and Other Developmental Disabilities, 19*(4), 194–204.

Sereno, S. C., & Rayner, K. (2003). Measuring word recognition in reading: Eye movements and event-related potentials. *Trends in Cognitive Sciences, 7*(11), 489–493.

Sitnikova, T., West, W. C., Kuperberg, G. R., & Holcomb, P. J. (2006). The neural organization of semantic memory: Electrophysiological activity suggests feature-based segregation. *Biological Psychology, 71*(3), 326–340.

Sitnikova, T., Holcomb, P. J., Kiyonaga, K. A., & Kuperberg, G. R. (2008). Two neurocognitive mechanisms of semantic integration during the comprehension of visual real-world events. *Journal of Cognitive Neuroscience, 20*(11), 2037–2057.

Sivaratnam, C. S., Cornish, K., Gray, K. M., Howlin, P., & Rinehart, N. J. (2012). Brief report: Assessment of the social-emotional profile in children with autism spectrum disorders using a novel comic strip task. *Journal of Autism and Developmental Disorders, 42*(11), 2505–2512.

van de Meerendonk, N., Kolk, H. H. J., Vissers, C. T. W. M., & Chwilla, D. J. (2010). Monitoring in language perception: Mild and strong conflicts elicit different ERP patterns. *Journal of Cognitive Neuroscience, 22*(1), 67–82.

Van Petten, C., & Luka, B. J. (2012). Prediction during language comprehension: Benefits, costs, and ERP components. *International Journal of Psychophysiology, 83*(2), 176–190. https://doi.org/10.1016/j.ijpsycho.2011.09.015

Võ, M. L. H. (2021). The meaning and structure of scenes. *Vision Research, 181*, 10–20. https://doi.org/10.1016/j.visres.2020.11.003

Wechsler, D. (2008). *Wechsler adult intelligence scale – Fourth edition (WAIS-IV)*. Pearson Education, Incorporated.

Wechsler, D. (2014). *Wechsler intelligence scale for children-fifth edition (WISC-V): Administration and scoring manual*. Pearson Education, Incorporated.
West, W. C., & Holcomb, P. J. (2002). Event-related potentials during discourse-level semantic integration of complex pictures. *Brain Research. Cognitive Brain Research, 13*, 363–375.
Yang, J., & Small, S. L. (2015). Language processing, functional magnetic resonance imaging of. In *International encyclopedia of the social & behavioral sciences: Second edition* (Vol. 13, Second ed., pp. 368–380). Elsevier. https://doi.org/10.1016/B978-0-08-097086-8.54033-1
Zhang, J., Jiang, C., Zhou, L., & Yang, Y. (2016). Perception of hierarchical boundaries in music and its modulation by expertise. *Neuropsychologia, 91*, 490–498. https://doi.org/10.1016/j.neuropsychologia.2016.09.013

Chapter 20
Gestational Exposure to Di-*n*-Butyl Phthalate Induces Autism-Like Behavior Through Inhibition of Neuro-Steroidogenesis

Françoise Sidime, Meriem Bendaoud, Maisara Abdelgawad, and Abdeslem El Idrissi

20.1 Introduction

Phthalates are liquid plasticizers used in the industrial production of consumer goods to impart flexibility in plastic products, food packaging materials, cosmetics and children's toys (Lottrup et al., 2006; Mylchreest et al., 2000; Petersen & Breindahl, 2000). As a result of their wide use, the potential for human exposure is significantly high, especially among individuals using polyvinyl chloride based devices like drinking bottles and plastic plates. Furthermore, young women of reproductive age using phthalate based cosmetic products have been reported to have high levels of phthalate metabolites in their urine (Blount et al., 2000; Kohn

F. Sidime
Helene Fuld School of Nursing, New York, NY, USA

The Center for Developmental Neuroscience, College of Staten Island, Staten Island, NY, USA

The Graduate Center, Program in Biology – Neurosciences, The City University of New York, New York, NY, USA

M. Bendaoud
Department of Biology, New Jersey City University, Jersey City, NJ, USA

M. Abdelgawad
The Center for Developmental Neuroscience, College of Staten Island, Staten Island, NY, USA

A. El Idrissi (✉)
Center for Developmental Neuroscience and Department of Biology, College of Staten Island, City University of New York, New York, NY, USA

Graduate Center, Program Biology-Neuroscience, City University of New York, New York, NY, USA
e-mail: Abdeslem.elidrissi@csi.cuny.edu

A. El Idrissi, D. McCloskey (eds.), *Neurobiology of Autism Spectrum Disorders*,
https://doi.org/10.1007/978-3-031-42383-3_20

et al., 2000; Koch et al., 2003; Thompson et al., 2004). Since phthalates are widely used, they are considered ubiquitous low level environmental contaminates.

Phthalates are endocrine-disrupting compounds (EDCs). The deleterious effects of Di-*n*-butyl phthalate (DBP), one of the most dominant phthalate esters affecting the endocrine system in human and animal models, is well documented (Crisp et al., 1998; Swan et al., 2005; Thompson et al., 2004). However, there is increasing concern over the potential impact of DBP on neurodevelopment (Li et al., 2013, 2014). A study by Kim and colleagues (2009) reported the association of DBP metabolites with neuropsychological impairment in children aged 8–11 years (Kim et al., 2009). Recently, significant progress was made on the effect of DBP *in utero*. Prenatal exposure to a high dose of DBP (500 mg/kg BW) induces hippocampal structural damage and neuronal loss in the developing rat brain. In addition, exposure to DBP induces impairment in spatial learning and memory (Li et al., 2013) through a downregulation of estradiol and its receptors (Li et al., 2014).

While steroids are synthesized by peripheral organs such as testes, ovaries, and the adrenal gland, neurosteroids are synthesized de novo by the brain (Akwa et al., 1991; Baulieu & Robel, 1990; Baulieu, 1991). An interesting finding in the field of endocrine steroids was the discovery that the brain expressed the steroidogenic acute regulatory protein (StAR) and the mitochondrial enzyme P450 side chain cleavage (P450scc). The biosynthesis of neurosteroids estrogen and pregnanolone proceeds through the transfer of cholesterol from the outer to the inner mitochondrial membrane through the steroidogenic acute regulatory protein (StAR). The cholesterol side chain is cleaved by cytochrome P450 side chain cleavage (P450scc), a rate limiting mitochondrial enzyme originally found in peripheral steroidogenic glands and subsequently in the brain (Baulieu & Robel, 1990). The cleavage results in cholesterol conversion to pregnanolone or estrogen, depending on the activation of other enzymes responsible for their synthesis (Baulieu & Robel, 1990; Baulieu, 1991).

De novo synthesis of active neurosteroids is elevated during embryogenesis and decreases early in development (King & Stocco, 2011). This spike in neurosteroid synthesis increases during neuronal growth, dendritic spine formation and brain development (Sierra et al., 2003) and coincides with peak expression of StAR and P450scc mRNA (King & Stocco, 2011). Later in life, neurosteroids regulate the physiology of neurons while providing neuro protection (McCarthy, 2008; King & Stocco, 2011). Neuroactive steroids regulate neuronal function through their concomitant influence on neuronal excitability and gene expression. There is now considerable evidence that neuroactive steroids have modulatory effects on the release of multiple neurotransmitters and are potent positive allosteric modulators of neurotransmitters. Pregnanolone is a positive allosteric modulator of $GABA_A$ receptors (Mensah-Nyagan et al., 1996; Akwa et al., 1991). Patch-clamp recording of inhibitory postsynaptic currents (IPSCs) in Purkinje cells from cerebellar slice showed that pregnenolone increased, dose-dependently, the frequency of IPSCs within 1 min of application (Mtchedlishvili & Kapur, 2003), indicating that this effect is unlikely to be induced through gene transcription. The modulation of neurotransmitter receptors by steroids is not limited to $GABA_A$ receptors. All members within the family of ligand-gated ion channels have been shown to be a target for steroid modulation (Lambert et al., 2003). Within the glutamate receptor family, N-methyl-d-aspartate (NMDA) receptors, α-amino-3-hydroxy-5-methyl-4-isoxazolepropionic acid (AMPA) receptors and kainate receptors have also been shown to be a substrate for steroids allosteric modulation (Lambert et al., 2003). At the molecular level, neuroactive steroids can bind to intracellular receptors that act as transcription factors and regulate gene expression. Therefore, active neurosteroids play an important role in neuronal development and function through their genomic and non-genomic effects.

The aim of this study was to determine the effect of gestational exposure to relatively low levels of DBP (1 mg/kg) on neurosteroidogenesis and neuronal development. Specifically, the effects of DBP on the expression levels of the rate-limiting enzymes in neurosteroids biosynthesis and the functional consequences on downstream target genes, $GABA_A$ and GAD.

20.2 Materials and Methods

20.2.1 Animals

All mice used in this study were two-month-old FVB/NJ males. Mice were reared in a controlled environment with a 12h:12h LD cycle and food and water available ad libitum. Use of animals was approved by the College of Staten Island's Institutional Animal Care and Use Committee (IACUC) according to NIH guidelines. Pregnant dams were injected with 1 mg/kg s.c. DBP at gestational day 10. For euthanasia, animals were injected with 125 mg/kg Nembutal i.p.

20.2.2 High-Performance Liquid Chromatography (HPLC)

Whole brains from euthanized pregnant dams and their pups (+/−) DBP were homogenized in hexane using 100 mg of brain tissue in 10 ml of hexane. The samples were lyophilized for 24 hours to remove any traces of hexane. The lyophilized tissues were then dissolved in acetonitrile and injected into the HPLC machine. The DBP standard curve was generated for the quantification of DBP tissue levels. Peaks generated in all the tissues (+/−) DBP were quantified and recorded by measuring the area under the peak and calculating the respective concentration.

20.2.3 Western Blotting

Brains from euthanized mice were removed and dissected into the cortex, hippocampus, cerebellum, brainstem and diencephalon and were frozen on dry ice within 3 min of the sacrifice. Total soluble and membrane-bound proteins were extracted from each region (Benke et al., 1999). Protein concentrations of the extracts were determined using a BCA protein assay kit (Sigma Aldrich) and a microplate spectrophotometer (Spectra MAX 340PC – Molecular Devices, CA, USA). After separation SDS-PAGE, the separated proteins were electrophoretically transferred to a PVDF membrane (Fisher Scientific, USA) in Tris-buffered saline with 20% methanol. The membranes were probed with antibodies directed against P450 (1:10,000 dilution – EMD, Millipore, USA), anti-StAR (1:500 dilution – Abcam, USA), $GABA_AR$ β2/3 (1:100 dilution – Chemicon International, USA), GAD67 (1:5000 dilution - EMD, Millipore, USA), or β actin (1:10,000 – dilution, Sigma, USA) and followed by a secondary antibody incubation using HRP conjugated goat anti-mouse or goat anti -rabbit IgG (1:1000 dilution - Santa Cruz Biotechnology, USA). An HRP enhanced chemiluminescence (ECL) detection kit (Pierce, USA) was used to detect immunoreactive proteins. Densitometry was used to measure the intensity of the protein bands. The relative density of the protein bands was calculated and normalized to the density of β-actin using Image J (NIH, USA).

20.2.4 Immunohistochemical Analysis

Two month old male mice (WT, WT-DBP) were Transcardially exsanguinated using phosphate buffer saline followed by 4% paraformaldehyde (PFA), and brains were removed and post-fixed in 4% PFA for 24 hours at 4 °C. Brains were cryoprotected in 30% sucrose, embedded in OCT (ProScitech, USA) and cyrosectioned at 30 μm. Free-floating sections were blocked and permeabilized with 2% nonfat dry milk (NFDM) in PBS containing 0.02% Triton X-100 and 10% normal goat serum (NGS) on a shaker for 1 hour. Samples were incubated overnight at 4 °C in primary antibodies: rabbit anti-P450 (1:8000 dilution – EMD, Millipore, USA) and mouse anti-StAR (1:500 dilution – Abcam, USA). Following immunolabeling, samples were washed in PBS and incubated with Cy 5-conjugated goat anti-mouse (Jackson Immuno-Research laboratories, USA) or Cy 3-conjugated goat anti-rabbit IgG (Jackson Immuno-Research laboratories, USA) secondary antibodies for 1 hour at RT. After washing with PBS, floating sections were mounted on slides and counterstained with Prolong Gold antifade containing DAPI (Invitrogen, USA) and cover slipped. The expression and distribution of the proteins were visualized and analyzed using the SP2 AOBS confocal microscope (Leica, USA). For all image acquisitions, the gain and offset were constant.

20.2.5 cDNA Preparation and Quantitative Real-Time PCR Analysis

Total RNA was extracted from frozen brain tissue using TRIZOL ® (Invitrogen, USA) and cleaned to eliminate any potential genomic DNA contamination using Qiagen RNeasy mini kit (Valencia, CA) according to the manufacturer's instruction. OD readings using the Nanodrop 1000 Spectrophotometer (Thermo Scientific, USA) at 260 nm and 280 nm were conducted to determine the RNA purity and concentration. Reverse transcription was conducted using the USB First Strand cDNA synthesis kit (Affymetrix, USA) in a reaction mixture that contained 10X transcription buffer, ribonuclease (RNase) inhibitor, 10 mM deoxynucleotides (dNTPs), and Moloney murine leukemia virus reverse transcriptase (M-MLV RT). Total RNA (1 μg) was reverse transcribed for 60 minutes at 44 °C; transcription was terminated by increasing the temperature to 92 °C for 10 minutes and finally kept at 4 °C. A volume of 1 μl of cDNA was used for quantitative real-time PCR analysis. cDNA was mixed in a 12.5 μl reaction mix containing SYBR® Green Master Mix (Applied Biosystem, USA) and 0.5 μl of 10 nM primer (Table 20.1). In order to activate SYBR® Green Master Mix that contains Hotstart DNA Taq polymerase, a 10 min heat activation step at 95 °C was conducted. Quantitative RT-PCR was performed using a thermal cycler (Biorad, USA) employing 40 cycles of denaturation for 15 seconds and 1 minute of annealing at 60 °C. Samples were run in triplicate and the mean Ct values were quantified to determine relative expression. The data were normalized to glyceraldehyde 3-phosphate dehydrogenase (GAPDH). GAPDH gene was chosen as the reference for normalizing target genes because it was consistently and reproducibly expressed in all samples.

Table 20.1 Primers employed in RT-qPCR (Integrated DNA Technologies)

Gene ID: StAR - Gene Name: Stard3nl - NM_024270 Forward: 5′-TTC GAT TCA AAG TGC TGA TAC TTG-3′ Reverse: 5′-GCT AAT AGA AAG GCA CTG GTC A-3′
Gene ID: P450scc - Gene Name: Cyp11b1 - NM_001033229 Forward: 5′-CAG GCT AAC TCA ATG GAA CTC A-3′ Reverse: 5′-GCC AGC TCA AAG AAA GTC AT-3′
Gene ID: GABAARB3 - Gene Name: Gabrb3 - NM_001038701 Forward: 5′-GCT CCC ACA GTT CTC CAT TG-3′ Reverse: 5′-ACC CGA TAT TTC TCT TCA ACC G-3′
Gene ID: GAD67 - Gene Name: GAD1 - NM_008077 Forward: 5′-CTT GCC TAC TAT CAC CAC GAC-3′ Reverse: 5′-ACA ATA ACC ACC AGA GAC ACA G-3′
Gene ID: GADPH - Gene Name: Slc12a5 - NM_008084 Forward: 5′-CCA CCT CTG CTG TCT ACA TC-3′ Reverse: 5′-AGC CTA TGA CGA TGA CCC A-3′

20.2.6 Scanning Electron Microscopy

After perfusion of mice with 4% paraformaldehyde, testes were dissected, post-fixed for 2 h at 4 °C, cross-sectioned in the midline and fixed separately in 2.5% glutaraldehyde at 4 °C for 12 h. Testes were dehydrated through an ascending ethanol series (70, 80, 85, 90 and 95%; 10 min at each concentration) and then completely dehydrated twice in 100% ethanol solution (for 15 min each time). The prepared testes were stored in a glass container with a desiccant and dried for 24 h. After drying, the specimens were mounted on a holder and sputter coated with gold-palladium. Testes samples were then observed and photographed with an SU-8020 (Hitachi Ltd., Tokyo, Japan) at 5 kV.

20.2.7 Statistical Analysis

SPSS 18.0 (SPSS, Chicago, IL, USA) was used for statistical analysis. Experimental results were expressed as mean ± standard deviation. The Levene's test was employed to check the normality and homogeneity of variances in the data. The Student's t test was used for paired comparisons. For the comparison of three or more groups, one-way ANOVA was used, which was followed by Duncan's post hoc test. Values of $p < 0.05$ were considered statistically significant.

20.2.8 Results

20.2.8.1 DBP Levels in the Brain of Offspring

In order to confirm that gestational exposure to low levels of DBP targets the brain of offspring mice, we used HPLC to measure the levels of DBP in both the injected pregnant dam and offspring's brain tissue. The subcutaneous delivery of a single dose of DBP (1 mg/kg BW) at gestational day 10 to the pregnant dam resulted in detectable traces of DBP in the brain of both the mother and offspring. Both dams and offspring were sacrificed on post natal day 1 (PND 1) when the pups were born and the levels of DBP in their brains were examined. DBP was injected into the HPLC and the retention peak was used as a reference. We found that pregnant dams had a DBP concentration of 0.5 ug/g of brain weight, whereas the offspring had a concentration of 0.05ug/g of brain weight. These data indicate that DBP was metabolized by the injected dams and crossed the placenta and blood brain barrier of the offspring and dams respectively.

20.2.8.2 Gestational Exposure to DBP Disrupts Key Proteins in Steroidogenesis

Expression of P450scc and StAR is critically important for the synthesis of neurosteroids that may regulate gene transcription (McCarthy, 2008). Neuroactive steroids regulate neuronal function through their concurrent influence on neuronal excitability and gene expression. To analyze the endocrine disrupting properties of DBP on neurosteroidogenesis, pregnant mice were injected subcutaneously with a single dose of DBP (1 mg/kg BW) at gestational day 10. Pregnant age-matched mice were injected with PBS and served as controls. Using quantitative real-time PCR analysis, ~~and~~ we investigated the effects of DBP on P450scc and StAR transcription. We found that DBP significantly reduced the mRNA levels for both StAR and P450scc. We also examined the transcripts of the β3 subunit of the $GABA_A$ receptors and GAD67 and found that the transcription levels of both of these genes was significantly reduced (Fig. 20.1). $GABA_A$ receptors are critically important for the development of immature neuronal networks. We hypothesized that DBP interfered with the biosynthesis of the neuroactive steroids by reducing the expression of P450scc and StAR. As a result, DBP induced a developmental delay in the maturation of the GABAergic system through inhibition of neuroactive steroid synthesis. Dysregulation in GABA-mediated inhibition has been shown as a critical link to

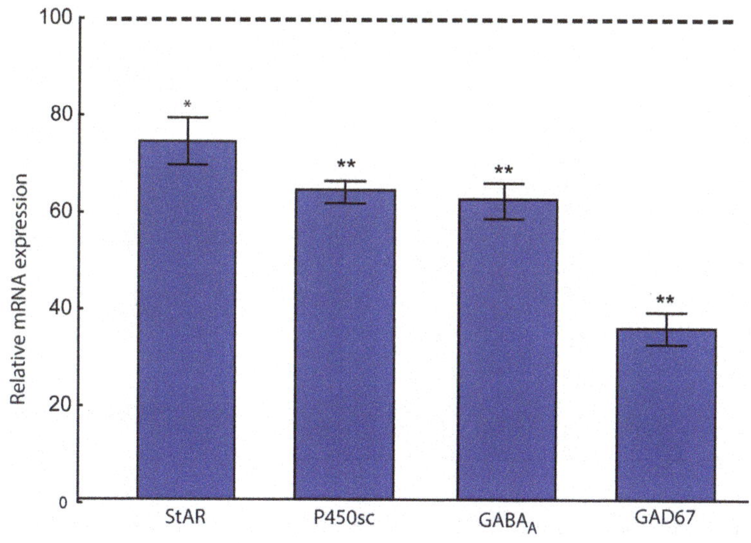

Fig. 20.1 mRNA expression of StAR, P450scc, $GABA_A$ β3 and GAD67 in the hippocampus. In utero exposure to DBP at gestational day 10 resulted in a significant decrease in the mRNA levels of both steroidogenic and GABAergic gene mRNAs that lasted into adulthood. Bars represent the relative mRNA expression normalized to GADPH mRNA and to the respective mRNA from control hippocampi. Statistical values are representative of an ANOVA with Duncan's HSD post hoc test alpha level 0.05% ($p < 0.05$*). Data represent mean ± SEM of at least three triplicated independently repeated experiments

neurological psychiatric disorders, most likely reflecting changes in GABAergic circuits, including axonal sprouting and formation of novel synapses (D'Hulst et al., 2006; El Idrissi et al., 2005; Selby et al., 2007).

20.2.8.3 Gestational Exposure to DBP Reduces the Expression of Steroidogenic and GABAergic Proteins

Consistent with this reduction in mRNA expression of StAR, P450scc, GAD67 and the β3 subunit of the $GABA_A$ receptors, protein expression was also significantly reduced. Western blot analysis of hippocampi from 2 months old male offspring of DBP injected dams showed a significant reduction of both steroidogenic and GABAergic proteins (Fig. 20.2). These data suggest that gestational exposure to endocrine disruptors induces long-lasting biochemical and neurological changes that persist into adulthood. Reduced expression of steroidogenic proteins in the brain will probably lead to reduced level of neuroactive steroids and this effect is

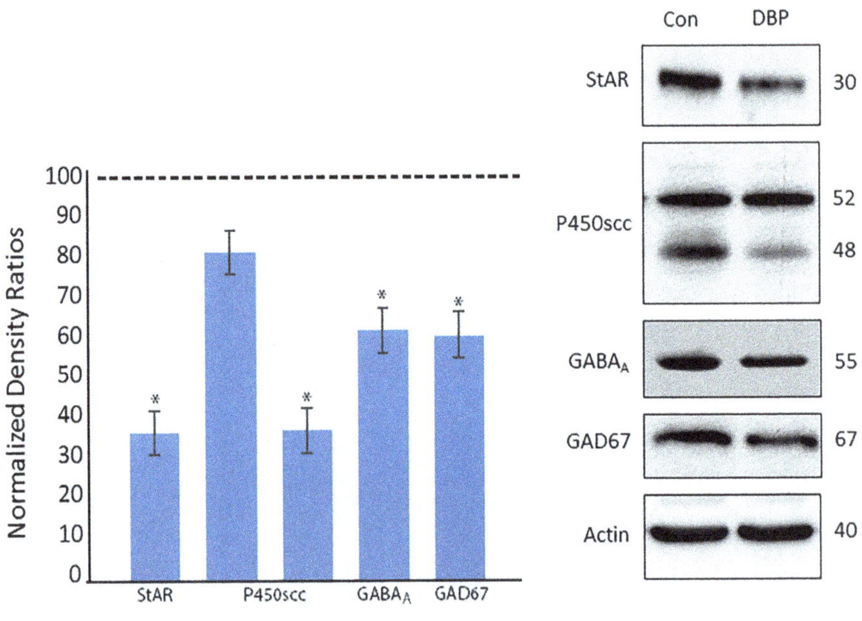

Fig. 20.2 Gestational exposure to DBP reduces the expression of steroidogenic and GABAergic proteins in the hippocampus. Representative image of a western blot showing immunoreactivity of StAR, P450scc, β3 of the $GABA_A$, GAD67 and Actin in the hippocampus from PND60 mice. Densitometry analysis of the western blots revealed that gestational exposure to DBP caused a significant reduction in the expression levels of StAR, P450scc, β3 of the $GABA_A$ and GAD67. P450 western blot showed two bands of 52 and 48 KDa. The high molecular weight isoform was not significantly affected by DBP treatment, whereas the 48 KDa was significantly reduced. Bars represent the densitometric ratios of StAR, P450scc, β3 of the $GABA_A$ and GAD67 to actin. Data represent mean ± SEM of at least three independently repeated experiments

further exacerbated by the reduced expression of steroidogenic proteins in the testes during adulthood. Western blot analysis of testis from 2 month old offspring of DBP injected dams showed a significant reduction of both steroidogenic proteins examined (StAR and P450scc) (Fig. 20.3). Further analysis of the spatial distribution of StAR and P450scc with immunofluorescence and confocal microscopy revealed a significant reduction in the expression and co-localization of both antigens (Fig. 20.4). Interestingly, both StAR and P450scc were expressed in endothelial cells of capillaries, suggesting a role for steroids in angiogenesis (Fig. 20.4e).

Gestational exposure to elevated levels of DBP has been shown to induce abnormal development of the testis and reproductive tract in male offspring. Although there is insufficient data for DBP effects on human reproduction, some studies in rodents have reported the influence of DBP on the male reproductive system (Alam et al., 2010). Previous studies have shown that in utero exposure to DBP leads to testicular histopathology and diminished testosterone production by the fetal rat testis (Foster et al., 2001; Shultz et al., 2001). Here we examined the potential toxic effect of DBP on testicular histology using scanning electron microscopy. We found that exposure to low doses of DBP *in utero* did not induce any marked histological abnormalities in the testes of the offspring (Fig. 20.5). We also examined the expression levels of StAR and P450scc in the testes using immunofluorescence and

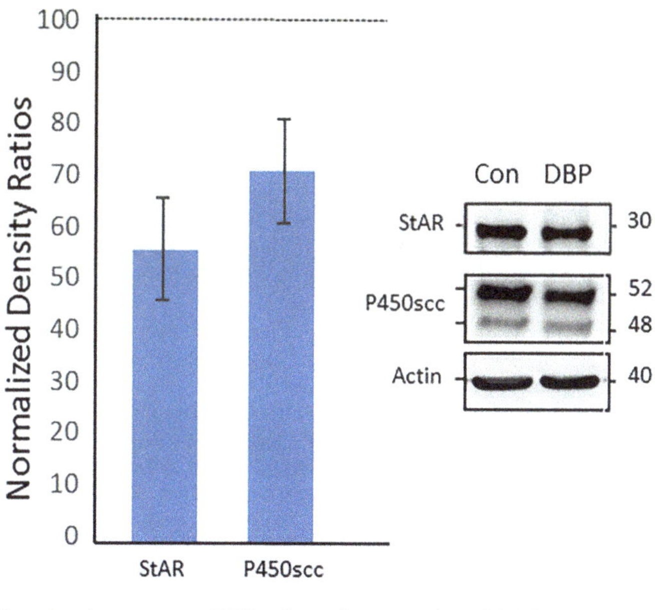

Fig. 20.3 Gestational exposure to DBP reduces the expression of StAR and P450 in the Testis. Representative image of a western blot showing immunoreactivity of StAR, P450scc and Actin in the testes of PND60 mice. Densitometry analysis of the western blots revealed that gestational exposure to DBP caused a significant reduction in the expression levels of StAR and P450scc proteins in testes of DBP exposed mice. Bars represent the densitometric ratios of StAR and P450scc to actin. Data represent mean ± SEM of at least three independently repeated experiments

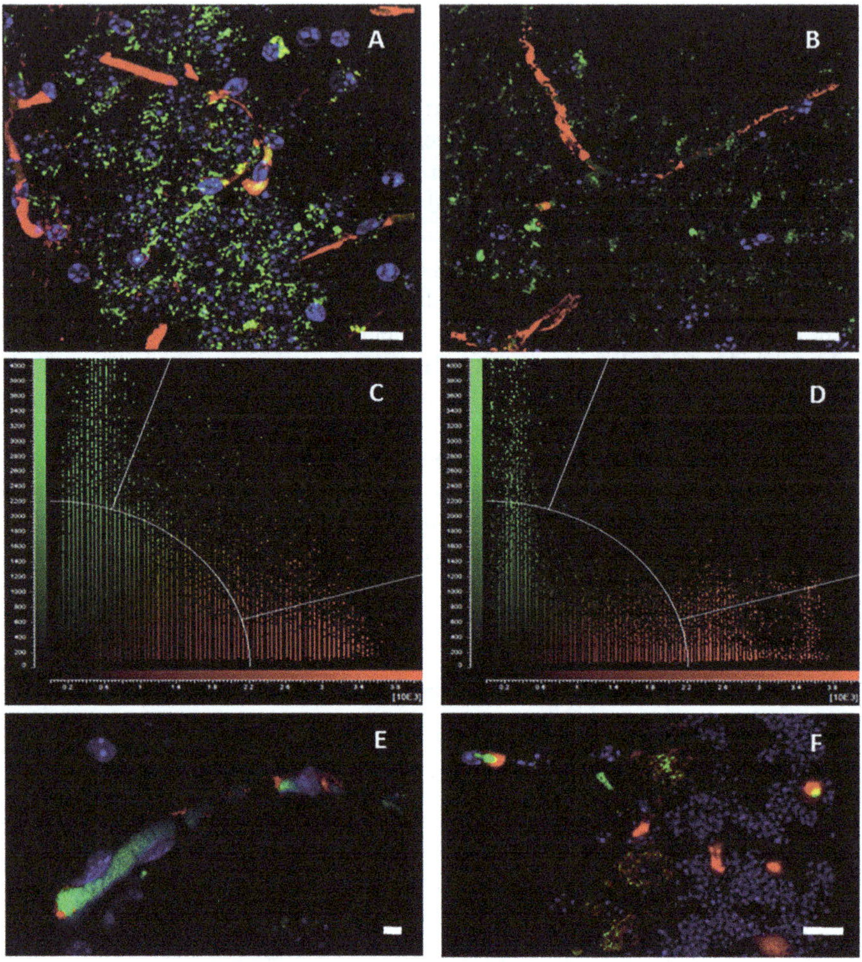

Fig. 20.4 Expression of P450scc and StAR in the hippocampus. Representative images of StAR (red) and P450scc (green) in CA3 region of the hippocampus from control (**a**) and DBP exposed brain (**b**). Images were reconstructed from a Z stack obtained with a confocal microscope and processed using Imaris software. (**c, d**) are Scatter plots showing the quantification of the immunoreactivity of StAR (red) and P450scc (green) from hippocampal slices of PND 60 male. Dots represent individual immunofluorescence puncta of StAR (red) and P450scc (green) from control (**c**) and DBP exposed brains (**d**). The arced lines represent the threshold of detection above background. The area between the oblique lines represent the degree of overlap or spatial co-localization of the two antigens. In (**c**) there is approximately 13% co-localization with distinct labeling for StAR along the X axis and P450 along the Y axis. Gestational exposure to DBP induced a drastic reduction in the expression of StAR and P450scc as well as their co-localization (**b**). (**e**) is a maximum projection of a Z stack images showing co-localization of StAR and P450scc immunoreactivity in endothelial cells of capillaries (**a**). In the cerebellum, immunoreactivity for StAR and P450scc were co-localized mainly to the Purkinje cells cytoplasm (**f**) with some distinct expression in the endothelial cells for both antigens

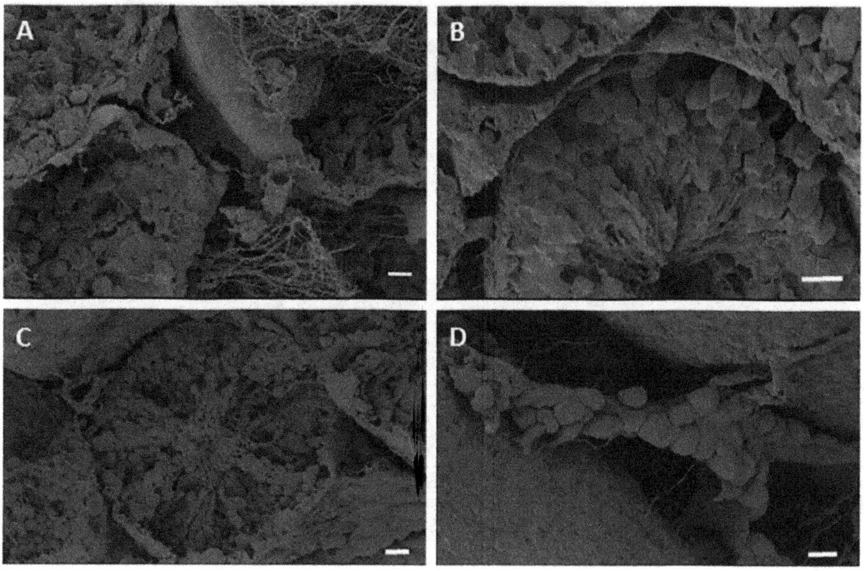

Fig. 20.5 Representative SEM images of the testes from a 2 months old mouse exposed to DBP in utero (**a**–**d**). No marked histopathology was observed with exposure to low level of DBP. Semineferous tubules had normal histology showing healthy spermatogonia (**b**), spermatocytes (**b**) and spermatids with visible intact flagella (**a**). Interstitial and Leydig cells showed normal histology and number (**d**). Calibration bars are 10 μm

confocal microscopy. We found that these two steroidogenic proteins showed high levels of expression and co-localization in Leydig cells and DBP reduced their level of expression (Fig. 20.6).

20.2.9 Discussion

Our studies revealed that *in utero* exposure of mice to a single dose of DBP (1 mg/kg BW s.c) at gestational day 10 significantly reduced the mRNA levels for both steroidogenic proteins StAR and P450scc. We also examined the transcription level of the β3 subunit of the $GABA_A$ receptors and GAD67 gene and found that these genes were equally affected by gestational exposure to DBP. Given the low levels of StAR expression in the brain we chose the hippocampus since this brain region has been reported to express slightly higher levels of StAR (King & Stocco, 2011; Kimoto & Tsurugizawa, 2001). Protein expression levels correlated with the mRNA data, we found that the expression of StAR and P450scc proteins was significantly reduced in the hippocampus by both western blot (Fig. 20.2) and immunofluorescence (Fig. 20.4). We also found that the expression of these steroidogenic proteins in the testes was significantly reduced (Fig. 20.3) with no apparent testicular histopathology. Testicular decrease in the expression of StAR and P450scc would further

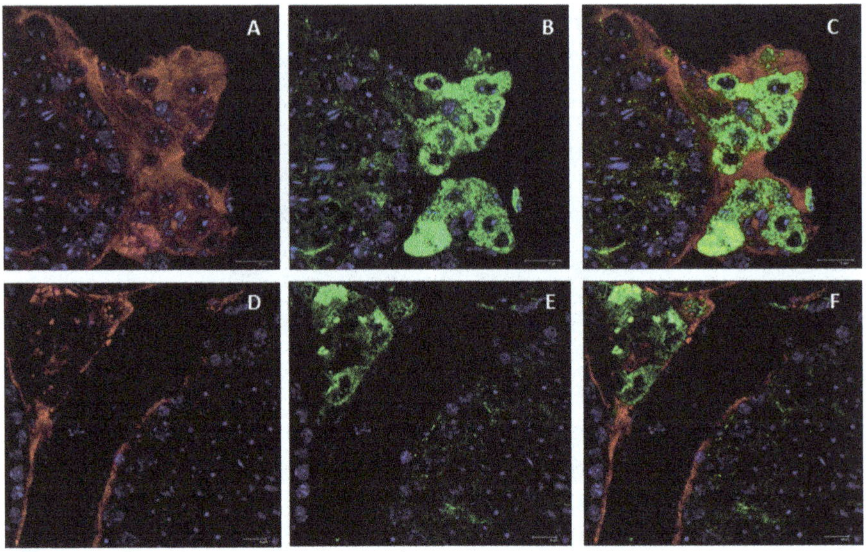

Fig. 20.6 Effect of DBP on the expression of P450scc and StAR in the testis. Representative images of StAR (red) and P450scc (green) in seminiferous tubules and Leydig cells of the testis from controls (**a**–**c**) and DBP exposed mice (**d**–**f**). Images were reconstructed from a Z stack obtained with a confocal microscope. Sections from control mice testis showed marked expression of StAR (**a**) and P450scc (**b**) in Leydig cells and throughout the cells of the seminiferous tubules. Exposure to DBP *in utero* resulted in a uniform reduction of the expression levels of both antigens in the testis (**d**–**f**). Calibration bars are 10 μm

decrease the bioavailability of neuroactive steroids in DBP injected offspring. Interestingly, the endocrine disrupting properties of DBP persisted into adulthood strongly suggesting that in utero exposure to endocrine disruptors can have a long lasting biochemical alterations. Recently, significant progress has been made in investigating the cellular and molecular mechanisms of neurosteroids in the brain. Most importantly, studies have focused on how neurosteroids like estrogen synergistically works with the GABAergic system in the brain (King & Stocco, 2011; McCarthy, 2008). Steroid hormones like pregnanolone and estrogen have been reported to influence the GABAergic system both early and later in development by exerting genomic and non-genomic effects (Baulieu, 1991; McCarthy, 2008). The neurosteroid pregnanolone allosterically modulates GABAARs and increases the influx of chloride ions into the cells (Baulieu, 1991), increasing therefore the efficacy of the GABAergic system. Additionally, early in development, steroid hormones like estrogen have been reported to augment the excitatory actions of GABA by increasing the amount of calcium influx, increasing the magnitude of the response to GABA. The vital role of estradiol orchestrating the excitatory actions of GABA aids in the regulation of dendritic growth, spine generation and synaptogenesis (McCarthy, 2008). In concordance with these activities, estrogen increases the expression of GAD, which consequently increases the synthesis of the neurotransmitter GABA by neurons (Mong & Blutstein, 2006). It is not clear however if gestational exposure to DBP directly affect the expression of GABAergic proteins or indirectly through downregulation of steroidogenic proteins and neuroactive steroids. Neuroactive steroids regulate neuronal function through their concomitant influence on neuronal excitability and gene expression. Thus, a reduced expression in steroidogenic proteins would have a long-lasting effects on neurotransmitter receptors and their function. GABAA receptor-mediated neurotransmission as a dynamic and highly flexible process is controlled by multiple mechanisms operating at the molecular, cellular, and systemic level. Neurosteroids most likely represent the major endogenous modulator of GABAA receptor function and represent a powerful homeostatic mechanism to enhance GABAergic function in response to increased excitability. Alterations of GABAA receptor expression and function is a key aspect of multiple neurological and psychiatric disorders, most likely reflecting changes in GABAergic circuits, including axonal sprouting and formation of novel synapses.

Alterations in P450scc and StAR by DBP early in development can affect the synthesis of neurosteroids. As a result, they can affect maturation of the GABAergic system. Studies have reported that a reduction in the expression of these proteins in the brain led to decreased synthesis of neurosteroids, which in turn resulted in decreased cell proliferation, increased cell death, altered spine densities and a significant effect on axonal outgrowth in vitro (King & Stocco, 2011). These supporting data point to the evidence that interference with these critical neurosteroidogenic proteins may affect or compromise the efficacy of the GABAergic system early in development. Later in development, the continued expression of these neurosteroids is also critical for the homeostatic allosteric modulation of neurotransmitter receptors. Neurosteroids have neuroprotective roles in the brain (King & Stocco, 2011; McCarthy, 2008) and their age-dependent decline has been postulated to be responsible for the decline in cognitive functions during aging (King & Stocco, 2011).

Acknowledgments We would like to thank the CSI vivarium staff for their care and wellbeing of our animals. We also thank the many undergraduate students who assisted in this research. This work was supported by a grant from PSC-CUNY and LSAMP Scholarship to F.S.

Conflict of Interest Statement The authors declare that the research was conducted in the absence of any commercial or financial relationships that could be construed as a potential conflict of interest.

References

Akwa, Y., Young, J., Kabbadj, K., Sancho, M. J., Zucman, D., Vourc'h, C., Jung-Testas, I., Hu, Z. Y., Le Goascogne, C., Jo, D. H., Corpéchot, C., Simon, P., Baulieu, E. E., & Robel, P. (1991). Neurosteroids: Biosynthesis, metabolism and function of pregnenolone and dehydroepiandrosterone in the brain. *The Journal of Steroid Biochemistry and Molecular Biology, 40*, 71–81.

Alam, M. S., Ohsako, S., Matsuwaki, T., Zhu, X. B., & Tsunekawa, N. (2010). Induction of spermatogenic cell apoptosis in prepubertal rat testes irrespective of testicular steroidogenesis: A possible estrogenic effect of di(n-butyl)phthalate. *Reproduction, 139*(2), 427–437.

Baulieu, E. E. (1991). Neurosteroids: A new brain function. *Biology of the Cell, 71*, 3–10.

Baulieu, E. E., & Robel, P. (1990). Neurosteroids: A new brain function? *The Journal of Steroid Biochemistry and Molecular Biology, 37*, 395–403.

Benke, D., Honer, M., Michel, C., Bettler, B., & Mohler, H. (1999). γ-Aminobutyric acid type B receptor splice variant proteins GBR1a and GBR1b are both associated with GBR2 in situ and display differential regional and subcellular distribution. *The Journal of Biological Chemistry, 274*, 27323–27330.

Blount, B. C., Silva, M. J., Caudill, S. P., Needham, L. L., Pirkle, J. L., Sampson, E. J., Lucier, G. W., Jackson, R. J., & Brock, J. W. (2000). Levels of seven urinary phthalate metabolites in a human reference population. *Environmental Health Perspectives, 108*, 979–982.

Crisp, T. M., Clegg, E. D., Cooper, R. L., Wood, W. P., Anderson, D. G., Baetcke, K. P., et al. (1998). Environmental endocrine disruption: An effects assessment and analysis. *Environmental Health Perspectives, 106*(Suppl. 1), 11–56.

D'Hulst, C., De Geest, N., Reeve, S. P., Van Dam, D., De Deyn, P. P., Hassan, B. A., & Kooy, R. F. (2006). Decreased expression of the GABAA receptor in fragile X syndrome. *Brain Research, 1121*(1), 238–245.

El Idrissi, A., Ding, X. H., Scalia, J., Trenkner, E., Brown, W. T., & Dobkin, C. (2005). Decreased GABA(A) receptor expression in the seizure-prone fragile X mouse. *Neuroscience Letters, 377*(3), 141–146.

Foster, P. M., Mylchreest, E., Gaido, K. W., & Sar, M. (2001). Effects of phthalate esters on the developing reproductive tract of male rats. *Human Reproduction Update, 7*, 231–235. https://doi.org/10.1093/humupd/7.3.231

Kim, B. N., Cho, S. C., Kim, Y., Shin, M. S., Yoo, H. J., Kim, J. W., Yang, Y. H., Kim, H. W., Bhang, S. Y., & Hong, Y. C. (2009). Phthalates exposure and attention-deficit/hyperactivity disorder in school-age children. *Biol Psychiatry*, 15;66(10):958–63. https://doi.org/10.1016/j.biopsych.2009.07.034. Epub 2009 Sep 12. PMID: 19748073.

Kimoto, T., & Tsurugizawa, T. (2001). Neurosteroid synthesis by cytochrome P450-containing systems localized in the rat brain hippocampal neurons: N-methyl-D-aspartate and calcium-dependent synthesis. *Endocrinology, 142*, 3578–3389.

King, S. R., & Stocco, D. M. (2011). Steroidogenic acute regulatory protein expression in the central nervous system. *Frontiers in Endocrinology, 2*(72), 1–6.

Koch, H. M., Rossbach, B., Drexler, H., & Angerer, J. (2003). Internal exposure of the general population to DEHP and other phthalates--determination of secondary and primary phthalate monoester metabolites in urine. *Environmental Research, 93*, 177–185.

Kohn, M. C., Parham, F., Masten, S. A., Portier, C. J., Shelby, M. D., Broch, J. W., & Needham, L. L. (2000). Human exposure estimates for phthalates. *Environmental Health Perspectives, 108*, A440–A442.

Lambert, J. J., Belelli, D., Peden, D. R., Vardy, A. W., & Peters, J. A. (2003). Neurosteroid modulation of $GABA_A$ receptors. *Progress in Neurobiology, 71*, 67–80.

Li, X. J., Jiang, L., Chen, L., Chen, H. S., & Li, X. (2013). Neurotoxicity of dibutyl phthalate in brain development following perinatal exposure: A study in rats. *Environmental Toxicology and Pharmacology, 36*, 392–402.

Li, X., et al. (2014). Dibutyl phthalate-induced neurotoxicity in the brain of immature and mature rat offspring. *Brain Dev, 36*(8), 653–660.

Lottrup, G., Andersson, A. M., Leffers, H., Mortensen, G. K., Toppari, J., Skakkebaek, N. E., & Main, K. M. (2006). Possible impact of phthalates on infant reproductive health. *International Journal of Andrology, 29*, 172–180.

McCarthy, M. M. (2008). Estradiol and the developing brain. *Physiological Reviews, 88*, 91–134. https://doi.org/10.1152/physrev.00010.2007

Mensah-Nyagan, A. G., Feuilloley, M., Do-Rego, J. L., Marcual, A., Lange, C., Tonon, M. C., Pelletier, G., & Vaudry, H. (1996). Localization of 17β-hydroxysteroid dehydrogenase and characterization of testosterone in the brain of the male frog. *Proceedings of the National Academy of Sciences of the United States of America, 93*, 1423–1428.

Mong, J. A., & Blutstein, T. (2006). Estradiol modulation of astrocytic form and function: Implications for hormonal control of synaptic communication. *Neuroscience, 138*(3), 967–997.

Mtchedlishvili, Z., & Kapur, J. (2003). A presynaptic action of the Neurosteroid Pregnenolone sulfate on GABAergic synaptic transmission. *Molecular Pharmacology, 64*, 857–864.

Mylchreest, E., Wallace, D. G., Cattley, R. C., & Foster, P. M. (2000). Dose-dependent alterations in androgen-regulated male reproductive development in rats exposed to Di(n-butyl) phthalate during late gestation. *Toxicological Sciences, 55*, 143–151.

Petersen, J. H., & Breindahl, T. (2000). Plasticizers in total diet samples, baby food and infant formulae. *Food Additives and Contaminants, 17*, 133–141.

Selby, L., Zhang, C., & Sun, Q. Q. (2007). Major defects in neocortical GABAergic inhibitory circuits in mice lacking the fragile X mental retardation protein. *Neuroscience Letters, 412*(3), 227–232.

Shultz, V. D., Phillips, S., Sar, M., Foster, P. M., & Gaido, K. W. (2001). Altered gene profiles in fetal rat testes after in utero exposure to di(n-butyl) phthalate. *Toxicological Sciences, 64*, 233–242. https://doi.org/10.1093/toxsci/64.2.233

Sierra, A., Lavaque, E., Perez-Martin, M., Azcoitia, I., Hales, D. B., & Garcia- Segura, L. M. (2003). Steroidogenic acute regulatory protein in the rat brain: Cellular distribution, develop- mental regulation and overexpression after injury. *The European Journal of Neuroscience, 18*, 1458–1467.

Swan, S. H., Main, K. M., Liu, F., Stewart, S. L., Kruse, R. L., Calafat, A. M., et al. (2005). Decrease in anogenital distance among male infants with prenatal phthalate exposure. *Environmental Health Perspectives, 113*, 1056–1061.

Thompson, C. J., Ross, S. M., & Gaido, K. W. (2004). Di (n-Butyl) phthalate impairs cholesterol transport and steroidogenesis in the fetal rat testis through a rapid and reversible mechanism. *Endocrinology, 145*, 1227–1237.

Index

A. El Idrissi, D. McCloskey (eds.), *Neurobiology of Autism Spectrum Disorders*, https://doi.org/10.1007/978-3-031-42383-3

GPSR Compliance
The European Union's (EU) General Product Safety Regulation (GPSR) is a set of rules that requires consumer products to be safe and our obligations to ensure this.

If you have any concerns about our products, you can contact us on

ProductSafety@springernature.com

In case Publisher is established outside the EU, the EU authorized representative is:

Springer Nature Customer Service Center GmbH
Europaplatz 3
69115 Heidelberg, Germany

www.ingramcontent.com/pod-product-compliance
Lightning Source LLC
Chambersburg PA
CBHW071738271025
34597CB00003B/116
* 9 7 8 3 0 3 1 4 2 3 8 2 6 *